Engineering Materials

This series provides topical information on innovative, structural and functional materials and composites with applications in optical, electrical, mechanical, civil, aeronautical, medical, bio- and nano-engineering. The individual volumes are complete, comprehensive monographs covering the structure, properties, manufacturing process and applications of these materials. This multidisciplinary series is devoted to professionals, students and all those interested in the latest developments in the Materials Science field, that look for a carefully selected collection of high quality review articles on their respective field of expertise.

Indexed at Compendex (2021) and Scopus (2022)

Jai Prakash · Junghyun Cho · Olim Ruzimuradov ·
Dong Fang
Editors

Titanium Dioxide-Based Multifunctional Hybrid Nanomaterials

Application on Health, Energy and Environment

Springer

Editors
Jai Prakash
Department of Chemistry
National Institute of Technology
Hamirpur, Himachal Pradesh, India

Junghyun Cho
Department of Mechanical Engineering
Binghamton University (State University of New York)
Vestal, NY, USA

Olim Ruzimuradov
Turin Polytechnic University in Tashkent
Tashkent, Uzbekistan

Dong Fang
Faculty of Materials Science and Technology
Kunming University of Science and Engineering
Kunming, China

ISSN 1612-1317 ISSN 1868-1212 (electronic)
Engineering Materials
ISBN 978-3-031-80624-7 ISBN 978-3-031-80625-4 (eBook)
https://doi.org/10.1007/978-3-031-80625-4

This Springer imprint is published by the registered company Springer Nature Switzerland AG
The registered company address is: Gewerbestrasse 11, 6330 Cham, Switzerland

Contents

TiO_2 Nanomaterials: Synthesis, Properties, Modifications, and Applications

Rupam Sharma and Jai Prakash

Abstract This chapter presents study of titanium dioxide (TiO_2) nanomaterials, highlighting their various properties, and synthesis strategies. It has been described how TiO_2's intrinsic restrictions, especially its broad bandgap, limit its photocatalytic activity to ultraviolet light. The chapter discusses several modification techniques that are intended to improve the performance of TiO_2 based nanomaterials in visible and solar light. Furthermore, it outlines the broad spectrum of applications for TiO_2, including its roles in renewable energy, environmental remediation, and biomedical fields. The chapter concludes by underscoring the significance of ongoing research to overcome the challenges associated with TiO_2 and to harness its full potential in addressing global energy and environmental issues.

Keywords TiO_2 · Modifications · Properties · Energy · Environment · Biomedical applications

1 Introduction

Due to its low cost, nontoxic behaviour, and effective photocatalytic capacity, titanium dioxide (TiO_2) has been the subject of significant research. As a result, TiO_2 has several uses in fields such as photocatalysts, Li-ion batteries, solar cells, and biotechnology [1]. Since its widespread application in several fields, including the production of sustainable energy and the elimination of environmental contaminants, TiO_2 nanoparticles have inspired a great deal of scholarly attention. TiO_2 has bandgap of (3.0–3.2 eV). Global worry has grown over the past few decades because there is a growth in environmental pollution and urgency in energy crisis. The exploration of environmentally friendly technologies that may be applied to clean energy sources and pollution recovery is vital in order to achieve sustainable growth of civilization. Since solar energy is an endless supply of renewable energy

R. Sharma · J. Prakash (✉)
Department of Chemistry, National Institute of Technology Hamirpur, 177005, Hamirpur, Himachal Pradesh, India
e-mail: jaip@nith.ac.in

J. Prakash et al. (eds.), *Titanium Dioxide-Based Multifunctional Hybrid Nanomaterials*, Engineering Materials, https://doi.org/10.1007/978-3-031-80625-4_1

in the long run, creating materials and methods to maximize solar energy usage is essential for environmental stewardship as well as energy security [2]. The first work on the hydrogen production on TiO_2 photoelectrodes under UV light illumination was published by Honda and Fujishima in 1972, and it attracted attention from all over the world [3]. The titanium minerals brookite, ilmenite, anatase, and rutile are significant from an economic standpoint. And some of the less frequent titanium minerals are titanite, Leucoxene, loparite, and perovskite, despite the fact that there are many titanium minerals. The minerals that contain the most titanium are anatase (90–95%), rutile (94–99%), and brookite (90–100%). There are 62 million tons of rutile and 832 million tons of ilmenite among the world's reserves of titanium. The bulk of the world's rutile reserves are found in South Africa (8.3 million tons), Kenya (13 million tons) and Australia (29 million tons), but the majority of ilmenite reserves are found in China (220 million tons) and Australia (250 million tons) [4]. TiO_2's distinct characteristics are closely linked to its crystal structure, which is also influenced by how it is processed. Three phases make up the crystal structures of TiO_2 are anatase, rutile and brookite. According to Hwu et al. [5], the processing temperature and method of preparation had a noteworthy impact on the crystal structure of TiO_2 nanoparticles. At temperatures over 973 K, Anatase showed a transformation to rutile for tiny TiO_2 nanoparticles (less than 50 nm). The TiO_2 generated at a temperature lower than 873 K exhibits an anatase crystal structure, according to XRD measurements [5]. Because of its small size and unique characteristics, TiO_2 nanoparticles have drawn attention from a wide range of scientific fields and have become a focus point for technological advancement. Nanoparticles with less then 100 nm diameter are incredibly reactive due to their remarkable surface area to volume ratio. Their ability to accelerate chemical reactions in the presence of light is one of their most remarkable characteristics [6]. This ability is often used in environmental remediation procedures including air and water purification. Moreover, TiO_2 nanoparticles are widely used in commonplace items like sunscreens and cosmetics because of their potent UVA and UVB absorption capabilities, which provide strong protection against damaging sun exposure. Beyond skincare, they are used in many other industries, including as energy (where they help create solar cells of the future) and biomedicine (where they could be used in medical imaging and drug delivery systems). Nevertheless, despite their enormous potential, questions remain concerning TiO_2 nanoparticle safety, especially in light of their possible toxicity when swallowed or inhaled. However, continued study into their synthesis, modification, and application promises to overcome safety concerns and uncover even greater utility, opening the door to creative solutions and developments [7].

This chapter highlights the basics of TiO_2 including its various properties, synthesis strategies and multifunctional applications in the field of energy, environment and biomedical/health applications.

2 TiO_2: Structural, Optical, and Electrical Properties

There are various crystalline phases of TiO_2 but if we talk about three most common crystalline phases which are brookite, anatase, and rutile. These phases have unique characteristics and crystal formations. Because of their distinctive features, anatase and rutile have been investigated the most. Whereas rutile displays a more symmetric and compact arrangement of atoms, anatase has a tetragonal structure. The variations in structure across different phases impact multiple attributes, including as stability, reactivity, and surface area. In photocatalytic applications, rutile is less reactive than anatase because anatase has a higher surface area [8]. Since TiO_2 has a broad bandgap energy that is located in the ultraviolet (UV) portion of the electromagnetic spectrum, it has intriguing optical features. The absorption and reflection of light by TiO_2 is dependent on the bandgap energy. The bandgaps of rutile TiO_2 and anatase TiO_2, which have varied sensitivity to different light wavelengths, are approximately 3.0 eV and 3.2 eV, respectively [9]. Other varieties of TiO_2 include ramsdellite, collumbite, and hollandite. Although these types of titanium are not found in nature, they may be created in a lab setting by applying high pressure to the anatase and rutile phases of the metal [10] (Fig. 1).

TiO_2 nanoparticles are used in sunscreens and UV-blocking coatings because of their ability to scatter light and absorb UV rays. In general, TiO_2 is categorized as a semiconductor with a large bandgap (Fig. 2).

Its bandgap energy determines its electrical conductivity qualities. TiO_2's large bandgap causes it to act as an insulator under typical circumstances. Nonetheless, the conductivity of the crystal lattice can be increased by doping or adding flaws [12].

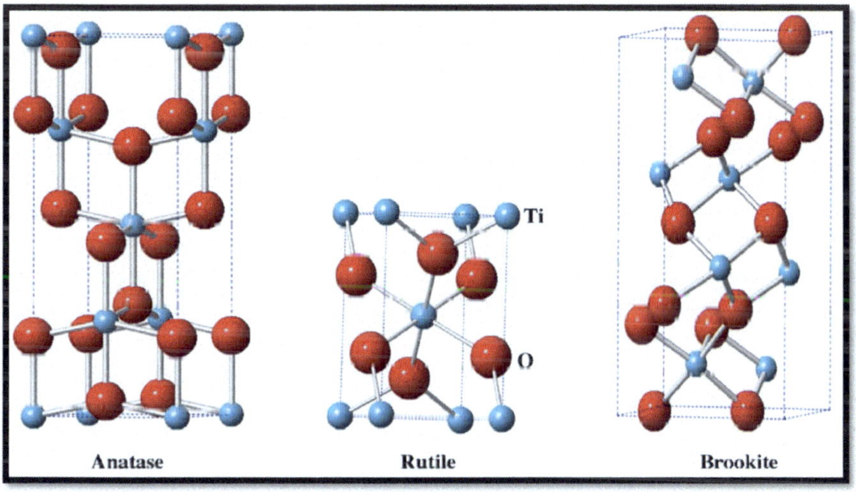

Fig. 1 TiO_2 crystal structures: anatase, rutile and brookite. Reproduced with permission from Etacheri et al. [11], Copyright (2015) Elsevier

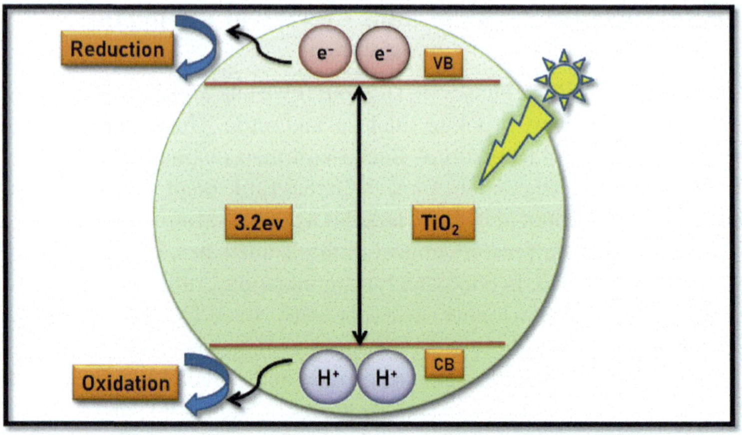

Fig. 2 The schematic illustration of photocatalysis process in TiO_2

TiO_2's semiconducting qualities are used in electronic devices like gas sensors and solar cells to facilitate effective charge transport and sensing [13].

3 Synthesis of TiO_2 Nanomaterials

TiO_2 is a highly significant nanomaterial that garners a lot of interest because of its special qualities. It has significant benefits for solar energy transmission, hydrogen synthesis from water through photochemical processes, and photocatalysis for the breakdown of harmful chemicals. TiO_2 is an effective photocatalyst due to its non toxic nature, strong oxidizing capacity of photogenerated holes and chemical inertness is also one of the reasons behind its effective photocatalyst property [14]. The key to using materials based on TiO_2 is their phase stability under various conditions (temperature, pressure, and media) [15]. Depending on the type of TiO_2 crystal, the material's chemical and physical characteristics might vary greatly. For many uses, the brookite phase is practically unfavorable. Although it can persist at low temperatures, it is unstable at room temperature. The most stable and widely utilized form of TiO_2 is rutile, which is especially useful as a white pigment in paint. TiO_2, a photocatalyst, mostly possesses an anatase-type crystalline structure with a band gap of 3.2 eV [14–16]. There are various methods for synthesizing TiO_2 nanomaterials as shown in Fig. 3.

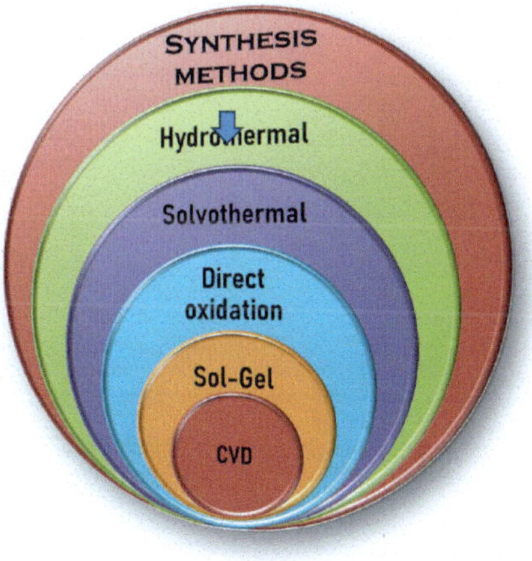

Fig. 3 Different methods for synthesizing TiO_2 nanomaterials

3.1 Hydrothermal Method

Typically, steel pressure autoclaves are used to carry out the hydrothermal reaction, which are containers either with or without teflon liners that are kept at controlled pressures or temperatures while the reaction takes place in aqueous solutions as shown in Fig. 4 [17]. It is feasible to reach vapour saturation pressure and elevate the reaction temperature over the boiling point of water in a hydrothermal process. The temperature and amount of solution supplied have a major influence on the internal pressure that builds up in the autoclave. The hydrothermal technique is any heterogeneous interaction with water-based solvents or mineralizers at high pressure and temperature to break down and recrystallize compounds that are barely insoluble under normal conditions [14, 18]. In hydrothermal process we use solvent which can handle the pressure typically ranges between 1 and 10,000 atm where as if we talk about temperature it ranges between 100 and 1000 °C, which promotes precursor contact during synthesis. If water is used as the solvent, the procedure is referred to as hydrothermal synthesis [19].

3.2 Solvothermal Method

In the presence of a particular solvent the hydrothermal process is known as the solvothermal method as shown in Fig. 5 [20]. This procedure uses a specific type of non-aqueous solvent. Considering that the solvothermal approach allows for the use of a broad range of organic solvents which have high boiling point. Temperatures can be raised more in solvothermal processes than in hydrothermal approaches [14, 21]. Using this method gives us more control over the TiO_2 nanoparticles' size, shape, and distribution of crystallinity. This versatile technique may be used to create

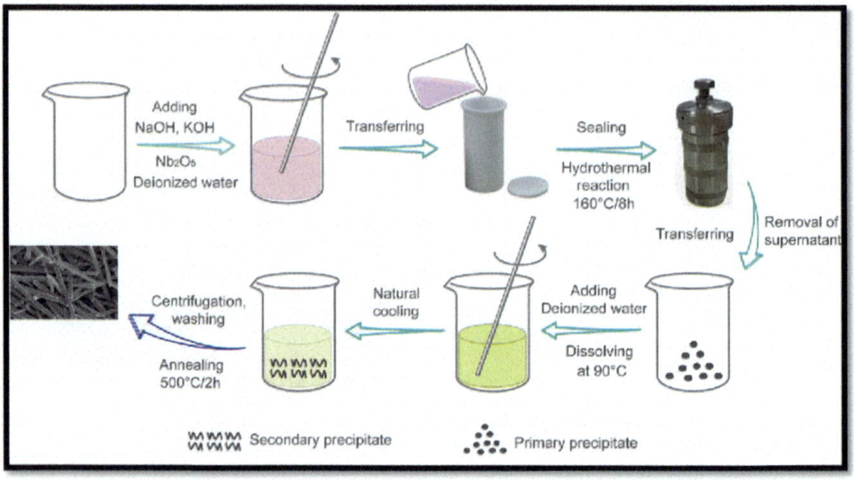

Fig. 4 Process flow of the hydrothermal method to prepare nanomaterials. Reproduced with permission from Zhang et al. [17], Copyright (2015) Elsevier

a range of nanoparticles that have restricted variation in size and dispersity. It is used to synthesize nanoscale metals, semiconductors, ceramics, and polymers at moderate to high pressure while using solvent. When H_2O is utilized as the solvent in a solvothermal process than it is termed as hydrothermal process. Under hydrothermal conditions, synthesis often takes place below water's supercritical temperature [22].

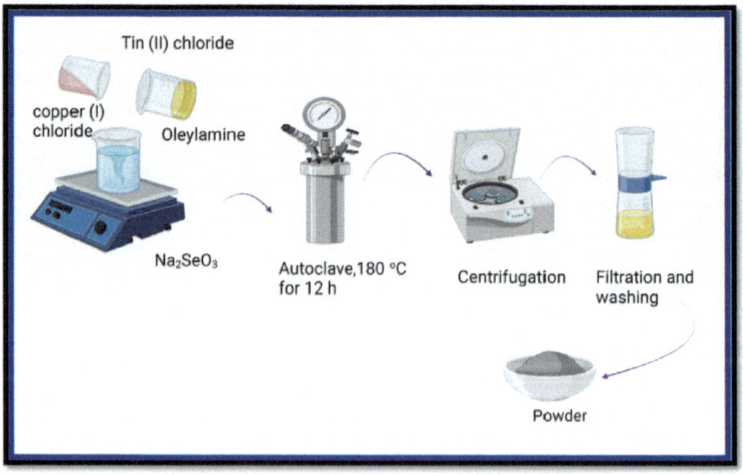

Fig. 5 Process flow of the solvothermal method to prepare nanomaterials. Reproduced with permission from Suriakarthick et al. [20], Copyright (2022) Elsevier

3.3 Direct Oxidation Method

Titanium metal is employed as oxidants or under anodization in the direct oxidation process. The oxidation process can be used to create nanostructured titania. For example, the crystalline TiO_2 nanorods were created by directly oxidizing a titanium metal plate with hydrogen peroxide [23, 24]. When a cleaned Ti plate is immersed in 50 mL of a 30% hydrogen peroxide solution at 353 °K for 72 h, TiO_2 nanorods are often detected on the plate. It is the dissolution precipitation mechanism that allowed to obtain crystalline TiO_2. By adding inorganic salts of NaX here X can be $F^-$and Cl^- and SO_4^{2-}, the crystalline phase of TiO_2 nanorods can be manipulated. Pure anatase can be produced in its synthesis by the addition of F and SO_4, whereas rutile is favoured by the addition of Cl [24].

3.4 Sol–gel Method

In the industrial sector, the sol–gel methodology is used more frequently than other current approaches. This method can produce superior nanomaterials of the identical size on a larger scale due to its unique features. By combining two or more metal or metal oxide reactant in certain proportions, alloy products can be generated in a single step due to the capability of creating two or more types of nanoparticles simultaneously during this process [25–29]. We can also control the final morphology in sol–gel method by having variation in progress as shown in Fig. 6 [30]. One of the tried-and-true synthetic methods for creating superior Metal Oxide NPs and mixed oxide composites is the sol–gel technique. The surface characteristics and texture of the materials may be precisely controlled with this technology. In general, the method of sol–gel involves five key processes: hydrolysis, polycondensation, aging, drying out, and heat breakdown [31]. For example TiO_2 NPs were made by Vijayalakshmi et al. [32] by adding nitric acid to ethanol and Titanium IV isopropoxide (TTIP) to facilitate the hydrolysis process. The study showed that, under the same reaction conditions, TiO_2 NPs generated by the sol–gel method had smaller crystallite diameters and were more crystalline than those prepared by the hydrothermal method.

3.5 Chemical Vapour Deposition

Chemical vapour deposition produces (CVD) vapours that condense further on substrate materials' surfaces to build solid phase materials. These procedures are typically employed to create thin films on material surfaces in order to modify the attributes of different substrates, including wear resistance, mechanical, electrical, optical, thermal, and corrosion resistance. The deposition reaction is started and the

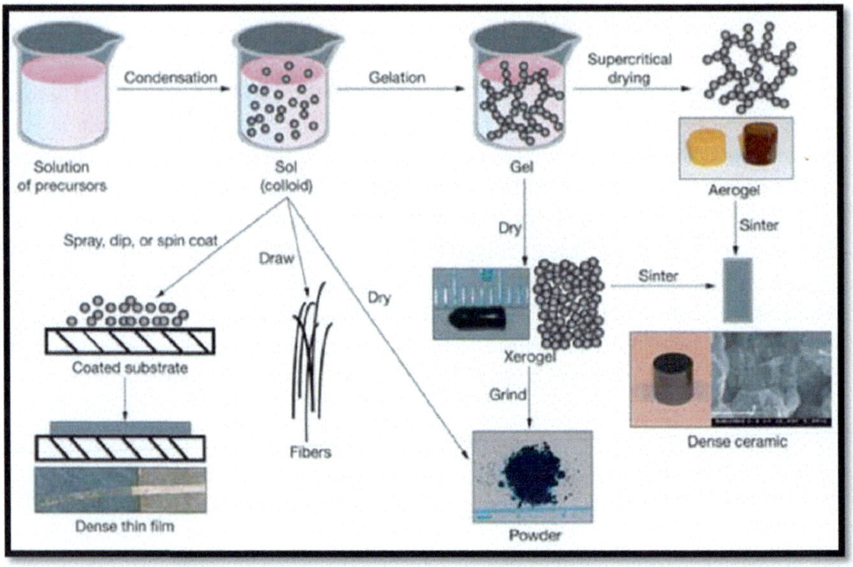

Fig. 6 Different stages of the sol–gel manufacturing process to control the final form of the product [30] (open access)

gasses in the CVD chamber are heated using thermal energy. A number of variables, including deposition temperature, pressure, flow rate, gas composition, and deposition chamber shape, influence CVD processes [14]. These are the variables that aid in regulating the target material's nanoforms. Using titanium tetrachloride as a precursor, Djerdja et al. [33] produced nanocrystalline TiO_2 films formed by CVD on several substrates at a very low temperature of 320 °C. They discovered that the substrates' characteristics affected the size and distribution of nano grains in the films [34]. The second CVD method uses a substrate for CVD deposition that is generated from a template [35] and Fig. 7 displays the block diagram of the CVD system [36].

4 Need of Modifications

TiO_2 is the most significant because of its superior chemical and physical characteristics. However, visible light makes them inactive since the TiO_2 band edge lies in the UV range. Considerable things have been done in this direction to alter the optical and electrical characteristics of TiO_2 so there must be enhancement in visible light activity. One of the common method which has been used significantly is cations and anions doping on TiO_2. An other method which is used is to combine narrow bad gap semiconductor with TiO_2 for example Bi_2S_3/TiO_2 [37]. Since a straight forward modification can quickly change the mechanism and quicken the kinetics of photocatalysis, Numerous attempts have been explored to alter TiO_2's surface in an

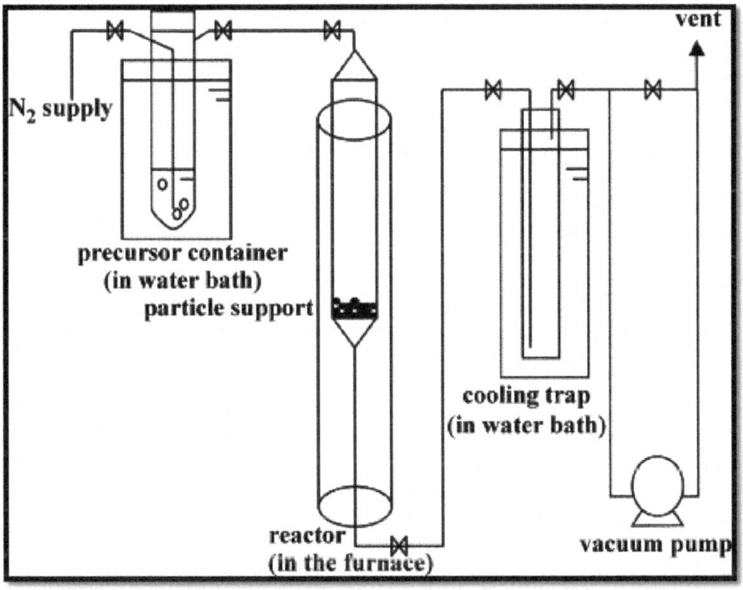

Fig. 7 An illustration of the CVD schematic system. Reproduced with permission from Ding et al. [36], Copyright (2001) Elsevier

effort to enhance photocatalytic activity [38]. Conversely, CO_2 is thought to be the most powerful greenhouse gas and the primary root cause of global warming. One technique that has been recommended as a potential remedy for the worldwide issue of industrial CO_2 emissions is photocatalysis. A process known as photocatalysis converts CO_2 into a variety of significant worth fuels, such as formic acid, methanol, methane and formaldehyde by using a type of semiconductor and sustainable solar energy. Methods for enhancing CO_2 adsorption on semiconductor surfaces are also offered. The semiconductor TiO_2 is now the focus of much study because of its effective photoactivity, high stability, low cost, and safety when adjusted properly. Research is also being done on its potential as a photocatalyst for the reduction of CO_2. We address the recently developed methods for producing and modifying TiO_2, which can improve the CO_2 photoreduction process efficiency. These modification methods, which include metal deposition, metal/non-metal doping, loading materials made from carbon, semiconductor hybrid structures, and dispersion on large surface area, are meant to improve the product's light sensitivity, separation of charged particles, and TiO_2 active surface area. They also seek to increase the yield and selectivity of the final product [39]. Ever since it was initially shown that TiO_2 could produce the photocatalytic splitting of water into hydrogen and oxygen, researchers have been concentrating their efforts on this material. Modified TiO_2 NPs are thought to be more successful than their unmodified fine particle analogs because of their distinct physicochemical features. However, there appears to have been a recent spike in research interest utilizing NPs. Various tactics have been utilized to alter TiO_2 in order to lower the rates of recombination of photogenerated charge carriers and improve TiO_2's ideal performance. The most often used modification methods are coupling to an additional semiconductor and doping with anions or cations [40].

4.1 Types of Modification in TiO_2

Since its initial demonstration of its capabilities in a wider variety of applications, TiO_2 has been the center of interest for researchers.. But still there are some areas where modification of TiO_2 nanomaterials are is possible as shown in Fig. 8 which can enhance the properties of TiO_2 and can give better results.

4.1.1 Doping

In band gap engineering, TiO_2 doping has been a crucial tactic for modifying the optical response of semiconductor photocatalysts. Doping is mostly done to introduce intra-band gap states or reduce the band gap in order to produce a bathochromic shift, or an increase in visible light absorption. Doping may lead to photocatalytic systems that are more effective [41]. To change the electrical structure of the photocatalyst without affecting its crystal structure, doping is the preferred method. O_2^- with another anion is more difficult to replace than Ti_4^+ in TiO_2 with a cation because

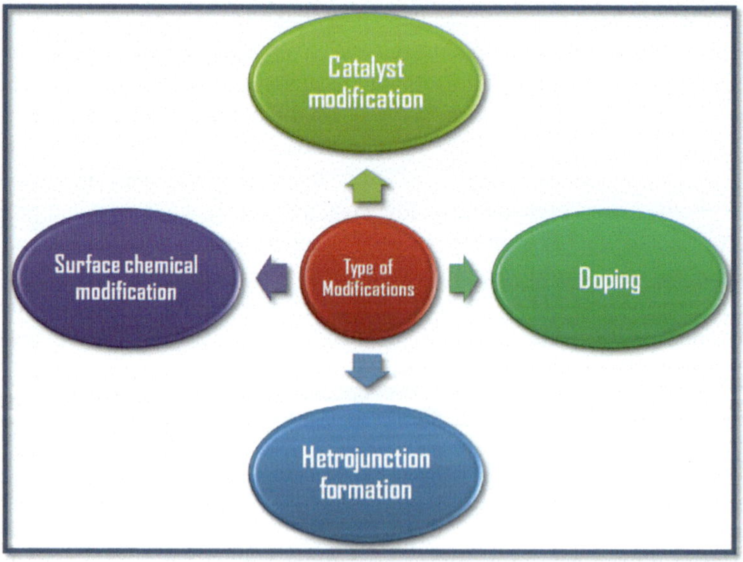

Fig. 8 Types of modification of TiO_2 nanomaterials

to variations in ionic radii and charge states [42]. Because of their inherent lattice strain, nanomaterials exhibit a higher tolerance to structural distortion than bulk materials. Consequently, it seems that surface modification of TiO_2 nanoparticles is more advantageous than bulk TiO_2 modification [43].

Metal Doping

Researchers investigated the impacts of metal doping (M = Cu, Ag, and Eu) in nanostructured (M–TiO_2) to ascertain the placement of metal ions in the host matrix, their stable valence states and surroundings, and the impact on the structural, electrical, and optical properties of M–TiO_2. In an experimental technique, metal-doped TiO_2 nanoparticles (with metal ratios of 1%, 2%, and 3 mol.%) were synthesized using TTIP and ethanol as the solvent. The procedure was run with intense magnetic stirring for fifteen minutes. The doping method comprised adding the proper amounts of copper or silver acetate to the solution and agitating it until a clear solution emerged, as stated above. The solution was hermetically sealed, ultrasonicated for two hours, heated to sixty degrees Celsius in an air oven for twelve hours, and then ultrasonicated again for two hours before the acetic acid was added. Throughout the air-calcined Cu or Ag doped TiO_2, the anatase phase was stable around 500 °C. Up to 800 °C of annealing preserves the anatase polymorph in TiO_2 doped with europium [44].

Transition Metal Doping

TiO_2 only uses a small portion of the solar spectrum, consequently up until now, its ability to absorb light into the range of visible light has been boosted via doping with transition metals [45]. The photoreactivity for both the oxidation of $CHCl_3$ and the reduction of CCl_4 was found to be greatly increased by doping quantum-sized TiO_2 with Fe_3^+, Mo_5^+,Ru_3^+, Os_3^+,Re_5^+,V_4^+, and Rh_3^+ at concentrations of 0.1–0.5 at.% [46]. The addition of WO_3 and MoO_3 greatly enhanced the photocatalytic activity of TiO_2 toward the oxidation of 1–4 dichlorobenzene [47]. When exposed to solar radiation, TiO_2 with Cr and V ions implanted has demonstrated three to four times the photocatalytic activity of TiO_2 when it comes to the breakdown of NO [48]. The catalyst's characteristics are obviously determined by the doping technique, Therefore, same experimental conditions for sample preparation and the use of identical bare TiO_2 as starting material are required for direct comparisons on the effects of metal doping. Because transition metal doped TiO_2 photocatalysts have higher photocatalytic activity than pure TiO_2, they are important for the decolorization of methylene blue dye. The photocatalytic characteristics of TiO_2 may be enhanced by transition metal doping in a number of ways. For example it can lower the bandgap energy of TiO_2, improving its ability to absorb visible light and enhancing its photocatalytic activity when exposed to it. Higher surface area materials can have more active sites accessible for photocatalytic processes as a result of transition metal doping. By lowering the rate at which electron–hole pairs recombine, transition metal doping can improve the efficiency with which these charge carriers

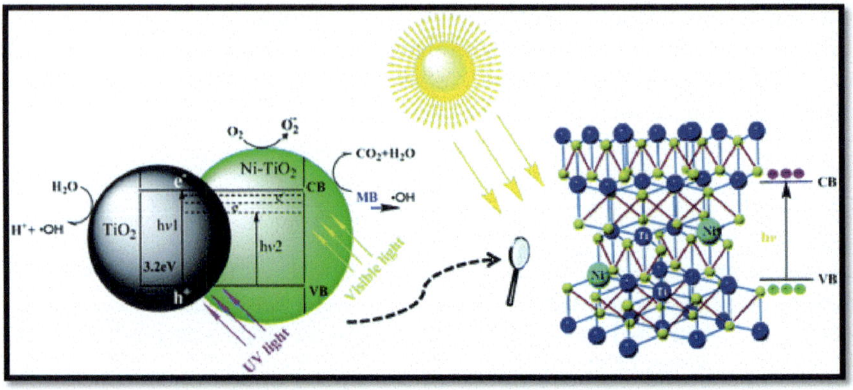

Fig. 9 Mechanism of MB degradation via Ni doped TiO_2 [50] (open access)

are used in photocatalytic processes. The production of electron–hole pairs, which are essential for the photocatalytic degradation of organic dyes like methylene blue, may be more efficiently shown by transition metal doped TiO_2 materials. All things considered, through improving separation of charge, absorption of light, and ultimate photocatalytic performance, the use of transition metal doped TiO_2 photocatalysts presents a viable strategy to increase the efficiency of photocatalytic degradation processes, such as the decolorization of methylene blue dye [49, 50], Fig. 9. Shows the mechanism of Methylene blue degradation mechanism through Ni doped TiO_2.

Rare Earth Metals Doping

Rare earth metal doping has a significant impact on pure TiO_2, especially in terms of its optical, electrical, and structural properties. Here are some effects of rare earth metal doping on pure TiO_2. Doping with rare earth metals alters the absorption threshold of TiO_2, leading to increased absorption of visible light radiation. This modification lowers the energy required for electron transitions, allowing for absorption of visible and even near-infrared (NIR) light [T1]. TiO_2's photocatalytic effectiveness may be raised by this improvement in optical characteristics. The band structure of TiO_2 is enhanced by the introduction of sub energy bands by rare earth metal ions beneath the conduction band. These sub-energy levels trap charge carriers, which helps different processes by enabling their separation from the bands and encouraging diffusion towards the surface. Better absorption of visible light is made possible by a red shift in the absorption spectra and a decrease in the band gap energy. Doping TiO_2 with rare earth metals enhances its photocatalytic activity by using a wider range of the solar spectrum, from UV to NIR. Doped TiO_2 nanostructures show improved photocatalytic efficiencies compared to undoped counterparts, making them suitable for applications such as photocatalytic CO_2 reduction, water splitting, and solar cell devices. Doping with rare earth metals enhances the structural and optoelectronic properties of TiO_2 nanostructures. Co-doping with other elements can further improve these properties and mitigate issues such as recombination of charge carriers. Co-doping strategies have been found to be effective in enhancing the photocatalytic properties of TiO_2 nanostructures [51]. According to certain reports, doping TiO_2 with rare earth metals, such as Gd, La, Ce, Pr, Nd, Sm, and Er, could effectively reduce the width of the forbidden band gap, resulting in a specific red-shift absorption. Furthermore, it was discovered that the charge density distribution surrounding TiO_2 was impacted by the presence of rare earth metals, with localized changes in electron loss being noted at the location of rare earth element doping. The investigation also showed that Gd-doped TiO_2 had a clear delocalized intrinsic electric field, which helped to enhance charge separation and photocatalytic activity [52]. In summary, rare earth metal doping in TiO_2 leads to improved optical absorption, band gap engineering, enhanced photocatalytic activity, and better structural and optoelectronic properties. These modifications make rare earth metal-doped TiO_2 nanostructures promising materials for various applications in energy, environment, and beyond.

Non Metal Doping

TiO_2 is one of the many transition metal oxides that is frequently utilized as a photocatalyst because of its special qualities, which include its low cost, good optical activity, high chemical stability, and non-toxicity [53, 54]. TiO_2-based nanoparticles have been used in a variety of fields, such as medicine delivery, energy storage, antimicrobial materials, and self-cleaning applications. This catalyst is specifically used extensively for the chemical polutents photodegradation which are found in wastewater; when exposed to UV–Vis light, it accelerates the breakdown of organic compounds [55, 56]. Despite this, the photocatalytic efficacy of unblended TiO_2 is yet relatively poor, because of its wide bandgap which makes it additionally vulnerable to UV light excitation, and the rapid recombination of photogenerated electron–hole (e^-/h^+) pairs. Nonetheless, a model photocatalyst need to have a bandgap that permits stimulation by visible light [57, 58]. Effective charge separation, charge carrier and transport, and appropriate valence band (VB) and conduction band (CB) locations for redox reactions are all desirable characteristics of these nanostructured photocatalysts. The UV region comprises a very small portion of the solar spectrum, hence visible-light active nano photocatalysts are favored in this situation [59, 60]. Therefore, altering the inner band structure of TiO_2 by substituting oxygen in the TiO_2 lattice to lower the optical energy bandgap is one way to cause visible light absorption in the material [61]. Numerous structural and chemical alterations have been employed to increase the photocatalytic activity of TiO_2 and solve some of the previously listed problems, notably bandgap energy and recombination rate. Figure 10 symbolysis the change in band gap after doping with metal and non metal. As non-metal doping inhibits electron–hole recombination and increases the material's absorption of visible light, it is believed to be a valuable method for increasing TiO_2's photocatalytic efficiency. Non-metal atoms in the first row have the ability to penetrate through lattice interstices and form bonds with other lattice atoms because

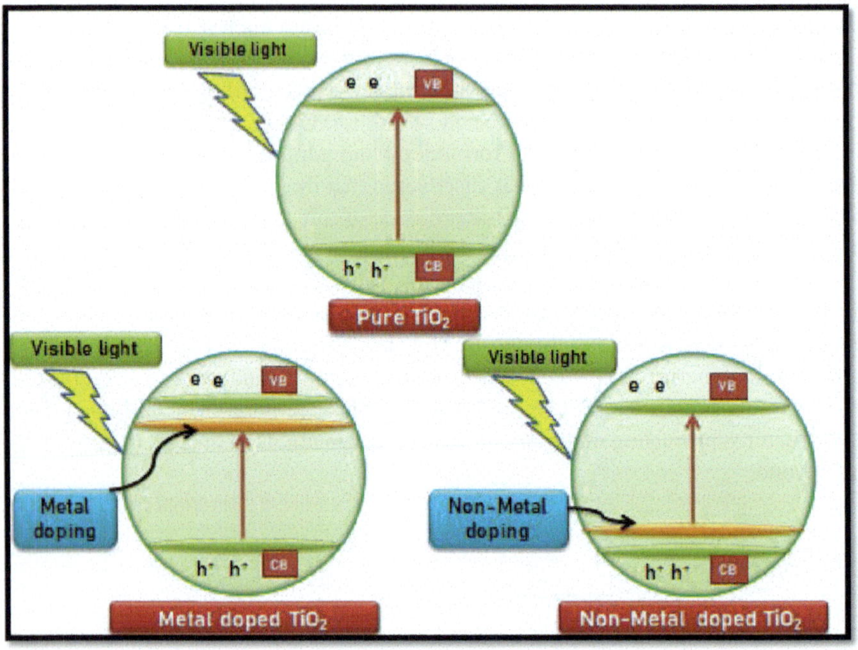

Fig. 10 Shift in band gap due to doping of metals and non metals

of their minuscule size, especially when they are neutral. Depending on their redox potential, they can donate one or more electrons to the lattice and become oxidized [62]. To change the properties of TiO_2, impurities are added to the crystal structure by a process called doping. Non-metal doping is commonly used to produce p-type or n-typed doping.

4.1.2 Heterojunction Formation

It has been shown that creating surface heterojunctions is another effective way to encourage the photocatalytic properties of TiO_2-based nanomaterials. It is widely accepted that highly active and stable photogenerated electron and hole recombination can be suppressed and charge transfer facilitated by well-defined junctions [63, 64]. n–n and p–n junctions are two types of TiO_2 semiconductor heterojunctions that are frequently built to enhance charge separation and increase the lifetime of photo-excited charge carriers. For example, TiO_2-based n–n heterojunctions have been made accessible for a variety of photocatalytic applications, such as CdS/TiO_2, CeO_2/TiO_2, WO_3/TiO_2, ZnO/TiO_2, TiO_2/In_2O_3 and Ag_2CO3/TiO_2. Because these semiconductor heterojunctions generate interfacial electrical fields, they can lower the recombination rate of e–h^+ pairs [65–70]. On the other hand, TiO_2-based p–n heterojunctions, like MoS_2/TiO_2, $BiOI/TiO_2$, NiO/TiO_2 and Cu_2O/TiO_2 have been extensively studied for their ability to effectively separate charges due to the photo-induced generated p–n junction interface and to extend the visible light absorption of TiO_2 semiconductors with wider band gaps [71–74].

4.1.3 Surface Chemical Modification

From a conceptual standpoint, one of the main limitations on the photocatalytic efficacy of TiO_2 nanoparticles is their limited absorption of visible light. TiO_2 nanoparticles can only utilise 3–4% of UV radiation due to their broad band gap, and they are unable to properly utilise the visible light spectrum, which contains around 48% of the total solar energy. In order to overcome this problem, research has concentrated on making TiO_2 nanoparticles more visible light active by moving the beginning of their light response from the ultraviolet to the visible spectrum. As in Fig. 11 [75] there is doping of diborane in TiO_2 which results in higher visible light senstivity.

Non-metal-doped and metal-doped TiO_2 nanomaterials have been created and extensively used in a variety of photocatalytic sectors over the past few decades. There are some strategies by which we can make changes in our material for example impurity doping, structural defects insertion, surface sensitization, surface plasmon resonance effect generation by this we can enhance the visible light activity of our materials [76].

4.1.4 Catalytic Doping

TiO_2 is one of the many semiconductors used, and because of its broad band gap and photosensitivity, it is well known to be an effective photocatalyst for the degradation of many environmental pollutants. Although this photocatalyst has several advantages, using it has certain disadvantages [77]. Some of the disadvantage in TiO_2 is that Visible light cannot be used due to the band edge absorption threshold [78] and charge Carrier recombination happens in a matter of nanoseconds [79]. Many methods have been put forth to extend the catalyst's carrier life time and enhance its light absorption properties in order to get over these restrictions. The long wavelength absorption can be achieved via dye sensitization, band gap alterations, or exterior surface alteration. By introducing electrons from the dye's excited level into the semiconductor conduction band, a variety of organic dyes can be used to dye TiO_2 and increase its sensitivity into the visible range. Doping is the most popular and effective method for bandgap tailoring. In certain cases, doping can extend the lifetime of charge carriers if the dopants have energy levels that are either slightly above or below TiO_2 valence band, which results in shallow charge carrier entrapment. If the dopant and semiconductor have a strong enough electronic coupling effect to alter the band structure, bandgap narrowing may also be achievable. The photo-redox potential energy of TiO_2 may be somewhat decreased as a result of the band gap reduction if the redox potential of the H_2O/OH pair (−2.8 V) is inside it. Nevertheless, a rise in photon absorption capacity or the rate at which charge

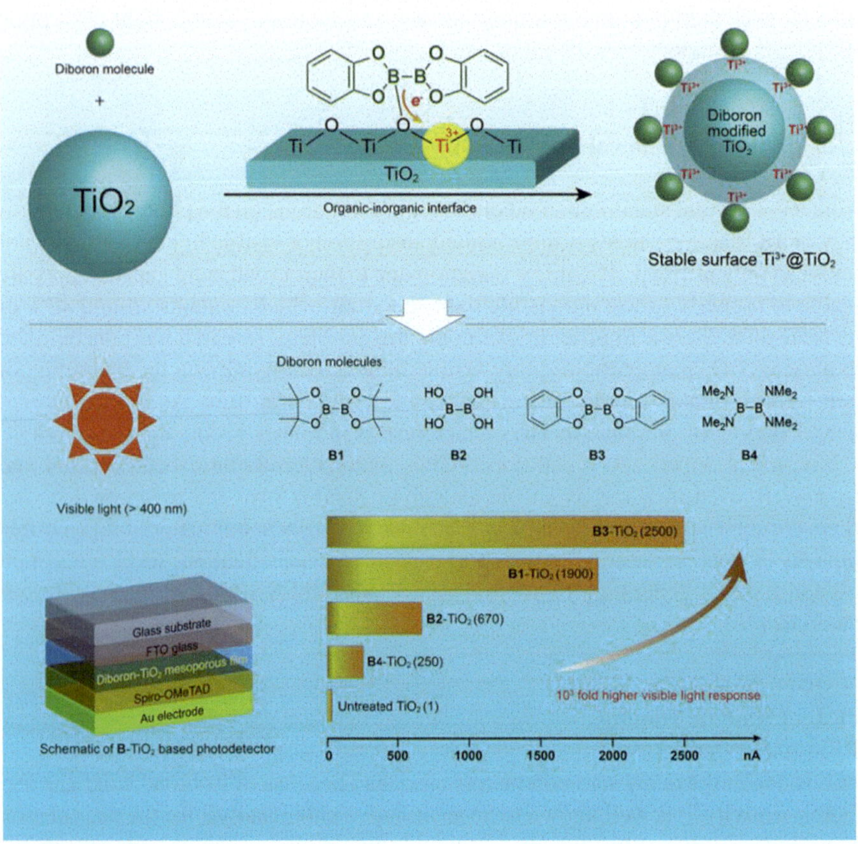

Fig. 11 Diborane doped TiO_2 results in better visble light response [75] (open access)

carriers are produced might counteract this detrimental impact and greatly raise the total photo-reactivity. To increase TiO_2 optical absorption and photocatalytic activity, doping with different metal and non-metal ions has been tried. In order to improve charge separation efficiency, connecting semiconductor particles with other semiconductors is a major strategy. TiO_2 quantum yield will rise when the photocatalyst is supported on an adsorbent such as zeolites [77].

5 Application of TiO_2 Nanomaterials

In the environmental and energy domains, TiO_2 photocatalysis is widely applied in a wide range of products and applications as shown in Fig. 12 [80] such as photoelectrochemical conversion, hydrogen evolution, water and air cleansing systems,

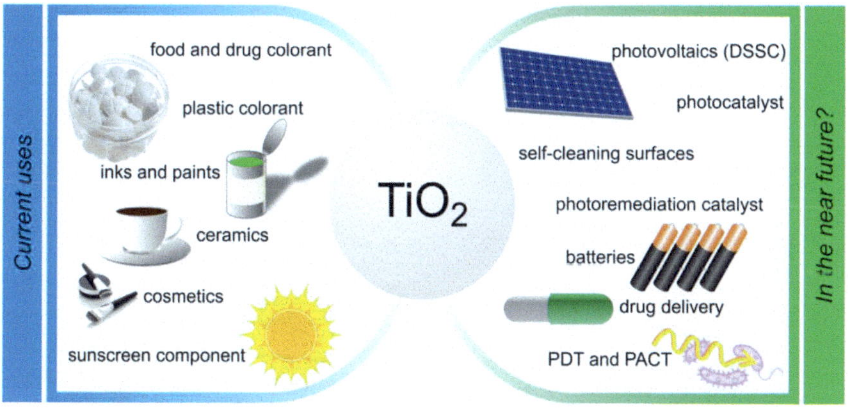

Fig. 12 Application of TiO_2 nanomaterials and some future aspects of TiO_2 nanomaterials [80] (open access)

sterilization, and self-cleaning surfaces. However, new material research is essential to enhance photocatalytic activity and find new uses for TiO_2 photocatalysis [81].

In the early 1900s, TiO_2 was first produced in large quantities as a non-toxic paint dye substitute. At the moment, the yearly production of TiO_2 exceeds 4 M tons. This compound is employed in a wide range of everyday products, which include sun cream production, white plastics, medication, and affordable, harmless food pigmentation that has been approved by the relevant European Union authorities for the appropriateness of food additives. It is additionally employed as a colorant in white plastics [41]. Research into the required applications of TiO_2 NPs began in 1985 with the publication of one of the first articles on photocatalytic disinfection [82]. Ever since, there has been a steady rise in the application of TiO_2 NPs in photodynamic therapy research. It has been associated with the photodynamic inactivation of bacteria that resist antibiotics and the use of TiO_2 NPs as photosensitizing compounds in the cancer treatment. TiO_2 NPs and their composites, combinations, and hybrids with other molecules have been thoroughly studied as photosensitizers in photodynamic therapy. TiO_2 can be used to treat cancerous tumors by creating bioconjugates with cell-specific monoclonal antibodies [83].

5.1 *Environmental Applications*

When exposed to UV light, TiO_2 NPs act as a photocatalyst, generating holes, electrons, and charge carriers in the VB and CB, respectively. As Moradeeya et al. proposed a model where with the help of TiO_2 photocatalysis process we can treat waste water and purify it and can overcome waste water problems which are present due to waste water in coming through factories in environment as shown in Fig. 13 [84]. Once at the surface, these separated charge carriers combine with ambient O_2 or

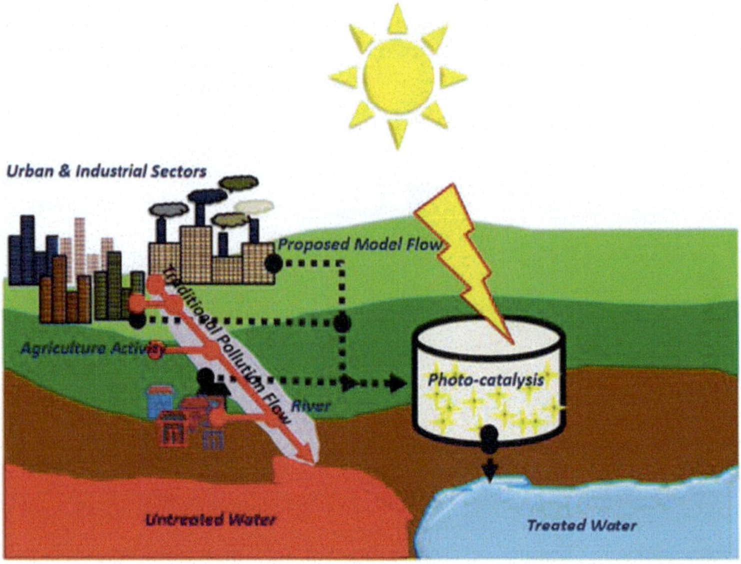

Fig. 13 Model signifies the usefulness of TiO_2 photocatalysis in treating waste water. Reproduced with permission from Ni et al. [84], Copyright (2022) Elsevier

adsorbed H_2O molecules to create several reactive oxygen species (ROS), including $O_2^{-\bullet}$ and $OH^{\bullet}$. For dyes to be destroyed by photocatalysis, $OH^{\bullet}$ must be created [85, 86]. Because the photogenerated charge carriers recombine quickly, there is less ROS produced, which lowers photocatalytic activity.

This could be enhanced by doping TiO_2 nanospheres with rare earth (RE) metals, which produce separation and charge imbalance by making more electrons and holes available to take part in O_2 and H_2O oxidation and reduction reactions [85, 87, 88]. By this there is production of more reactive oxygen species (ROS) and accelerates the photocatalytic degradation of dyes adsorbed on TiO_2 surfaces. Visible light radiation is absorbed more readily when RE metal is doped because it alters the absorption threshold to a lower energy or a larger wavelength [88]. The TiO_2 band gap is engineered by doping with RE metal ions, which introduce sub-energy levels below the CB [89, 90]. Electrons from VB require less energy to transfer to the sub energy level than does the transition from VB to CB of TiO_2 because of the presence of varied energy levels caused by the inclusion of RE metal ions. This transition can occur when visible light is absorbed, and it can be adjusted to even absorb near-infrared light by changing the shape of TiO_2 nanosheets alone or in combination with other functional nanomaterials [85, 87, 90]. All of these changes brought about by RE metal ion doping in TiO_2 improve the optical characteristics and photocatalytic efficiency of photocatalysis [88]. The modification of TiO_2 was investigated to boost the feasibility of photocatalytic oxidation for use in large-scale applications. In order to increase TiO_2's UV and visible photoactivity and remove impurities from interior environments that include volatile organic compounds, C-doped TiO_2 photocatalysts were created. The findings demonstrated that carbon atoms are found in the TiO_2 lattice at both interstitial and substitutional sites. On the surface of the photocatalyst, carbon atoms are also dispersed as carbonaceous species. Above the TiO_2 valence band, the doped carbon creates a new electronic site that promotes charge carrier separation, prevents their recombination, and causes the light to be absorbed into the visible spectrum. The C-doped TiO_2 photocatalysts shown an improvement over bare TiO_2 in terms of both the reaction rate in the visible- and UV-photocatalytic oxidation processes and the degradation of gaseous methyl ethyl ketone. Additionally, the effects of carbon content and relative humidity on the removal of methyl ethyl ketone were investigated under both UV and visible light irradiations. The findings demonstrated that C-TiO_2-0.1% had the highest rate of methyl ethyl ketone degradation under all studied settings [91].

5.2 Energy Applications

Due to their improved qualities over pure TiO_2, doped TiO_2 materials have demonstrated considerable potential in a variety of energy applications. Considerable research has been done on doped TiO_2 for application in dye-Sensitized Solar Cells. TiO_2 nanomaterials have many By doping TiO_2 with elements like gadolinium, one can increase its capacity to absorb light and increase the efficiency of turning solar energy into electrical power [92]. Additionally, dopants can decrease recombination rates and enhance charge transfer, which will enhance the efficiency of dye-Sensitized Solar Cells [93, 94]. Numerous studies have been conducted on doped TiO_2 nanoparticles for photocatalytic uses. TiO_2 can be doped with elements such as lanthanides to increase its absorption range into the visible spectrum, which will increase its efficiency for photocatalytic reactions in the presence of sunlight [95–98]. Because of their great theoretical and practical energy density, Li–S batteries are attracting attention. When elemental sulfur combines with Li^+ ions during charge/discharge cycles, soluble lithium polysulfides (LiPS) are produced in the electrolyte, which is the main cause of capacity deterioration. These LiPS diffuse toward the anode and cause the shuttle effect. One reasonable strategy to improve the electrochemical performance of the cell is to regulate the reactions on the coated separator surface in order to reduce LiPS shuttle between electrodes. In this study, the sol–gel method was used to manufacture Fe-doped TiO_2 [1 5% of weight]. The glass fiber separator's new coating enhanced polysulfide trapping and electrical conductivity [99]. There are so much of potential in TiO_2 nanomaterials in various fields. These versatile material can be applied in biomedical areas like disinfection, imaging, sensing, and drug delivery. Additionally, they hold promise for energy applications which are explained in next section such as environmental remediation, enhancing battery performance, producing hydrogen through water splitting, and improving solar cell efficiency (Fig. 14).

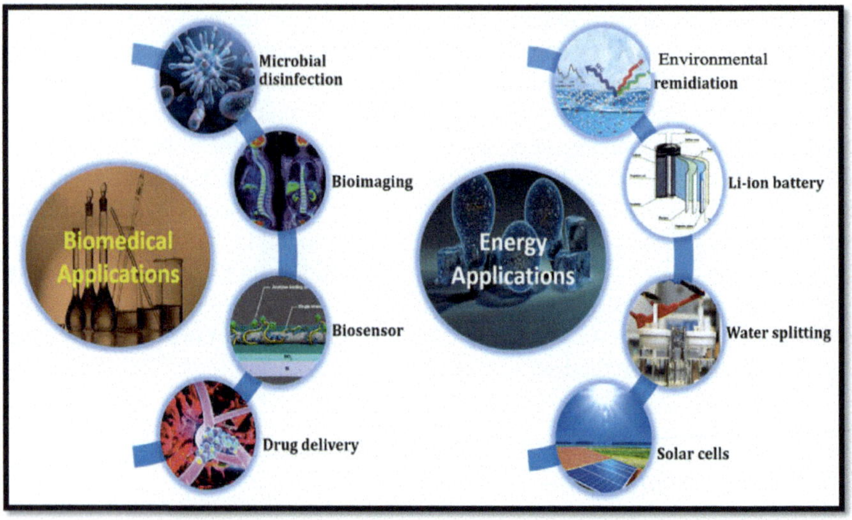

Fig. 14 Biomedical and energy application of TiO_2 nanomaterials [100] (open access)

5.3 *Biomedical Applications*

One of the most common materials to appear in a variety of technological domains, including biosensing, energy, and medicine, is TiO_2 nanostructures. Several TiO_2 nanostructures, such as nanowires and nanotubes, have been used in photoelectrochemical (PEC) biosensing applications, significantly improving target detection. By offering nanoporous surfaces and hierarchical structures, TiO_2 nanostructures are effective additive materials that can make up for the drawbacks of titanium implants, such as poor surface contact with surrounding tissues. They are used as coatings or strengthened materials for the exposed titanium implant surface. Additionally, a variety of medications, including those for osteoporosis, cancer, and infections, can be placed into these nanostructures and utilized as local drug delivery systems. TiO_2 nanostructures and their derivatives are also newly developed antibacterial agents that can combat pathogenic germs that pose a threat to humans [25]. Numerous research lines have demonstrated that the therapeutic efficacy of drugs used in the treatment of cancer may be enhanced by the targeted medication delivery of cytotoxic chemotherapy agents loaded on TiO_2-based nanostructures. Li et al. [101] examined for the first time the possible use of one-dimensional TiO_2 whiskers for drug delivery in cancer treatment, as well as the combination of these whiskers impacts on the uptake and buildup of daunorubicin in SMMC-7721 cells [101]. Because nanomaterials are so small, they can enter and be absorbed by human cells and tissue through injections, inhalation, skin penetration, and ingestion. Furthermore, the toxicity of nanostructures is frequently manifested by membrane damage, protein malfunction,

DNA mutation, cell death, proliferative behavior, and other mechanism. TiO_2 nanomaterials are very use full in bone implants, dental implants, biosensors, treatment of brain tumours and in cardiovascular stents [102].

6 Conclusion

TiO_2 becomes a key component in the search for long-term answers to the energy and environmental problems of the modern world. Because of its special qualities such as its low cost, non-toxicity, and potent photocatalytic capabilities it is now at the forefront of research into renewable energy sources and pollution mitigation. The report emphasizes the intrinsic limits of TiO_2, in particular how its broad bandgap limits its photocatalytic activity to UV light. On the other hand, investigating different modification methods—like doping and linking with other semiconductors offers viable paths for improving its visible light performance. These modifications improve the light absorption characteristics of TiO_2, enable better separation of charges and lessen the photogenerated electron–hole pairs recombination, all of which increase TiO_2 efficiency in applications like photocatalysis. The wide range of uses for TiO_2, from antimicrobial treatments and solar cells to water purification and lithium-ion batteries, highlight its adaptability and significance across numerous industries. The increasing understanding of TiO_2 nanoparticles' potential to surpass conventional materials in a range of applications is demonstrated by the continuous study conducted on them, especially on those that have undergone modifications to improve their physicochemical qualities. This trend is vital since the world is facing serious challenges including energy scarcity and environmental deterioration, which call for the creation of novel and useful technology.

References

1. Wang, J., et al.: Synthesis, modification and application of titanium dioxide nanoparticles: a review. Nanoscale **14**(18), 6709–6734 (2022)
2. Long, L. L., et al.: A green approach for preparing doped TiO_2 single crystals. ACS Appl. Mater. Interfaces **6**(19), 16712–16720 (2014)
3. Fujishima, A., Honda, K.: Electrochemical photolysis of water at a semiconductor electrode. Nature **238**(5358), 37–38 (1972)
4. Lakshmanan, V., Bhowmick, A., Abdul Halim, M.: Titanium dioxide: production, properties and applications. In: Titanium Dioxide: Chemical Properties, Applications and Environmental Effects. Nova Science Publishers, New York (2014)
5. Hwu, Y., et al.: X-ray absorption of nanocrystal TiO_2. Nanostruct. Mater. **9**(1–8), 355–358 (1997)
6. Samat, M., et al.: Hubbard U calculations on optical properties of 3d transition metal oxide TiO_2. Results Phys. **6**, 891–896 (2016)
7. Dreno, B., Alexis, A., Chuberre, B., Marinovich, M.: Safety of titanium dioxide nanoparticles in cosmetics. J. Eur. Acad. Dermatol. Venereol. **33**, 34–46 (2019)

8. Kavan, L., et al.: Electrochemical and photoelectrochemical investigation of single-crystal anatase. J. Am. Chem. Soc. **118**(28), 6716–6723 (1996)
9. Diebold, U.: The surface science of titanium dioxide. Surf. Sci. Rep. **48**(5–8), 53–229 (2003)
10. Kavan, L.: Titania in diverse forms as substrates. In: Dye-Sensitized Solar Cells, pp. 45–81. CRC Press Taylor & Francis, Boca Raton (2010)
11. Etacheri, V., et al.: Visible-light activation of TiO_2 photocatalysts: advances in theory and experiments. J. Photochem. Photobiol., C **25**, 1–29 (2015)
12. Li, C., et al.: Effect of sinter temperature on the electrical properties of TiO_2-based capacitor–varistors. Mater. Lett. **57**(8), 1400–1405 (2003)
13. Grätzel, M.: Solar energy conversion by dye-sensitized photovoltaic cells. Inorg. Chem. **44**(20), 6841–6851 (2005)
14. Kumar, A., Pandey, G.: Different methods used for the synthesis of TiO_2 based nanomaterials: a review. Am. J. Nano Res. Appl. **6**(1), 1 (2018)
15. Zhang, H., Banfield, J.F.: Structural characteristics and mechanical and thermodynamic properties of nanocrystalline TiO_2. Chem. Rev. **114**(19), 9613–9644 (2014)
16. Noman, M.T., Ashraf, M.A., Ali, A.: Synthesis and applications of nano-TiO_2: a review. Environ. Sci. Pollut. Res. **26**, 3262–3291 (2019)
17. Zhang, D., et al.: Modified hydrothermal synthesis and structural characterization of monoclinic (KxNa1− x) NbO3 ($0.05 \leq x \leq 0.15$) rods. Ceram. Int. **41**(7), 8837–8842 (2015)
18. Kretzschmar, A.L., Manefield, M.: The role of lipids in activated sludge floc formation. AIMS Environ. Sci. **2**(2), 122–133 (2015)
19. Makwana, N.M., et al.: Pilot plant scale continuous hydrothermal synthesis of nano-titania; effect of size on photocatalytic activity. Mater. Sci. Semicond. Process. **42**, 131–137 (2016)
20. Suriakarthick, R., et al.: Solvothermal synthesis, structural and transport properties of polycrystalline copper tin selenide for thermoelectric applications. Inorg. Chem. Commun. **140**, 109491 (2022)
21. Zdravkov, A., et al.: A new solvothermal route to efficient titania photocatalyst. Mater. Chem. Phys. **160**, 73–79 (2015)
22. Kurajica, S., et al.: The utilization of modified alkoxide as a precursor for solvothermal synthesis of nanocrystalline titania. Mater. Chem. Phys. **196**, 194–204 (2017)
23. Albano, C., Sarmiento, Y., González, G.: Synthesis and Characterization of Nanostructures: MWCNTf/TiO2 and MWCNTf/TiO2/HAp. in Macromolecular Symposia. Wiley Online Library (2012)
24. Wu, J.-M., et al.: Large-scale preparation of ordered titania nanorods with enhanced photocatalytic activity. Langmuir **21**(15), 6995–7002 (2005)
25. Jafari, S., et al.: Biomedical applications of TiO_2 nanostructures: recent advances. Int. J. Nanomed. 3447–3470 (2020)
26. Owens, G.J., et al.: Sol–gel based materials for biomedical applications. Prog. Mater. Sci. **77**, 1–79 (2016)
27. Tian, N., et al.: Synthesis of tetrahexahedral platinum nanocrystals with high-index facets and high electro-oxidation activity. Science **316**(5825), 732–735 (2007)
28. Catauro, M., et al.: Influence of the heat treatment on the particles size and on the crystalline phase of TiO_2 synthesized by the sol-gel method. Materials **11**(12), 2364 (2018)
29. Rahman, I.A., Padavettan, V.: Synthesis of silica nanoparticles by sol-gel: size-dependent properties, surface modification, and applications in silica-polymer nanocomposites—a review. J. Nanomater. **2012**(1), 132424 (2012)
30. Aguilar, G.V.: Introductory chapter: a brief semblance of the sol-gel method in research. In: Sol-Gel Method-Design and Synthesis of New Materials with Interesting Physical, Chemical and Biological Properties (2018)
31. Gupta, S., Tripathi, M.: A review on the synthesis of TiO_2 nanoparticles by solution route. Open Chem. **10**(2), 279–294 (2012)
32. Vijayalakshmi, R., Rajendran, V.: Synthesis and characterization of nano-TiO_2 via different methods. Arch. Appl. Sci. Res. **4**(2), 1183–1190 (2012)

33. Djerdj, I., et al.: Transmission electron microscopy studies of nanostructured TiO_2 films on various substrates. Vacuum **80**(4), 371–378 (2005)
34. Lee, J.-H., et al.: Fabrication of titania nanotubular film with metal nanoparticles. J. Cryst. Growth **311**(3), 638–641 (2009)
35. Wu, J.-M., Hayakawa, S., Tsuru, K., Osaka, A.: Nanocrystalline titania made from interactions of Ti with hydrogen peroxide solutions containing tantalum chloride. Cryst. Growth Des. **2**(2), 147–149 (2002)
36. Ding, Z., et al.: Synthesis of anatase TiO_2 supported on porous solids by chemical vapor deposition. Catal. Today **68**(1–3), 173–182 (2001)
37. Bessekhouad, Y., Robert, D., Weber, J.: Bi2S3/TiO_2 and CdS/TiO_2 heterojunctions as an available configuration for photocatalytic degradation of organic pollutant. J. Photochem. Photobiol., A **163**(3), 569–580 (2004)
38. Ali, I., Suhail, M., Alothman, Z.A., Alwarthan, A.: Recent advances in syntheses, properties and applications of TiO_2 nanostructures. RSC Adv. **8**(53), 30125–30147 (2018)
39. Al Jitan, S., Palmisano, G., Garlisi, C.: Synthesis and surface modification of TiO_2-based photocatalysts for the conversion of CO_2. Catalysts **10**(2), 227 (2020)
40. Ghosh, S., Das, A.: Modified titanium oxide (TiO_2) nanocomposites and its array of applications: a review. Toxicol. Environ. Chem. **97**(5), 491–514 (2015)
41. Carp, O., Huisman, C.L., Reller, A.: Photoinduced reactivity of titanium dioxide. Prog. Solid State Chem. **32**(1–2), 33–177 (2004)
42. Mor, G.K., et al.: A review on highly ordered, vertically oriented TiO_2 nanotube arrays: fabrication, material properties, and solar energy applications. Solar Energy Mater. Solar Cells **90**(14), 2011–2075 (2006)
43. Burda, C., et al.: Enhanced nitrogen doping in TiO_2 nanoparticles. Nano Lett. **3**(8), 1049–1051 (2003)
44. Hernández, J.V., et al.: Effects of metal doping (Cu, Ag, Eu) on the electronic and optical behavior of nanostructured TiO_2. J. Alloys Compd. **710**, 355–363 (2017)
45. Wilke, K., Breuer, H.: The influence of transition metal doping on the physical and photocatalytic properties of titania. J. Photochem. Photobiol. A: Chem. **121**(1), 49–53 (1999)
46. Choi, W., Termin, A., Hoffmann, M.R.: The role of metal ion dopants in quantum-sized TiO_2: correlation between photoreactivity and charge carrier recombination dynamics. J. Phys. Chem. **98**(51), 13669–13679 (2002)
47. Papp, J., Soled, S., Dwight, K., Wold, A.: Surface acidity and photocatalytic activity of TiO_2, WO_3/TiO_2, and MoO_3/TiO_2 photocatalysts. Chem. Mater. **6**(4), 496–500 (1994)
48. Choi, J., Park, H., Hoffmann, M.R.: Combinatorial doping of TiO_2 with platinum (Pt), chromium (Cr), vanadium (V), and nickel (Ni) to achieve enhanced photocatalytic activity with visible light irradiation. J. Mater. Res. **25**(1), 149–158 (2010)
49. Karuppasamy, P., et al.: An investigation of transition metal doped TiO_2 photocatalysts for the enhanced photocatalytic decoloration of methylene blue dye under visible light irradiation. J. Environ. Chem. Eng. **9**(4), 105254 (2021)
50. Guan, B., et al.: Porous nickel doped titanium dioxide nanoparticles with improved visible light photocatalytic activity. Nanoscale Adv. **2**(3), 1352–1357 (2020)
51. Prakash, J., et al.: Novel rare earth metal–doped one-dimensional TiO_2 nanostructures: fundamentals and multifunctional applications. Mater. Today Sustain. **13**, 100066 (2021)
52. Tang, X., et al.: DFT and experimental study on visible-light driven photocatalysis of rare-earth-doped TiO_2. Vacuum **200**, 110972 (2022)
53. Reddy, K.R., et al.: Enhanced photocatalytic activity of nanostructured titanium dioxide/polyaniline hybrid photocatalysts. Polyhedron **120**, 169–174 (2016)
54. Guo, G.-S., et al.: Synthesis of titania and titanate nanomaterials and their application in environmental analytical chemistry. Talanta **72**(5), 1687–1692 (2007)
55. de Lima, J.F., et al.: Control of chemical state of cerium in doped anatase TiO_2 by solvothermal synthesis and its application in photocatalytic water reduction. J. Mater. Chem. A **3**(18), 9890–9898 (2015)

56. Zhu, et al.: Hydrogen titanate nanofibers covered with anatase nanocrystals: a delicate structure achieved by the wet chemistry reaction of the titanate nanofibers. J. Am. Chem. Soc. **126**(27), 8380–8381 (2004)
57. Reddy, K.R., et al.: Facile fabrication and photocatalytic application of Ag nanoparticles-TiO_2 nanofiber composites. J. Nanosci. Nanotechnol. **11**(4), 3692–3695 (2011)
58. Zhou, W., et al.: Ag_2O/TiO_2 nanobelts heterostructure with enhanced ultraviolet and visible photocatalytic activity. ACS Appl. Mater. Interfaces **2**(8), 2385–2392 (2010)
59. Priyanka, K., et al.: Influence of La doping on structural and optical properties of TiO_2 nanocrystals. Mater Charact **113**, 144–151 (2016)
60. Salehifar, N., Nikfarjam, A.: Improvement the visible light photocatalytic activity of gold nanoparticle, Co_2O_3 and nitrogen doped TiO_2 nanofibers. Mater. Lett. **188**, 59–62 (2017)
61. Lázaro-Navas, S., Prashar, S., Fajardo, M., Gómez-Ruiz, S.: Visible light-driven photocatalytic degradation of the organic pollutant methylene blue with hybrid palladium–fluorine-doped titanium oxide nanoparticles. J. Nanopart. Res. **17**, 1–14 (2015)
62. Finazzi, E., Di Valentin, C., Pacchioni, G.: Boron-doped anatase TiO_2: pure and hybrid DFT calculations. J. Phys. Chem. C **113**(1), 220–228 (2009)
63. Li, X., et al.: Engineering heterogeneous semiconductors for solar water splitting. J. Mater. Chem. A **3**(6), 2485–2534 (2015)
64. Yuan, Y.-P., et al.: Hetero-nanostructured suspended photocatalysts for solar-to-fuel conversion. Energy Environ. Sci. **7**(12), 3934–3951 (2014)
65. Li, X., et al.: Synthesis and photoactivity of nanostructured CdS–TiO_2 composite catalysts. Catal. Today **225**, 64–73 (2014)
66. Li, G., Zhang, D., Jimmy, C.Y.: Thermally stable ordered mesoporous CeO_2/TiO_2 visible-light photocatalysts. Phys. Chem. Chem. Phys. **11**(19), 3775–3782 (2009)
67. Yang, L., Si, Z., Weng, D., Yao, Y.: Synthesis, characterization and photocatalytic activity of porous WO_3/TiO_2 hollow microspheres. Appl. Surf. Sci. **313**, 470–478 (2014)
68. Wang, Y., et al.: One-step template-free fabrication of mesoporous ZnO/TiO_2 hollow microspheres with enhanced photocatalytic activity. Appl. Surf. Sci. **307**, 263–271 (2014)
69. Li, C., et al.: Ultrasonic aerosol spray-assisted preparation of TiO_2/In_2O_3 composite for visible-light-driven photocatalysis. J. Catal. **310**, 84–90 (2014)
70. Yu, C., et al.: Enhancing the photocatalytic performance of commercial TiO_2 crystals by coupling with trace narrow-band-gap Ag_2CO_3. Ind. Eng. Chem. Res. **53**(14), 5759–5766 (2014)
71. Dai, G., Yu, J., Liu, G.: Synthesis and enhanced visible-light photoelectrocatalytic activity of p−n junction $BiOI/TiO_2$ nanotube arrays. J. Phys. Chem. C **115**(15), 7339–7346 (2011)
72. Yu, J., Wang, W., Cheng, B.: Synthesis and enhanced photocatalytic activity of a hierarchical porous flowerlike p–n junction NiO/TiO_2 photocatalyst. Chem. Asian J. **5**(12), 2499–2506 (2010)
73. Liu, L., Yang, W., Sun, W., Li, Q., Shang, J.: Creation of Cu_2O@ TiO_2 composite photocatalysts with p–n heterojunctions formed on exposed Cu_2O facets, their energy band alignment study, and their enhanced photocatalytic activity under illumination with visible light. ACS Appl. Mater. Interfaces. **7**, 1465–1476 (2015)
74. Xiong, D., et al.: Preparation and characterization of $CuCrO_2/TiO_2$ heterostructure photocatalyst with enhanced photocatalytic activity. Appl. Surf. Sci. **347**, 747–754 (2015)
75. Cao, Y., et al.: Modification of TiO_2 nanoparticles with organodiboron molecules inducing stable surface Ti^{3+} complex. IScience **20**, 195–204 (2019)
76. Chen, X., Mao, S.S.: Titanium dioxide nanomaterials: synthesis, properties, modifications, and applications. Chem. Rev. **107**(7), 2891–2959 (2007)
77. Anandan, S., Ikuma, Y., Niwa, K.: An overview of semi-conductor photocatalysis: modification of TiO_2 nanomaterials. Solid State Phenom. **162**, 239–260 (2010)
78. Kumar, S.G., Devi, L.G.: Review on modified TiO_2 photocatalysis under UV/visible light: selected results and related mechanisms on interfacial charge carrier transfer dynamics. J. Phys. Chem. A **115**(46), 13211–13241 (2011)

79. Lettieri, S., et al.: Charge carrier processes and optical properties in TiO_2 and TiO_2-based heterojunction photocatalysts: a review. Materials **14**(7), 1645 (2021)
80. Ziental, D., et al.: Titanium dioxide nanoparticles: prospects and applications in medicine. Nanomaterials **10**(2), 387 (2020)
81. Nakata, K., Fujishima, A.: TiO_2 photocatalysis: design and applications. J. Photochem. Photobiol. C: Photochem. Rev. **13**(3), 169–189 (2012)
82. Matsunaga, T., Tomoda, R., Nakajima, T., Wake, H.: Photoelectrochemical sterilization of microbial cells by semiconductor powders. FEMS Microbiol. Lett. **29**(1–2), 211–214 (1985)
83. Ni, W., et al.: 808 nm light triggered black TiO_2 nanoparticles for killing of bladder cancer cells. Mater. Sci. Eng. C. **81**, 252–260 (2017)
84. Moradeeya, P.G., Sharma, A., Kumar, M.A., Basha, S.: Titanium dioxide based nanocomposites—current trends and emerging strategies for the photocatalytic degradation of ruinous environmental pollutants. Environ. Res. **204**, 112384 (2022)
85. Prakash, J., Sun, S., Swart, H.C., Gupta, R.K.: Noble metals-TiO_2 nanocomposites: from fundamental mechanisms to photocatalysis, surface enhanced Raman scattering and antibacterial applications. Appl. Mater. Today **11**, 82–135 (2018)
86. Đorđević, V., Milićević, B., Dramićanin, M.D.: Rare earth-doped anatase TiO_2 nanoparticles. Titanium Dioxide 24–60 (2017)
87. Jai, P., et al.: Synthesis, characterization and multifunctional properties of plasmonic Ag–TiO_2 nanocomposites. Nanotechnology **27**(35), 355707 (2016)
88. Reszczyńska, J., et al.: Photocatalytic activity and luminescence properties of RE^{3+}–TiO_2 nanocrystals prepared by sol–gel and hydrothermal methods. Appl. Catal. B **181**, 825–837 (2016)
89. Singh, N., et al.: Dual functional Ta-doped electrospun TiO_2 nanofibers with enhanced photocatalysis and SERS detection for organic compounds. ACS Appl. Mater. Interfaces **9**(34), 28495–28507 (2017)
90. Mazierski, P., et al.: Enhanced photocatalytic properties of lanthanide-TiO_2 nanotubes: an experimental and theoretical study. Appl. Catal. B **205**, 376–385 (2017)
91. Shayegan, Z., Haghighat, F., Lee, C.-S.: Carbon-doped TiO_2 film to enhance visible and UV light photocatalytic degradation of indoor environment volatile organic compounds. J. Environ. Chem. Eng. **8**(5), 104162 (2020)
92. Wei, H., Wu, Y., Lun, N., Zhao, F.: Preparation and photocatalysis of TiO_2 nanoparticles co-doped with nitrogen and lanthanum. J. Mater. Sci. **39**, 1305–1308 (2004)
93. Sayama, K., Arakawa, H.: Effect of carbonate salt addition on the photocatalytic decomposition of liquid water over Pt–TiO_2 catalyst. J. Chem. Soc., Faraday Trans. **93**(8), 1647–1654 (1997)
94. Xu, A.-W., Gao, Y., Liu, H.-Q.: The preparation, characterization, and their photocatalytic activities of rare-earth-doped TiO_2 nanoparticles. J. Catal. **207**(2), 151–157 (2002)
95. Li, R., et al.: Preparation of Gd-doped TiO_2 nanotube arrays by anodization method and its photocatalytic activity for methyl orange degradation. Catalysts **10**(3), 298 (2020)
96. Cheng, X., et al.: Structural, morphological, optical and photocatalytic properties of Gd-doped TiO_2 films. Thin Solid Films **615**, 13–18 (2016)
97. Zhou, Y., et al.: Preparation and performance of ordered porous TiO_2 film doped with Gd^{3+}. J. Mater. Eng. Perform. **20**, 1319–1322 (2011)
98. El-Bahy, Z.M., Ismail, A.A., Mohamed, R.M.: Enhancement of titania by doping rare earth for photodegradation of organic dye (Direct Blue). J. Hazard. Mater. **166**(1), 138–143 (2009)
99. Eroglu, O., Kiai, M.S., Kizil, H.: Performance enhancement of Li-S battery with the anatase nano structured Fe doped TiO_2 as a robust interlayer. J. Alloy. Compd. **838**, 155607 (2020)
100. Bokare, A., Chinnusamy, S., Erogbogbo, F.: Tio_2-graphene quantum dots nanocomposites for photocatalysis in energy and biomedical applications. Catalysts **11**(3), 319 (2021)
101. Li, Q., et al.: The incorporation of daunorubicin in cancer cells through the use of titanium dioxide whiskers. Biomaterials **30**(27), 4708–4715 (2009)
102. Cheng, Y., et al.: Progress in TiO_2 nanotube coatings for biomedical applications: a review. J. Mater. Chem. B **6**(13), 1862–1886 (2018)

Synthesis and Characterizations of Hybrid TiO_2 Based Multifunctional Nanomaterials

Tao Tao and Junghyun Cho

Abstract Aqueous precursor solution is often employed to produce metal oxide nanostructures and thin films through the hydrolysis of metal ions. Due to the constraints imposed by being far away from an equilibrium condition of synthetically grown oxides in solution, the degree of crystallinity is low, and the controllability of specific phases and structures can be challenging. It also requires very low supersaturation and a seed crystal, which often makes the process impractical and limited despite its potential as an alternative to vapor/plasma-based processes. A hydrothermal solution processing technique can, however, offer a more precise control of process parameters as a means of tailoring nanostructure developments of TiO_2 and its hybrid architecture. Especially, when it establishes the aggregation-driven growth model from oxide nanoparticles precipitated in supersaturated solution, it has potential to produce highly crystalline structures along with a high growth rate even at low temperature and cost. Hydrothermal process is carried out at various temperatures (between 80 and 220 °C) using mixed solvents and various fill levels to gauge the effect of autogenous pressure and liquid density during processing. In particular, there exists a supersaturation level of precursor solutions, below which the growth of high-quality, anisotropic crystals of TiO_2 is feasible. For a low degree of supersaturation, nanoparticles or nanowires with high crystallinity can aggregate to form the secondary nanostructures depending on their oriented attachment. Such particle-based crystallization pathways have the potential to produce a single crystalline nanorod array of TiO_2 from solution. The surface of these nanorods can be further treated via another hydrothermal method or ion exchange with other oxide phases to form the hybrid structure consisting of a core and a shell. It improved its functional properties compared to those of the original core phase, thereby making the hybrid TiO_2 nanostructures well suited for light or pathogen stimuli applications that require excellent photocatalytic, photoelectric, and antimicrobial properties. Importantly, the hydrothermal process provides more *sustainable* synthetic paths by offering low temperature, low energy consumption, and less toxic precursors.

T. Tao · J. Cho (✉)
Materials Science and Engineering Program, Department of Mechanical Engineering, Thomas J. Watson College of Engineering and Applied Science, Binghamton University (State University of New York), Binghamton, NY 13902, USA
e-mail: jcho@binghamton.edu

J. Prakash et al. (eds.), *Titanium Dioxide-Based Multifunctional Hybrid Nanomaterials*, Engineering Materials, https://doi.org/10.1007/978-3-031-80625-4_2

Keywords Rutile · Anatase · Supersaturation · Nucleation · Growth · Aggregation · Aqueous precursor · Solution · Non-classical crystallization · Sustainable

1 Introduction

Oxide materials have received great attention for use as a functional ceramic film [1–13], as well as a protective and barrier coating [14–18]. For these ceramic films and coatings, high-temperature processing is normally required to achieve sufficiently dense and high crystalline structures, which limits their use in many applications. Furthermore, those processes involve consumption of high energy and resources [19]. More attention has been recently paid to alternatives to such high-temperature, high-energy production methods. Therefore, low-temperature processing that can lead to the development of highly functional ceramic films will offer innovative solutions for numerous technical issues that neither metals nor polymers can solve. Current developments in organic and flexible electronics, which can be portable and rollable, present another opportunity to have low temperature processed ceramic films used as a hybrid or a standalone structure in those devices [20, 21].

One possible way of fabricating ceramic films at low temperatures is to take advantage of *in-situ* precipitated nanoparticles and nanostructures grown from *aqueous* solution [18, 22–29]. These nanostructures can tailor ceramic film formation and the subsequent microstructure development [28]. This approach is sufficiently general that it can be applied to a wide range of ceramic systems and does not require high-temperature post-annealing except a drying step. Not only does it provide '*environment-friendly*' processing, but aqueous solution processing also consumes much less energy compared to vapor/plasma based processes [30].

A good example of such aqueous-based processing can indeed be found in nature as seen in formation of biominerals such as nacre, sea urchin spine, and sponge spicule [31–33]. These aspects have been extensively studied in calcite- or silica-based systems, but little explored for formation of metal oxide films such as titanium oxide from the corresponding aqueous solution. The biominerals, however, form in ambient conditions over an extended period while synthetic oxides need to be processed in a reasonable time scale (within a few minutes to hours, as opposed to weeks to years). Due to the difficulty in reaching an equilibrium state under laboratory experimental conditions, synthetic oxide films processed at low temperatures usually contain defects and a high degree of non-crystallinity, which often make functional properties inferior to those of the films generated from conventional vapor-phase deposition, or sol–gel methods followed by high-temperature annealing.

In order to provide enhanced functional properties from solution-grown oxide films, it is, therefore, imperative to have a means of precisely controlling hydrolysis of metal ions in aqueous solution processing which strongly depends on reaction conditions such as pH, concentration, and temperature [34–36]. Good examples on this aspect have been published on three consecutive papers of TiO_2 films [28, 37, 38].

Especially, a low temperature processing has shown versatility to generate various nanostructures as well. The growth of low-dimensional nanostructures (1-D, 2-D) provides a possible way to enhance the crystallinity of the solution-prepared films that can ultimately benefit many applications including, for example, photoelectrochemical cells [39–41], photodetection and biosensing devices [42], chemical sensors [43], and piezoelectric energy harvesting devices [44–46]. Having them organized into micro- and macroscale structures in a *desirable shape* will be a key to scale-up their functional properties at the device levels.

Hydrothermal synthetic routes have received a particular attention as a means of producing various forms of inorganic materials such as thin films, single crystals, nanocrystals, nanorods, nanoplates, and bulk powders [47–51]. It utilizes the solubility of inorganic substances in water at elevated temperatures and pressures and the subsequent crystallization of the dissolved species from the aqueous solution. Hydrothermal processing is carried out in an autoclave that is sealed so that it can withstand high pressure and temperature for an extended time. It also offers the opportunity to produce oxide nanostructures in a variety of shapes and sizes from the nano- to micro-scale [52–54]. Careful control of the hydrothermal processing parameters is needed for successfully preparing for high-quality inorganic nanostructures. Hence, temperature and pressure are two important thermodynamic parameters while other reaction conditions such as liquid fill level, stirring of liquid, and temperature gradient inside the reactor will play an important role. Key concerns will be the solubility and diffusion characteristics of precursor liquid that tailors the quality of the crystals grown [49]. The former can be increased by adding a complexing agent (mineralizer) that includes inorganic ions such as OH^-, Cl^-, F^-, H^+, CO_3^{2-}, and SO_4^{2-} as it increases the solubility and promotes the recrystallization of minerals. Diffusion problems can be partly resolved in hydrothermal crystallization with decreased viscosity of the liquid.

Among various binary oxide systems, TiO_2 thin films have shown the great potential as high-*k* dielectrics [5, 55], photocatalysts for decomposition of environmentally hazardous compounds [56, 57], antimicrobial surfaces [58], and photoelectrodes for dye-sensitized solar cells [6, 7, 59]. Having a capability to fabricate nanostructured TiO_2 films at low temperature and low cost will provide a significant benefit for those fields provided that their functional properties are also improved. In this case, exact roles of solution parameters from hydrothermal processing need to be clarified to trigger nucleation of the nanoparticles and their consolidation mechanism(s) to form nanostructured TiO_2. Surface of these nanostructures can be further hybridized using the same hydrothermal system with other oxide phases to create the nanocomposite structures that should exert multifunctional properties by taking advantage of both oxide structures. Two rutile-based nanocomposite systems were widely studied: TiO_2 (rutile)–TiO_2 (anatase) [60–62], TiO_2 (rutile)–ZnO [63–65] that have been reported to show the effective charge separation and better response under visible light. Both *classical* and *non-classical* crystallization theories have been employed to establish the quantitative relationship between processing parameters and nucleation, growth, and aggregation dynamics [36]. The latter may also involve organic additives or oriented attachment to mimic the hierarchical organization of oxide nanostructures [66].

By gaining the knowledge on tailoring the structure and properties of these oxide films, such processes have the potential of spawning major impact on the application of these oxide materials for devices that can be fabricated on either rigid (e.g., Si, glass, metal) or flexible substrates (e.g., polymer, thin glass, metal foil). This synthetic pathway consumes very low energies with low-temperature processing conditions, which is also more sustainable while having minimal impact on environmental concerns in terms of resources. It is expected that the knowledge obtained from TiO_2 can be extended to other solution grown oxide systems such as ZrO_2, ZnO, SnO_2, In_2O_3, and $BaTiO_3$. This review therefore offers recipes for overcoming the challenges found in solution-based ceramic thin film processing and leading to revolutionary advances in the solution growth of nanostructures for enhanced functional properties.

2 Solution Chemistry of TiO_2 Formation

The formation of synthetic oxides shares many characteristics with biominerals that naturally form under ambient conditions over long periods [31]. These oxides in aqueous environment are, however, formed under an accelerated hydrolysis environment for a relatively short period. For this synthetic pathway, quantitative understanding of the hydrolysis process of precursor species is essential to more precisely tailor nucleation, growth, and aggregation of nanoparticles via solution parameters such as pH, concentration and temperature. Since the solubility of the titanium oxides and their hydroxides are often unknown, thermodynamics data are used to calculate equilibrium solubilities of the stable phases. This allows the calculation of the degree of supersaturation (*S*), which is the ratio of the solution ion activity product to the equilibrium solubility product of the precipitated phase [28, 37].

$TiCl_4$, one of the common precursors for TiO_2, undergoes a hydrolysis in aqueous solution to generate hydroxo-, chloro-, and chloro-hydroxo-titanium species:

$$TiCl_4 + xH_2O = TiCl_{4-x}(OH)_y^{4-x-y} + xH^+ + xCl^- \quad (1)$$

It turned out that the extent of association of chlorine ions is of little significance especially at low concentration of Cl^- and low pH, in which the predominance of hydroxotitanium species ($Ti(OH)_y^{(4-y)+}$) has been reported [67]. Depending on pH of solution that controls availability of OH^-, the extent of hydrolysis may, therefore, vary. At a low pH range (1–2), the predominance of $Ti(OH)_2^{2+}$ or TiO^{2+} has been confirmed by different spectroscopic techniques [68]. In addition, the dominance of such species are supported by theoretical calculations including Gibbs free energy based chemical speciation models and partial charge models, especially at room temperature [28, 69, 70]. $Ti(OH)_3^+$ is another species reported to be dominating in a pH range of 1–2 when the temperature is raised to 70 °C [28]. A higher degree of *supersaturation (S)* is obtained either by *increasing pH (more OH^-)* of the solution or *increasing [$TiCl_4$]*, thereby prompting subsequent precipitation.

In order to correlate the particle dynamics in precursor solutions with the degree of supersaturation (S), the solutions with a wide range of [$TiCl_4$] and pH were investigated *in-situ* using a dynamic light scattering (DLS) technique [28]. As seen in this study, the typical nucleation and growth of nanoparticles can be divided into three periods, i.e., induction, initial growth, and saturated growth. The induction time (t_{ind}) is defined as the time, where growth of the nuclei is more dominant than formation of new nuclei that makes the number of nuclei formed constant. Thus, this time represents for kinetics of nucleation. Using the classical nucleation theory [71, 72], the activation energy is expressed as:

$$\Delta G^* = \frac{\beta v^2 \sigma^3}{(kT\text{In}S)^2}, \tag{2}$$

where β is a shape factor ($16\pi/3$ with an assumption that the nuclei have a spherical shape), v is the molecular volume (3.402×10^{-29} m^3 from the bulk density and molar mass of an anatase phase of TiO_2), σ is the interfacial energy of the nucleus assumed to be independent upon the radius of the nucleus, k is the Boltzmann's constant, T is the absolute temperature, and S is the degree of supersaturation. By using the Arrhenius rate theory, the nucleation rate, J, is obtained from the above activation energy and a constant, Ω:

$$J = \Omega \cdot \exp\left(-\frac{\Delta G^*}{kT}\right). \tag{3}$$

Rearranging the above equations yields the induction time that is defined as the number of nuclei, N^* divided by the nucleation rate, which becomes a function of the supersaturation, as shown in:

$$\text{In } t_{\text{int}} = \text{In}\left(\frac{N^*}{\Omega}\right) + \frac{\beta v^2 \sigma^3}{(kT)^3}(\text{In}S)^{-2}. \tag{4}$$

Experimental results on TiO_2 particle analyses showed a good correlation with this prediction, suggesting the importance of introducing the degree of supersaturation for predicting the kinetics of nucleation [28].

After the induction period, there was a rapid particle growth period, in which the TiO_2 particles grow from critical nucleus size. This growth can be achieved either by single particle growth or by particle aggregation, but the latter seems to be more responsible for the size range above 20 nm. DLS techniques cannot differentiate two different mechanisms. As the typical size of primary particles is about 5–10 nm, as shown in the TEM work [18, 26–28], the initial growth rate is defined as the one that reaches this size after critical nuclei form. This rate increases with [Ti^{4+}] and pH. Figure 1 shows the growth rates with supersaturation (different [$TiCl_4$]), which have fallen into a single line with supersaturation that implies that the growth rate

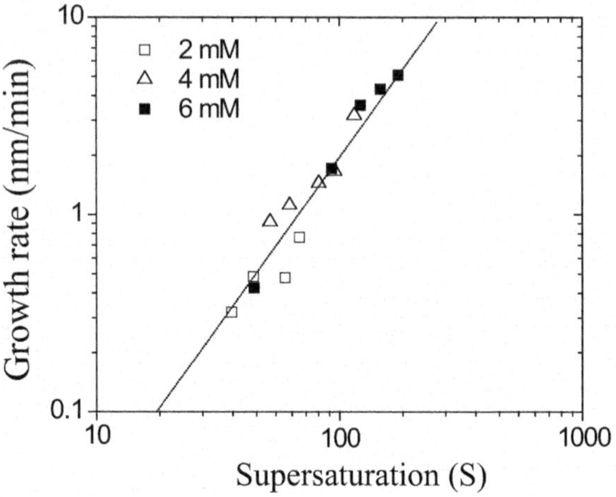

Fig. 1 Initial growth rate of TiO_2 nanoparticles measured from DLS for the precursor solutions with the degree of supersaturation [Reprinted with permission from Zhang et al. [28]. Copyright 2008, American Ceramic Society]

only depends on the degree of supersaturation. In this case, the growth kinetics can be related to the following power law relationship:

$$\dot{d} = K \cdot S^p, \tag{5}$$

where $\dot{d}$ is the growth rate, p is the growth exponent and K is the constant. In Fig. 1, $p \approx 1.8$ suggesting almost a parabolic function of supersaturation [73].

3 Hierarchical Organization of Solution-Grown TiO_2 Nanostructures

With the aforementioned definition of S, it is feasible to simultaneously incorporate the effects from precursor concentration, pH, and solution temperature. With no surface nucleation (i.e., no seed layer), nucleation and initial growth can occur in the supersaturated bulk solution that results in nanoparticle formation. The subsequent growth of nanoparticles can then be governed by a ***non-classical crystallization pathway*** **through nanoparticle aggregation**, rather than ion-by-ion growth (Fig. 2), in which a wide range of microstructures was observed as the supersaturation level increased (from the lower-left region to the upper-right region). Such a particle assembly can be directed with the aid of either organic additives or self-assembly of nanoparticles (oriented attachment). This control in transition periods has an important implication as its aggregation behavior will determine the crystal quality of those clusters/units. Furthermore, it will directly impact final products of those structural developments by having a capacity to produce any complex shape (e.g., thin films, membranes, sensing surfaces, etc.). A hierarchical organization of nanostructures also enables a near-ambient processing with highly crystalline structure (possibly,

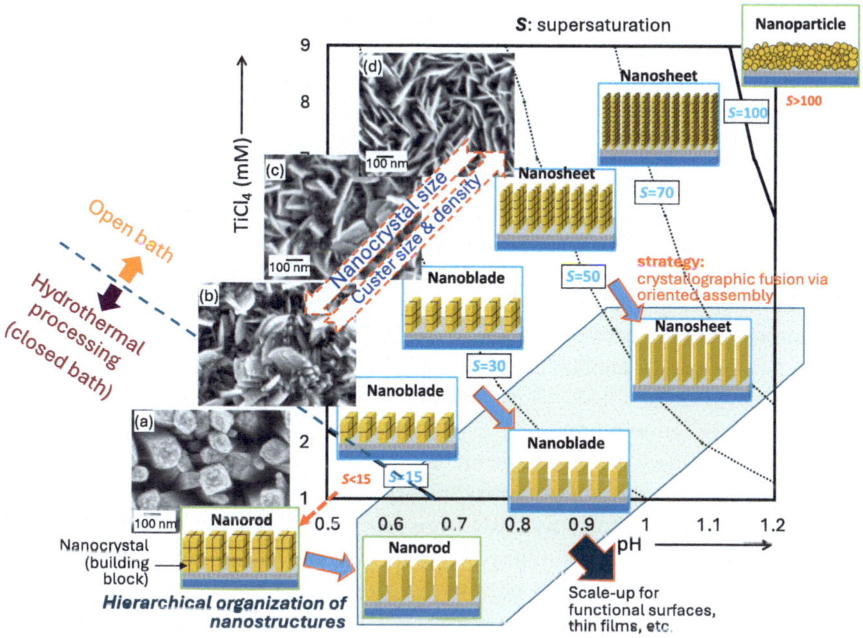

Fig. 2 Hierarchical structural organization map (in a plot of [$TiCl_4$] vs. pH) of solution grown TiO_2 nanostructures in thin films originated from nanoparticles. The degree of supersaturation increases with higher concentration of [$TiCl_4$] and higher pH and the corresponding microstructures are displayed at each supersaturation

as a means of forming a single crystal) for enhanced functional properties that are normally obtained from high-energy, vapor-phase based processes.

All the SEM images shown here are obtained from the samples that have the same deposition time (6 h). Since each sample had a different deposition rate, the thicknesses of TiO_2 thin films processed with various conditions were different; that is, for $S < 15$ (hydrothermal method), $S = 15$ (chemical bath deposition (CBD), 2 steps), 30 (CBD, 12 steps), 50 (CBD, 12 steps), 70 (CBD, 12 steps), the thicknesses were 1510 nm, 543 nm, 767 nm, 1532 nm, and 2182 nm, respectively. For the hydrothermal case, the supersaturation level (<15) was assumed as its precursor solution was clearer after 3 h than the CBD at $S = 15$. In an earlier study [37], the film formation was examined with the supersaturation down to ~70, at which the dendritic growth of the tree-like and leaf-like nanostructures became dominant. It indicates that minor fluctuation in concentration of the solution ahead of the growing interface can lead the crystal to form the protuberances.

By further decreasing the supersaturation (less than 70), well-defined nanostructures such as 1-D nanorods or 2-D nanoblades began to show up in the growing film. For instance, using the hydrothermal method, the 1-D nanorod structures with the square-shaped growth front (Fig. 2a) grew up with its width around 100–150 nm.

Since the segmented nanorod structure offers a great amount of the inter-rod void spaces, these TiO_2 nanorod films could be a good candidate for a photoelectrode or a photocatalyst [59, 68, 74]. An ethanol addition also plays an important role in the hydrothermal process as it makes a precipitation process in the precursor solution more controllable. There was a report that ethanol slows down hydrolysis of Ti(IV) species in the precursor solution at 50 °C [75]. Another study was processed at 140 °C, during which the solution did not become hazy that indicates the hydrolysis reaction even at this high temperature was still relatively slow [76].

On the other hand, the 2-D nanoblade structures with rectangular-shaped growth front (i.e., the width much larger than the thickness) grew up by CBD with the supersaturation (S) ranging between 15 and 70 (Fig. 2b–d). With lower supersaturations close to $S = 15$, the nanoblades got thicker in the thickness direction and more separated between nanablades (i.e., the density of the film decreased). The width dimension of the growing faces of nanoblades was ranged from 100 to 300 nm. Below $S = 15$ in the CBD process, the precipitation was so slow, so there was no film deposition unless the hydrothermal process was employed. While the top portion of all those TiO_2 films grown by CBD from low supersaturations ($S = 15$–70) was segmented and porous, the bottom portion appeared to be denser. Such densification of the bottom portion may be related to the lateral thickening effect of the nanoblades during the multistep deposition [77]. In particular, as the supersaturation of the precursor solution during the film growth could not be maintained at a constant level for a long continuous deposition, a *step-wise, multistep deposition* was needed for CBD, which could slow down the vertical growth of nanostructures and fill the gaps between the nanoblades due to the transition right after each step.

It seems that variations in the thickness, size and density of nanostructured thin films are attributed to changes in process conditions, particularly due to the degree of supersaturation. Very low supersaturation ($S < 15$) from hydrothermal process at which the precursor solution was clearer, however, made the deposition rate quite comparable, unlike the one shown under $S = 15$–70. It suggests the growth rate of nanostructure clusters may also depend on the shapes of those clusters (for example, nanoblades from $S = 15$–70 and nanorods from $S < 15$), and that those clusters may form resulting from the oriented attachment (OA) of nanoparticles from non-classical crystallization pathway [66, 76]. Complete fusion between those nanoparticles, however, did not seem to occur to make a single crystal blade or rod, as confirmed by TEM study [78]. In this non-classical pathway, the shape of nanoclusters depends on surface energies of the nanocrystal faces formed in the specific solution conditions (esp. under various pH levels). Therefore, various levels of supersaturation of the precursor offers a means of not only controlling the shapes and sizes of TiO_2 nanostructures in solution but also of tailoring their growth rates. The scale-up process of such fusions will lead to formation of surfaces and thin films with improved functional properties.

4 Anisotropic Nanostructure Formation

The film growth based upon particle attachment at high *S* was shown to be very slow and contain a large amount of amorphous phase [37, 79]. This will limit functional properties of titania films. On the other hand, when *S* is very low (i.e., closer to the equilibrium condition), a preferential growth of rutile occurs, and its growth rate can be quite rapid due to anisotropic growth. It also increases the crystallinity that will lead to enhanced properties. Zhang et al. demonstrated that rutile grows along the [001] direction (Fig. 3) [37]. This sample was obtained from low supersaturation in solution ($S < 80$), and had cracked areas, from which the inward growth of nanostructures could be observed in this TEM image (Fig. 3a). Initial TiO_2 particles or layers deposited earlier seemed to template the subsequent growth of nanostructures. The crystal shape and growth direction appear to be consistent with the surface energy argument as (001) surface is the highest energy surface for a rutile phase [80]. This 'leaf'-like shape also has some resemblance to an equilibrium shape of rutile crystal in a vacuum (Fig. 3b) [80]. The 'tree' shaped nanostructures of oriented growth also appeared in the film [37]. When supersaturation got even lower (<50), it was possible to obtain the film structure only with nanosheets and nanoblades grown from the leaf-like structure (see Fig. 2b–d). It shows the potential to grow the 1-D nanostructures (that can be developed from 'tree'-like needle) or 2-D nanostructures (from 'leaf'-like sheet) by carefully controlling solution conditions.

There are two possible scenarios once nuclei form at the critical supersaturation level: (i) *classical* anisotropic growth driven by, for example, a screw dislocation [81, 82]; (ii) *non-classical* growth mechanisms through oriented attachment (OA) of the nanoparticles. It was shown that the former mechanism could trigger the growth of single crystal ZnO nanowires and nanotubes from solution phase [82]. Figure 4a indeed shows the evidence of growth spiral (marked by a circle) at the tip of one of the square-shaped nanorods, an indicative of screw dislocation-driven growth. However, this low supersaturation case showed only a few of such single crystal nanorods while most of the nanowires were randomly arranged. It suggests that this growth mechanism may not be practical to scale up to form a thin film. As

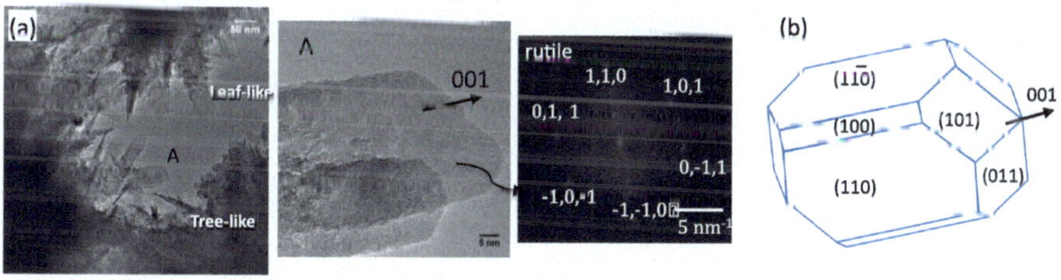

Fig. 3 **a** TiO_2 "leaf" and "tree"-like structure grown at low *S* (<80). Inset A shows the growth of nanosheet ("leaf-like") and its Fourier-transformed electron diffraction pattern that confirms a rutile phase and the [001] growth direction [Reprinted with permission from Zhang et al. [37]. Copyright 2008, American Ceramic Society]; **b** equilibrium shape of a rutile crystal in vacuum

for the second mechanism, the aggregation of several rutile nanowires with the same crystallographic orientation was also shown to form nanorod bundles, which shows that OA process was the dominant nanorod growth [83, 84]. Figure 4b–d show the examples of such oriented assemblies (marked by circles) exhibiting a square shaped tip [(002) plane, as confirmed by XRD] whose size increases with supersaturation. Internal boundaries did not seem to be fused, so single crystalline nanorods were not produced. This aggregation mechanism should provide more efficient way to cover the surface with single crystal TiO_2 nanowires provided that the crystallographic fusion of the high-energy side surfaces of nanowires could occur. In addition to those nanorods, CBD-produced nanosheets and nanoblades have not been a single crystalline nature (see Fig. 2), suggesting a perfect alignment among the nanoparticles or nanowires be a key.

For TiO_2 nanorod growth, there is no common seed layer to trigger heterogeneous nucleation of the rutile nanorod from a substrate unlike that of ZnO nanorod growth [85, 86]. Initial growth of TiO_2 films will then start with ***homogeneous nucleation*** in supersaturated solution to form **primary particles**. It is conjectured that the subsequent growth of these nanoparticles is governed by ***non-classical crystallization pathways*** **through nanoparticle aggregation,** rather than classical ion-by-ion growth. Figure 5a shows the Penn and Banfield model [84] that proposes the

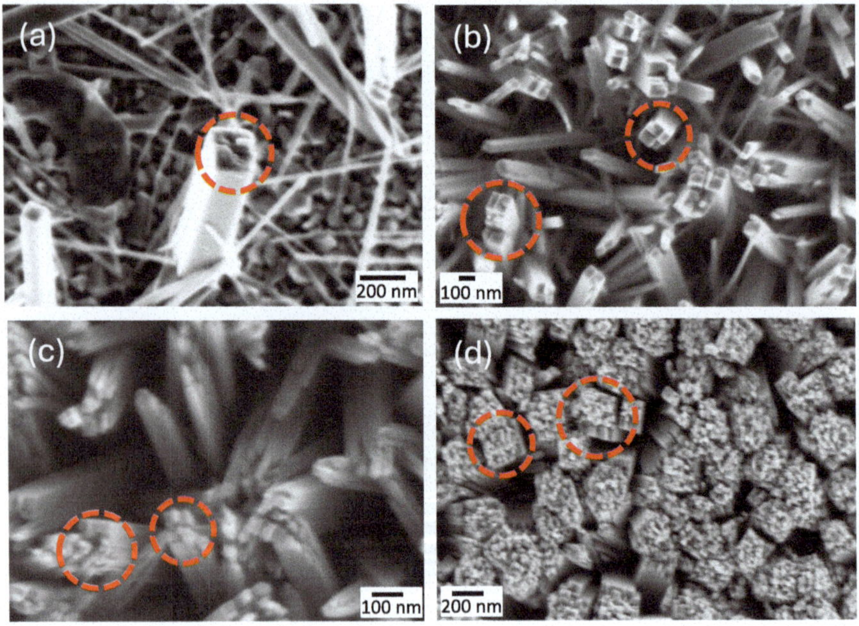

Fig. 4 Hydrothermally grown 1D nanostructure of TiO_2 using mixed solvent (water, ethanol), processed at 180 °C: **a** single crystal nanowires and nanorods; **b**, **c**, **d** nanorod clusters via side-by-side attachment of nanowires. FTO coated glass was used as a substrate. The degree of supersaturation increases from (**a**) through (**d**) as the corresponding HCl concentration decreases)

imperfect oriented attachment among the three particles (particles 2 and 3 are at the same plane while particle 1 on top of those two), thereby creating edge dislocations (between 2 and 3) and a screw dislocation (between 1 and 3). The latter can provide a spiral growth motif in the arrow direction. Our previous observation on various sizes of TiO_2 nanorods made by joining nanowires sideways supports this model. Once a few nanoparticles aggregate each other on the substrate, imperfect attachment creates screw dislocations that trigger to grow into the vertical direction (shown by 'arrows) responsible for the growth of individual nanowires. While nanowires grow vertically along [001] direction via spiral growth mechanism, they will begin to attach each other from large (110) side planes to minimize total energy of the system, which will create steps on the growth front as well as on the side surfaces (Fig. 5b). These steps can accommodate other particles precipitated from the precursor solution, which will grow in parallel to form interweaved nanorod structure as shown in Fig. 5c. Here, traces of original nanowires and their bundled domains are still noticeable on the growth surface. It will result in poor functional properties due to high defect density (dislocations, low angle grain boundaries, vacancies).

Hence, initial particle attachment needs to be *near* perfect to grow a single crystalline nanorod. Such control during an initial transition period has important implication as its aggregation behavior will determine the crystal quality of those clusters/units. Given that, vertical growth of nanowires is regulated by the spiral growth following nanoparticle aggregation while the lateral growth is progressed with the attachment of nanoparticles and nanowires via a non-classical mechanism. As a

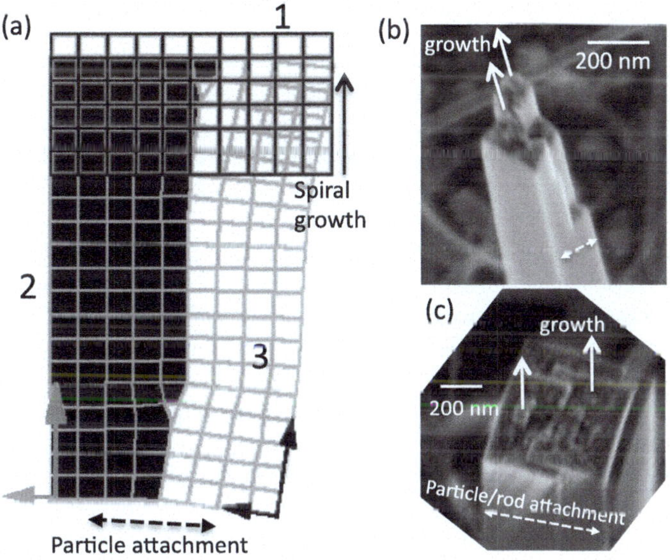

Fig. 5 Particle attachment mechanism: **a** Penn and Banfield model [Reprinted with permission from Ref. [84]. Copyright 1998, The American Association for the Advancement of Science]; **b** nanorod growth (intermediate stage); **c** nanorod growth (later stage)

result, maintaining a very low degree of supersaturation is critical to better control imperfections during initial particle attachment, as well as aiding for spiral growth [87, 88]. **The hierarchical organization of nanostructures** will then depend on the control of these growth mechanisms involved.

For the oriented assembly, organic additives provide an effective means of changing the morphology of a crystal by face-selective adsorption. Low molecular weight additives are advantageous over polymeric ones because the former has only a few potential binding sites in contrast to the latter additive cases. Anisotropic growth of low-dimensional nanostructures of anatase TiO_2 has been attempted through non-classical crystallization pathway to form mesocrystals [89–91]. For anatase nanoparticles, an organic ligand such as trizma ($(HOCH_2)_3CNH_2$) and serinol ($(HOCH_2)_2CNH_2$) was demonstrated to be effective for oriented attachment of nanoparticles [90, 91]. Such anisotropic assembly occurs due to the unattached organic ligand and dissociative adsorption of water on the (001) surface that induce the preferred attachment of the anatase particle along the [001] direction.

5 Processing of Hybrid TiO_2 via Hydrothermal Method

TiO_2 is a well-studied material [56, 74, 92, 93] but functional properties of solution grown TiO_2 films are not at its full capacity yet, compared to those from high-temperature, high-energy processed counterparts. This is partly because mechanistic understanding of complex growth regimes is lacking due to difficulty in controlling process parameters independently. It results in *inferior* crystalline quality, along with extra defects trapped inside the structure. Thus, process optimizations and detailed structural analyses are critically needed to advance *hydrothermally* grown rutile nanorod structure to the next level. In particular, vertically grown TiO_2 (rutile) nanorod arrays, once structural optimization is achieved, can be further treated for diverse *functional* properties with other oxides to form hybrid nanostructure with, for examples, TiO_2 (anatase), ZnO, or SnO_2.

The main goal is to use oxide nanostructures as a very efficient outer functional surface that complements more robust TiO_2 nanorod framework that can be grown on the surface of any base materials. This surface will not only better respond to UV light, but it is also possible to make the nanocomposite coating *smarter* so it can work under visible light or room light. It was reported that the rutile-anatase interface is a key to extend its enhanced photoactivity into visible wavelengths [94]. This is attributed to a possibility of creating some sub-band levels within the band structure of rutile due to impurities, defects, and/or disordered interfacial regions.

There are several advantages to go with the hybrid structure consisting of **rutile (inside)** and **ZnO or anatase (outside)** structure, or **SnO_2 (inside) and anatase (outside)**, as shown in Fig. 6. It will provide (i) more efficient charge separation under light illumination between electrons and holes by moving electrons from TiO_2 (anatase) (or ZnO) nanostructures to TiO_2 (rutile) core, and holes from TiO_2 (rutile) to anatase (or ZnO). If it is for TiO_2 (anatase) to SnO_2 core, electrons move from

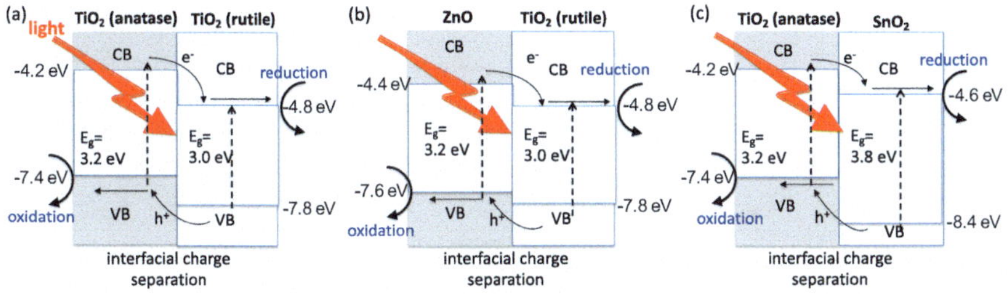

Fig. 6 Schematic illustration of electron–hole separation from the hybrid system: **a** anatase shell—rutile core; **b** ZnO shell—rutile core; **c** anatase shell—SnO_2 core

anatase to SnO_2 and holes from SnO_2 to anatase. This is achieved because the bottom of conduction band (LUMO) of anatase (−4.2 eV) or ZnO (−4.4 eV) is higher than that of TiO_2 (rutile) (−4.8 eV), and in addition, the top of their valence band (HOMO) (−7.4 eV and −7.6 eV, respectively) is slightly higher than that of TiO_2 (rutile) (−7.8 eV) [61, 64, 95–97]. The LUMO and HOMO levels are also favorably matched for the TiO_2 (anatase) (−4.2 eV, −7.4 eV) and the SnO_2 core (−4.6 eV, −8.4 eV).

This configuration of the core–shell structure also (ii) avoids the recombination of electrons with the oxidized species in the environment while holes accumulated in the outer layer generates the *hydroxyl radicals* in the presence of moisture, which are a very strong oxidative agent. Furthermore, (iii) high oxidative potential of the holes in the outer shell allows the direct oxidation of organic species. The bandgap energies of TiO_2 (rutile) (3.0 eV), ZnO or TiO_2 (anatase) (3.2 eV), and SnO_2 (3.8 eV) will enable the hybrid structure (iv) to cover a wider range of wavelength of the light due to the creation of the sub-bands within the band structure owing to impurities (e.g., Zn in TiO_2, Ti in SnO_2) and defects (e.g., oxygen vacancies, cation interstitials, interfaces), thus making it more efficient. Other benefits include (v) increased external surface area in contact with environment, compared to that of a simple nanorod array made with TiO_2 or SnO_2 alone. Further, various shapes and sizes of nanowires and nanoclusters are possible with oxide nanostructures grown by the hydrothermal method. All these aspects make the 'TiO_2/ZnO', 'TiO_2 (rutile)/TiO_2 (anatase)', 'TiO_2 (anatase)/SnO_2 nanocomposites very efficient achieving the strong photocatalytic performance, as well as serving as a photoanode for DSSC, which do not solely depend on UV light or hydroxyl radicals. SnO_2, which has the same rutile structure as TiO_2, shows the potential for nanorod formation through hierarchical organization of nanoparticles [98–100]. SnO_2 can serve as an efficient surface for gas sensing [100].

Recent progresses made from the nanostructured coatings of TiO_2 (rutile, anatase), TiO_2 (anatase)/TiO_2 (rutile), and ZnO for the photocatalytic behavior are shown in Fig. 7, in which the dye degradation kinetics defined by:

$$\ln\left(\frac{C}{C_0}\right) = -k \cdot t \tag{6}$$

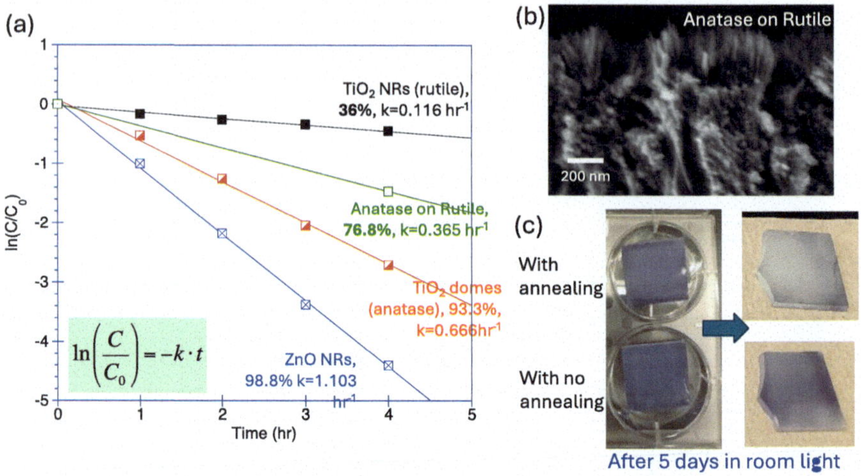

Fig. 7 Photocatalytic activities of nanostructured oxide films; **a** MB dye degradation rates measured from UV/Vis spectroscopy; **b** cross-sectional SEM image of anatase on rutile; **c** MB degradation on anatase on rutile under room light after 5 days (dye solution made with 0.025 mM of dye mixed in DI water)

where C is the concentration of the solution taken at each step, C_0 is the initial concentration ($t = 0$), and k-value (per hour) is the degradation reaction constant that is used to compare the photocatalytic properties of the materials. These nanostructured coatings based upon TiO_2 (anatase) nanoclusters and ZnO nanorod arrays displayed very high photocatalytic deactivation of methylene blue (MB) dyes under UV light (Fig. 7a). Strong bleaching of the MB dye was seen after short exposure under UV irradiation, where the degradation rate depended upon specific microstructures and crystalline planes generated from those secondary nanostructures [101]. However, TiO_2 (rutile) nanorods did not show the same level of the photocatalytic performance until it is functionalized with the anatase phase to form the hybrid core–shell structure (Fig. 7b). Interestingly, this hybrid structure even displayed a visible light response for the MB dye degradation when the dye-soaked coupons were exposed to room light, as shown in Fig. 7c. The photocatalytic activity also gives a good indication for the antimicrobial activity required for prevention of the biofilm formation. Previously, TiO_2 (anatase) and ZnO surfaces displayed promising antibacterial behavior under UV light as well [102–104].

Dye-sensitized solar cells (DSSC) [6] with TiO_2 nanostructured films were assembled with ruthenium dye and iodide-based liquid electrolytes for photovoltaic (PV) measurements. TiO_2 films were designed to maximize the dye adsorption in the internal surface of the films that also provides the pathway for the liquid electrolyte infiltration. For the TiO_2 film case, the PV performance was enhanced when the cells were assembled with the *vertically* aligned nanorod structures (rutile) prepared by hydrothermal processing at 140 °C, compared to nanoparticle or nanoblade-based

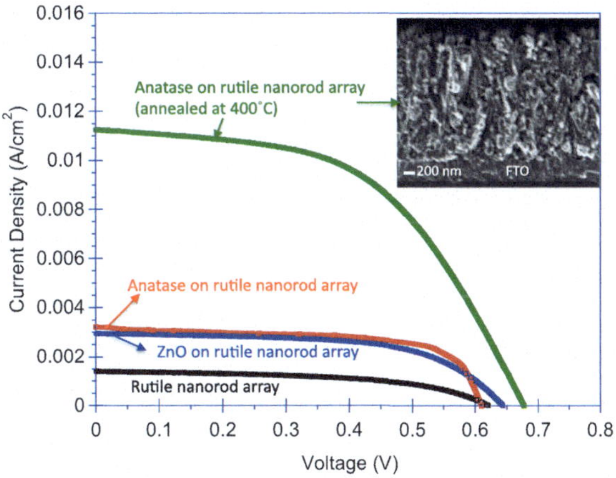

Fig. 8 I–V curves obtained from the rutile nanorod film and its hybrid films on FTO substrate (under solar simulator at 100 mW/cm^2)

TiO_2 films [38]. The rutile nanorod film alone still has shown a poor power conversion efficiency (η) of ~0.45% with a fill factor (FF) of 0.53 (Fig. 8). When it was hybridized with anatase or ZnO nanostructured layer, the efficiencies got more than doubled. It indicates that its improvement was obtained from the hybrid structure through more efficient electron–hole separation, as well as creating more effective surface areas through covering the surface of inactive rutile nanorods. More improvements were achieved when the rutile-anatase hybrid structure was annealed at high temperatures (up to 400 °C), for which the efficiency rose to ~4.0% with a fill factor of 0.52. Cross-sectional microstructure observation showed denser structure than the corresponding rutile-only sample while the inter-rod gaps were effectively filled with anatase nanocrystals (Fig. 8, inset). Annealing may also have improved a crystalline quality of the rutile nanorod-bundled structures due to its high temperature exposure. It indicates that, with the process optimization, good crystalline quality of the rutile nanorod structure combined with the help of an anatase or ZnO surface layer, could allow the fabrication of efficient, multi-functional surfaces. Thus, the potential of enhanced functional properties of the rutile nanorod arrays warrants a further study to better understand the effects of hybrid structure.

6 Processing of Hybrid TiO_2 via Ion Exchange

Alkaline solution etching has shown to be effective in generating various morphologies of ceramic materials for the preparation of nanotubes, nanoflakes and nanofibers. Tsai and Teng obtained a sodium titanate nanotube through NaOH etching [105]. Furthermore, TiO_2 nanoparticles were transformed to nanoflakes and nanofibers through NaOH solution treatment with different temperatures and concentrations [106]. In order to investigate such an effect of the core–shell configurations on the photocatalytic properties, a vertical array of rutile nanorods was first prepared through the hydrothermal method, followed by the surface etching and ion exchange, both in low-temperature solution conditions [107]. By controlling the etching temperature and time, the resultant morphologies varied widely. The ion exchange process resulted in transforming the skin layer of a rutile nanorod into anatase that showed an improved photocatalytic performance, as explained in the previous section. It provides an effective way of creating a phase junction between the remaining rutile core and the newly formed anatase shell. The influences from ion exchange time and the resultant surface morphologies were systematically studied to evaluate its capacity as a photocatalyst for decomposing the organic substances.

In this process, TiO_2 begins to be dissolved in concentrated NaOH solution and the reaction rate depends on solution temperature. This reaction is generally represented by [108]:

$$y\mathrm{TiO_2} + x\mathrm{NaOH} \rightarrow \mathrm{Na}_x\mathrm{Ti}_y\mathrm{O}_{2y+\frac{x}{2}} + \frac{x}{2}\mathrm{H_2O}. \qquad (7)$$

The reaction results in forming sodium titanate products consisting of variable Na to Ti ratios (x: y), which could be influenced by many factors [108, 109]. Sodium titanate is generally considered a wide band gap semiconductor, and its band gap energy was reported to be ~3.30 eV to 3.38 eV [110–112]. Additionally, sodium titanate can react with acid and generate hydrogen titanate through an ion exchange process, which is unstable and immediately converted into an anatase TiO_2 phase [110]. This process is shown in the following two reactions:

$$\mathrm{Na}_x\mathrm{Ti}_y\mathrm{O}_{2y+\frac{x}{2}} + x\mathrm{HNO_3} \rightarrow x\mathrm{NaNO_3} + \mathrm{H}_x\mathrm{Ti}_y\mathrm{O}_{2y+\frac{x}{2}}. \qquad (8)$$

$$\mathrm{H}_x\mathrm{Ti}_y\mathrm{O}_{2y+\frac{x}{2}} \rightarrow \frac{x}{2}\mathrm{H_2O} + y\mathrm{TiO_2}(\text{anatase})\ldots \qquad (9)$$

These reactions provide the pathway for rutile surface to be transformed into an anatase layer through the titanate transition phases. Subsequently, the rutile core is unaltered and only the shell layer is covered by an anatase particle layer, from which the anatase–rutile phase junction is generated. This surface conversion reaction is schematically explained in Fig. 9.

Figure 10a shows the SEM plan view of the hydrothermally grown rutile nanorods that were vertically aligned, where the square growth faces can be seen. These nanorods formed by having a lateral oriented attachment of smaller, 001-oriented nanowires, as reported in our previous study [113]. In addition, these rutile nanorods were present mainly as Ti^{4+} instead of Ti^{3+} based upon the previous X-ray photoelectron spectroscopy study [114]. Etching conditions become the critical parameter to obtain an anatase shell while maintaining the 1-D feature of the rutile core. When the hydrothermal temperature is too high or when the treatment time is too long, the rutile nanorod could start to lose its orientation or be completed dissolved.

From Fig. 10b, c it was found that the nanorod framework remains for the etching treatment at 100 °C and 140 °C for 2 h, while for the former case, the surface of

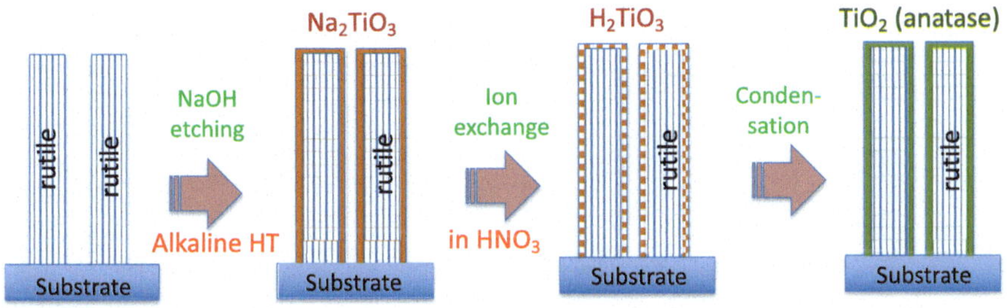

Fig. 9 Microstructure development into the core–shell configuration after ion exchange of vertically aligned rule nanorods (HT: hydrothermal)

nanorods remains almost intact, presenting little trace of surface etching. However, as the etching temperature increased to 140 °C, the surface of the nanorods was modified and covered with a layer of nanoparticles. When the temperature increased to 180 °C, as shown in Fig. 10d, those vertical nanorods started to get bundled together. The interspace between the nanorods was mostly eliminated and the cracks began to appear on a microscale. Such a surface morphology is not preferable for the photocatalytic applications which highly depend on the internal surface area. This will also potentially result in poor adhesion of the rutile coating to FTO surface as the nanorods are tilted or physically displaced to be entangled with other surrounding nanorods. The surface dissolution process appeared to get these nanorods easily bundled together through the reprecipitation process. Therefore, etching at 180 °C was considered to be too high, and 140 °C was used as an etching temperature for the subsequent surface functional process.

Various etching times were also tested for the 140 °C condition. The result for 0, 2, 4 and 6 h etched sample were shown in Fig. 10a, c, e, f, respectively. The nanorod core feature still remained for 0 and 2 h durations with no obvious change in the rod surface morphology and array density. On the other hand, for the 4 and 6 h durations, the original shape of nanorods was distorted and the coating began to get delaminated from the substrate since the prolonged etching treatment affected the adhesion between the nanorods and substrate. In addition, with more TiO_2 dissolved in NaOH solution, sodium titanate particles tended to aggregate not only on the nanorod surfaces, but they could also precipitate in the solution, making the precursor solution very unstable. Based upon these observations, the etching condition at 140 °C for 2 h in 10 M NaOH solution was selected for the subsequent ion exchange process.

To further verify the ion exchange effects on the surface morphology and the remaining sodium, SEM and EDS were used for characterizing the nanorod coatings with different ion exchange times. The microstructure developments are shown in Fig. 11, where surface morphologies of the nanorods change dramatically during ion exchange. The surface of nanorods was smooth after the etching process was applied; however, a layer of nanoparticles began to appear after 1 h ion exchange and was very distinctively particulate when ion exchange time increased to 3 and 4 h (Fig. 11e, f, respectively), compared to those from pure rutile nanorods and their

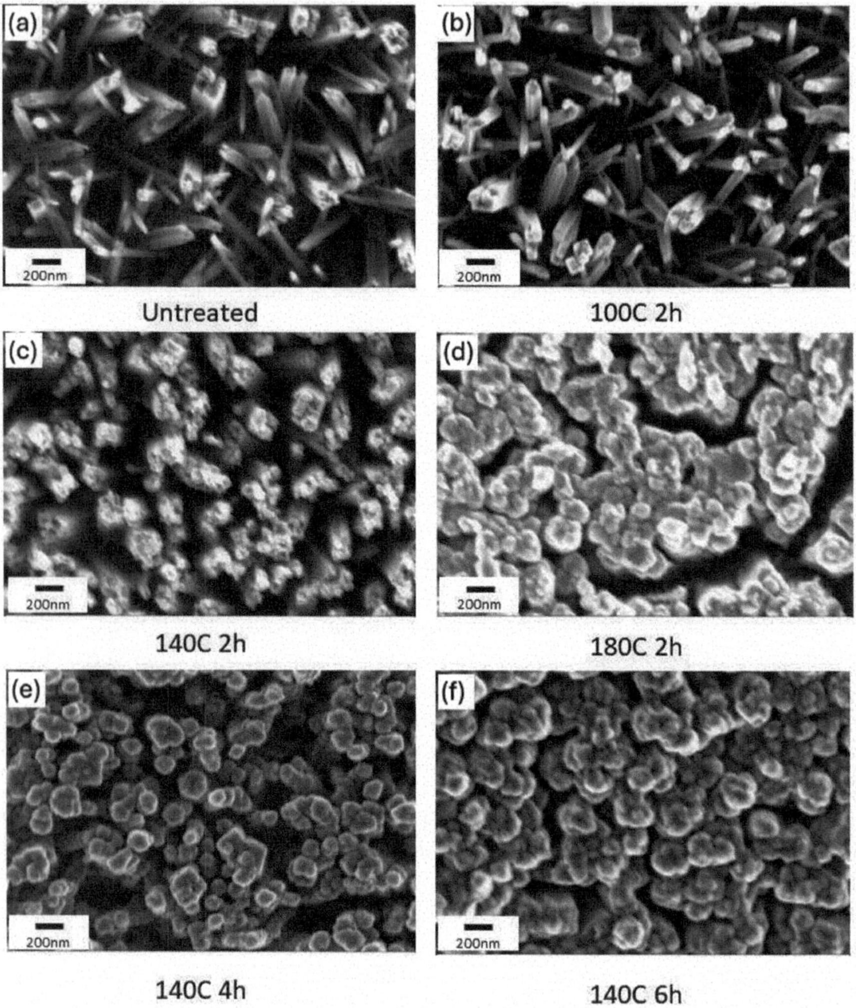

Fig. 10 Plan view SEM images for rutile TiO_2 nanorods after different etching processes, showing the impacts from etching temperature: **a** control sample of untreated rutile nanorods; **b** 100 °C 2 h; **c** 140 °C, 2 h; **d** 180 °C, 2 h. Impacts from the etching time were compared: **c** 140 °C, 2 h; **e** 140 °C, 4 h; **f** 140 °C, 6 h. All these samples were treated with 1 h ion exchange process

etched nanorods. The nanoparticle size was around 30 nm for 1 h ion exchange case and with more ion exchange time, nanoparticle size increased gradually to around 65 nm.

The sodium atomic percentage was found to be ~15.5% after the etching process, which decreased with ion exchange time from 15.5% to ~0% after 3 h (within the detectability limit of EDS). This result is consistent with the XRD and Raman spectra

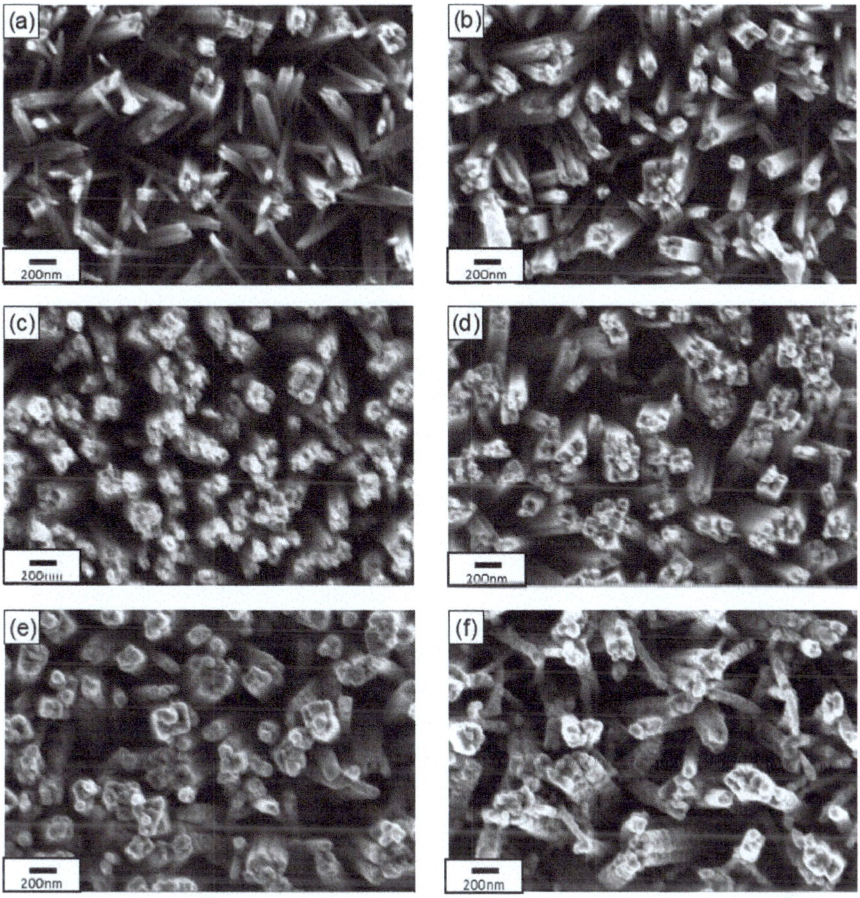

Fig. 11 Plan view SEM images of **a** rutile nanorods with no treatment, along with etched rutile nanorods with ion exchange for **b** 0 h; **c** 1 h; **d** 2 h; **e** 3 h; **f** 4 h

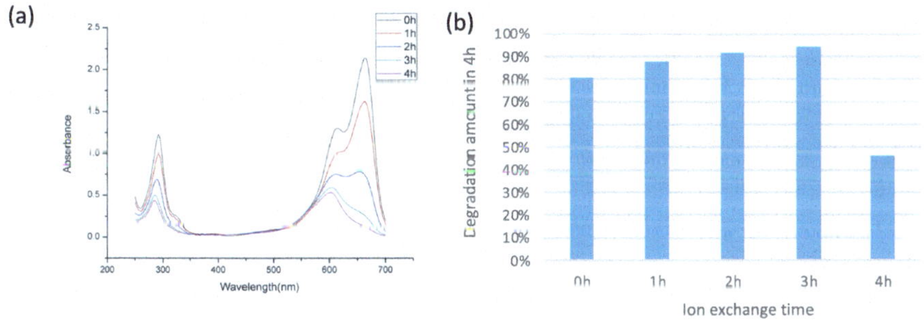

Fig. 12 **a** UV–Vis absorbance spectra of MB dye solution immersed with the 3-h ion exchanged TiO_2 nanorods for various exposure times under UV light. **b** Dye degradation efficiency with ion exchange time for the 4 h-UV exposure test

[107]. Furthermore, the basic nanorod framework of the rutile coating remained during the ion exchange process.

In order to verify the surface functionalization of the nanorod core for its photocatalytic property, the dye degradation tests were carried out on different ion exchange time samples as shown in Fig. 12. The dye degradation efficiency of 3-h ion exchange structure was significantly improved to 94.2% compared to that of an original rutile nanorod array that had 25.9% dye degradation efficiency during four hours of UV exposure for the given volume of the dye. Such a photocatalytic performance from the rutile core—anatase shell configuration is very similar to that of 'highly photocatalytic' anatase coatings [103] or of ZnO nanorod-based coatings [114]. Bare FTO was used as a reference and only had a 2.5% degradation efficiency for the same 4 h experiment. This improvement is attributed to the generation of sodium titanate as an intermediate product, which is also a wide bandgap semiconductor that has been proven to be an effective photocatalyst [111]. However, the shell layer made of sodium titanate has shown 80% of the dye degradation after 4 h exposure under UV. This layer is very reactive in aqueous condition, so it can be easily converted into another functional layer by other ions.

Figure 13 represents the dye degradation behavior of the ion exchanged TiO_2 due to its strong photocatalytic property compared to those of other oxide coatings. *k*-value (0.6721/h) from 3 h ion exchange is now comparable to that of nanocrystalline anatase with the [001]-textured (004) surface (about 93%; $k = 0.662/h$) [103] and of ZnO nanorod based coatings [114]. The aforementioned anatase/rutile hybrid coating, however, showed much less dye degradation percentage (about 76.8%; $k = 0.365/h$) [103], so such an improvement from ion exchanged anatase on the rutile nanorods should be attributed to the effectiveness of ion exchange on tailoring the surface functionality.

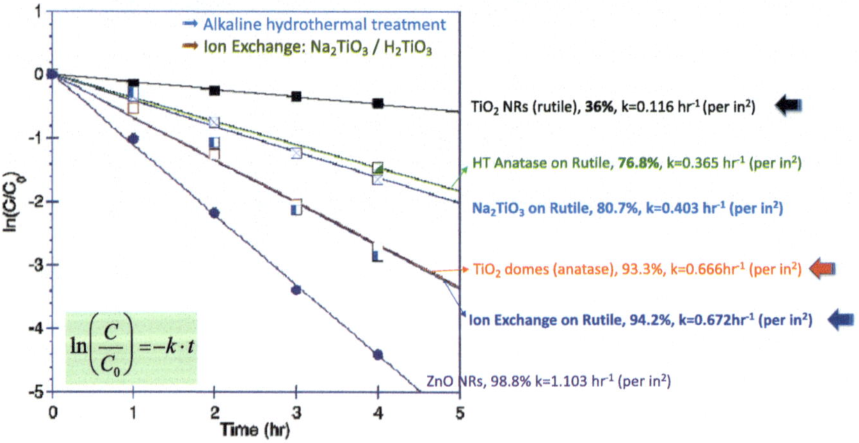

Fig. 13 Photocatalytic degradation kinetics of MB dye solution under UV light irradiation to show the effect of engineered surface functionality through ion exchange on rutile nanorods

During the gradual transition of the surface layer from sodium titanate to anatase, the transformation led to the creation of a phase junction interfaced between the remained rutile as a core and the outer anatase layer as a shell. Once charge carriers are separated upon light absorption, photoelectrons could be rapidly transferred to the rutile phase instead of staying with holes within the single anatase phase region [115–117]. The holes that move to the anatase phase and are released from the surfaces will be an effective oxidizing agent, as well as generate the hydroxyl radicals to decompose the organic species, as seen in Fig. 6. While the photocatalytic performance shown here is based upon a UV activation mechanism, hybridizing the rutile-anatase nanostructures may result in self-structural modification near the interface created between nanostructured rutile and anatase phases, as shown in Fig. 7. It could make some band narrowing effects via creation of sub-bands from the associated defects to be activated under visible light.

Since the shell layer contains residual Na ions, the resultant shell layer will represent a Na-doped TiO_2 (anatase), which provides a means of effectively surface doping by controlling the etch and ion exchange times. One example of such ion exchange with cobalt (Co) is explained in the reaction below.

$$2Na_2TiO_3 + 2Co(NO_3)_2 \rightarrow 4NaNO_3 + Co_2TiO_4 + TiO_2. \tag{10}$$

The surface layer can be Co-doping or Co titanate depending on its concentration in the shell layer of the rutile nanorods. After this conversion, the band gap energy (2.8 eV) and sheet resistance (49 kΩ/sq) were reduced compared to those of untreated rutile (3.0 eV, 93.8 MΩ/sq) [107]. It shows the potential for designing magnetic properties by having the surface impregnated with magnetic elements (Fig. 14).

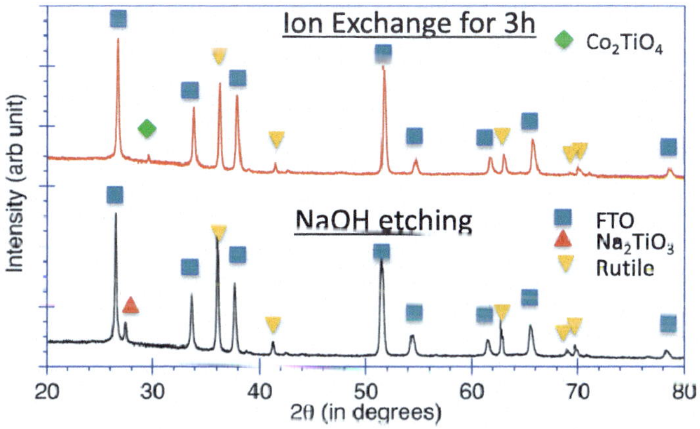

Fig. 14 An example of hybrid oxide system through ion exchange to form a shell layer containing a dopant element

7 Conclusion

Metal oxide nanostructures and thin films are often produced by aqueous precursor solution. Its synthetic principle is based upon hydrolysis of metal ions, where the degree of supersaturation (S), determined from equilibrium solubilities, is a key driving force for nucleation and growth of oxide phases. This processing is similar to the formation of certain biominerals created at ambient conditions that possesses very high crystallinity. Unlike such biomineralization, however, the controllability of specific phases and structures can be quite challenging in solution-based processing, which results in poor crystallinity. As a result, the solution process generally requires very low supersaturation and/or a seed crystal to form high crystallinity for enhanced functional properties, which often makes the process impractical and limited.

A hydrothermal processing technique can overcome such limitation to offer more precise control of process parameters to prepare the precursor solution with very low degrees of supersaturation. Not only does it produce highly crystalline quality at low temperature and cost, along with a high growth rate, a hydrothermal method also enables to tailor 1D and 2D nanostructure developments of TiO_2 (rutile). In particular, the latter aspect can be achieved through the *aggregation-driven* growth based upon *non-classical crystallization* from oxide nanoparticles precipitated in supersaturated solution. For example, well-oriented 1-D *nanowires* can be consolidated into a larger scale 1-D *nanorod*. Depending on the orientation between attached nanoparticle or nanowires, the particle-based crystallization will also have the potential to produce a single crystalline nanorod array of TiO_2 from solution.

Oriented attachment of nanoparticles and nanowires facilitates the hierarchical organization to form high quality nanorods that can be further promoted with the aid of organic additives or seed crystals. An array of vertically oriented TiO_2 (or SnO_2) nanorods can then work as a platform to introduce further advancement of photo-responsive properties by mediating its surface functionality via more 'photoefficient' anatase or ZnO nanostructures. Furthermore, general principles involved in structure evolution of TiO_2 and its hybrids in solution can be applied to the formation of other hybrid systems by selecting the phases for desirable band gap energy along with LUMO and HOMO levels with respect to those of a core phase.

Among them, anatase—rutile hybrid nanocomposite structures have been extensively studied. This core–shell configuration can also be prepared by an etching and ion exchange process on as-grown rutile nanorods. A control of the depth of an etching layer on the surface of rutile nanorod is critical to form a nanoparticulate anatase layer through ion exchange of the titanate layer in HNO_3. With an optimum ion exchange condition, the overall photocatalytic property reached to >94%, comparable to that of nanotextured anatase or ZnO nanorods.

This hybrid nanocomposite in the core–shell configuration effectively inhibits the recombination of charge carriers while increasing the surface area, thereby dramatically improving the photocatalytic activity. It is expected that further improvement of the photocatalytic and other functional properties will be possible depending on surface phases and morphologies, as well as texture and chemistry of nanoparticles on the surface layer. Therefore, TiO_2-based hybrid structures provide numerous opportunities in various applications including wastewater treatment systems, self-cleaning, photovoltaics, and gas sensing devices.

Acknowledgements This article was written partly based upon our previous studies and supported by the Small Scale Systems Integration and Packaging (S^3IP) Center, NASA grant (#NNX09AT3OG), and NASA SBIR Program (Contract #: NNX16CJ42P) through N2 Biomedical LLC, Bedford, MA (Arash Aslani, Ward Halverson). Some of the characterization work presented here was performed at the Analytical and Diagnostics Lab (ADL) of the S^3IP Center. I would also like to acknowledge contributions of the research work from my former graduate students: Jong Hyun Shim, Biplab Roy, Guangneng Zhang, and Sunghee Lee.

References

1. Robertson, J.: High dielectric constant gate oxides for metal oxide Si transistors. Rep. Prog. Phys. **69**, 327–396 (2006)
2. Chang, J.P., Lin, Y.-S.: Dielectric property and conduction mechanism of ultrathin zirconium oxide films. Appl. Phys. Lett. **79**(22), 3666–3668 (2001)
3. Roy, B.K., Zhang, G., Cho, J., Magnuson, R., Poliks, M.: High-k ceramic thin films for embedded capacitors. In: Proceedings of the US-Korea Conference on Science, Technology, and Entrepreneurship (UKC 2008). San Diego, CA (2008)
4. Martinez, A.I., Acosta, D.R.: Effect of fluorine content on the structural and electrical properties of SnO_2 and ZnO–SnO_2 thin films prepared by spray pyrolysis. Thin Solid Films **483**, 107–113 (2005)
5. Zhou, L., Hoffmann, R.C., Zhao, Z., Bill, J., Aldinger, F.: Chemical bath deposition of thin TiO_2-anatase films for dielectric applications. Thin Solid Films **516**, 7661–7666 (2008)
6. Grätzel, M.: Photoelectrochemical Cells. Nature **414**, 338–344 (2001)
7. Grätzel, M.: Solar energy conversion by dye-senstitized photovoltaic cells. Inorg. Chem. **44**, 6841–6851 (2005)
8. Maex, K., Baklanov, M.R., Shamiryan, D., Lacopi, F., Brongersma, S.H., Yanovitskaya, Z.S.: Low dielectric constant materials for microelectronics. J. Appl. Phys. **93** (11), 8793–841 (2003)
9. Morgen, M., Ryan, E.T., Zhao, J.H., Hu, C., Cho, T.H., Ho, P.S.: Low dielectric constant materials for ULSI interconnects. Annu. Rev. Mater. Sci. **30**, 645–680 (2000)
10. Ding, S.-J., Chen, L., Wan, X.-G., Wang, P.-F., Zhang, J.-Y., Zhang, D.W., Wang, J.-T.: Structure characterization of carbon and fluorine-doped silicon oxide films with low dielectric constant. Mater. Chem. Phys. **71**, 125–130 (2001)
11. Hoffman, R.L., Norris, B.J., Wager, J.F.: ZnO-based transparent thin-film transistors. Appl Phys. Lett. **82**(5), 733 (2003)
12. Carcia, P.F., McLean, R.S., Reilly, M.H., Malajovich, I., Sharp, K.G., Agrawal, S., Nunes, J.G.: ZnO thin film transistors for flexible electronics. In: MRS Sympossium Proceedings, vol. 769. Warrendale, PA (2003)
13. Roy, B.K., Cho, J.: Dielectric properties of solution-deposited crystalline barium titanate thin films. J. Am. Ceram. Soc. **95**(4), 1189–1192 (2012)
14. Goward, G.W.: Protective coatings-purpose, role, and design. Mater. Sci. Tech. **2**, 194–200 (1986)
15. Miller, R.A.: Thermal barrier coatings for aircraft engines: history and directions. J. Therm. Spray Technol. **6**(1), 35–42 (1997)
16. Maluf, N.: An Introduction to Microelectromechanical Systems. Artech House, Norwood, MA (2000)

17. Kleiman, J.I., Iskanderova, Z.A., Tennyson, R.C., Pérez, F.J.: Protective coatings for LEO environments in spacecraft applications. Surf. Coatings Tech. **76–77**(2), 827–834 (1995)
18. Zhang, G., Cho, J.: Tailoring of nano- and microstructure in biomimetically synthesized ceramic films. In: Proceedings of the 25th Army Science Conference. Orlando, FL (2006)
19. Yoshimura, M., Livage, J.: Soft processing for advanced inorganic materials. MRS Bull. **25**(09), 12–13 (2000)
20. Lee, H., Zhang, G., Cho, J.: Inorganic-Organic barrier coatings for flexible OLED applications. In: Flexible Electronics and Displays Conference and Exhibition (IEEE Conf. Proc.)
21. Wright, M., Uddin, A.: Organic—inorganic hybrid solar cells: a comparative review. Sol. Energy Mater. Sol. Cells **107**, 87–111 (2012)
22. Bunker, B.C., Rieke, P.C., Tarasevich, B.J., Campbell, A.A., Fryxell, G.E., Graff, G.L., Song, L., Liu, J., Virden, J.W., McVay, G.L.: Ceramic thin-film formation on functionalized interfaces through biomimetic processing. Science **264**, 48–55 (1994)
23. Agarwal, M., Guire, M.R.D., Heuer, A.: Synthesis of ZrO_2 and Y_2O_3-Doped ZrO_2 thin films using self-assembled monolayers. J. Am. Ceram. Soc. **80**(12), 2967–2981 (1997)
24. Supothina, S., Guire, M.R.D.: Characterization of SnO_2 thin films grown from aqueous solutions. Thin Solid Films **371**, 1–9 (2000)
25. Niesen, T.P., Guire, M.R.D.: Review: deposition of ceramic thin films at low temperatures from aqueous solutions. J. Electroceramics **6**, 169–207 (2001)
26. Zhang, G., Howe, J.Y., Coffey, D.W., Blom, D.A., Allard, L.F., Cho, J.: A biomimetic approach to the deposition of ZrO_2 films on self-assembled nanoscale templates. Mater. Sci. Eng. C **26**(8), 1344–1350 (2006)
27. Zhang, G., Cho, J.: Structural evolution and mechanical behavior of bio-inspired oxide films on self-assembled organic layers. In: Mechanics of Biological and Bio-Inspired Materials (MRS Proceedings), vol. 975, pp. 153–59. MRS, Warrendale, PA (2007)
28. Zhang, G., Roy, B.K., Allard, L., Cho, J.: Titanium oxide nanoparticles precipitated from low-temperature aqueous solutions: I. Nucleation, growth, and aggregation. J. Am. Ceram. Soc. **91**(12), 3875–3882 (2008)
29. Cai, N., Cho, J.: Low temperature processed SnO_2 films using aqueous precursor solutions. Ceram. Int. **39**(1), 143–151 (2013)
30. Yoshimura, M., Suchanek, W.L., Byrappa, K.: Soft solution processing: a strategy for one-step processing of advanced inorganic materials. MRS Bull. **25**(09), 17–25 (2000)
31. Imai, H.: Self-Organized formation of hierarchical structures. Top. Curr. Chem. **270**, 43–72 (2007)
32. Oaki, Y., Imai, H.: The hierarchical architecture of nacre and its mimetic material. Angew. Chem. Int. Ed. **44**, 6571–6575 (2005)
33. Sethmann, I., Hinrichs, R., Wörheide, G., Putnis, A.: Nano-cluster composite structure of calcitic sponge spicules—a case study of basic characteristics of biominerals. J. Inorg. Biochem. **100**, 88–96 (2006)
34. Calvert, P., Mann, S.: Review: synthetic and biological composites formed by in-situ precipitation. J. Mater. Sci. **23**, 3801–3815 (1988)
35. Mann, S.: Biomineralization and Biomimetic Materials Chemistry. In: Mann S. (ed.) Biomimetic Materials Chemistry, pp. 1–40. VCH Publishers, New York, NY (1996)
36. Cölfen, H.: Bio-inspired mineralization using hydrophilic polymers. Topp. Curr. Chem. **271**, 1–77 (2007)
37. Zhang, G., Roy, B.K., Allard, L., Cho, J.: Titanium oxide nanoparticles precipitated from low-temperature aqueous solutions: II. Thin film formation and microstructure developments. J. Am. Ceram. Soc. **93**(7), 1909–1915 (2010)
38. Roy, B.K., Zhang, G., Cho, J.: Titanium oxide nanoparticles precipitated from low-temperature aqueous solutions: III. Thin film properties. J. Am. Ceram. Soc. **95**(2), 676–683 (2012)
39. Feng, X., Shankar, K., Varghese, O.K., Paulose, M., Latempa, T.J., Grimes, C.A.: Vertically aligned single crystal TiO_2 nanowire arrays grown directly on transparent conducting oxide coated glass: synthesis details and applications. Nano Lett. **8**(11), 3781–3786 (2008)

40. Liu, B., Aydil, E.S.: Growth of oriented single-crystalline rutile TiO_2 nanorods on transparent conducting substrates for dye-sensitized solar cells. J. Am. Chem. Soc. **131**(11), 3985–3990 (2009)
41. Kumar, A., Madaria, A.R., Zhou, C.: Growth of aligned single-crystalline rutile TiO_2 nanowires on arbitrary substrates and their application in dye-sensitized solar cells. J. Phys. Chem. C **114**(17), 7787–7792 (2010)
42. Zhou, R., Chang, H.-C., Protasenko, V., Kuno, M., Singh, A. K., Jena, D., Xing, H.: CdSe nanowires with illumination-enhanced conductivity: Induced dipoles, dielectrophoretic assembly, and field-sensitive emission, vol. 101, pp. 073704. AIP (2007)
43. Chen, P.-C., Shen, G., Zhou, C.: Chemical Sensors and electronic noses based on 1-D metal oxide nanostructures. IEEE Trans. Nanotech. **7**(6), 668–682 (2008)
44. Im, B., Joshi, U.A., Lee, K.H., Lee, J.S.: Growth of single crystalline barium tatanate nanowires from TiO_2 seeds deposited on conducting glass. Nanotechnology **21**, 5, 425601 (2010)
45. Sun, C., Shi, J., Wang, X.: Fundamental study of mechanical energy harvesting using piezoelectric nanostructures, vol. 108, pp. 034309. AIP (2010)
46. Wang, Z., Hu, J., Suryavanshi, A.P., Yun, K., Yu, M.-F.: Voltage generation from individual $BaTiO_3$ nanowires under periodic tensile mechanical load. Nano Lett. **7**(10), 2966–2969 (2007)
47. Bamiduro, F., Ward, M.B., Brydson, R., Milne, S.J.: Hierarchical growth of ZnO particles by a hydrothermal route. J. Am. Ceram. Soc. **97**(5), 1619–1624 (2014)
48. Drummond, S.E., Ohmoto, H.: Chemical evolution and mineral deposition in boiling hydrothermal systems. Econ Geol Lancaster, Pa. **80**(1), 126–147 (1985)
49. Laudise, R.A.: Hydrothermal synthesis of crystals. Chem Eng News Arch **65**(39), 30–43 (1987)
50. Bhosale, P.N., Kondalkar, V.V., Mane, R.M., Choudhury, S., Khot, K.V.: Hydrothermal assisted synthesis of hierarchical nanostructured metal oxide thin film. Nanomed Nanotechnol **6**(4), 1–2 (2015)
51. Gupta, T., Samriti, Cho, J., Prakash, J.: Hydrothermal synthesis of TiO_2 nanorods: formation chemistry, growth mechanism, and tailoring of surface properties for photocatalytic activities. Mater. Today Chem. **20**, 100428 (2021)
52. Yang, Q., Lu, Z., Liu, J., Lei, X., Chang, Z., Luo, L., Sun, X.: Metal oxide and hydroxide nanoarrays: hydrothermal synthesis and applications as supercapacitors and nanocatalysts. Prog. Nat. Sci. Mater. Int. **23**(4), 351–366 (2013)
53. Dong, L., Cheng, K., Weng, W., Song, C., Du, P., Shen, G., Han, G.: Hydrothermal growth of rutile TiO_2 nanorod films on titanium substrates. Thin Solid Films **519**(15), 4634–4640 (2011)
54. Noh, H.W., Jeong, S.M., Cho, J., Hong, J.-I.: Ultrahigh photosensitivity of the polar surfaces of single crystalline ZnO nanoplates. Nanoscale **10**(15), 6801–6805 (2018)
55. Roy, B.K., Zhang, G., Magnuson, R., Poliks, M., Cho, J.: Electrodeposited titania thin films grown on a Cu electrode for high-k dielectric applications. J. Am. Ceram. Soc. **93**(3), 774–781 (2010)
56. Carp, O., Huisman, C.L., Reller, A.: Photoinduced reactivity of titanium dioxide. Prog. Solid State Chem. **32**, 33–177 (2004)
57. Chakraborty, A., Samriti, Ruzimuradov, O., Gupta, R.K., Cho, J., Prakash, J.: TiO_2 nanoflower photocatalysts: synthesis, modifications and applications in wastewater treatment for removal of emerging organic pollutants. Environ. Res. **212**, 113550 (2022)
58. Prakash, J., Cho, J., Mishra, Y.K.: Photocatalytic TiO_2 nanomaterials as potential antimicrobial and antiviral agents: scope against blocking the SARS-COV-2 spread. Micro Nano Eng. **14**, 100100 (2022)
59. Ye, M., Wen, X., Wang, M., Iocozzia, J., Zhang, N., Lin, C., Lin, Z.: Recent advances in dye-sensitized solar cells: from photoanodes, sensitizers and electrolytes to counter electrodes. Mater. Today **18**(3), 155–162 (2015)

60. Jiang, J., Zhou, H., Zhang, F., Fan, T., Zhang, D.: Hydrothermal synthesis of core–shell TiO_2 to enhance the photocatalytic hydrogen evolution. Appl. Surf. Sci. **368**, 309–315 (2016)
61. Pei, C.C., Leung, W.W.-F.: Enhanced photocatalytic activity of electrospun TiO_2/ZnO nanofibers with optimal anatase/rutile ratio. Catal. Commun. **37**, 100–104 (2013)
62. Hurum, D.C., Agrios, A.G., Grey, K.A., Rajh, T., Thurnauer, M.C.: Explaining the enhanced photocatalytic activity of degussa P25 mixed-phase TiO_2 using EPR. J. Phys. Chem. B **107**, 4545–4549 (2003)
63. Tian, J., Chen, L., Yin, Y., Wang, X., Dai, J., Zhu, Z., Liu, X., Wu, P.: Photocatalyst of TiO_2/ZnO nano composite film: preparation, characterization, and photodegradation activity of methyl orange. Surf. Coat. Technol. **204**(1–2), 205–214 (2009)
64. Zhao, Y., Li, C., Liu, X., Gu, F., Du, H.L., Shi, L.: Zn-doped TiO_2 nanoparticles with high photocatalytic activity synthesized by hydrogen–oxygen diffusion flame. Appl. Catal. B **79**(3), 208–215 (2008)
65. Lin, C., Song, Y., Cao, L., Chen, S.: TiO_2 nanotubes/ZnO/CdS ternary nanocomposites: preparation, characterization and photocatalysis. J. Chin. Adv. Mater. Soc. **1**(3), 188–199 (2013)
66. Cölfen, H., Mann, S.: Higher-Order organization by mesoscale self-assembly and transformation of hybrid nanostructures. Angew. Chem. Int. Ed. **42**, 2350–2365 (2003)
67. Nabivanets, B., Kudritskaya, L.: Determination of the composition and dissociation constants of titanyl chloro-complexes. Russ. J. Inorg. Chem. **12**(6), 789–792 (1967)
68. Grätzel, M., Rotzinger, F.P.: Raman spectroscopic evidence for the existence of titanyl (TiO^{2+}) in acidic aqueous solutions. Inorgan. Chem. **24**(14), 2320–2321 (1985)
69. Henry, M., Jolivet, J.P., Livage, J.: Aqueous chemistry of metal cations, hydrolysis, condensation, and complexation. In: Reisfeld, R., Jorgensen, C.K. (eds.) Structure and Bonding, vol. 77, pp. 155–206. Springer-Verlag, Berlin, Germany (1992)
70. Lencka, M.M., Riman, R.E.: Thermodynamic modeling of hydrothermal synthesis of ceramic powders. Chem. Mater. **6**, 61–70 (1993)
71. Mullin, J.W.: Crystallization, 4th Ed. Butterworth-Heinemann (2001)
72. Rieke, P.C., Marsh, B.D., Wood, L.L., Tarasevich, B.J., Liu, J., Song, L., Fryxell, G.E.: Aqueous-solution deposition kinetics of iron oxyhydroxide on sulfonic-acid terminated self-assembled monolayers. Langmuir **11**(1), 318–326 (1995)
73. Chiang, P.-P., Donohue, M.D.: A kinetic approach to crystallization from ionic solution. I. Crystal growth. J. Colloid Int. Sci. **122**(1), 230–250 (1988)
74. Chen, X., Mao, S.S.: Titanium dioxide nanomaterials: synthesis, properties, modifications, and applications. Chem. Rev. **107**(7), 2891–2959 (2007)
75. Wang, Y., Zhang, L., Deng, K., Chen, X., Zou, Z.: Low temperature synthesis and photocatalytic activity of rutile TiO_2 nanorod superstructures. J. Phys. Chem. C **111**(6), 2709–2714 (2007)
76. Shim, J.H., Cho, J.: Hierarchical organization of TiO_2 nanostructures in low temperature solution processes. J. Am. Ceram. Soc. **99**(2), 431–39 (2016). https://doi.org/10.1111/jace.13972
77. Wu, W.-Q., Xu, Y.-F., Su, C.-Y., Kuang, D.-B.: Ultra-long anatase TiO_2 nanowire arrays with multi-layered configuration on FTO glass for high-efficiency dye-sensitized solar cells. Energy Environ. Sci. **7**(2), 644–649 (2014)
78. Shim, J.H., Bae, I.-T., Cho, J.: Microstructure development of hydrothermally grown TiO_2 thin films with vertically aligned nanorods. J. Am. Ceram. Soc. **101**, 50–60 (2018)
79. Roy, B.K., Zhang, G., Yoo, M., Bae, I.-T., Cho, J.: Developments of low-temperature solution processing for nanostructured titania dielectric films. Sci. Adv. Mater. **2**(1), 90–101 (2010)
80. Ramamoorthy, M., Vanderbilt, D.: First-principles calculations of the energetics of stoichiometric TiO_2 surfaces. Phys. Rev. B **49**(23), 16721–16727 (1994)
81. Bierman, M.J., Lau, Y.K.A., Kvit, A.V., Schmitt, A.L., Jin, S.: Dislocation-driven nanowire growth and eshelby twist. Science **320**(5879), 1060–1063 (2008)
82. Morin, S.A., Bierman, M.J., Tong, J., Jin, S.: Mechanism and kinetics of spontaneous nanotube growth driven by screw dislocations. Science **328**(5977), 476–480 (2010)

83. Ribeiro, C., Vila, C., Stroppa, D.B., Mastelaro, V.R., Bettini, J., Longo, E., Leite, E.R.: Anisotropic growth of oxide nanocrystals: insights into the rutile TiO_2 phase. J. Phys. Chem. C **111**(16), 5871–5875 (2007)
84. Penn, R.L., Banfield, J.F.: Imperfect oriented attachment: dislocation generation in defect-free nanocrystals. Science **281**(5379), 969–971 (1998)
85. Greene, L.E., Law, M., Tan, D.H., Montano, M., Goldgerger, J., Somorjai, G., Yang, P.: General route to vertical ZnO nanowire arrays using textured ZnO seeds. Nano Lett. **5**(7), 1231–1236 (2005)
86. Lee, S., Roy, B.K., Cho, J.: Vertically aligned ZnO nanorods grown by low-temperature solution processing. Jpn J. Appl. Phys. **52**(5S1), 05DA09 (2013)
87. Pina, C.M., Enders, M., Putnis, A.: The composition of solid solutions crystallising from aqueous solutions: the influence of supersaturation and growth mechanisms. Chem. Geol. **168**, 195–210 (2000)
88. Pina, C.M., Putnis, A., Astilleros, J.M.: The growth mechanisms of solid solutions crystallising from aqueous solutions. Chem. Geol. **204**, 145–161 (2004)
89. Niederberger, M., Cölfen, H.: Oriented attachment and mesocrystals: non-classical crystallization mechanisms based on nanoparticle assembly. Phys. Chem. Chem. Phys. **8**, 3271–3287 (2006)
90. Polleux, J., Pinna, N., Antonietti, M., Niederberger, M.: Ligand-directed assembly of preformed titania nanocrystals into highly anisotropic nanostructures. Adv. Mater. **16**(5), 436–439 (2004)
91. Polleux, J., Pinna, N., Antonietti, M., Hess, C., Wild, U., Schlögl, R., Niederberger, M.: Ligand functionality as a versatile tool to control the assembly behavior of preformed titania nanocrystals. Chem. Eur. J. **11**, 3541–3551 (2005)
92. Diebold, U.: The surface science of titanium dioxide. Surf. Sci. Rep. **48**(5–8), 53–229 (2003)
93. Liu, K., Cao, M., Fujishima, A., Jiang, L.: Bio-inspired titanium dioxide materials with special wettability and their applications. Chem. Rev. **114**(19), 10044–10094 (2014)
94. Liu, L., Chen, X.: Titanium dioxide nanomaterials: self-structural modifications. Chem. Rev. **114**(19), 9890–9918 (2014)
95. Yu, X., Zhao, Z., Zhang, J., Guo, W., Qiu, J., Li, D., Li, Z., Mou, X., Li, L., Li, A., Liu, H.: Rutile nanorod/anatase nanowire junction array as both sensor and power supplier for high-performance, self-powered, wireless UV photodetector. Small **12**(20), 2759–2767 (2016)
96. Sutiono, H., Tripathi, A.M., Chen, H.-M., Chen, C.-H., Su, W.-N., Chen, L.-Y., Dai, H., Hwang, B.-J.: Facile synthesis of [101]-oriented rutile TiO_2 nanorod array on FTO substrate with a tunable anatase-rutile heterojunction for efficient solar water splitting. ACS Sustain. Chem. Eng. **4**(11), 5963–5971 (2016)
97. Xiong, Z., Wu, H., Zhang, L., Gu, Y., Zhao, X.S.: Synthesis of TiO_2 with controllable ratio of anatase to rutile. J. Mater. Chem. A **2**(24), 9291–9297 (2014)
98. Zhang, W., Feng, L., Chen, H., Zhang, Y.: Hydrothermal synthesis of SnO_2 nanorod as anode materials for lithium-ion battery. Nano **14**(09), 1950109 (2019)
99. Liang, Y., Fang, B.: Hydrothermal synthesis of SnO_2 nanorods: morphology dependence, growth mechanism and surface properties. Mater. Res. Bull. **48**(10), 4118–4124 (2013)
100. Kuang, X., Liu, T., Shi, D., Wang, W., Yang, M., Hussain, S., Peng, X., Pan, F.: Hydrothermal synthesis of hierarchical SnO_2 nanostructures made of superfine nanorods for smart gas sensor. Appl. Surf. Sci. **364**, 371–377 (2016)
101. Cho, J.: Application of nanostructured metal oxides and its hybrids for inactivation of bacteria and viruses. In: Prakash, J., Cho, J., Campos Janegitz, B., Sun, S. (eds.) Multifunctional Hybrid Semiconductor Photocatalyst Nanomaterials: Application on Health, Energy and Environment, pp. 53–80. Springer International Publishing, Cham (2023)
102. Tao, T., Shim, J.H., Joshi, J., Sauer, K., Cho, J.: Tailoring of solution grown TiO_2 and ZnO nanostructures for photoactive performances. In: 2016 MRS Fall Meeting (Poster Presentation) (2016)
103. Tao, T., Bae, I.-T., Woodruff, K.B., Sauer, K., Cho, J.: Hydrothermally-grown nanostructured anatase TiO_2 coatings tailored for photocatalytic and antibacterial properties. Ceram. Int. **45**(17, Part B), 23216–24 (2019)

104. Prakash, J., Krishna, S.B., Kumar, P., Kumar, V., Ghosh, K.S., Swart, H.C., Bellucci, S., Cho, J.: Recent advances on metal oxide based nano-photocatalysts as potential antibacterial and antiviral agents. Catalysts **12**(9), 1047 (2022)
105. Tsai, C., Teng, H.: Structural features of nanotubes synthesized from NaOH treatment on TiO_2 with different post-treatments. Chem. Mater. **18**, 367 (2006)
106. Peng, C.-W., Richard-Plouet, M., Ke, T.-Y., Lee, C.-Y., Chiu, H.-T., Marhic, C., Puzenat, E., Lemoigno, F., Brohan, L.: Chimie douce route to sodium hydroxo titanate nanowires with modulated structure and conversion to highly photoactive titanium dioxides. Chem. Mater. **20**(23), 7228–7236 (2008)
107. Tao, T.: Solution Processing of Ceramic Nanocomposite Coatings for Improved Functional Properties, State University of New York at Binghamton (Ph.D. dissertation) (2018)
108. Tanaka, S.-I., Hirose, N., Tanaki, T.: Effect of the Temperature and Concentration of NaOH on the Formation of Porous TiO_2, vol. 152, pp. C789-C94 (2005)
109. Huang, J., Cao, Y., Deng, Z., Tong, H.: Formation of titanate nanostructures under different NaOH concentration and their application in wastewater treatment. J. Solid State Chem. **184**(3), 712–719 (2011)
110. Yang, D., Liu, H., Zheng, Z., Yuan, Y., Zhao, J.-C., Waclawik, E.R., Ke, X., Zhu, H.: An efficient photocatalyst structure: TiO_2(B) nanofibers with a shell of anatase nanocrystals. J. Am. Chem. Soc. **131**(49), 17885–17893 (2009)
111. Muniyappan, S., Solaiyammal, T., Sudhakar, K., Karthigeyan, A., Murugakoothan, P.: Conventional hydrothermal synthesis of titanate nanotubes: systematic discussions on structural, optical, thermal and morphological properties. Mod. Electron. Mater. **3**(4), 174–178 (2017)
112. Ali, S., Granbohm, H., Ge, Y., Singh, V.K., Nilsén, F., Hannula, S.-P.: Crystal structure and photocatalytic properties of titanate nanotubes prepared by chemical processing and subsequent annealing. J. Mater. Sci. **51**(15), 7322–7335 (2016)
113. Shim, J.H., Bae, I.-T., Cho, J.: Microstructure development of hydrothermally grown TiO_2 thin films with vertically aligned nanorods. J. Am. Ceram. Soc. **101**(1), 50–60 (2018)
114. Shim, J.H.: Multifunctional Ceramic Nanostructure Thin Films Prepared By Solution Processing, Binghamton University (Ph.D. dissertation) (2017)
115. Kawahara, T., Konishi, Y., Tada, H., Tohge, N., Nishii, J., Ito, S.: A patterned TiO_2(Anatase)/TiO_2(Rutile) bilayer-type photocatalyst: effect of the anatase/rutile junction on the photocatalytic activity. Angew. Chem. Int. Ed. **41**(15), 2811–2813 (2002)
116. Miyagi, T., Kamei, M., Mitsuhashi, T., Ishigaki, T., Yamazaki, A.: Charge separation at the rutile/anatase interface: a dominant factor of photocatalytic activity. Chem. Phys. Lett. **390**(4), 399–402 (2004)
117. Zhu, Y., Deng, K., Sun, H., Gu, B., Lu, H., Cao, F., Xiong, J., Li, L.: TiO_2 phase junction electron transport layer boosts efficiency of planar perovskite solar cells. Adv. Sci. **5**(3), 1700614 (2018)

TiO_2 Sol–Gel Nanomaterials: Synthesis, Properties and Applications

Dmitry Kovalenko, Vladimir Gaishun, Vasily Vaskevich, Alina Semchenko, and Olim Ruzimuradov

Abstract The present study examines the synthesis, properties, and applications of nanostructured materials based on titanium dioxides (TiO_2) via the sol–gel method. The sol–gel method makes it possible to vary the physicochemical properties of nanostructured materials over a wide range and achieve significant interaction between components at the stage of the initial sol. Simultaneously, sol–gel synthesis of nanostructured materials is more energy efficient compared to vacuum and plasma methods. The main physical and chemical processes that occur throughout the preparation and maturation stages of sol are determined. The influence of synthesis conditions on the dispersion, phase composition, and properties of nanostructured materials based on titanium dioxide is established. A description of the sol–gel method for producing bulk nanostructured materials is provided. Extensive studies have been conducted to examine the optical and structural properties of nanostructured materials based on TiO_2. Numerous investigations have been carried out on the photocatalytic activity of nanostructured films and xerogels obtained through the sol–gel method. The study findings indicate that nanostructured films and xerogels derived from TiO_2 possess photocatalytic properties, making them suitable for application in water purification systems.

Keywords TiO_2 · Sol–gel method · Nanostructured films · Optical properties · Photocatalytic activities

D. Kovalenko (✉) · V. Gaishun · V. Vaskevich · A. Semchenko
Francisk Skorina Gomel State University, Sovetskaya 104, 246028 Gomel, Belarus
e-mail: dkov@gsu.by

O. Ruzimuradov
National University of Uzbekistan Named After Mirzo Ulugbek, Student's Campus, University Street 4, Tashkent, Uzbekistan 100174

J. Prakash et al. (eds.), *Titanium Dioxide-Based Multifunctional Hybrid Nanomaterials*, Engineering Materials, https://doi.org/10.1007/978-3-031-80625-4_3

1 The Main Stages of Photocatalytic TiO_2 Sol–Gel Nanomaterials Synthesys

1.1 Stable Sols Based on Organic Titanium Compounds

The hydrolysis of various titanium compounds (alkoxides and inorganic titanium salts, mainly $TiCl_4$) in aqueous solutions leads to primary products at low pH values such as basic salts of variable composition. At higher pH values, hydrated forms of TiO_2 are formed, which correspond to the formula $Ti(OH)_2$ or titanium dioxide $(TiO_2)\cdot nH_2O$, where *n* depends on aging and drying conditions.

Freshly precipitated hydrated TiO_2 has a high adsorption capacity for both cations and anions; the content and nature of impurities in TiO_2 depend on the pH of the medium during precipitation, the nature of the precipitant, and the initial titanium compound.

The specific surface area of the hydrated TiO_2 generated during hydrolysis under hydrothermal conditions also significantly depends on the temperature and duration of the process. The available data indicates that the degree of crystallization of the product resulting from hydrothermal hydrolysis is greatly affected by the initial concentration of $TiCl_4$, which determines the acidity of the reaction environment.

The photocatalysis mechanisms by which certain types of organic compounds are decomposed completely to carbon dioxide and water are well known. New materials have been developed based on titania, and the sensitivity to visible light has improved [1]. The generation of new catalytic active sites either on the silica surface or in the silica matrix is discussed and the insights obtained from these studies allow a fundamental understanding of the relationships between the structural characteristics and the physico-chemical/reactivity properties of titania–silica catalysts [2]. Titania–silica materials represent a novel class of catalysts and have been widely applied in photocatalysis, acid catalysis and oxidation catalysis. The intimate interaction of TiO_2 and SiO_2 has been shown to result in new structural characteristics and physicochemical/reactivity properties. The degree of interaction, in other words, the homogeneity or dispersion when TiO_2 is mixed with or supported on SiO_2, largely depends on the preparation methods and synthesis conditions. Also, the surface enrichment of either Ti or Si in TiO_2–SiO_2 mixed oxides has been shown to depend on the preparation conditions and chemical compositions [2].

The possibilities for the purposeful control and stabilization of properties of TiO full control and stabilization of properties of TiO_2 nanopowders and sols are very wide [3]. Particles of titanium dioxide and titanium dioxide-hydroxypropyl cellulose organic–inorganic composite can be easily synthesized by the sol–gel method [4].

A simple and accessible method for obtaining pure and stable nanosized TiO_2 (anatase) under mild hydrothermal conditions is well known. Performing solvothermal hydrolysis of titanium ethoxide in anhydrous ethanol with the addition of a strictly defined amount of ultrapure water under mild conditions also results in the production of pure ultrafine dispersed nanocrystalline anatase [5, 6].

TiO_2 exists in the form of several crystalline modifications. Anatase, rutile, and brookite can be found in nature. Note that brookite is seldom manufactured on an industrial scale and is rarely encountered in nature. The anatase form is also significantly inferior in production volume to rutile, since rutile dioxide scatters light approximately 30% better (better hiding power) and is more weather-resistant. When heated, both anatase and brookite are irreversibly converted to rutile (transition temperatures, respectively, (400–1000) °C and about 750 °C).

During heat treatment of gels, hydrated forms of TiO_2, they crystallize with the formation of anhydrous TiO_2. Depending on the calcination temperature, polymorphic transformations of hydrated TiO_2 into anatase, rutile, or brookite are possible. These transformations are followed by changes in the specific surface area and porous structure. At temperatures below 600 °C, the crystallization process occurs with the formation of anatase with practically unchanged pore volume and specific surface area. At higher temperatures, a transition of anatase to rutile is observed, accompanied by a sharp decrease in both pore volume and specific surface area. The temperature of polymorphic transformation can also be greatly influenced by mineral impurities. Numerous data have shown that, in addition to the phase composition, the conditions of heat treatment of TiO_2 gels determine other important performance characteristics of TiO_2—morphology and particle size [7].

The most common methods for producing TiO_2 are sulfate and chloride (from $TiCl_4$). The processing of titanium tetrachloride is carried out according to three different schemes: hydrothermal hydrolysis, vapor-phase hydrolysis or combustion in a stream of oxygen. Recently, the sol–gel method, which is suitable for obtaining nanosized TiO_2 particles with a given structure and properties, has become increasingly important and, therefore, is of interest in connection with its application in the field of nanotechnology. Materials based on TiO_2 have extremely diverse applications: as photocatalytic, antibacterial, antireflective coatings [8–21]; coatings to protect the surface from various influences [18, 20–28]; coatings for solar cells [29–31].

The hydrolysis of various titanium compounds (alkoxides and inorganic titanium salts, mainly ethoxides) in aqueous and anhydrous solutions leads to primary products at low pH values such as basic salts of variable composition. This research employs titanium ethoxide as the initial alcohol compound to obtain film-forming solutions. This is because the hydrolysis of titanium ethoxide in anhydrous isopropyl alcohol with the addition of a strictly defined amount of ultrapure water under mild conditions leads to the formation of a stable solution.

1.2 Formation of Thin Films of Titanium Oxide.

The spin-coating procedure is the optimal technique for obtaining coatings on flat substrates.

The centrifugal application method involves the formation of a deposited layer by spreading the film-forming solution under centrifugal forces. The excess film-forming solution is forced out of the substrate in a radial flow due to the effect of centrifugal force.

It is well-known that the main parameters influencing the thickness of sol–gel coatings produced via centrifugation are the viscosity of the film-forming solution and the rotation frequency of the substrate. Experiments were conducted to determine the optimal range of rotation speeds for the workpieces in relation to the selected sol. These experiments involved the deposition of sol–gel films on substrates of various sizes by centrifugation. A series of experiments established a correlation between the diameter of the substrate and the rotation frequency at which the resulting films have a minimum width of the brim formed by the coating at the edge of the substrate surface.

Ready-made film-forming solutions were deposited on previously prepared substrates made of silicon, quartz glass, and metal using centrifugation and dipping methods. After deposition, the resulting samples were heat-treated step by step in air from 100 to 800 °C.

1.3 Formation of Bulk Materials Based on TiO_2

The method for obtaining bulk photocatalytic materials includes the following stages:

1. Obtaining a sol by hydrolysis in a water solution in the presence of nitric acid as a catalyst. The hydrolysis of titanium-based sols was carried out in an anhydrous environment. The starting components were mixed in a special polymer reactor with a paddle stirrer for 30 min;
2. Fine dispersion under ultrasonic conditions;
3. Centrifugal separation of solid particles and impurities at a centrifuge rotation speed of 3000 rpm for 60 min;
4. For neutralizing the acidic environment and accelerating the gelation process, drops of a 0.1 N alkali solution are added to the mixture to increase the pH to 4–6. Casting the sol into molds and gelling for 15–30 min at room temperature;
5. Maturation of gels during the day;
6. Formation of xerogels by drying the gels in an oven at a temperature of 40–50 °C for 3–5 days, depending on the geometric dimensions of the gels.

Figure 1 presents a schematic diagram illustrating the process of creating xerogels doped with noble metals and rare earth elements.

The hydrolysis rate is influenced by many factors, in particular the presence of a catalyst, temperature, solvent concentration, etc.

Controlling the structure of TiO_2 at the nanoscale can lead to significant improvements in the functional properties of the resulting material. By employing the sol–gel technique, it becomes feasible to not only regulate and reproduce the ratio of TiO_2

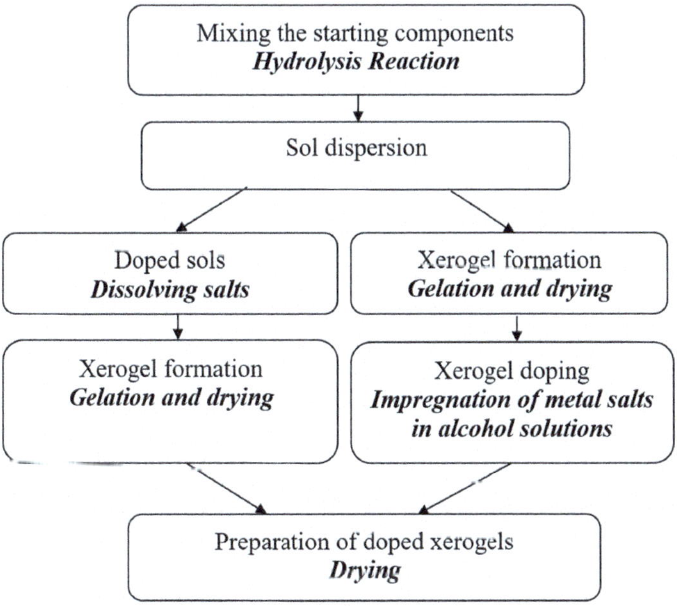

Fig. 1 Schematic diagram illustrating the formation of xerogels doped with metals

crystalline phases (anatase, rutile, and brookite) in the resulting product, but also to obtain each phase in its pure form [5, 6].

Preparing the sols to obtain gels and xerogels from titanium isopropoxide and titanium ethoxide as detailed below. The synthesis was carried out in a glass vessel. Previously, a hydrolysis mixture was placed in the flask, which consisted of isopropyl alcohol and concentrated hydrochloric or nitric acid. Titanium isopropoxide or titanium ethoxide was added to the flask while stirring continuously. The resulting mixture was homogenized. The hydrolysis of titanium alkoxides can be represented by the following equations:

$$\mathrm{Ti-OR + H_2O \rightarrow Ti-OH + ROH,}$$
$$\mathrm{Ti-OH + Ti-OR \rightarrow Ti-O-Ti + ROH,}$$
$$\mathrm{Ti-OH + Ti-OH \rightarrow Ti-O-Ti + HOH.}$$

Sols of various compositions containing titanium isopropoxide or titanium ethoxide were prepared since the resulting structure, which determines the structural characteristics, strongly depends on the preparation conditions, including the composition of the initial components.

1.3.1 Formation of Doped Xerogels by Gelation

Gelation in alkoxide sol–gel processes occurs spontaneously and continues for several days. Nevertheless, with a rise in temperature, the gelation time experiences a substantial decrease.

To create homogeneous matrices with optimal parameters (formation of solid gel blanks, obtaining a given xerogel porosity, etc.), the type and amount of the introduced solid filler, TiO_2 powder, play a decisive role. It was found that introducing a filler, for example, directly into the hydrolyzate product leads to its clumping and, consequently, to the presence of large agglomerates of particles in the final ash. Therefore, the dispersion of the filler in an aqueous environment is separated into a distinct process [7].

Depending on the calcination temperature, polymorphic modifications of hydrated TiO_2 into anatase, rutile or brookite are possible, accompanied by changes in the specific surface area and porous structure. The sol–gel method allows for regulation and reproduction of the crystalline phase ratio of TiO_2 (anatase, rutile, and brookite) in the resulting product while also obtaining each phase in its pure form [8–21].

1.3.2 Formation of Xerogels by Pressing

After obtaining a xerogel for photocatalytic purification based on TiO_2, a porous structure (tablets with a diameter of 1.6 cm and a thickness of 1 cm) was created by compression molding in order to enhance the working surface, which will interact with the solution during the photocatalysis process. The resulting xerogels, after mixing with TiO_2 powder, were placed in a mold to form tablets. Initial solutions based on titanium compounds were used as a binder and a plasticizer when forming the workpieces.

Subsequently, the resulting tablets were subjected to a step-by-step heat treatment process in order to produce the photocatalytically active polymorphic modification rutile or brookite.

The materials were placed in an oven and heat-treated at 600 and 800 °C for 60 min, followed by cooling in air.

The materials obtained via this method can be used as targets to form coatings from the plasma of a pulsed cathode-arc discharge. In order to accomplish this, the dried xerogels are annealed at a temperature of 900–950 °C for 6 h, which causes the formation of a ceramic material. During annealing, numerous complex physical and chemical processes occur which are related to the degradation of organic fragments embedded in the inorganic gel network, the removal of solvents, volatile degradation products, and chemically bound water. The inorganic polymer undergoes a restructuring process known as sintering, which may also involve crystallization in certain instances.

2 Properties of TiO_2 Sol–Gel Nanomaterials

2.1 Calculation of the Refractive Index of the Resulting Thin Films

The refractive index of a TiO_2 coating can be determined by using the absorption maximum observed in the transmission spectrum. For a weak absorptive coating, assuming normal light incidence, the minimum transmittance is calculated as follows:

$$T_{\min} = 4n_f^2 n_s / \left(n_f^2 + n_s^2\right).$$

The coating refractive index

$$n_f = \left[n_s(2 - T_{\min}) + 2n_s(1 - T_{\min})^{1/2} / T_{\min}\right]^{1/2},$$

where n_f is the refractive index of the coating, n_s is the refractive index of the substrate [32].

Table 1 shows that as the heat treatment temperature is raised, the refractive index of the coating exhibits a gradual increase. This phenomenon is explained by the fact that with the refractive index of the anatase structure of 2.23, rutile is denser, i.e., its refractive index is 2.60. The coating samples after heat treatment at 700 °C represent an anatase structure. At 800 °C, the samples represent a mixed structure of anatase and rutile. When temperature reaches 900 °C, the coating structure changes to the rutile phase, showing the maximum refractive index. Achieving precise tuning of the optical properties of the TiO_2 coating requires careful control of the ratio between the two structural structural types, anatase and rutile. This ratio is susceptible to alteration during heat treatment, thereby causing a modification in the refractive index.

Table 2 illustrates that increase in the annealing temperature is accompanied by the gradual increase in the refractive index of the coating. This phenomenon is explained by the fact that when the refractive index of the mixed structure of anatase and rutile is 2.10, rutile is denser, i.e., its refractive index is 2.55. Following heat treatment at 500 °C, the coating samples represent a mixed structure of anatase and rutile; after heat treatment at 900 °C, the structure completely transforms into the rutile phase, showing the maximum refractive index.

Table 1 Refractive index of TiO_2 (isopropoxide) coatings

Annealing temperature	600 °C	700 °C	800 °C	900 °C
Refractive index	2.17	2.23	2.55	2.60

Table 2 Refractive index of the TiO_2 coatings

Annealing temperature	500 °C	600 °C	800 °C	900 °C
Refractive index	2.10	2.20	2.36	2.55

2.2 *Structural Properties of Synthesized Sol–Gel Coatings Using Raman Spectroscopy*

The Raman spectrum was recorded using a SENETRA express Raman spectrometer (Bruker).

The analysis revealed that the TiO_2 coating captures oxygen from the atmosphere during the annealing process. Additionally, the annealing temperature promotes surface modification, grain growth, grain bonding, and the growth and development of the crystalline structure (Figs. 2 and 3).

Annealing at a temperature of 600 °C leads to the emergence of Raman peaks at 639, 513, 392 and 148 cm^{-1}. These vibrations correspond to the structure of the TiO_2 coating, i.e., anatase. The transition from the amorphous structure to the crystalline phase of anatase begins at an annealing temperature of 600 °C.

Increasing the annealing temperature up to 700 °C leads to a gradual increase in the intensity of the lines in the TiO_2 spectrum and a corresponding rise in the degree of crystallization of the coating. However, it is important to note that the rutile phase does not occur at these temperatures. The peaks of the coating, which includes titanium ethoxide, at frequencies of 148, 392, 513 and 639 cm^{-1} are much more intense than the peaks of the coating, which includes titanium isopropoxide. It can be seen that

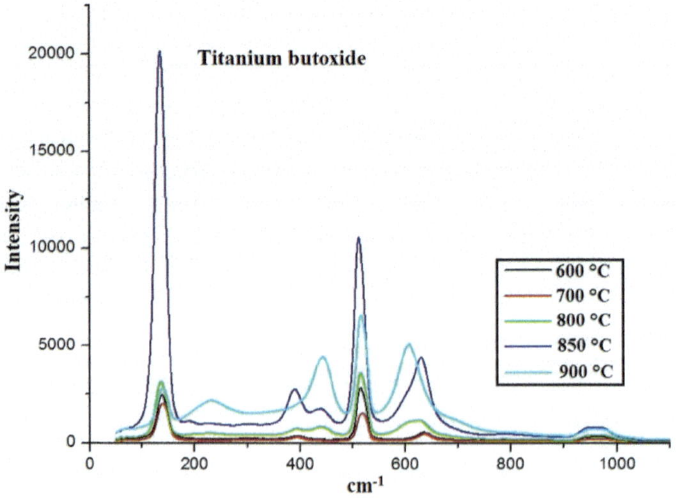

Fig. 2 Raman spectrum of TiO_2 coatings at different annealing temperatures

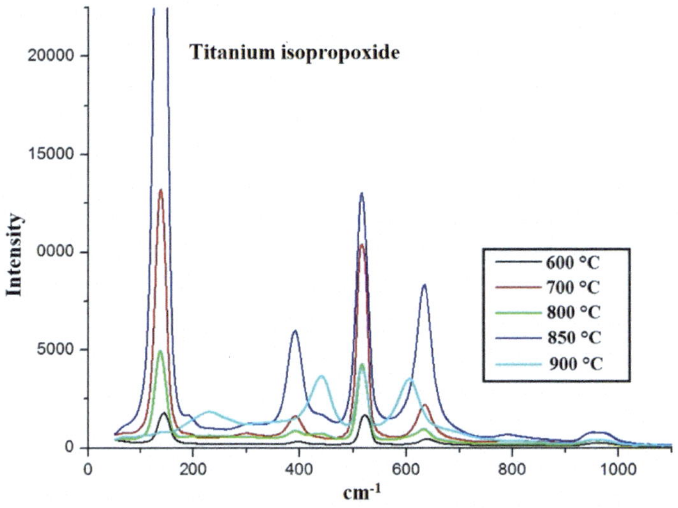

Fig. 3 Raman spectrum of TiO_2 coatings based on titanium isopropoxide at different annealing temperatures

at annealing temperatures of 600~700 °C, the TiO_2 coating predominantly exhibits a crystalline structure within the anatase phase.

The intensity of the coating peaks characterizing the anatase phase experienced a substantial increase after annealing at a temperature of 800 °C. The coating, including ethoxide, exhibits a novel distinctive peak at a frequency of 441 cm^{-1}, which corresponds to the structure of rutile TiO_2. The intensity of the peak is weak. This spectrum indicates that the structure of the TiO_2 coating is in the early stages of transitioning from the anatase phase to the rutile phase.

At 850 °C, the peaks of the coating, which contained isopropoxide, representing the anatase phase at frequencies of 398 and 640 cm^{-1}, nearly vanished. The intensity of anatase peaks at frequencies of 513, 148, 392 and 639 cm^{-1} is significantly reduced. Intensity peaks appeared at 441 and 607 cm^{-1}, which characterize the rutile phase.

At an annealing temperature of 900 °C, the intensity of the peaks in the Raman spectrum of the coating at frequencies of 607 cm^{-1} and 441 cm^{-1} exhibited a substantial increase, whereas the peaks at frequencies of 392 cm^{-1} and 639 cm^{-1} disappeared. Furthermore, in the case of the coating containing isopropoxide, the peak at 148 cm^{-1} vanished. The peak at 513 cm^{-1} decreased significantly. A new intensity peak appeared at a frequency of 235 cm^{-1}, which characterizes the rutile structure. The TiO_2 coating almost completely crystallizes into the rutile phase.

The above-described analysis demonstrates that as the annealing temperature is raised, the structure of the TiO_2 coating transforms from amorphous to rutile. The rutile phase exhibits superior thermal stability in comparison to the anatase phase. This property is shared by anatase and rutile, despite the fact that they both represent a tetragonal system. The rutile phase represents an octahedron with a shared face and vertex, while the anatase phase is an octahedron with a shared vertex. Such a structure is unstable. Therefore, from a thermodynamic point of view, the rutile phase is a stable structure at high temperatures of TiO_2. In addition, due to the fact that the density of the amorphous coating (3.2~3.65 g cm^{-3}) compared to the density of rutile (4.26 g cm^{-3}) is much lower, the structure of the coating is converted from an amorphous state to rutile. This is accompanied by a process of significant compaction of the substance, while the binding energy is greater than during the formation of anatase (3.2 g cm^{-3}). Similarly, when the annealing temperature increases, the phase transition of TiO_2 proceeds as follows: amorphous–anatase–rutile.

2.3 Structural Properties via X-ray Phase Analysis

The study of structural properties by X-ray phase analysis was carried out on samples of titanium oxide that underwent heat treatment from 400 to 800 °C. Previously, when analyzing Raman spectra, it was found that at annealing temperatures above 800 °C, the content of the anatase phase in the sample decreases, due to which the materials have pronounced photocatalytic properties.

The results of X-ray phase analysis indicate that low-temperature annealing ($T \leq 400$ °C) does not lead to the formation of active TiO_2 particles in the resulting films. The diffraction patterns of films, fabricated on the basis of titanium ethoxide (Fig. 4), show that TiO_2 crystals transforming into the brookite phase and subsequently functioning as active centers in photocatalytic reactions are formed at temperatures $T \geq 600$ °C.

With a further increase in the heat treatment temperature to 800 °C, there is an increase in the intensity of the lines due to the presence of brookite in the coatings ($2\theta = 25.24°$). This modification of TiO_2 is superior in photocatalytic activity to anatase and rutile [4]. The intensity of the Ti_2O_3 peaks ($2\theta = 61.55°$) decreases with increasing the annealing temperature from 600 to 800 °C. Thus, the presence of brookite in the thin-film systems under study makes it possible to use them as coatings with a self-cleaning surface, as well as to create water purification systems based on filters made from a set of metal meshes with an active photocatalytic coating [7].

2.4 Surface Properties of the Resulting Sol–Gel Coatings via AFM Spectroscopy

Figure 5 shows the morphology of TiO_2 (isopropoxide) coatings obtained through the sol–gel method after annealing at different temperatures. It is evident from the figure that during annealing at 400 °C, the surface is composed of particles that have an irregular shape and blurred boundaries. This may be due to the low annealing

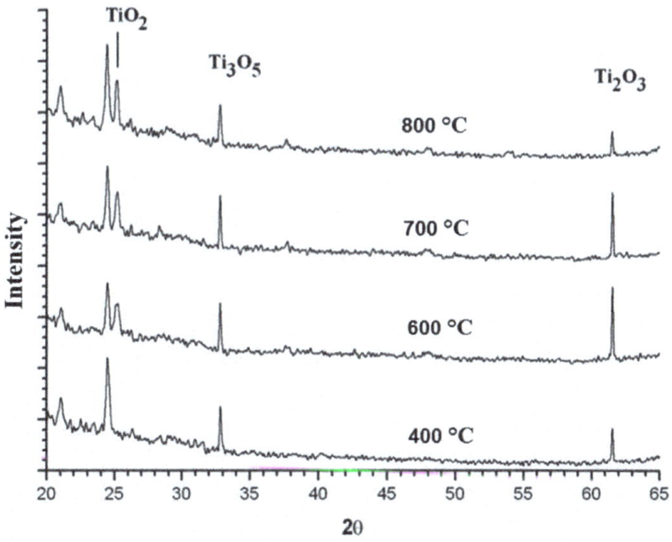

Fig. 4 Diffraction patterns of sol–gel coatings, formed on the basis of titanium ethoxide

temperature. The degree of crystallization is low, resulting in the blurred boundary between grains and a decrease in the grain size.

Following the annealing process at 600 °C, it is obvious that the particles present on the coating's surface exhibit an even distribution, a considerable degree of crystallinity, and distinct boundaries between grains.

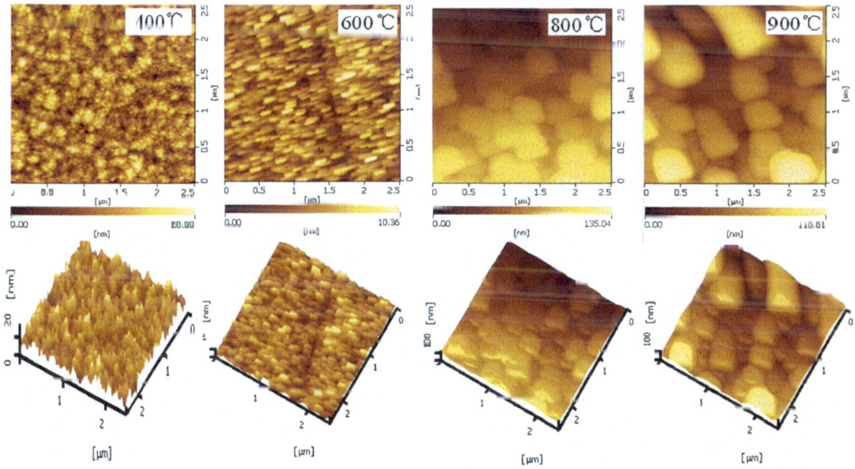

Fig. 5 AFM images of TiO_2 coatings based on titanium isopropoxide obtained at different annealing temperatures

Table 3 Surface roughness parameters

Annealing temperature	400 °C	600 °C	800 °C	900 °C
Relief microroughness R_z, nm	385.1	421.3	61.5	56.9
Mean absolute error from profile points to the center line R_a, nm	5.8	9.3	2.6	4.4
Roof-mean-square deviation from profile points to the center line R_q, nm	18.2	24	4.5	6.4

Upon annealing at 800 °C, the grain size of the coating surface grew larger, the particles became densely packed, and the grain morphology was accurate. The surface structure changed significantly compared to the structure of the coating when subjected to annealing at a temperature of 600 °C. The structural analysis of the TiO_2 coating shows that the coating contains two phases, anatase and rutile.

An elevation in the annealing temperature generates sufficient energy to facilitate complete crystallization within the rutile phase, thereby promoting an increase in the grain size. After annealing at 900 °C, the grain size reached 200 nm in the TiO_2 coating.

According to the data presented in Table 3, the relief microroughness (R_z) is equal to 385.1 nm when annealed at 400 °C. It implies that the surface is composed of irregularly shaped particles. This may be due to the low annealing temperature. The degree of crystallization is low, which blurs the boundary between grains and reduces the grain size.

Upon annealing at 600 °C, the relief microroughness R_z exhibited an increase to 421.3 nm. This phenomenon can be attributed to the formation of the anatase phase, which takes place at this specific annealing temperature. This is evident from the structural analysis of the TiO_2 coating.

After subjecting the coatings to annealing at 800 °C, there was a notable alteration in the surface structure, in contrast to the structure observed at the annealing temperature of 600 °C. The relief microroughness decreased to 61.5 nm. When temperature reached 900 °C, the R_z index reduced to 56.9 nm. which indicates that the coating structure underwent a transition to the rutile phase.

2.5 *Photocatalytic Properties of TiO_2 Sol–Gel Materials*

Heterogeneous photocatalysis using titanium dioxide-based materials is considered a promising and innovative solution to the water pollution problem. The challenge is to convert the basic research into a successful innovation, leading to the implementation of this process into wastewater treatment [13, 14]. Currently, TiO_2-based photocatalytic membranes are at the forefront of photodegradation research and technical readiness. The membrane setup provides a high contact surface area for effective filtration and degradation, without the necessary hassle of photocatalyst recovery

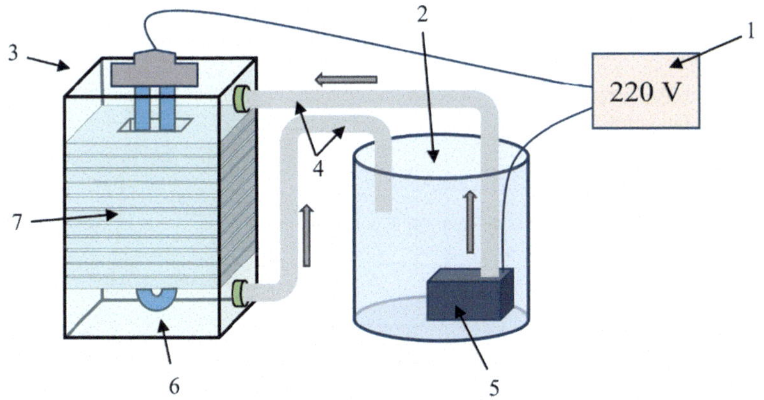

1 – power supply; 2 – flask; 3 – chamber;
4 – rubber hoses; 5 – pump; 6 – UV lamp;
7 – filter made of a series of brass meshes with a photocatalytic coating

Fig. 6 Diagram illustrating the setup for studying the photocatalytic properties of the resulting coatings

after water and wastewater treatment [15, 16]. The other way is the using of TiO_2 and Ag/TiO_2 nanoparticles to influence on the oxidation of organic matter of domestic wastewater was evaluated using a heterogeneous photocatalytic system [17, 18].

The photocatalytic activity of the materials was studied using model reactions of photocatalytic oxidation of methylene blue ($C_{16}H_{18}ClN_3S{:}H_2O$) in an aqueous solution subjected to ultraviolet irradiation. The dye concentration was $5 \cdot 10^{-3}$ wt.%. This was sufficient for the initial solution to possess a rich blue color. Xerogels were placed in the chamber of the setup, and the chamber was then filled with a solution of methylene blue.

For these purposes, an experimental setup was constructed, consisting of a chamber with an ultraviolet light emitter and a brass mesh filter with a photocatalytic coating (Fig. 6).

The volume of the chamber was designed considering the dimensions of the UV radiation source and the maximum capacity of the setup. The chamber parts were assembled using screw connections. A small-sized fluorescent U-shaped lamp was used as a radiation source, emitting in the ultraviolet and visible range of the radiation spectrum from 300 to 450 nm. The solution was supplied to the working chamber of the installation using SOBO Internal Filter WP–330F.

In order to investigate the photocatalytic properties of bulk materials, ceramic blanks that had been manufactured in advance were substituted for plates and the experiment was re-conducted.

The photocatalytic properties of the films and bulk xerogels were assessed by the decomposition of an organic dye, methylene blue ($C_{16}H_{18}ClN_3S{:}H_2O$), in an aqueous solution subjected to ultraviolet radiation. The dye concentration was $5 \cdot 10^{-3}$ wt.%. This was sufficient for the initial solution to have a rich blue color.

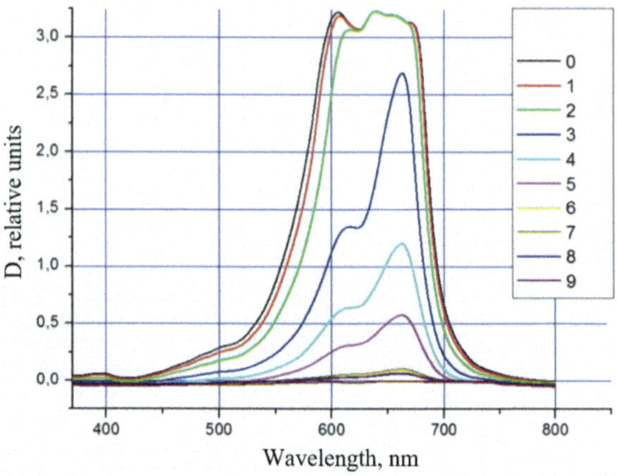

Fig. 7 Absorption spectra of the methylene blue solution for a filter made from a series of meshes deposited with photocatalytic films

Brass meshes were used to study the photocatalytic properties of the films. The film samples were deposited by dipping. Then the meshes were placed in the chamber. The irradiation time was 9 h. Water sampling were performed every 60 min. Additionally, a sample was taken that contained a dye solution that had not been exposed to radiation. The decomposition of the dye was monitored by changes in its optical transmission spectra in the visible wavelength range.

The photocatalytic properties of the resulting coatings were controlled by observing changes in the optical absorption spectra of a methylene blue solution within the visible wavelength range. The absorption spectra are presented in Fig. 7.

The photocatalytic properties of bulk xerogels were studied in a similar way. The xerogels were placed in the chamber of the setup, and the chamber was filled with a solution of methylene blue.

The absorption spectra of the methylene blue solution for a filter made of bulk xerogels are provided in Fig. 8.

Analyzing the collected absorption spectra, it becomes evident that the intensity of the absorption peak decreases over time. Within nine hours, the absorption peak, which was localized at a wavelength of $\lambda = 659$ nm, reduced from 3.5 to 0.2 rel. units for films and bulk xerogels, which is a consequence of methylene blue photocatalytic oxidation. Consequently, the concentration of methylene blue in the solution also decreases due to its decomposition in the solution.

According to the results, it can be claimed that samples of coatings and xerogels containing TiO_2 exhibit photocatalytic activity and can be employed for the purpose of water purification.

After analyzing the findings of investigations concerning the photocatalytic properties of the resulting materials (Figs. 9 and 10), it can be concluded that samples subjected to heat treatment at 700 and 800 °C exhibit less pronounced photocatalytic abilities in contrast to samples subjected to heat treatment at temperatures ranging from 400 to 600 °C. The outcome is predominantly influenced by the anatase phase

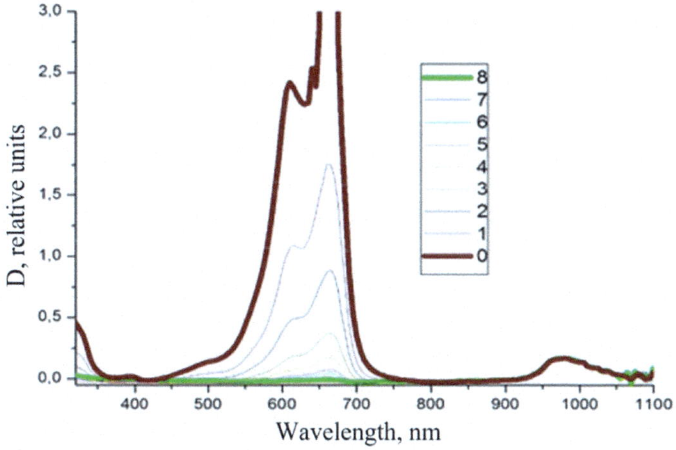

Fig. 8 Absorption spectrum of solution samples

concentration in titanium oxide, which imparts more prominent photocatalytic properties. Additionally, elevated temperatures (approximately 700 °C) result in a greater rutile phase content, which diminishes the photocatalytic properties.

The study's findings revealed a decrease in the methylene blue concentration in the solution to 1–2% of its initial concentration, which provides further evidence for the photocatalytic activity of coatings and xerogels based on TiO_2.

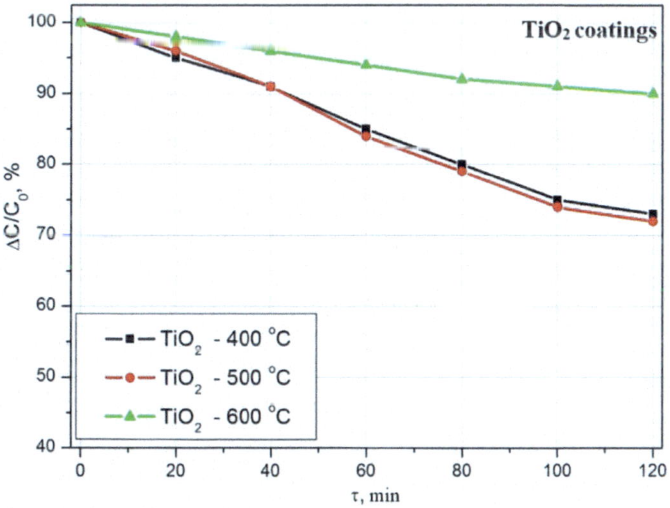

Fig. 9 Change in dye concentration—methylene blue during photocatalytic degradation on samples of TiO_2 coatings

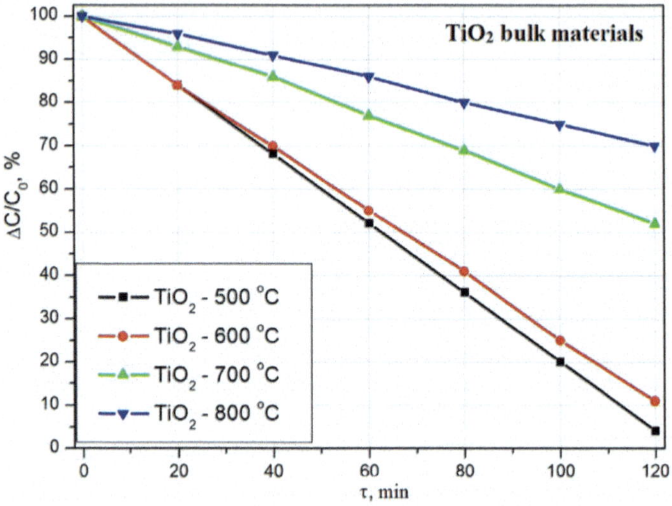

Fig. 10 Change in dye concentration—methylene blue during photocatalytic degradation on samples of TiO_2 bulk materials

3 Conclusion

An examination of the structural properties of the TiO_2 films generated at temperatures exceeding 600 °C through X-ray phase analysis revealed the presence of brookite-shaped TiO_2 crystals, which functioned as active centers in photocatalytic reactions. The influence of the concentration of cerium ions on the surface morphology of the resulting coatings was studied via the AFM method. It has been established that titanium films doped with cerium are characterized by the presence of pores, the size and number of which depend on the concentration of cerium salt in the original film-forming ash. The research results on the photocatalytic properties of cerium-doped TiO_2 films, using the example of the photocatalytic oxidation of methylene blue, show a decrease in the concentration of methylene blue in solution to 1–2% of its initial concentration. Coatings that have been developed can find their application as oxidative photocatalytic materials in small-scale cyclic water purification systems.

References

1. Fujishima, A., Zhang, X., Tryk, D.A.: TiO photocatalysis and related surface phenomena. Surf. Sci. Rep. **63**, 515–582 (2008). https://doi.org/10.1016/j.surfrep.2008.10.001
2. Xingtao, G., Wachs, I.E.: Titania-silica as catalysts: molecular structure characteristics and physico-chemical properties. Catal. Today. **51**, 233–254 (1999). https://doi.org/10.1016/S0920-5861(99)00048-6

3. Ismagilov, Z.R.: Synthesis and stability of nanosized titanium dioxide. Adv. Chem. **9**, 942–953 (2009). https://doi.org/10.1070/RC2009v078n09ABEH004082
4. Kraev, A.S., Agafonov, A.V., Davydova, O.I., et al.: Sol-gel synthesis of titanium dioxide and titanium dioxide-hydroxypropyl cellulose hybrid material and electrorheological characteristics of their dispersions in poly(dimethylsiloxane). Colloid J. **69**, 620–626 (2007). https://doi.org/10.1134/S1061933X07050122
5. Jia, Z.M., Zhao, Y.R., Shi, J.N.: Adsorption kinetics of the photocatalytic reaction of nano-TiO_2 cement-based materials: a review. Constr. Build. Mater. **370**, 130462 (2023). https://doi.org/10.1016/j.conbuildmat.2023.130462
6. Zhu, Z., Lu, Y., Zhou, M.: Surface modification of recycled tire rubber powders with Tannic acid and Nano-TiO_2 for enhanced performance and photocatalytic properties of Rubberized Cement-Based materials. Constr. Build. Mater. **399**, 132607 (2023). https://doi.org/10.1016/j.conbuildmat.2023.132607
7. Kovalenko, D.L., Gaishun, V.E., Vaskevich, V.V., Khakhomov, S.A., Khudaverdyan, S.K., Ayvazyan, G.Y.: Formation and research of properties of photocatalytic materials on the basis of TiO_2 for water treatment. In: Várkonyi-Kóczy, A. (eds.) Engineering for Sustainable Future. INTER-ACADEMIA 2019. Lecture Notes in Networks and Systems, vol. 101. Springer, Cham (2020). https://doi.org/10.1007/978-3-030-36841 8_4
8. Madkhali, N. et al.: Recent update on photocatalytic degradation of pollutants in waste water using TiO_2-based heterostructured materials. Results Eng. **17**, 100920 (2023). https://doi.org/10.1016/j.rineng.2023.100920
9. Wei, P., Zhang, Y., Huang, Y., Chen, L.: Structural design of SiO_2/TiO_2 materials and their adsorption-photocatalytic activities and mechanism of treating cyanide wastewater. J. Mol. Liq. **377**, 121519 (2023). https://doi.org/10.1016/j.molliq.2023.121519
10. McCormick, W.J., McCrudden, D., Skillen, N., Robertson, P.K.J.: Electrochemical monitoring of the photocatalytic degradation of the insecticide emamectin benzoate using TiO_2 and ZnO materials. Appl. Catal. A: Gen. **660**, 119201 (2023). https://doi.org/10.1016/j.apcata.2023.119201
11. Shi, Y., Zhang, Z., Dai, Y. et al.: Influence of current density on the photocatalytic activity of Nd:TiO_2 coatings. J. Wuhan Univ. Technol.-Mat. Sci. Edit. **39**, 32–38 (2024). https://doi.org/10.1007/s11595-024-2851-4
12. Jia, L., Yang, L.M., Wang, W., et al.: Preparation and characterization of Rb-doped TiO_2 powders for photocatalytic applications. Rare Met. **43**, 555–561 (2024). https://doi.org/10.1007/s12598-019-01241-2
13. Musial, J., Mlynarczyk, D.T., Stanisz, B.J.: Photocatalytic degradation of sulfamethoxazole using TiO_2-based materials—perspectives for the development of a sustainable water treatment technology. Sci. Total Environ. **856**(2), 159122 (2023). https://doi.org/10.1016/j.scitotenv.2022.159122
14. Yu, Y., Wana, J., Parr, J.F.: Preparation and properties of TiO_2/fumed silica composite photocatalytic materials. Procedia Eng. **27**, 448–456 (2012). https://doi.org/10.1016/j.proeng.2011.12.473
15. Wang, Q., et al.: Axial suspension plasma sprayed Ag-TiO_2 coating for enhanced photocatalytic and antimicrobial properties. Surf. Interfaces. **45**, 103856 (2024). https://doi.org/10.1016/j.surfin.2024.103856
16. Kirk, C.H., et al.: TiO_2 photocatalytic ceramic membranes for water and wastewater treatment: technical readiness and pathway ahead. J. Mater. Sci. Technol. **183**, 152–164 (2024). https://doi.org/10.1016/j.jmst.2023.09.055
17. Estrada-Vázquez, R., Vaca-Mier, M., Bustos-Terrones, V., et al.: Assessment of TiO_2 and Ag/TiO_2 photocatalysts for domestic wastewater treatment: synthesis, characterization, and degradation kinetics analysis. Reac Kinet Mech Cat (2023). https://doi.org/10.1007/s11144-023-02557-y
18. Dontsova, T., Kyrii, S., Yanushevska, O., et al.: Physicochemical properties of TIO_2, ZrO_2, Fe_3O_4 nanocrystalline adsorbents and photocatalysts. Chem. Pap. **76**, 7667–7683 (2022). https://doi.org/10.1007/s11696-022-02433-4

19. Shilova, O.A., Glebova, I.B., Voshchikov, V.I., Ugolkov, V.L., Dolmatov, V.Yu., Komarova, K.A., Ivanova, A.G.: Environmentally friendly antifouling transparent coatings based on sol-gel "epoxy/titanium tetrabutoxide" composition modified with detonation nanodiamond. J. Adv. Mater. Technol. **7**(3), 201–218 (2022). https://doi.org/10.17277/jamt.2022.03
20. Kovalenko, D.L., Gaishun, V.E., Vaskevich, V.V., Aleshkevich, N.A.: Formation of sol-gel by the method of antireflective two-layer TiO_2-SiO_2 coatings. Vesti, Gomel State University named after F. Skorina, Gomel. **72**(3), 63–68 (2012)
21. Shilova, O.A., Vlasov, D.Y., Zelenskaya, M.S., Ryabusheva, Y.V., Khamova, T.V., Glebova, I.B., Sinelnikov, A.A., Marugin, A.M., Frank-Kamenetskaya, O.V.: Sol-gel derived TiO_2 and epoxy-titanate protective coatings: Structure, property, fungicidal activity and biomineralization effects. In: Frank-Kamenetskaya, O.V., Vlasov, D.Yu., Panova, E.G., Lessovaia, S.N. (eds.) Processes and Phenomena on the Boundary between Biogenic and Abiogenic Nature, pp. 619–638. Swizerland, Springer. (Scopus) (2020). ISBN 978-3-030-21613-9. https://doi.org/10.1007/978-3-030-21614-6_33
22. Shapovalov, V.I., Shilova, O.A., Smirnova, I.V., Zav'yalov, A.V., Lapshin, A.E., Magdysyuk, O.V., Panov, M.F., Plotnikov, V.V., Shutova, N.S.: Modification of the glass surface by titanium dioxide films synthesized through the sol–gel method. Glass. Phys. Chem. **37**, 150–156 (2011). https://doi.org/10.1134/S1087659611020143
23. Bouyarmane, H., et al.: Photocatalytic degradation of emerging antibiotic pollutants in waters by TiO_2/Hydroxyapatite nanocomposite materials. Surf. Interfaces. **24**, 101155 (2021). https://doi.org/10.1016/j.surfin.2021.101155
24. Bayan, E.M., Pustovaya, L.E., Volkova, M.G.: Recent advances in TiO_2-based materials for photocatalytic degradation of antibiotics in aqueous systems. Environ. Technol. Innovation. **24**, 101822 (2021). https://doi.org/10.1016/j.eti.2021.101822
25. Michael, M.P., Singh, S.K., Sahini, M.G.: Facile conglomeration of guar gum/TiO_2/Fe_3O_4 composite materials for photocatalytic antimicrobial activities. J. Indian Chem. Soc. **99**(10), 100688 (2022). https://doi.org/10.1016/j.jics.2022.100688
26. Pandele-Cusu, J., Petrescu, S., Preda, S., et al.: Comparative study of the TiO_2 nanopowders prepared from different precursors and chemical methods for heterogeneous photocatalysis application. J. Therm. Anal. Calorim. **147**, 13111–13124 (2022). https://doi.org/10.1007/s10973-022-11544-9
27. Dobromir, M., Konrad-Soare, C.T., Stoian, G., Semchenko, A., Kovalenko, D., Luca, D.: Surface wettability of ZnO-loaded TiO_2 nanotube array layers. Nanomaterials **2020**, 10 (1901). https://doi.org/10.3390/nano10101901
28. Lukong, V.T., Ukoba, K., Jen, TC.: Review of self-cleaning TiO_2 thin films deposited with spin coating. Int. J. Adv. Manuf. Technol. **122**, 3525–3546 (2022). https://doi.org/10.1007/s00170-022-10043-3
29. Ayvazyan, G.Y., Kovalenko, D.L., Matevosyan, L.A., Semchenko, A.V.: Investigation of the structural and optical properties of silicon-perovskite structures with a black silicon layer. J. Contemp. Phys.**57**(3), 274–279. https://doi.org/10.1134/S1068337222030069
30. Kovalenko, D.L., et al.: Development of a combined method for the formation of ZnO/TiO_2 nanotubes. Prob. Phys. Math. Tech. **49**(4), 11–16 (2021). https://doi.org/10.54341/20778708_2021_4_49_11
31. Semchenko, A.V., et al.: Characteristics of nanocomposite sol-gel films on black silicon surface. In: Khakhomov, S., Semchenko, I., Demidenko, O., Kovalenko, D. (eds.) Research and Education: Traditions and Innovations. INTER-ACADEMIA 2021. Lecture Notes in Networks and Systems, vol. 422. Springer, Singapore (2022). https://doi.org/10.1007/978-981-19-0379-3_21
32. Sanchez-Gonzalez, J., Diaz-Parralejo, A., Ortiz, A.L., Guiberteau, F.: Determination of optical properties in nanostructured thin films using the Swanepoel method. Appl. Surf. Sci. **252**, 6013–6017 (2006). https://doi.org/10.1016/j.apsusc.2005.11.009

Recent Advances in Titanium Dioxide Nanotube Synthesis: A Comprehensive Review

Dilshod Boykobilov, Sherzod Jumagulov, Askar Parmanov, Ulugbek Shaislamov, Jamoliddin Razzokov, and Olim Ruzimuradov

Abstract In recent years, titanium dioxide (TiO_2) nanotubes have attracted much attention within both scientific and industrial domains, primarily owing to their exceptional photocatalytic properties and numerous other attributes. In particular, they are of wide interest including pharmaceuticals, gas detection sensors, renewable energy sources, biomedicine and other branches. However, their application presents certain challenges. Given that their inner cavities are confined to the visible light spectrum, the need for doping with metals arises to expand their physicochemical capabilities. Therefore, achieving precise nanotube synthesis tailored to industrial requirements becomes a paramount consideration. In this study, we introduce the proposed models for the synthesis of TiO_2 nanotubes, elucidate the influencing factors, and draw conclusions regarding the synthesis's impact and the mechanisms underlying TiO_2 nanotube formation. However, our primary focus lies on comprehensive analyses and the practical implications of advancing TiO_2 technology.

Keywords Anodization · TiO_2 · Nanotubes · Electrolyte · Anodization voltage · Temperature and time · pH · Anatase · Rutile

---

D. Boykobilov · S. Jumagulov · A. Parmanov · O. Ruzimuradov (✉)
Faculty of Chemistry, National University of Uzbekistan, Universitet 4, Tashkent, Uzbekistan 100174
e-mail: o.ruzimuradov@polito.uz

U. Shaislamov
Department of Physics, National University of Uzbekistan, University Str. 4, Tashkent, Uzbekistan 100174

Center for Development of Nanotechnology, National University of Uzbekistan, University Str. 4, Tashkent, Uzbekistan 100174

J. Razzokov
Institute of Fundamental and Applied Research, National Research University TIIAME, Kori Niyoziy 39, Tashkent, Uzbekistan 100000

O. Ruzimuradov
Department of Natural and Mathematic Sciences, Turin Polytechnic University in Tashkent, Kichik Halqa Yoli 17, Tashkent, Uzbekistan 100095

J. Prakash et al. (eds.), *Titanium Dioxide-Based Multifunctional Hybrid Nanomaterials*, Engineering Materials, https://doi.org/10.1007/978-3-031-80625-4_4

1 Introduction

TiO_2 is a highly valuable substance with numerous useful properties. The formation of nanotubes in TiO_2 phases, their interaction with other metals, the impact of biomolecules, and interfacial processes can yield promising results in the fields of renewable energy and medicine. However, there are still many unknown properties and applications of TiO_2 nanotubes (TNTs) in science, and much work needs to be done in this area. Industrialization and practical application of synthesized nanotubes pose significant challenges. Therefore, this paper aims to emphasize the importance of the formation and application of TNTs.

TiO_2 nanostructures play a crucial role in science and engineering, with semiconductors like TiO_2, ZnO, and Fe_2O_3 being widely used in industry. Anatase TiO_2, in particular, is a highly efficient photocatalyst due to its extensive surface area [1]. Additionally, titanium nanotubes exhibit exceptional properties such as low density, high mechanical strength, and corrosion resistance [2].

In today's world, the challenges of energy consumption and environmental pollution are escalating due to population and economic growth. To address these issues, it is imperative to explore new renewable energy sources for sustainable development. The extensive exploration of hydrogen energy derived from solar power, particularly through the utilization of advanced photocatalysts, holds promise in resolving the energy crisis and environmental pollution [3].

In 1972, Fujishima and Honda reported on water electrolysis at a TiO_2 electrode. Gerischer, Tributsch, D-Edwars, Grtzel, and Orega took the first steps in synthesizing nanotubes [4]. In 1999, Zwilling, Gong et al. first stated the formation of porous anodic oxides on TiO_2 in a hydrofluoric acid electrolyte, and in 2001, Grimes et al. demonstrated the nanotube structure [5]. Masak et al. studied the arrangement of hexagonal massifs of nanotubes (NTs) using a multistep approach [6, 7]. TiO_2 is a versatile material used in various fields, including paints, food, biomedicine, photocatalysis, water photo-splitting, cosmetics, and sun protection. Additionally, it can be used in light-sensitive materials, dye-sensitive solar batteries, gas sensors, biomedical coatings, and smart surface coatings [8–10].

The research landscape in nanotechnology is witnessing a steady expansion in last two decades. In the period from 2000 to 2021, a total of 18 958 scientific articles were published exclusively in Elsevier journals within the Scopus database (Fig. 1). Reflecting the increasing interest in this field, over 40,000 scientific works have been published across a diverse range of journals. This research endeavors to explore the pivotal role of TNTs, examining their diverse applications, unraveling the intricate mechanisms behind their formation, and scrutinizing the influential factors impacting their properties.

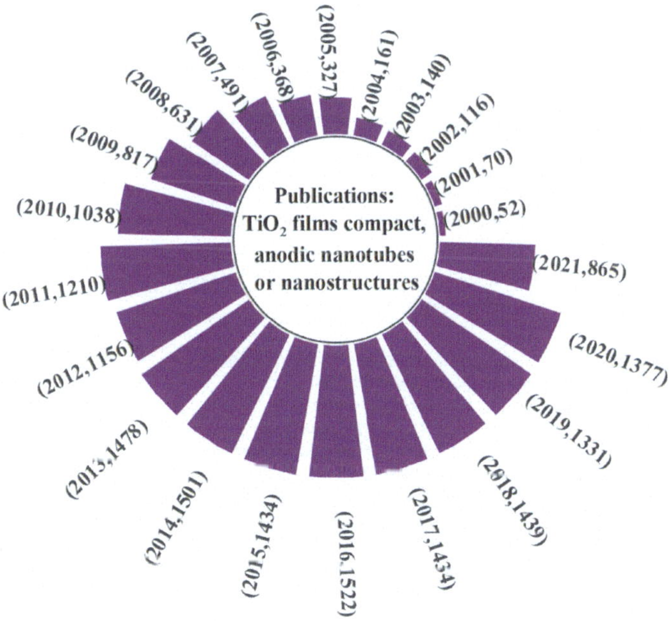

Fig. 1 The number of publications (Scopus-Elsevier) related to the synthesis of TNTs between 2000 and 2021. Adopted from Ref. [11]. (https://www.mdpi.com/2079-4991/11/9/2359)

2 Synthesis of TiO_2 Nanotubes

2.1 Methods of Synthesis of TiO_2 Nanotubes

TiO_2 NTs can be fabricated using a range of methods, including sol–gel, template-assisted methods, hydro/solvothermal approaches, electrochemical methods, sono-electrochemical, microwave irradiation, and magnetron deposition [12–16]. Electrochemical anodization is the oxidation process under the influence of an electric field. This is an electrochemical anodizing process, in which the thickness of the oxide layer on the metal surface increases with the appropriate electrolyte. In general, the process of electrochemical anodization can be described based on the following scheme (Fig. 2):

2.2 Synthesis of TiO_2 Nanotubes Based on Anodic Oxidation Method

The formation of nanotubes is a relatively straightforward process. Figure 3 illustrates a schematic diagram of a standard two-electrode anodizing device for creating

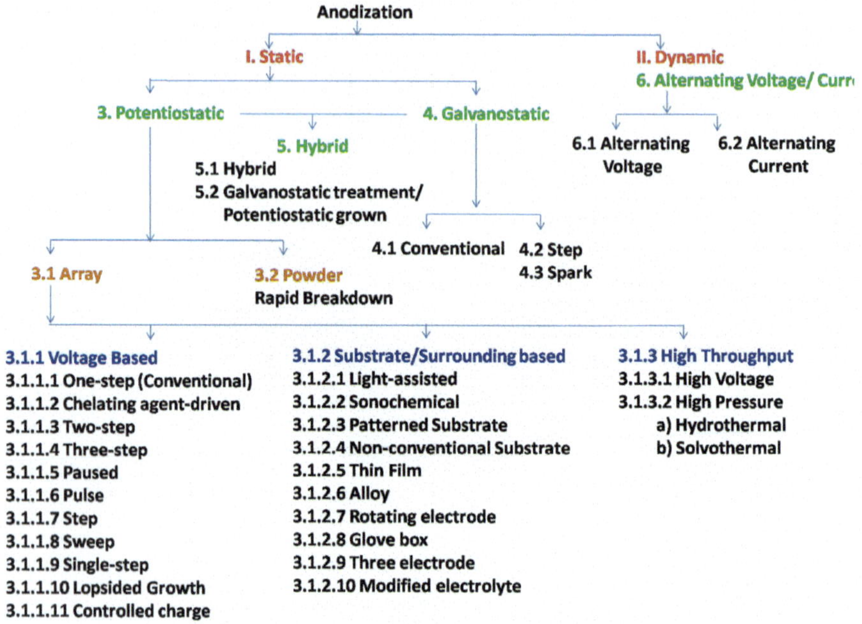

Fig. 2 Classification of anodization methods based on static and dynamic processes. Adopted from Ref. [17]. (https://chemistry-europe.onlinelibrary.wiley.com/doi/abs/)

titanium nanotube (TNT) layers. In the context of the electrochemical procedure described, the titanium foil assumes the role of the anode, where oxidation occurs, while the counter electrode, responsible for reduction reactions, can be composed of platinum or alternative materials suitable for electrode functions.

2.3 *Anodizing Stages*

The synthesis of nanotubes through the anodization method primarily progresses via three distinct phases. The imperative task during the initial stage is to purify the titanium foil [18, 19]. Certain investigations have explored the utilization of HF acid for the cleansing of the titanium foil surface. Nevertheless, it is crucial to highlight that employing HF may induce diverse anomalies due to an extra chemical "interaction" with the foil surface. This process could potentially have unfavorable consequences on the creation of nanotubes with a regular and ordered structure [20–22].

Stage I

To start the process, the foil undergoes a 15-min immersion in ultrasonic equipment containing acetone, ethanol (or isopropanol), and distilled water. Subsequently, it

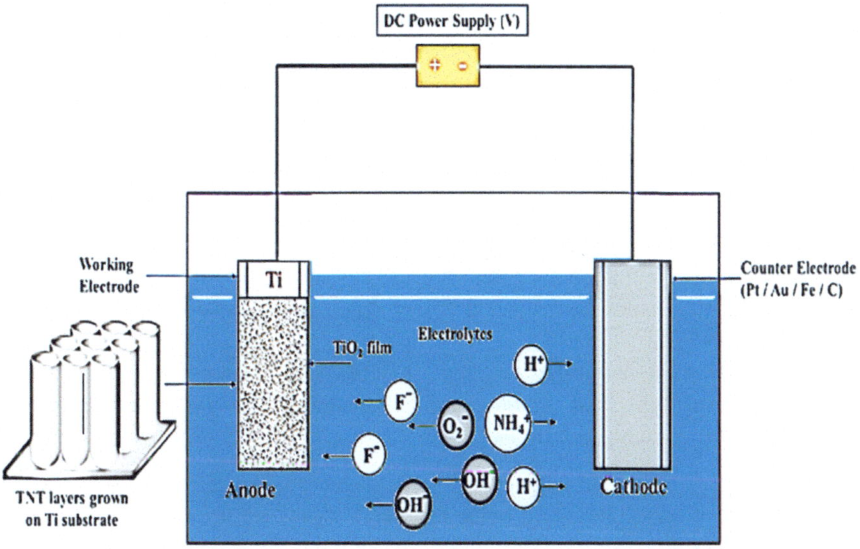

Fig. 3 Schematic representation of a typical two-electrode anodizing device for TNT layers formation

is left to air-dry for a duration of 1 h. Certain investigations propose an alternative method, suggesting the possibility of drying it using a stream of nitrogen [20–24].

Stage II

During the second phase of anodization, the titanium foil is submerged in an electrolyte solution composed of 0.3–0.5% NH_4F, 2–3% H_2O, and ethylene glycol. Additionally, platinum or graphite is employed as a counter electrode. To prevent electrolyte overheating and ensure a uniform distribution of ions, a water bath is established, and constant mixing is maintained using a magnetic stirrer. Anodization is conducted at room temperature, applying voltages of 40, 50, and 60 V, with varying anodizing durations of 2, 4, and 6 h, and so forth [25–29].

Stage III

Following the second anodization stage, the resulting foil undergoes a meticulous cleansing process in distilled water and is subjected to ultrasonic washing in ethanol and distilled water for a duration of 15 min. Subsequent to this, the foil experiences a heating process in a furnace, reaching temperatures between 450 and 500 °C at a rate of 2 °C/sec [8–17, 26–30].

2.4 *Synthesis and Organization Mechanism of Titanium Dioxide Nanotubes Employing Diverse Electrolytic Solutions*

The electrolytes utilized in the production of titanium nanotubes can be primarily categorized into 3 generations:

(a) The length of the synthesized nanotubes was extended to 500 nm through the use of HF acid and aqueous electrolytes [31];
(b) The length of nanotubes reached 5 μm when using F^- containing aqueous electrolytes [32];
(c) The length of the nanotubes reached 100–1000 μm when using F^- containing organic electrolytes, and smoother and better nanotubes were formed than those of the previous two generations [33].

Hence, employing a combination of F^- ions with organic electrolytes may lead to the generation of nanotubes with extended, well-organized structures and smooth walls. Despite comprehensive investigations into the fundamentals of nanotube formation, there remain lingering uncertainties within the field. Nevertheless, our attention will now shift towards examining the mechanisms involved in the formation of nanotubes. The processes of anodization and growth, resulting in the development of TNTs within organic electrolytes containing F^- ions, can be categorized into four distinct stages (see, Fig. 4):

Here, (a) development of a TiO_2 barrier layer; (b) initiation of pits formation in the oxide layer; (c) conversion of pits into pores; (d) expansion of pores and the establishment of space; (e) generation of TNTs; (f) top view showcasing fully grown TNTs. The chemical reactions occurring during anodic oxidation are represented by i-vi.

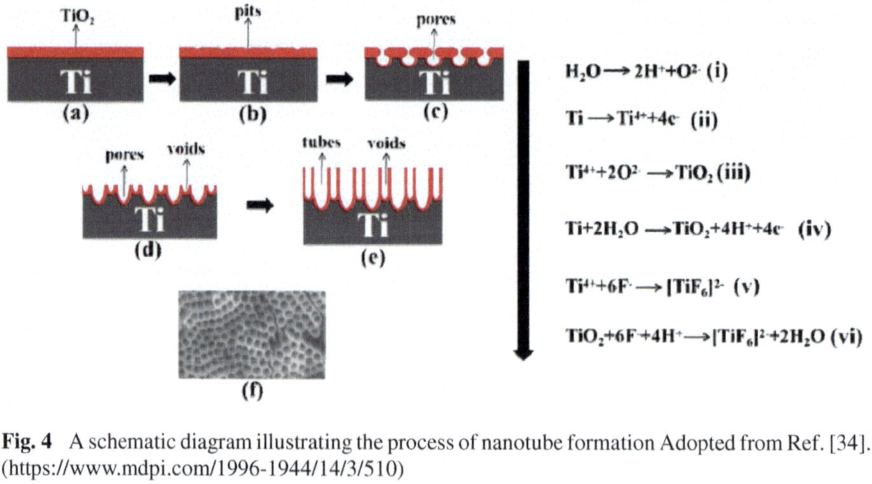

Fig. 4 A schematic diagram illustrating the process of nanotube formation Adopted from Ref. [34]. (https://www.mdpi.com/1996-1944/14/3/510)

This procedure can be classified into various phases: Upon the application of voltage, a dense (TiO_2) oxide layer forms on the anode. Initially, water undergoes dissociation (refer to Fig. 4, equation, (i) allowing Ti^{4+} cations to enter the solution (Fig. 4, equation (ii). Following this, Ti^{4+} cations react with O^{2-} anions released from water dissociation, leading to the formation of a dense oxide layer (Fig. 4, equations (iii–iv). Additionally, during this process, titanium cations combine with fluoride ions, resulting in the production of fluoride compounds (Fig. 4, equation (v). The overall progression of these processes is elucidated in Fig. 4, equation (vi) [34, 35].

Step 2 involves the dissociation of NH_4F in the electrolyte:

$$NH_4F = NH^{4+} + F^-$$

The F^- ions produced disperse across the oxide layer's surface, leading to its dissolution and the creation of numerous "pits" (refer to Figs. 4 and 5b).

As fluoride ions permeate the oxide (barrier) layer, the depressions in the layer expand, leading to the initiation of holes (see Fig. 4c). Subsequently, fluoride ions react with titanium cations, derived from the metal substrate, forming $[TiF_6]^{2-}$ complexes (as depicted in Fig. 4, equation v, and Fig. 5c). A fraction of fluoride ions interacts with the oxide layer, resulting in the "dissolution" of the oxide layer (refer to Fig. 4, equation vi).

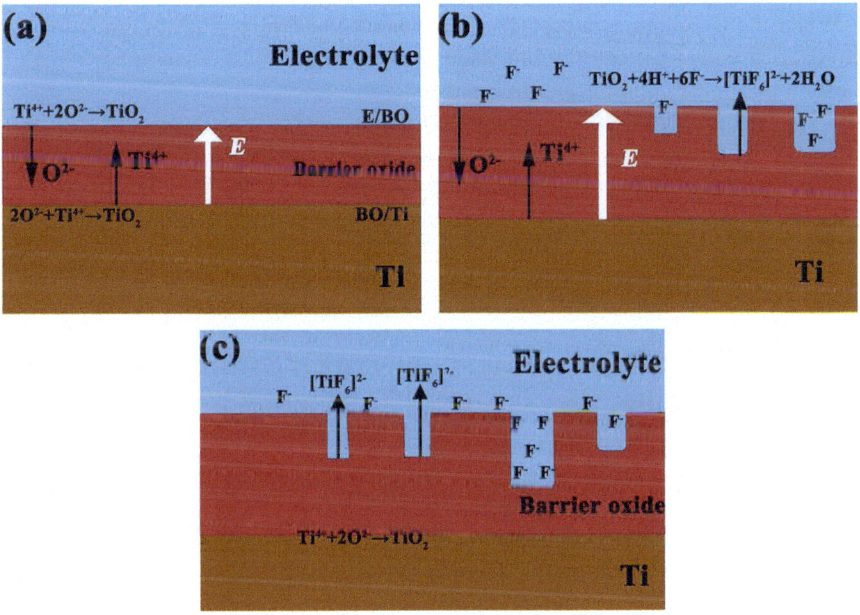

Fig. 5 A schematic representation depicting the formation process of TNTs. Reproduced with permission from Ref. [36] Copyright (2017) Elsevier

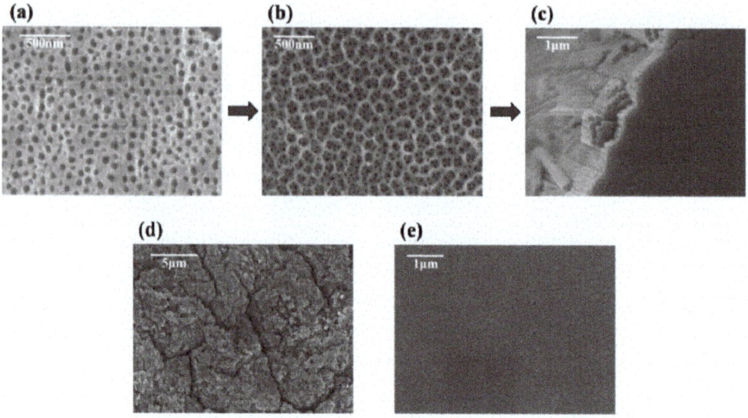

Fig. 6 SEM images display the growth process of TNTs: **a** pits formed in the oxide layer; **b** turning pits into nanopores; **c** conversion of nanopores into nanotubes; **d** top view of the nanotube; **e** bottom view of the nanotube. Adopted from Ref. [34]. (https://www.mdpi.com/1996-1944/14/3/510)

The generated electric field propagates from the oxide layer to the metal substrate through the formed holes, accelerating the melting and growth rates of the oxide layer. Concurrently, "stress corrosion" arises among the holes, creating voids in the spaces between them. As the holes enlarge, the interconnecting "joints" also expand, culminating in the formation of a wall composed of nanotubes [37, 38].

When the dissolution rate of TNTs reaches equilibrium with the growth rate, the reaction enters the equilibrium stage, and the length of the nanotubes remains constant during this phase (refer to Fig. 6).

It is imperative to highlight that the expansion of the oxide layer occurs at two distinct boundaries. Precisely, the oxide layer at the interface between the metal and oxide establishes the inner wall of the nanotube, whereas the oxide layer at the boundary between the oxide and electrolyte constitutes the outer wall of the nanotube [39].

3 Various Models Have Been Proposed for Synthesizing TiO_2 Nanotubes

There are different methods available for the synthesis of titanium nanostructures, including "Oxygen bubble", Equal strength model", "Droplet", "Plastic flow", "Equal power model", "FAD model (field assisted melting)" and others [20–26, 40, 41]. The objective of these proposed models is to elucidate the underlying mechanisms governing the formation of TNTs, with a particular emphasis on the fundamental models.

3.1 Oxygen Bubble Model

The O^{2-} ions traversing through the metal undergo electron donation, leading to the formation of oxygen molecules (O_2) and the generation of voids filled with oxygen gas within the oxide layer. Consequently, this induces a reduction in the oxide's conductivity. With the expansion of oxygen bubbles, the available area for ion migration diminishes, resulting in a decrease in flux. Sustaining the titanium foil at 25 V for 1 h prompts a decline in electron flow and rapid growth of oxygen bubbles. Altering the pressure and electric field reduces the "electrostatic" voltage within the oxide-electrolyte interface, causing the pressure within the bubbles to become adequate for dissolving the oxide.

Following the dissolution of the oxide, electrolytes (F^- ions) ingress the void created by the oxygen bubble rupture. Subsequently, as the electrolyte infiltrates the cavity, the electric field surpasses the value present in the oxide layer due to the proximity of the electrolyte to the oxide-metal interface. Consequently, the oxide layer attains the typical nm/V value, resulting in an increase in the thickness of the resultant oxide layer to sustain the electric field within. This, in turn, leads to the bursting of oxygen-filled bubbles and the subsequent re-growth of the oxide, restoring the applied voltage (refer to Fig. 7).

Therefore, the "oxygen bubble" model yields the following conclusions (a-f):

(a) The formation of an amorphous oxide layer is observed.
(b) Nucleation of nanocrystals within the oxide occurs, contributing to an increase in electron conductivity.
(c) The presence of oxygen-filled voids near crystal cores within the oxide leads to a reduction in current.

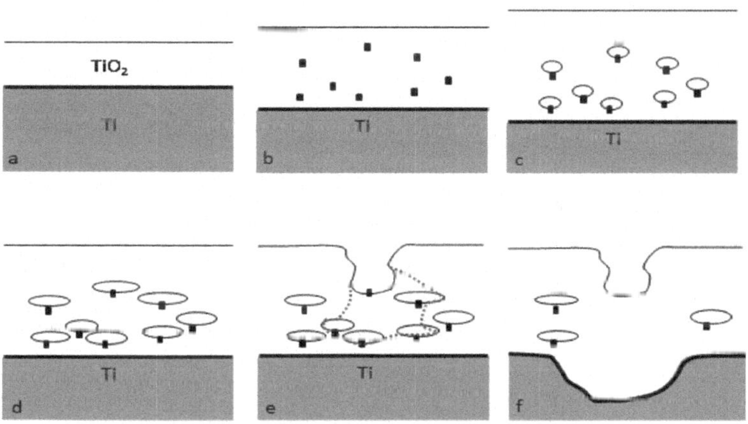

Fig. 7 Formation of nanotubes based on the oxygen bubble model. Reproduced with permission from Ref. [26] Copyright (2012) Elsevier

(d) Potentiostatic arrest induces an augmentation of oxygen-filled voids, consequently resulting in decreased current.
(e) The reduction of potentiodynamic polarization leads to the rupture of certain oxygen-filled voids, driven by the equilibrium between internal pressure and electrostriction force.
(f) Following the rupture of one of the oxygen-filled voids, there is a reorganization of the oxide structure and metal consumption, accompanied by the introduction of electrolytes into the oxide layer.

In this investigation, we elucidate the findings pertaining to the generation of TNTs via this approach. More precisely, we engage in discussions concerning the inception phases of nanopore formation utilizing the oxygen bubble method.

3.2 *Equal Strength Model*

The equal strength model reveals certain flaws in the oxide layer and serves as the foundation for pore formation. These flaws may arise due to variations in the speed of electrolytes and the presence of OH^- ions. During the pore formation process, the even distribution of the electric field results in the creation of hemispherical nanopores, causing the separated holes to merge. According to this model, the oxide layer is influenced by a balance between dissolution and growth processes [42]. When two pores are in close proximity, they adjust their distance to an optimal level. A pressure of 100 MPa induces uniform shrinkage and mechanical stress, leading to the adjustment of pore–pore distance and the formation of a hexagonal shape (Fig. 8) [43, 44].

The model (a–i) leads to the following conclusions:

(a) a uniform thickness flat oxide layer;
(b) a flat layer with a corrosive pit;
(c) the surface of the corrosive pit at the electrolyte/oxide interface is repeated at the oxide/metal interface;
(d) formation of hemispherical pore base and cylindrical wall of one pore;
(e) capable of increasing the diameter of two separated pores;
(f) self-enlarging pits that reach each other;
(g) pits approach with balanced curvature $2\theta < 180$;
(h) two pores are very close to each other;
(i) they adjust themselves to increase the wall thickness.

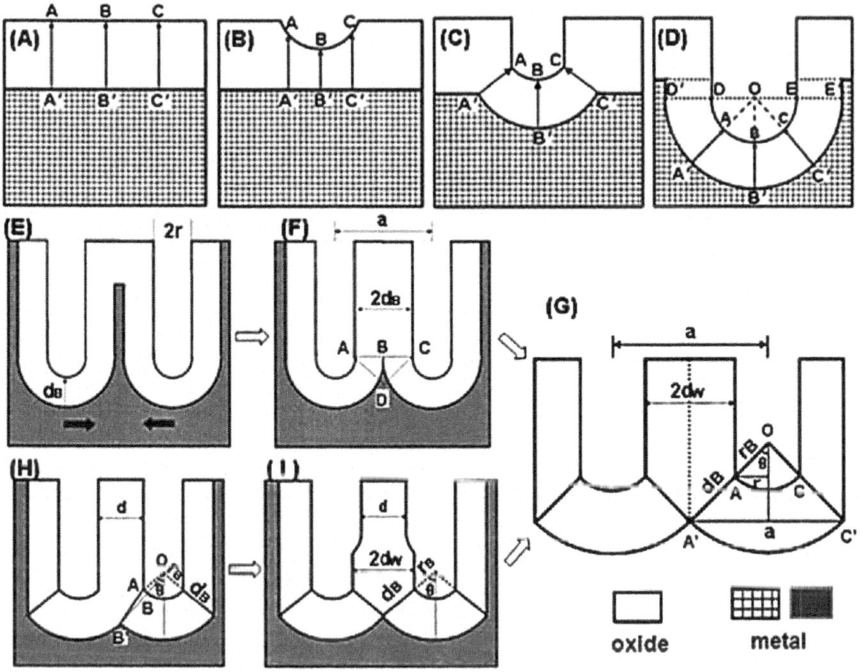

Fig. 8 Schematic diagram of the self-regulating process in anodic metal oxide formation based on equilibrium oxidation and dissolution with the help of an electric field described in the equal strength model. Reproduced with permission from Ref. [40] Copyright (2011) Royal Society of Chemistry

4 Factors Influencing the Formation of TiO_2 Nanotubes

Currently, many studies are focused on the morphology of titanium nanotubes. For example, several factors such as hole length, wall thickness, and distance between holes are taken into account. Such factors may include electrolyte type and concentration, pH, temperature, electrolyte wear, anodizing voltage, time, and type of electrodes (Fig. 9).

4.1 Impact of Electrolytes

Electrolytes play a crucial role in controlling the rate of formation and dissolution of oxide layers, thereby significantly affecting the morphology of (anodic titanium oxide) ATO structures [45]. The outcome of electrochemical processing is contingent upon the characteristics of the electrolyte, with due consideration to its composition.

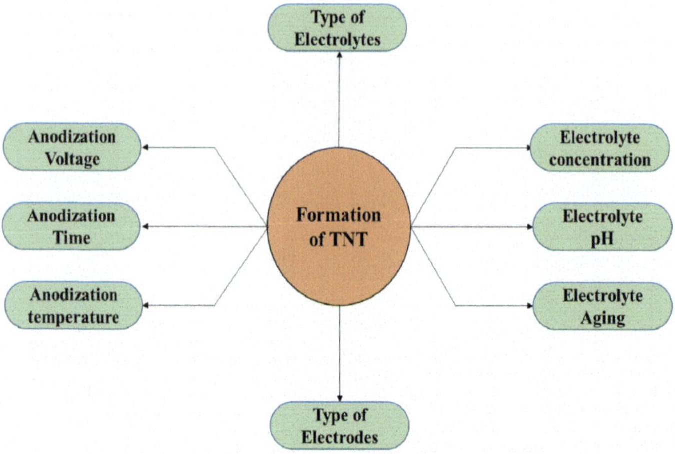

Fig. 9 Factors affecting TNTs

Within the realm of electrochemical anodization, electrolytes are categorically delineated into two primary groups: (a) aqueous electrolytes and (b) organic electrolytes. The subsequent sections explicate the pragmatic significance associated with both aqueous and organic electrolytes. This division is instrumental in elucidating the distinct roles and applications of these electrolytic mediums in the electrochemical anodization process.

4.2 Composition of Electrolytes

The electrolyte composition can be categorized as fluorinated or non-fluorinated, organic or inorganic, aqueous or non-aqueous. It is essential to minimize the water content in the electrolyte to control the expansion of anodized titanium nanotubes and the breakdown of the oxide. The electrolyte composition, which is a decisive factor in electrochemical processes, exhibits a fine classification based on several key attributes. In particular, electrolytes can be classified as fluorinated or non-fluorinated, where the presence of fluorine-containing compounds confers specific electrochemical properties. In addition, electrolytes are divided into aqueous or non-aqueous forms, which refers to their solubility in water or alternative solvents, thereby affecting their compatibility with various electrochemical applications. This comprehensive classification illuminates the multifaceted nature of electrolyte composition and highlights its key role in shaping the electrochemical landscape.

4.3 Fluoride Electrolytes

The electrolyte composition includes precursors such as NaF, KF, NH4F, and HF. Fluoride compounds dissociate during anodization and produce F^- ions, which are crucial in dissolving the oxide layer. To achieve a balance between nanotube growth and dissolution, the concentration of F^- ions must be maintained at the desired level. The amount of fluoride in the electrolyte generally has a significant impact on the nanotubes' morphology. Under optimal conditions, the fluoride concentration is maintained at approximately 0.3–0.6%, a factor carefully considered in experimental setups [5–8, 45–47]. Depending on the fluoride concentration, three different electrochemical properties can be studied. A very low amount of fluoride (0.05%) leads to the formation of a stable compact oxide layer, while high fluoride concentrations (about 1%) do not result in oxide formation because Ti^{4+} is too soluble to form $[TiF_6]$ immediately reacts with fluoride [2, 48].

The concentration of F^- ions plays a crucial role in the formation of nanotubes, leading to the following conclusions:

- Excessively high fluoride concentration leads to rapid formation of soluble complex ions $[TiF_6]^{2-}$, which further inhibits oxide layer formation and nanotube growth;
- Low concentration of F^- ions leads to insufficient dissolution of the formed barrier layer (oxide layer) and the formation of a dense oxide layer;
- As the concentration of F^- ions increases, nanotubes are formed in the Ti foil, leading to a further increase in the extended field capacity;
- F^- concentration does not affect the growth of nanotube length, it only increases the diameter of TiO_2 NT;
- The fluoride layer can form oxofluorides (TiO_xF_y) in addition to titanium and fluorine complexes;
- Increasing the fluoride concentration from 0.3 to 0.5% does not affect the nanotube length;
- Nanotubes can be obtained using 0.3–0.6% fluoride concentration;
- The fluoride concentration determines the dissolution rate of the oxide layer and should be kept as low as possible, but the fluoride ions should be at a high enough concentration to allow nanotube growth. The concentration of fluoride also affects the pH value of the solution.

4.4 NaF as an Electrolyte

When using NaF electrolyte, it results in the formation of self-assembled, vertically standing, and well-adhered nanotubes on a titanium substrate. An anodizing time of 2 h or more will result in a small amount of residue and defects. The addition of sulfuric acid to NaF limits the maximum length of the nanotubes to 2.4 μm [49, 50].

Thicker porous layers can be formed by adding NaF to weakly acidic electrolytes [51].

4.5 KF as an Electrolyte

The cross-sectional images of the self-assembled titanium nanotubes formed in the KF electrolyte reveal the presence of a precipitate layer in the composition, which disrupts the continuity and uniformity of the ion current. Nanotubes with a length of approximately 500 nm were obtained in an electrolyte containing 0.2 M K_2SO_4 + 0.5% KF, where the samples covered nearly the entire area (~99%) and the deposition of insoluble salts in the pores was observed [50].

4.6 HF as an Electrolyte

HF has been widely utilized as an electrolyte and in combination with other acids. In this instance, nanotubes with a length of only 5 μm were created [52]. The chemical dissolution of TiO_2 in HF electrolyte is essential for the formation of nanotubes rather than simple nanoporous structures. C.A. Grimes et al. provided detailed information on the use of hydrofluoric acid in combination with other acids [53]. The reaction between HF and TiO_2 during anodization can be expressed as:

$$TiO_2 + HF = TiF_6^2 + 2H^+ + 2H_2O \quad (1)$$

Some literature indicates the formation of thin-walled nanotubes in the presence of HF (see, Fig. 10) [54].

Synthesis of titanium nanotubes based on HF involves several key processes:

- During the anodization of TNTs, HF is capable of removing the initial layer [20];
- The efficiency of nanotube formation can be determined by immersing the foil in a 0.5% HF solution at 20 V for 2 h before and after anodization, resulting in the transformation of 2% titanium into nanotubes and a barrier layer;
- The addition of HF to various electrolyte mixtures can influence the morphology of nanotubes in different ways.

4.7 NH_4F as an Electrolyte

Cations with a larger structure, such as quaternary ammonium ions, enable thicker nanotube layers to grow and increase their formation rate. This may be attributed to the "infinitive" effect, which enhances ionic conductivity and restricts the thickness

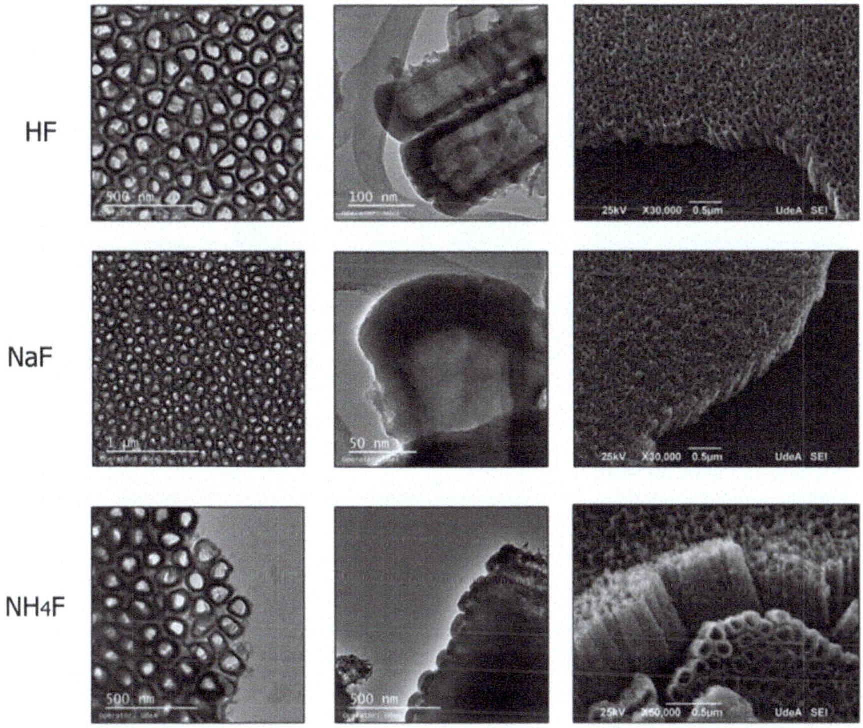

Fig. 10 TEM and SEM images of top view, barrier layer and side view of anode coatings. Reproduced with permission from Ref. [2] Copyright (2018) Elsevier

of the interactive (barrier) oxide layer [55]. This process can also be illustrated by the reaction of NH_4F with TiO_2 on the foil surface after the initiation of anodization.

Several practical considerations are crucial in the formation of titanium nanotubes using NH_4F. For instance:

- Uniform, smooth, and well-aligned nanotubes can be produced at a weight concentration of 0.25% NH_4F;
- As the concentration of NH_4F increases up to 1.5%, the nanotubes tend to adhere together and eventually "collapse". Nanotubes formed at 1.0% NH_4F have larger diameters and thicker walls than those formed at 0.25% NH_4F [56];
- Titanium foil can be corroded at concentrations above 0.3% of NH_4F in the absence of water in the electrolyte. This can also be seen in the following reaction:

$$TiO_2 + 6NH_4F = TiF_6^{2-} + 6NH_3 + 2H^+ + 2H_2O \quad (2)$$

NH_4F was used as an electrolyte in different concentrations. The formation of smooth nanotubes in the optimal state was observed in the NH_4F concentration range from 0.1 to 0.6% (Fig. 11) [57].

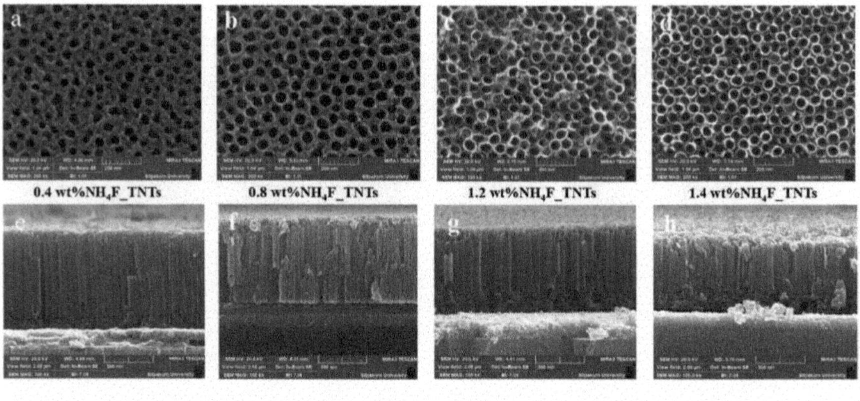

Fig. 11 Top views (**a–d**) and cross-sectional views of SEM images of titanium nanotubes obtained in the presence of different concentrations of NH_4F. Adopted from Ref. [27] (https://www.nature.com/articles/s41598-018-30668-3)

4.8 Impact of Ethylene Glycol

F^- ions have higher mobility in aqueous solutions compared to organic media. Typically, nanotube sidewalls exhibit a disordered state in water, while they grow smoothly in organic solutions. A low water content in organic electrolytes increases the nanotube growth rate [37, 58]. Therefore, it can be inferred that the presence of F^- ions and water content in the organic electrolyte are crucial factors. In most studies, ethylene glycol (EG) was utilized as the electrolyte. Depending on its water content, EG plays a role in various processes as an electrolyte. When anhydrous EG is electrolyzed at 20 V for 2 h or more, or when anhydrous EG is reused (3 times), compact nanotubes of various sizes can be obtained. It should be noted that, in the anhydrous state, a significant current was observed only from a voltage of 1.5 V, and the process was observed as follows [59]:

$$2(CH_2OH)_2 + TiO_2 \rightarrow Ti^{4+} + 2(COOH - CH_2OH) + 4H^+ + 4\,\bar{e}' \quad (3)$$

$$R - COOH \rightarrow (R - COO)^- + H^+ + \bar{e}' \quad (4)$$

$$Ti^{4+} + 4(R - COO)^- \rightarrow (RCOO)_4Ti + 4\,\bar{e}' \quad (5)$$

$$(RCOO)_4Ti \rightarrow TiO_2 + CO_2 + nH^+ + n\,\bar{e}' \quad (6)$$

In this process, the presence of the nucleophilic $RCOO^-$ ion in the solution leads to the formation of glycolic acid, which in turn facilitates oxidation reactions and creates a slightly acidic pH environment. This environment is favorable for the formation of nanotubes. However, in the EG-synthesized sample, a cracked surface with nanotubes can be observed (Fig. 12f), possibly due to the surface tension that occurs during slow oxide formation. As mentioned earlier, cracks are not observed on the surface of titanium nanotubes synthesized in aqueous electrolyte.

Ordered nanotubes based on EG are synthesized in aqueous solutions. In the scientific context presented, the focus is on the synthesis of ordered nanotubes utilizing ethylene glycol (EG) within aqueous solutions. The critical parameter is the quantity of water introduced to the EG, serving as an electrolyte. The significance of this lies in the hygroscopic nature of EG, where its properties are closely tied to the water content in the electrolyte. The text suggests subsequent chemical processes influenced by the precise amount of water, which contribute to the formation of the nanotubes. In this case, the following chemical processes occur (Table 1) [61]:

Table 1 The transformation of an aqueous solution containing EG into carbon dioxide (CO_2) and associated intermediates is investigated herein. The intermediates, namely glycolaldehyde (CH_2OH-CHO) and glycolic acid (CH_2OH-COOH), were discerned through the application of infrared (IR) spectroscopy [61]. Furthermore, an augmented current density of ions was observed in the aqueous medium as opposed

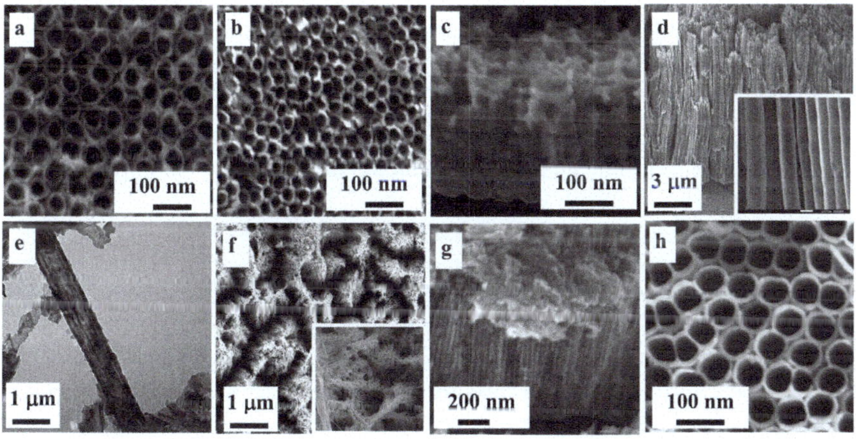

Fig. 12 Top view of FESEM images of titanium nanotubes prepared in **a** aqueous electrolyte **b** EG. Cross-sectional views of the sample of (**a**) and (**b**) in (**c**) and (**d**), respectively. TEM image of the sample in (**e**) (**d**). (**f**) Top view of FESEM image at 75 V as an anodized EG sample. Reproduced with permission from Ref. [60] Copyright (2011) Advanced Materials Research

Table 1 The transformation of an aqueous solution containing EG into carbon dioxide (CO_2) and associated intermediates

Ethylene glycol	Glycolaldehyde (CH_2OH-CHO)
	Glyoxal or glycol acid (CH_2OH-COOH)
	Glyoxyl acid ($C = O_2 (OH)$) → Oxalate acid (HOOC–COOH)
	Carbonate anhydride CO_2

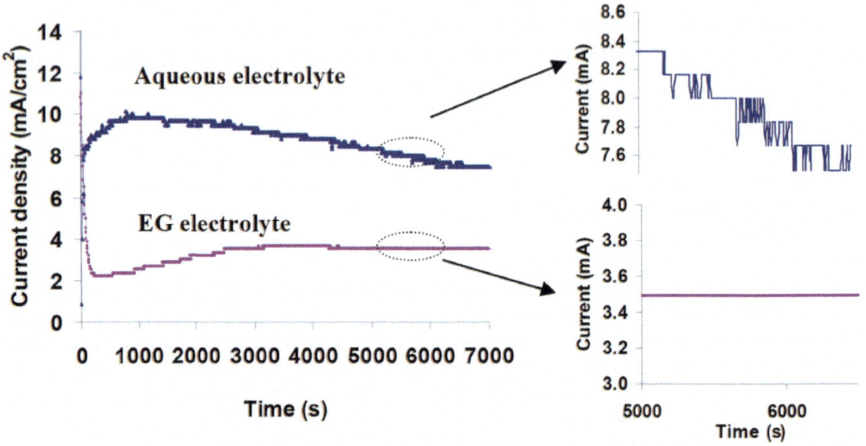

Fig. 13 The period of current passage recorded during anodizing in EG and aqueous electrolyte Reproduced with permission from Ref. [60] Copyright (2011) Advanced Materials Research

to the EG solution, ostensibly attributed to enhanced diffusivity within the aqueous electrolyte, as illustrated in Fig. 13.

The practical importance of forming EG nanotubes involves several important considerations:

1. The TiO_2 layer demonstrates diminished corrosion properties when juxtaposed with alternative organic and inorganic electrolytes.
2. The ion flow in EG manifests relatively typical behavior, with the diffusion rate of ions being contingent upon the temperature of the electrolyte.
3. In the absence of pH adjustment, the EG solution autonomously generates a mildly acidic environment, attributable to the presence of glyoxal or oxalic acid. This acidic milieu facilitates the formation of systematically ordered nanotubes.
4. Elevating the temperature of TNTs formed within EG to the range of 500–650 °C induces the diffusion of adsorbed carbon species into the nanotube structure during anodization. This thermal treatment leads to the transformation of the heated nanotubes into $TiO_{2x}C_x$, signifying the incorporation of carbon within the TiO_2 lattice. Notably, this process enhances the photoactivity of the material in the context of photoelectrochemical hydrogen production [62].
5. The viscosity of electrolytes plays a crucial role in influencing the mobility of ions. At 25 °C, EG exhibits a relatively low viscosity of 13.5 centipoise (cP), facilitating faster ion movement. In contrast, glycerol, with a considerably higher viscosity of 945 cP under the same conditions, hinders ion mobility to a greater extent [56].

4.9 Effect of Water

In the synthesis of titanium nanotubes, the presence of water in the electrolyte proves to be a critical factor, exerting a substantial influence on the morphology of the resulting nanotubes. Research suggests a close correlation between the diameter and length of TNTs and their photocatalytic activity. An investigation specifically focused on the direct impact of water quantity on nanotube length reveals noteworthy trends. The length of TNTs was observed to increase to 7.3 μm with a 1% water content. Further elevation of water content to 2% led to a peak in nanotube length at 10.6 μm. However, a subsequent increase in water content resulted in a decline in nanotube length. Specifically, when an electrolyte solution containing 10% water was employed, the length of TNTs decreased to 2.2 μm, as illustrated in Fig. 14. A comprehensive study conducted by Li Y., Yu H., Zhang C., et al. delves into the intricate details regarding the significance of water in the synthesis of nanotubes [3].

During the formation of nanotubes, water is initially dissociated and mainly serves as a source of oxygen [63]:

$$H_2O = H^+ + OH \quad (7)$$

$$OH^- \rightarrow O^{2-} + H^+ \quad (8)$$

or water splits heterolytically [64]:

$$2H_2O \rightarrow O_2 + 4\bar{e} + 4H^+ \quad (9)$$

An augmentation in the water content within the electrolyte induces the development of a dense and substantial oxide layer, thereby diminishing the growth rate of

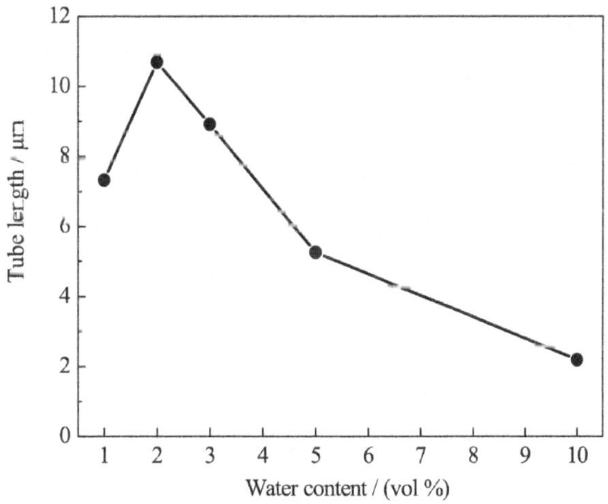

Fig. 14 The effect of water on the length of nanotubes. Reproduced with permission from Ref. [3] Copyright (2017) Elsevier

TNTs [23]. In a study conducted by Raja et al., the influence of water content in the electrolyte was examined, leading to the conclusion that a minimum of 0.18% water is imperative to achieve well-ordered nanotubes [59].

Furthermore, diverse concentrations of water were employed in the nanotube formation process. Notably, nanotubes of maximal length were synthesized when the water content reached 2%, achieved within a 30-min timeframe at a voltage of 60 V [63–66].

Moreover, EG employed as an electrolyte was initially utilized in anodization without the inclusion of water. Intriguingly, following the third re-anodization of EG, the emergence of nanotubes was observed. This phenomenon is attributed to the hygroscopic nature of EG. Indeed, the water content in the electrolyte was found to have escalated to 35% during this process. The progressive increase in water content over successive anodization cycles, in contrast to other ions, serves as additional evidence affirming the hygroscopic characteristics of EG, as detailed in Table 2 [23, 59].

The influence of water in the electrolyte on the formation of nanotubes is elucidated through the following key points:

1. Optimal nanotube formation:
    - To attain nanotubes of optimal quality, the electrolyte must incorporate a minimum of 0.18% water.
2. Highly organized nanotubes:
    - For the creation of highly organized and smooth-walled nanotubes, the water content in the electrolyte should be maintained at 2%.
3. Role of water as an oxygen source:
    - Water present in the electrolyte primarily serves as a source of oxygen, playing a crucial role in the nanotube formation process.
4. Impact of increased water content:
    - Beyond 3% water content, an escalation in water content leads to the erosion of the oxide layer (TiO_2), adversely affecting the length of the nanotubes.
5. Hygroscopic nature of EG:
    - The rise in water content during anodization serves as an indication of the hygroscopic nature of EG.

Table 2 Increase in water content after several times of anodizing [23]

Sample	0	1	2	3	4
F^- %	0.156	0.149	0.143	0.137	0.133
NH_4^+ mg L^{-1}	0.134	0.127	0.107	0.132	0.120
H_2O %	1.98	–	–	–	2.65

6. Effect on pore spacing and wall thickness:
   - Higher water concentration detrimentally affects pore spacing and wall thickness, accelerating chemical dissolution and reaction rates.
7. Stokes–Einstein relation and ion diffusion:
   - According to the Stokes–Einstein relation, the diffusion coefficients of H^+ and F^- ions increase with higher water concentration. This leads to fluctuations in ion concentration and pH levels during anodization, resulting in irregular melting rates and sidewall profiles [33].
8. Regulation of nanotube spacing:
   - The spacing between nanotubes can be controlled by adjusting the amount of water in the electrolyte.

In summary, the precise regulation of water content in the electrolyte emerges as a critical factor in determining the quality, organization, and characteristics of the synthesized nanotubes.

4.10 Effect of Solution Medium (pH)

The pH of the electrolyte plays a pivotal role in significantly influencing the formation of nanotubes, as the anodization products alter the solution environment, thereby impacting crucial nanotube characteristics such as length and pore diameter [67–69]. Observations by Wu et al. revealed the formation of nanotubes with lengths up to 6 μm at pH 4.5–5.0, while Feng et al. reported the formation of longer nanotubes at pH 8–9 [70–72].

It is evident that modifying the solution environment induces changes in the anodization rate. In the absence of precursors altering the pH value, anodization progresses slowly. Conversely, in a low acidic medium, the process is accelerated. The pH value undergoes changes during anodization due to an increase in hydrolysis products in an acidic environment. A shift in pH from strongly acidic to weakly acidic conditions results in an increase in nanotube length from 4 to 6 μm. In an acidic environment, soluble TiF_6^{2} complexes are formed, while in an alkaline environment, the formation of substances such as oxofluoride compounds (TiOF) leads to the disruption of the anodizing process [73].

A change in the solution medium (pH) leads to the following transformations:

- The manipulation of the solution environment enables precise control over parameters such as nanotube length, diameter, thickness, and the spacing between nanopores. Key considerations in this regard include:

1. Optimal pH Range for Nanotube Formation:
   - Maintaining the solution environment within the pH range of 4.5–5.0 (weakly acidic) is favorable to the formation of self-ordered, smooth-walled nanotubes.
2. pH level considerations:
   - The pH level of the solution must be carefully considered due to the accelerated dissolution of oxide in an acidic environment. When utilizing fluoride salts, some of which undergo basic hydrolysis, in lieu of hydrofluoric acid (HF), the solubility of the anodic oxide decreases.
3. Selection of Acid for Solution Modification:
   - Sulfuric acid (H_2SO_4) is identified as the most optimal option for altering the solution medium. The concentration of acid is a critical parameter in this context.

These considerations emphasize the importance of the solution environment in tailoring the characteristics of the synthesized nanotubes and highlight sulfuric acid as a strategic choice for achieving desired outcomes in nanotube synthesis.

4.11 Effect of Anodizing Voltage

The applied voltage also plays a vital role in influencing various parameters such as pore diameter, pore spacing, film thickness, and nanotube arrangement in the anodization process. As the applied potential increases, both the outer diameter and thickness of the NTs exhibit noticeable augmentation. Specifically, the NT diameter shows a linear increase from 85 nm at 10 V to 224 nm at 40 V. Conversely, the wall thickness experiences an increment from 14 to 26 nm within the applied potential range of 10–20 V. Remarkably, the anodized titanium NT sample demonstrates its thickest walls at an applied potential of 40 V [74].

At a lower voltage, such as 10 V, the porous layer forms solely on the surface of the oxide layer, resulting in barely visible pores and very small pits measuring 20 nm [75]. Masak et al. conducted research on the impact of applied voltage on nanotube formation, revealing that nanotube diameter was excessively small between 2 and 5 V. Figure 15 visually illustrates the substantial effect of potential on nanotube diameters: 20 nm at 2 V, 105 nm at 20 V, and nearly 300 nm at 40 V. These findings underscore the significant role of applied voltage in determining the morphology and dimensions of the synthesized nanotubes during anodization [76].

The synthesis of nanotubes with varying diameters at the same voltage is achievable through the modification of electrolyte type, anodizing time, and other pertinent parameters. For instance, nanotubes anodized at 20 V exhibit an average diameter of 30 nm, whereas nanotubes anodized at 25 V have an average diameter of 72 nm.

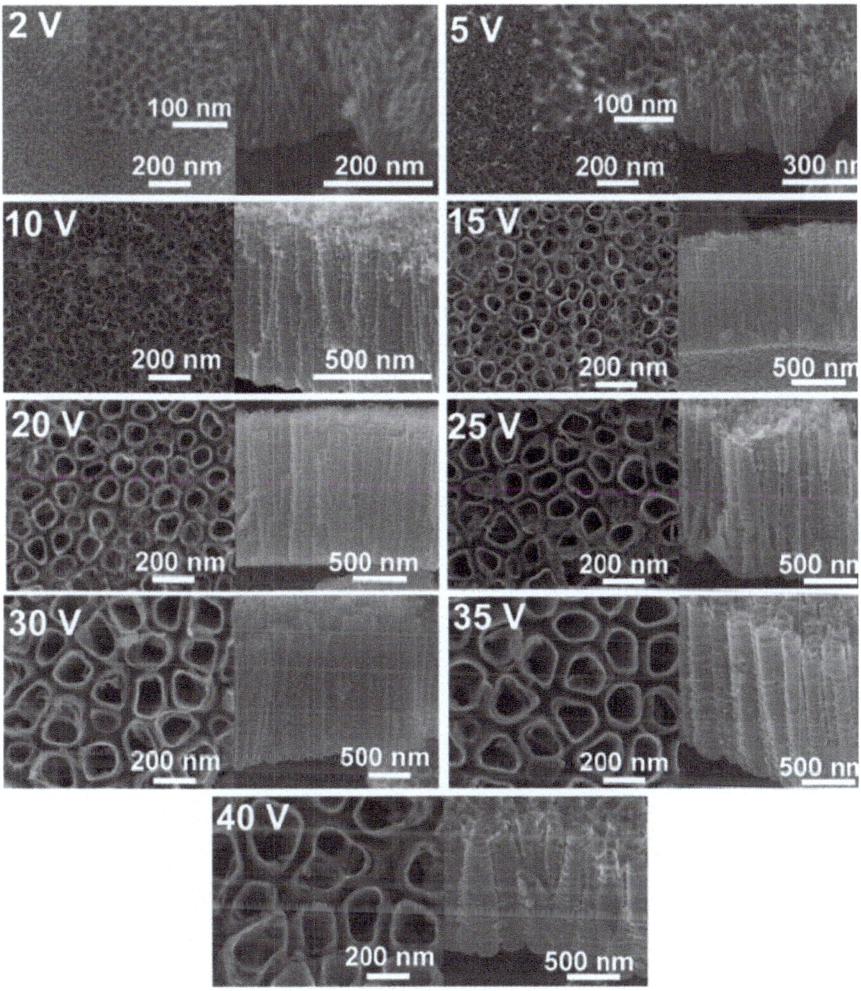

Fig. 15 SEM top and cross-sectional images of nanotubes formed at different voltages. Reproduced with permission from Ref. [76] Copyright (2008) Elsevier

Reports indicate that at low voltages (3 and 5 V), the metal oxide layer was porous, with pore sizes ranging from 15 to 30 nm, and no NTs were formed. As the anodizing potential increased (10 and 20 V), distinctive hollow tubular features with well-defined wall thickness and pore length began to emerge. However, when the voltage exceeded 23 V, disordered TNTs were formed [77].

The linear increase in nanotube diameter with anodizing voltage was mentioned earlier. In this context, the applied voltage primarily generates an electric field between the anode and the cathode, ensuring the formation of more fluoride ions. At 55 V, the pores grown were more compact, resulting in samples with a uniform diameter and wall thickness. Conversely, samples prepared at 35 and 45 V were compact

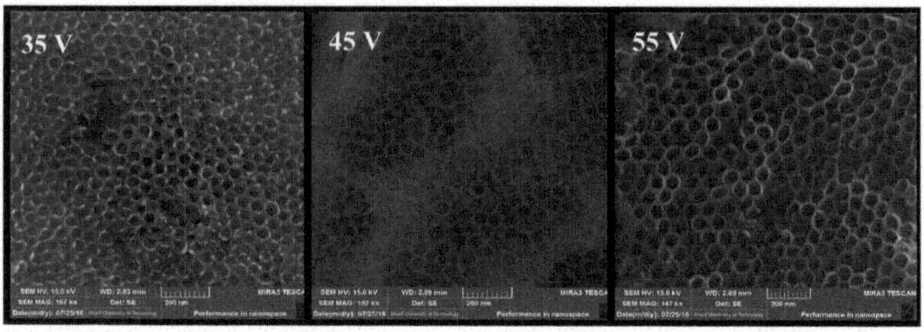

Fig. 16 SEM images of TiO_2 nanotube samples anodized at different voltages and heated at 450 °C 35, 45 and 55 V NH_4F (0.3%), distilled water (2%) electrolyte solution Adopted from Ref. [78] (https://link.springer.com/article/https://doi.org/10.1007/s40094-017-0257-9)

but did not form uniform pores. Greater uniformity was also observed in the samples prepared at a lower applied voltage of 35 V, as illustrated in Fig. 16. These observations emphasize the intricate relationship between anodizing parameters and the resulting morphology of TNTs.

The diameter and wall thickness of TNTs are intricately determined by the anodizing voltage [79].

$$d(nm) = kV$$

$$w(nm) = hV$$

Here, k and h are the growth ratios of the nanotube wall thickness and diameter, respectively, approximately equal to 0.13 (nm V^{-1}) and 1.2 (nm V^{-1}).

XRD images of samples grown at 35 and 45 V exhibit mixed crystalline phases of anatase and rutile in the (210) and (212) areas (Fig. 17) [78]. In contrast, the sample synthesized at 55 V predominantly shows the anatase phase in the (200) and (105) areas. Consequently, an increase in anodizing voltage results in an enhancement of the peak intensity of anatase, indicative of the formation of thicker and denser nanotubes in the metal layer [80, 81].

Simultaneously, as the anodizing voltage is raised from 40 to 60 V, there is an observable increase in the diameter of nanotubes, growing from 54 to 101 nm [82]. The relationship between anodizing voltage and nanotube diameter is illustrated in Fig. 18, depicting the dependency of nanotube dimensions on the applied voltage.

Nanotubes with different shapes were synthesized at voltages of 10, 20, 30, 40 V due to the partially uneven layer left after the initial treatment, while normally ordered nanotubes were obtained at voltages of 50 and 60 V. (Fig. 19).

Figure 20 illustrates the significance of applied voltage in the production of nanotubes. It depicts the SEM images acquired to demonstrate the impact of escalating voltage on the nanotube diameter.

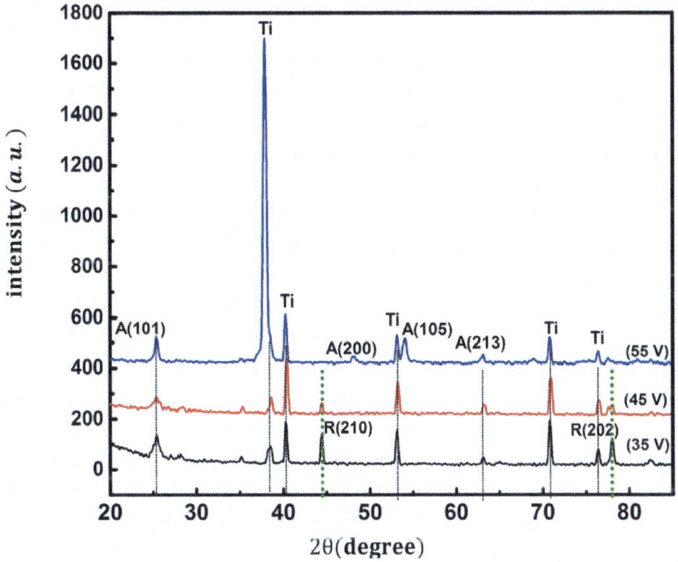

Fig. 17 XRD spectra for anodized samples at 35, 45 and 55 V. Adopted from Ref. [78] (https://link.springer.com/article/)

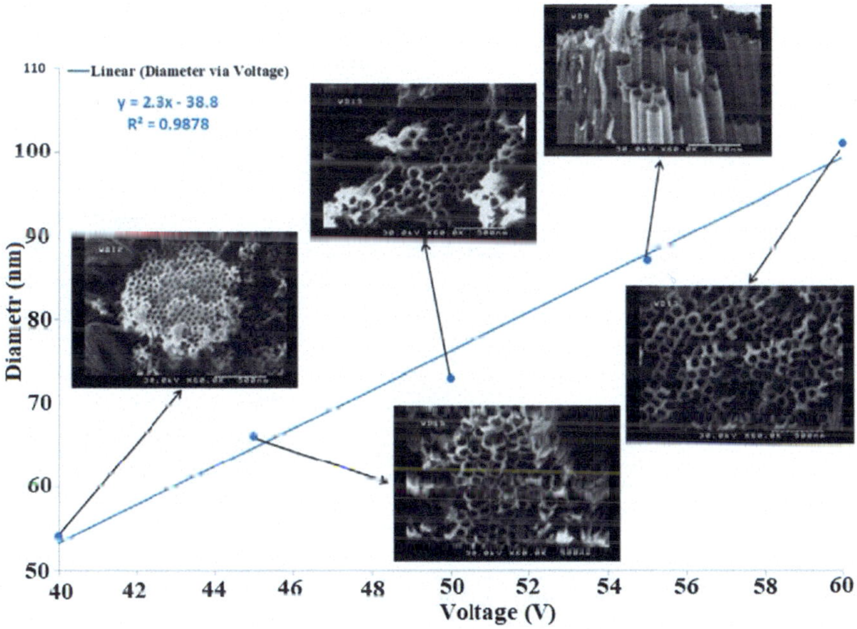

Fig. 18 Graph of dependence of nanotube diameter on anodizing Voltage. Reproduced with permission from Ref. [79] Copyright (2016) Elsevier

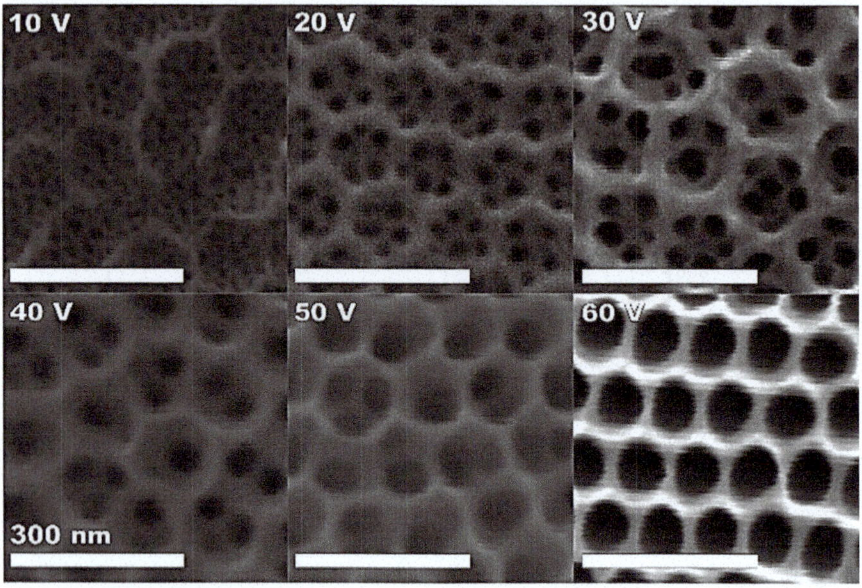

Fig. 19 SEM images of nanotubes formed at different voltages. Reproduced with permission from Ref. [83] Copyright (2018) Elsevier

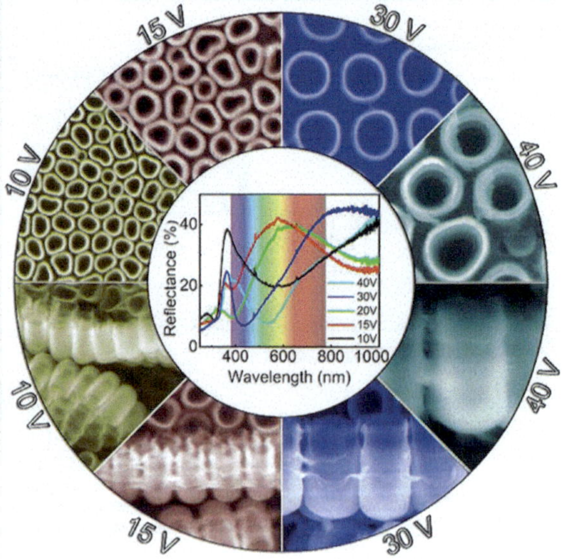

Fig. 20 Dependence of anodizing voltage on nanotube diameter. Adopted from Ref. [74] (https://pubs.acs.org/doi/abs/)

Jakub W, Zeng J, and their research team conducted a comprehensive investigation into the specific influence of anodizing voltage on the morphological characteristics of nanotubes [84–86]. Through a systematic analysis of varying anodizing voltages, the following findings have been elucidated:

Key observations arising from the study include:

1. The applied voltage exhibits a direct correlation with the nanotube diameter, demonstrating a linear increase as the voltage is elevated.
2. Anodizing voltage plays an essential role in initiating an electric field and facilitating ion movement during the initial stages.
3. Tailoring the anodizing voltage to the specific application requirements of nanotubes is feasible and recommended.
4. Voltage adjustment within the range of 55–60 V leads to a predominant transformation into the anatase phase, resulting in the formation of more ordered and optimized nanotubes.
5. Anodizing voltage has a noteworthy impact on reducing the overall anodizing time.
6. In the majority of titanium anodizing experiments, potentiostatic conditions are employed, maintaining a constant voltage and controlling the current. The applied voltage governs the strength of the electric field across the oxide layer, influencing the nanotube diameter obtained at higher voltage levels.

4.12 Thermal Effects on Anodization: A Study of the Effect of Electrolyte Temperature

The anodizing temperature emerges as a pivotal determinant in the formation of NT, particularly in the context of titanium TNTs. Temperature plays a crucial role in guiding the development of TNTs with ordered directional alignment. Additionally, electrolyte temperature stands out as a decisive factor influencing diverse surface morphologies and dimensions of nanotubes, as indicated in previous studies [87].

The evolution of pore profiles from a generalized "**V**" shape to a distinctive "**U**" shape is intricately linked to a gradual elevation in electrolyte temperature, a phenomenon particularly noteworthy in low-temperature electrolyte conditions [88]. These nuanced temperature-driven transformations provide valuable insights into the intricate dynamics governing the anodization process.

In order to elucidate the intricate relationship between anodizing temperatures and the resulting nanotube characteristics, a comprehensive analysis is presented in Fig. 21, highlighting the correlation among pore length, diameter, and wall thickness across varying anodizing temperatures. This graphical representation serves as a visual testament to the significant impact that anodizing temperature exerts on the dimensional attributes of nanotubes, contributing to a deeper understanding of the thermal intricacies inherent in the anodization process.

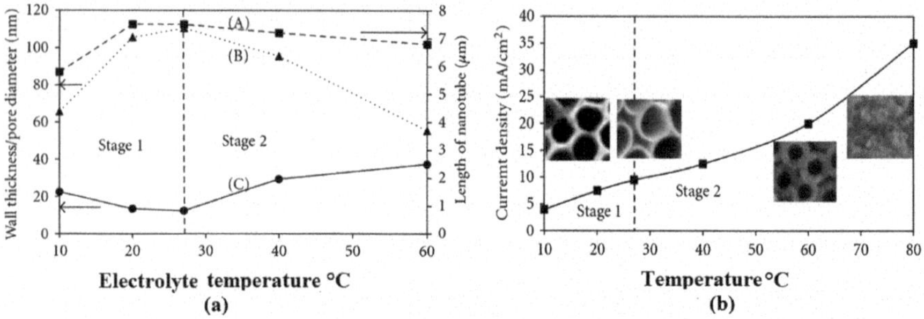

Fig. 21 **a** Effect of anodizing temperature of TNTs anodized at 60 V for 1 h (A) length, (B) pore diameter, and (C) wall thickness; **b** effect of anodizing temperature on current density. Adopted from Ref. [24] (https://onlinelibrary.wiley.com/doi/abs/)

In general, the effect of changing the electrolyte temperature on the anodization process and some properties of TNTs gives the following conclusions:

1. Low temperature (10 °C):
   - Inner pore diameter: 65.5 nm.
   - Length: 5.8 μm.
2. Moderate temperature (20 °C):
   - Inner pore diameter: 105.5 nm.
   - Length: 7.5 μm.
3. Room temperature (27 °C):
   - Smooth, defect-free TNTs.
   - Average inner pore diameter: 110.5 nm.
   - Length: 7.5 μm.
   - Wall thickness: 12.5 nm.
4. Higher temperature (40 °C):
   - Inner pore diameter: 95.5 nm.
   - Length: about 7.2 μm.

It's noted that as the electrolyte temperature increases to 40 °C, a negative impact on self-assembly was observed. This leads to the loss of the nanotube structure, and the anodic oxide transforms into an irregular porous layer.

This information suggests that the electrolyte temperature during the anodization process significantly influences the morphology and structural characteristics of the TNTs, with lower temperatures generally resulting in smaller dimensions and higher temperatures causing a negative impact on the self-assembly process (Fig. 22).

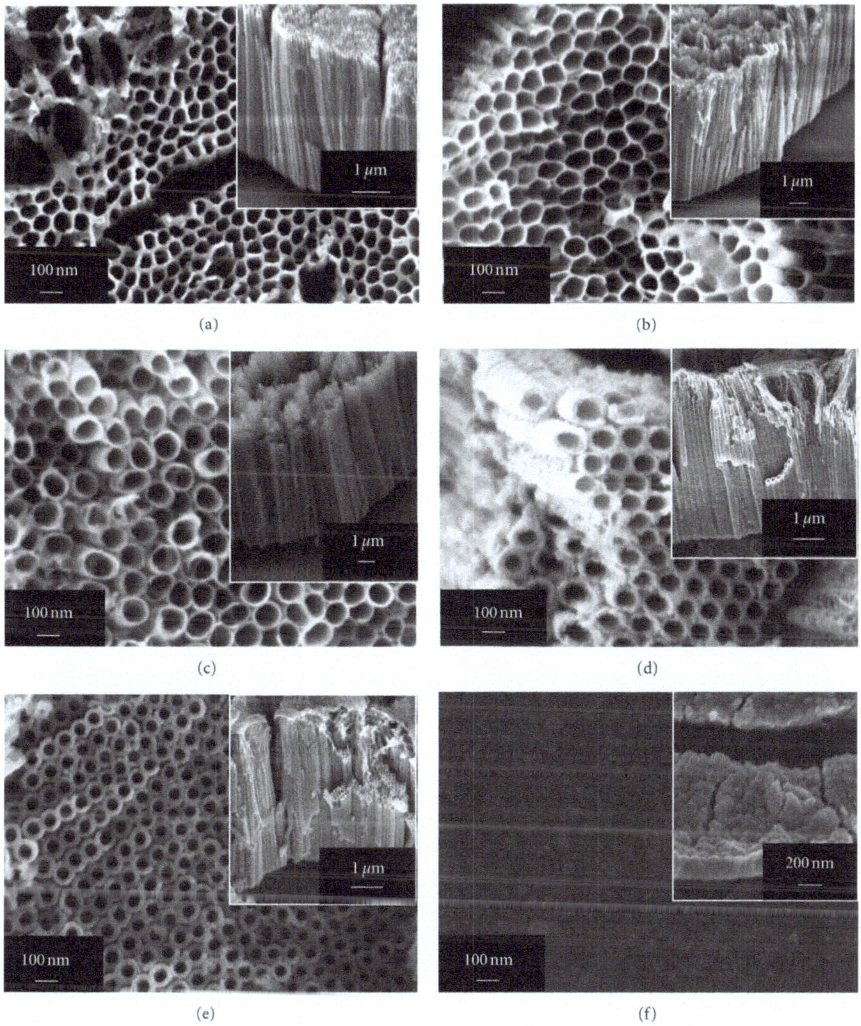

Fig. 22 FESEM images of TNTs synthesized at **a** 10 °C, **b** 20 °C, **c** room temperature, **d** 40 °C, **e** 60 °C and **f** 80 °C (1 h, 60 V, NH_4F). Adopted from Ref. [24] (https://onlinelibrary.wiley.com/doi/abs/)

Mohan et al. conducted a meticulous examination into the impact of electrolyte composition and temperature on the properties of nanotubes. The study encompassed an analysis of current density, time response, oxide layer growth, and melting processes employing HF and 1 M H_2SO_4 electrolytes at a fixed potential of 30 V, spanning a temperature range from 10 to 70 °C. At 10 °C, discernible structures with pore diameters of 35 nm and surface defects on the oxide layer were observed. Contrastingly, at 25 °C, the synthesis process yielded nanotubes characterized by an inner pore

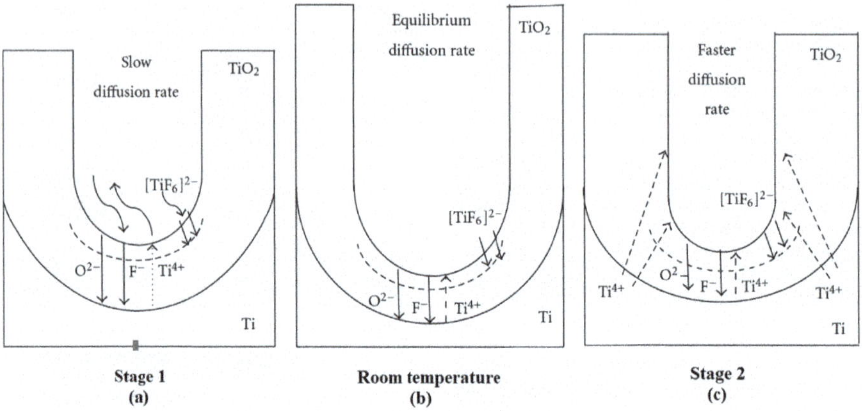

Fig. 23 Illustrates the dissemination rate of ionic species in metal oxide layers on Ti foil, with **a** representing stage 1 (cold state), **b** room temperature (27 °C) and **c** stage 2 (hot state). Adopted from Ref. [24] (https://onlinelibrary.wiley.com/doi/abs/)

diameter of 125 nm, a length of 250 nm, and a wall thickness of 30 nm. However, as the temperature exceeded 70 °C, the formation of self-organized nanotubes exhibited defects, disrupting the equilibrium necessary for proper self-organization [89, 90]. This underscores the direct influence of anodizing temperature on nanotube dimensions, particularly length and wall thickness. Notably, optimal nanotube formation was ascertained at room temperature, where the nuanced interplay of ion movement and diffusion rates at different temperatures facilitated the favorable conditions for self-organization, as depicted in Fig. 23. The findings underscore the intricate dynamics of temperature in governing nanotube morphology and provide valuable insights for optimizing anodization conditions for enhanced nanotube synthesis.

In the exploration of the anodization process, a critical thermal threshold is identified when the electrolyte temperature reaches the range of 70–80 °C, corresponding to approximately 1500 s. At this juncture, a discernible instability in the current at the anode emerges, precipitating an anomalous phenomenon characterized by a "burnt surface" evident on nearly every nanotube (refer to Fig. 24).

Fig. 24 Image of the electrode after heating the electrolyte at T > 70 °C. Reproduced with permission from Ref. [91] Copyright (2001) Elsevier

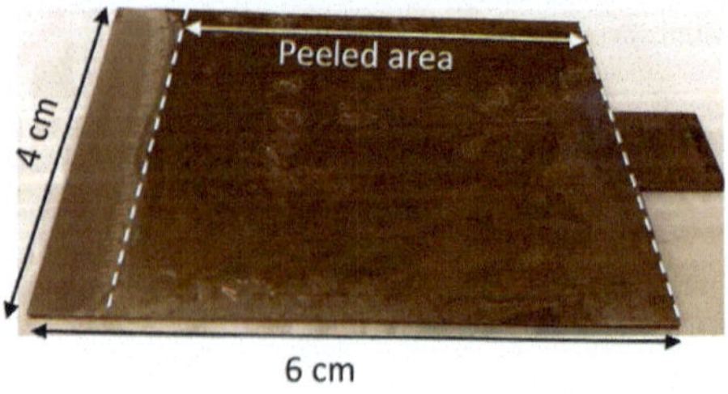

The study on the impact of electrolyte temperature on electrochemical anodization yields noteworthy conclusions elucidating distinctive phenomena at varying temperature regimes:

1. **Low temperatures (t < 10 °C)**:
   - The oxide layer surface exhibits the formation of structures termed "pits" due to constrained ion movement.
   - Limited ion mobility at lower temperatures impedes the dissolution of the oxide surface by F^- ions in fluoride compounds.
2. **Room temperature (25–27 °C)**:
   - At room temperature, the ion diffusion rate normalizes, fostering the optimal state for the formation of nanotubes.
3. **Moderate temperature (40 °C)**:
   - Accelerated ion movement at 40 °C disrupts the equilibrium between oxide layer growth and dissolution.
   - This disturbance results in the formation of thick-walled nanotubes and an increased separation distance between nanotubes.
4. **High temperatures (70–80 °C)**:
   - Electrode "burning" occurs due to the rapid melting of the oxide layer at the oxide-electrolyte interface.
5. **Dissolution rate and growth conditions**:
   - Electrolyte temperature significantly influences the dissolution rate.
   - Optimal nanotube growth is typically achieved at room temperature.
   - In an aqueous environment, nanotube growth occurs comparatively at a lower temperature (2 °C), while nanotube dissolution in an organic environment is viable in the temperature range of 0–40 °C.

These comprehensive findings shed light on the intricate interplay between electrolyte temperature and electrochemical anodization outcomes, providing valuable insights for the precise control and optimization of nanotube fabrication processes under varying temperature conditions.

4.13 *Effect of Time on Anodization*

Also, the time spent on anodization is very important in the formation of nanotubes. Changing the anodizing time has a significant effect on the morphology of the nanotubes. For example, Kadim et al., reported that the anodizing time increase leads to the raise in the nanotube diameter (Table 3 and Fig. 25) [92, 93].

Table 3 Impact of anodizing duration on nanotube diameter (EG, H_2O, NH_4F)

№	Anodizing time (minute)	Nanotube diameter (nm)
1	1	–
2	15	5–10
3	30	16–42
4	60	60–90
5	120	72–108

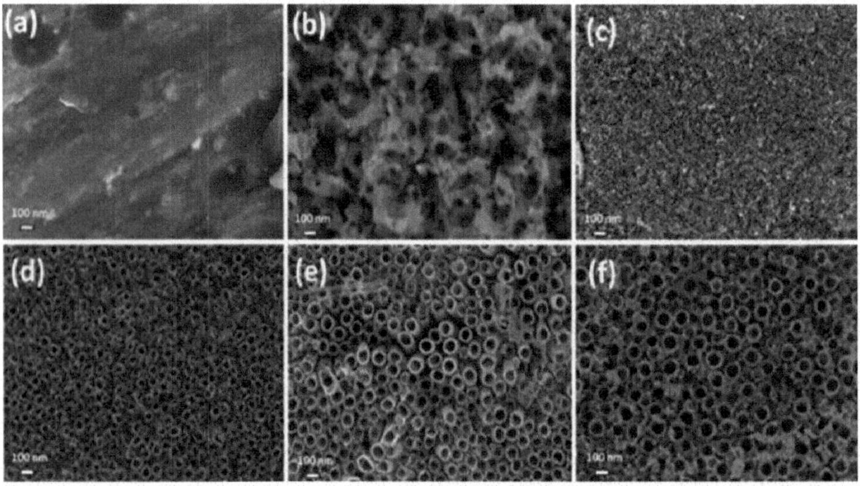

Fig. 25 SEM images of **a** Ti foil and $NH_4F/EG/H_2O$ solution at 40 V **b** 1 min **c** 15 min **d** 30 min **e** 60 min and **f** 120 min. Reproduced with permission from Ref. [93] Copyright (2015) Elsevier

An increase in anodization time, ranging from 1 to 6 h, corresponds to an expansion in the inner diameter of titanium nanotubes. Concurrently, the length of the nanotubes initially decreases but stabilizes over time. Despite some nanotube dissolution, the overall length remains constant throughout the anodization period [94]. A swift formation of titanium TNT occurs within the initial hour, yielding nanotubes approximately 1 μm in length.

It should also be noted that the oxidation and melting processes show a slower rate with increasing anodization time. This slowdown is due to the increased electrical resistance of the thicker oxide layer and the decrease in the diffusion rate of F^- ions at the base of the nanotube. As a result, as detailed in Table 4 [95–98], a dynamic equilibrium is reached between continuous formation and dissolution of nanotubes.

These insights contribute to a nuanced understanding of the temporal intricacies governing titanium nanotube evolution during extended anodization, providing valuable knowledge for tailoring nanotube morphology and optimizing anodization protocols for specific structural outcomes.

Table 4 Change in morphology of nanotubes during 2 and 6 h anodization at different temperatures. Reproduced with permission from Ref. [93] Copyright (2010) Elsevier

Anadization time, hour	Anadization temperature 28 °C				Anadization temperature 50 °C			
	Di^a, nm	t^b, nm	L^b, μm	$H_2{}^c$, Mmol/m^2	Di, nm	t, nm	L^b, μm	$H_2{}^c$, Mmol/m^2
2	96	27	1.570	47.0	107	19	0.84	32.6
6	105	20	7.707	37.2	158	23	0.80	31.3

Di^a = average inner diameter, t^b = average thickness, L^b = average length, $H_2{}^c$ = H_2 production

Increasing the anodization time has a minimal effect on the diameter and thickness of the nanotubes. This can be attributed to the effect of anodization time on the length of the nanotubes over a period of time, and as the anodization time increases, the length of the nanotubes increases [97, 98]. Some studies have shown that the length of TiO_2 nanotubes does not always increase with increasing oxidation time. In addition, the concentration of F^- ions in the electrolyte decreases continuously, and the pore length does not increase with the extension of anodic oxidation time [99, 100]. During this period, the growth rate of nanotubes decreases. When the dissolution rate of TNTs is equal to the growth rate, the anodization reaches equilibrium and the length of the nanotubes remains unchanged [101]. Therefore, the conclusions regarding the effect of anodization time on the morphology of nanotubes are as follows:

- until the anodization reaches equilibrium, the anodization time has almost no effect on the nanotube diameter and thickness, but it has a significant effect on the nanotube length. The length of nanotubes increases with the extension of anodization time;
- changes in anodization time does not affect the crystal structure of TNTs;
- it enables the control of the morphology of titanium nanotubes by adjusting the anodization time.

4.14 Advancements in Enhancing the Natural Oxide Layer on Titanium: Implications for Corrosion Resistance and Optical Interference Effects

Titanium, through exposure to atmospheric oxygen, naturally develops a thin protective oxide layer, typically only a few nanometers in thickness. In ongoing scientific endeavors, researchers are actively exploring methodologies to enhance both the quality and thickness of this intrinsic oxide layer. The primary objectives of such enhancements are to augment the corrosion resistance of the metal and improve the overall quality of the protective film.

One notable technique employed in this pursuit is the anodization process, which not only contributes to the fortification of the oxide layer but also involves a coloring process. This coloring effect arises from the optical interference phenomenon, a

Fig. 26 The range of colors obtained by anodizing titanium dioxide at different voltages. Adopted from Ref. [102] (https://pubs.acs.org/doi/abs/)

well-established technique widely employed in jewelry manufacturing for its ability to produce vibrant and visually appealing coatings [102]. Figure 26 illustrates the intricate relationship between the transparent oxide layer on the titanium surface, the optical interference effect, and the applied voltage during the anodization process. This visual representation serves as a valuable tool in comprehending the interplay of these factors and offers insights into optimizing anodization conditions for achieving desired optical effects and enhancing the overall protective properties of titanium surfaces. The ongoing research in this domain is poised to contribute significantly to advancements in titanium surface engineering and the broader fields of materials science and engineering applications.

5 Conclusion

This report comprehensively explores the formation mechanisms of titanium nanotubes synthesized through electrochemical anodization and delves into the factors influencing their morphology. Additionally, a brief overview of the applications of TNTs is provided. Recognizing the pivotal role TNTs are poised to play in the future, there is a pressing need to advance specific aspects within this field. In this study, a meticulous examination of the formation, synthesis, applications, influencing factors, and other pertinent aspects of TNTs has been conducted, yielding the following key conclusions:

- Anode TNTs have emerged as a focal point of research due to their formation mechanisms and exceptional performance. The structural elucidation of TNTs has been facilitated by the application of the equal capacity model and the oxygen bubble model. These models prove instrumental in explaining the attainment of uniformly polished highly ordered nanotubes.

- Recognizing the diverse industrial applications, it becomes imperative to tailor the morphology of nanotubes accordingly. A comprehensive study has been undertaken to understand the impact of various factors, including electrolyte composition and type, temperature, voltage, anodization time, and solution environment, on the electrochemical anodization process for nanotube formation. By strategically manipulating these factors, the synthesis of highly ordered nanotubes can be achieved, enhancing their applicability and performance in specific industrial contexts.

Acknowledgements This study was supported by the Agency of Innovative Development of the Republic of Uzbekistan within the Research Project (No. FZ-2020092325).

References

1. Boehme, M., Ensinger, W.: Mixed phase anatase/rutile titanium dioxide nanotubes for enhanced photocatalytic degradation of methylene-blue. Nano-Micro Letters. **3**, 236–241 (2011)
2. Aguirre, O.R., Echeverría, E.F.: Effects of fluoride source on the characteristics of titanium dioxide nanotubes. Appl. Surf. Sci. **445**, 308–319 (2018)
3. Sun, Y., Qiang, Z., Guixin, W., Kangping, Y.: Influence of water content on the formation of TiO_2 nanotubes and photoelectrochemical hydrogen generation. J. Alloy. Compd. **711**, 514–520 (2017)
4. Roy, P., Berger, S., Schmuki, P.: TiO_2 nanotubes: synthesis and applications. Angew. Chem. Int. Ed. **50**(13), 2904–2939 (2011)
5. Habazaki, H., Fushimi, K., Shimizu, K, Skeldon, P., Thompson, G.E.: Fast migration of fluoride ions in growing anodic titanium oxide. Electrochem. Commun. **9**(5), 1222–1227 (2007)
6. Macak, J.M., Albu, S.P., Schmuki, P.: Towards ideal hexagonal self-ordering of TiO_2 nanotubes. Physica Status Solidi (RRL)–Rapid Res. Lett. **1**(5), 181–183 (2007)
7. Albu, S.P., Ghicov, A., Macak, J.M., Schmuki, P.: 250 μm long anodic TiO_2 nanotubes with hexagonal self-ordering. Physica Status Solidi (RRL)–Rapid Res. Lett. **1**(2), R65–R67 (2007)
8. Regonini, D., Bowen, C.R., Jaroenworaluck, A., Stevens, R.: A review of growth mechanism, structure and crystallinity of anodized TiO_2 nanotubes. Mater. Sci. Eng. R. Rep. **74**(12), 377–406 (2013)
9. Huusko, J., Lantto, V., Torvela, H.: TiO_2 thick-film gas sensors and their suitability for NO_x monitoring. Sens. Actuators, B Chem. **16**(1–3), 245–248 (1993)
10. Huang, S.Y., Kavan, L., Exnar, I., Graetzel, M.: Rocking chair lithium battery based on nanocrystalline TiO_2 (anatase). J. Electrochem. Soc. **142**(9), L142 (1995)
11. Park, J., Cimpean, A., Tesler, A.B., Mazare, A.: Anodic TiO_2 nanotubes: tailoring osteoinduction via drug delivery. Nanomaterials **11**, 2359 (2021)
12. Xia, Y., Yang, P., Sun, Y., Wu, Y., Mayers, B., Gates, B., Yin, Y., Kim, F.: One-dimensional nanostructures: synthesis, characterization, and applications. Adv. Mater. **15**(5), 353–389 (2003)
13. Bavykin, D.V., Friedrich, J.M., Walsh, F.C.: Protonated titanates and TiO_2 nanostructured materials: synthesis, properties, and applications. Adv. Mater. **18**(21), 2807–2824 (2006)

14. Ghicov, A., Schmuki, P.: Self-ordering electrochemistry: a review on growth and functionality of TiO_2 nanotubes and other self-aligned MO_x structures. Chem. Commun. **20**, 2791–2808 (2009)
15. Kisumi, T., Tsujiko, A., Murakoshi, K., Nakato, Y.: Crystal-face and illumination intensity dependences of the quantum efficiency of photoelectrochemical etching, in relation to those of water photooxidation, at n-TiO_2 (rutile) semiconductor electrodes. J. Electroanal. Chem. **545**, 99–107 (2003)
16. Lin, Z.A., Francisco Quintero., Pedro, J.A., Dussan, A., Hugo, R.F.: fabrication and tuning of TiO_2 nanoarchitectured morphology using magnetron sputtering and its applications to photocatalysis. Ceram. Int. **40**(10), 15523–15529 (2014)
17. David, T.M., Priya, R.D., Wilson, P., Sagayaraj, P., Tom, M.: A critical review on the variations in anodization parameters toward microstructural formation of TiO_2 nanotubes. Electrochem. Sci. Adv. **2**(4), e202100083 (2022)
18. Zhu, K., Vinzant, T.B., Neale, N.R., Frank, A.J.: Removing structural disorder from oriented TiO_2 nanotube arrays: reducing the dimensionality of transport and recombination in dye-sensitized solar cells. Nano Lett. **7**(12), 3739–3746 (2007)
19. Kim, D., Ghicov, A., Schmůki, P.: TiO_2 Nanotube arrays: Elimination of disordered top layers ("nanograss") for improved photoconversion efficiency in dye-sensitized solar cells. Electrochem. Commun. **10**(12), 1835–1838 (2008)
20. Kant, K., Losic, D.: A simple approach for synthesis of TiO_2 nanotubes with through-hole morphology. Physica Status Solidi (RRL)–Rapid Res. Lett. **3**(5), 139–141 (2009)
21. Wang, J., Lin, Z.: Anodic formation of ordered TiO_2 nanotube arrays: effects of electrolyte temperature and anodization potential. J. Phys. Chem. C **113**(10), 4026–4030 (2009)
22. Mor, G.K., Carvalho, M.A., Varghese, O.K.: A room-temperature TiO_2-nanotube hydrogen sensor able to self-clean photoactively from environmental contamination. J. Mater. Res. **19**(2), 628–634 (2004)
23. Suhadolnik, L., Živa, M., Maja, P.S., Gasper, T., Janez, K., Miran, Č.: Influence of anodization-electrolyte aging on the photocatalytic activity of TiO_2 nanotube arrays. J. Phys. Chem. C. **124**(7), 4073–4080 (2020)
24. Lai, C.W., Sreekantan, S.: Photoelectrochemical performance of smooth TiO_2 nanotube arrays: effect of anodization temperature and cleaning methods. Int. J. Photoenergy **2012**, 11 (2012). https://doi.org/10.1155/2012/356943
25. Qin, S., Lizhen, W., Jinhua, Z., Ying, H., Hongxian, Z., Jing, M., Weihua, M.: A novel elaboration for various morphologies of three-layer TiO_2 nanotubes by oxygen bubble mould. Mater. Res. Bull. **106**, 220–227 (2018)
26. Mazzarolo, A., Curionib, M., Vicenzoa, A., Skeldon, P., Thompson, G.E.: Anodic growth of titanium oxide: electrochemical behaviour and morphological evolution. Electrochim. Acta **75**, 288–295 (2012)
27. Kim, J., Kim, B., Oh, C., Ryu, J., Kim, H., Park, E.: No Effects of NH_4F and distilled water on structure of pores in TiO_2 nanotube arrays. Sci. Rep. **8**(1), 12487 (2018)
28. Strnad, G., Cazacu, R., Chetan, P.: Optimized anodization setup for the growth of TiO_2 nanotubes on flat surfaces of titanium based materials. MATEC Web Conf. EDP Sci. **137**, 02011 (2017)
29. Doll, P.W., Wolf, M., Weichert, M., Ahrens, R.: Nanostructuring of titanium by anodic oxidation with sulfuric and hydrofluoric acid. Curr. Direct. Biomed. Eng. **4**(1), 641–644 (2018)
30. Hou, X., Aitola, K., Lund, P.D.: TiO_2 nanotubes for dye-sensitized solar cells—a review. Energy Sci. Eng. **9**(7), 921–937 (2021)
31. Raja, K.S., Misra, M., Paramguru, K.: Formation of self-ordered nano-tubular structure of anodic oxide layer on titanium. Electrochim. Acta **51**(1), 154–165 (2005)
32. Macák, J.M., Tsuchiya, H., Schmuki, P.: High-aspect-ratio TiO_2 nanotubes by anodization of titanium. Angew. Chem. Int. Ed. **44**(14), 2100–2102 (2005)
33. Macak, J.M., Tsuchiya, H., Taveira, L., Aldabergerova, S., Schmuki, P.: SmoothSmooth anodic TiO_2 nanotubes. Angew. Chem. Int. Ed. **44**(45), 7463–7465 (2005)

34. Cao, W., Chen, K., Xue, D.: Highly ordered TiO_2 nanotube arrays with engineered electrochemical energy storage performances. Materials. **14**(3), 510 (2021)
35. Yi, Z., Zeng, Y., Wu, H., Chen, X., Fan, Y., Yang, H., Tang, Y.: Synthesis, surface properties, crystal structure and dye-sensitized solar cell performance of TiO_2 nanotube arrays anodized under different parameters. Results in Physics. **15**, 102609 (2019)
36. Chen, S., Chen, Q., Gao, M., Yan, S., Jin, R., Zhu, X.: Morphology evolution of TiO_2 nanotubes by a slow anodization in mixed electrolytes. Surf. Coat. Technol. **321**, 257–264 (2017)
37. Rao, B.M., Torabi, A., Varghese, O.K.: Anodically grown functional oxide nanotubes and applications. MRS Commun. **6**(4), 375–396 (2016)
38. Jennings, J.R., Ghicov, A., Peter, L.M.: Dye-sensitized solar cells based on oriented TiO_2 nanotube arrays: transport, trapping, and transfer of electrons. J. Am. Chem. Soc. **130**(40), 13364–13372 (2008)
39. Valota, A., LeClere, D.J., Skeldon, P., Curioni, M., Hashimoto, T.: Influence of water content on nanotubular anodic titania formed in fluoride/glycerol electrolytes. Electrochim. Acta **54**(18), 4321–4327 (2009)
40. Su, Z., Zhou, W.: Formation, morphology control and applications of anodic TiO_2 nanotube arrays. J. Mater. Chem. **21**(25), 8955–8970 (2011)
41. Zhao, S., Ji, X., Haowen, F., Shaoyu, Z., Dongdong, L., Xufei, Z.: Derivation of a mathematical model for the growth of anodic TiO_2 nanotubes under constant current conditions. J. Electrochem. Soc. **164**(9), E187 (2017)
42. Houser, J.E., Hebert, K.R.: Modeling the potential distribution in porous anodic alumina films during steady-state growth. J. Electrochem. Soc. **153**(12), B566 (2006)
43. Houser, J.E., Hebert, K.R.: The role of viscous flow of oxide in the growth of self-ordered porous anodic alumina films. Nat. Mater. **8**(5), 415–420 (2009)
44. Su, Z., Hähner, G., Zhou, W.: Investigation of the pore formation in anodic aluminium oxide. J. Mater. Chem. **18**(47), 5787–5795 (2008)
45. Su, Z., Bühl, M., Zhou, W.: Dissociation of water during formation of anodic aluminum oxide. J. Am. Chem. Soc. **131**(24), 8697–8702 (2009)
46. Taveira, L.V., Macak, J.M., Sirotna, K.: Voltage oscillations and morphology during the galvanostatic formation of self-organized TiO_2 nanotubes. J. Electrochem. Soc. **153**(4), B137 (2006)
47. Habazaki, H.S., Nagata, P., Skeldon, G.E.: Thompson Ionic mobilities in amorphous anodic titania. J. Electrochem. Soc. **149**(3), B70 (2002)
48. Beranek, R., Hildebrand, H., Schmuki, P.: Electrochem. Solid State Lett. **6**(3), B12 (2003)
49. Macak, J.M., Taveira, L.V., Tsuchiya, H., Sirotna, K.: Influence of different fluoride containing electrolytes on the formation of self-organized titania nanotubes by Ti anodization. J. Electroceram. **16**, 29–34 (2006)
50. Balaur, E., Macak, J.M., Tsuchiya, H., Schmuki, P.: Wetting behaviour of layers of TiO_2 nanotubes with different diameters. J. Mater. Chem. **15**(42), 4488–4491 (2005)
51. Macak, J.M., Sirotna, K., Schmuki, P.: Self-organized porous titanium oxide prepared in Na_2SO_4/NaF electrolytes. Electrochim. Acta **50**, 3679–3684 (2005)
52. Liang, Z., Bai, X., Hao, P., Guo, Y., Xue, Y., Tian, J.: Full solar spectrum photocatalytic oxygen evolution by carbon coated TiO_2 hierarchical nanotubes. Appl. Catal. B **243**, 711–720 (2019)
53. Grimes, C.A., Mor, G.K.: TiO_2 Nanotube Arrays: Synthesis, Properties, And Applications Springer Science & Business Media (2009)
54. Jarosz, M., Kapusta-Kołodziej, J., Jaskuła, M., Sulka, G.D.: Effect of different polishing methods on anodic titanium dioxide formation. J. Nanomater. **16**(1), 86–86 (2015)
55. Shankar, K., Gopal, M.K., Adriana, F., Craig, A.: GrimesCation effect on the electrochemical formation of very high aspect ratio TiO_2 nanotube arrays in formamide—water mixtures. J. Phys. Chem. C **111**(1), 21–26 (2007)
56. Lim, Y.C., Zulkarnain, Z., Zuraida, K., Siti, S.I., Gliserol, G.: Electrochemical synthesis of ordered titania nanotubes in mixture of ethylene glycol and glycerol electrolyte. Malaysian J. Anal. Sci. **20**(2), 373–381 (2016)

57. Kojima, R., Kimura, Y., Bitoh, M., Abe, M., Niwano, M.: Investigation of influence of electrolyte composition on formation of anodic titanium oxide nanotube films. J. Electrochem. Soc. **159**(11), D629 (2012)
58. Huang, W., Xu, H., Ying, Z., Dan, Y., Zhou, Q., Zhang, J.: Split TiO_2 nanotubes—evidence of oxygen evolution during Ti anodization. Electrochem. Commun. **106**, 106532 (2019)
59. Raja, K.S., Gandhi, T., Misra, M.: Effect of water content of ethylene glycol as electrolyte for synthesis of ordered titania nanotubes. Electrochem. Commun. **9**(5), 1069–1076 (2007)
60. Ying, C.L., Zainal, Z., Hussein, M.Z.: Effect of electrolyte composition in electrochemical synthesis of self-organized TiO_2 nanotubes. Adv. Mater. Res. **364**, 298–302 (2012)
61. Wieland, B., Lancaster, J.P., Hoaglund, C.S., Holota, P., Tornquist, W.J.: Electrochemical and infrared spectroscopic quantitative determination of the platinum-catalyzed ethylene glycol oxidation mechanism at CO adsorption potentials. Langmuir **12**(10), 2594–2601 (1996)
62. Mohapatra, S.K., Misra, M., Mahajan, V.K., Raja, K.S.: A novel method for the synthesis of titania nanotubes using sonoelectrochemical method and its application for photoelectrochemical splitting of water. J. Catal. **246**(2), 362–369 (2007)
63. Ruzimuradov, O., Braglia, M., Vacandio, F., Knauth, P.: A humidity-sensitive nanocomposite solid ion conductor: sulfonated poly-ether-ether-ketone in nanotubular TiO_2 or ZrO_2 matrix. J. Solid State Electrochem. **22**, 3255–3260 (2018)
64. Tao, J., Zhao, J., Tang, C., Kang, Y., Li, Y.: Mechanism study of self-organized TiO_2 nanotube arrays by anodization. New J. Chem. **32**(12), 2164–2168 (2008)
65. Li, Y., Yu, H., Zhang, C., Song, W., Li, G., Shao, Z., Yi, B.: Effect of water and annealing temperature of anodized TiO_2 nanotubes on hydrogen production in photoelectrochemical cell. Electrochim. Acta **107**, 313–319 (2013)
66. Li, H., Ding, M., Jin, J., Sun, D., Zhang, S, Jia, C.: Effect of electrolyte pretreatment on the formation of TiO_2 nanotubes: an ignored yet non-negligible factor. ChemElectroChem. **5**(7), 1006–1012 (2018)
67. Paulose, M., Varghese, O.K., Mor, G.K., Grimes, C.A.: Unprecedented ultra-high hydrogen gas sensitivity in undoped titania nanotubes. Nanotechnology **17**(2), 398 (2005)
68. Indira, K., Mudali, U.K., Nishimura, T.: A review on TiO_2 nanotubes: influence of anodization parameters, formation mechanism, properties, corrosion behavior, and biomedical applications. J. Bio-and Tribo-Corrosion. **1**, 1–22 (2015)
69. Liu, Y., Zhou, B., Li, J., Gan, X., Bai, J., Cai, W.: Preparation of short, robust and highly ordered TiO_2 nanotube arrays and their applications as electrode. Appl. Catal. B **92**(3–4), 326–332 (2009)
70. Wu, H., Chen, Y, Liang, J., Shi, B., Wu, G., Zhang Y., Wang, D.: Hypomethylation-linked activation of PAX_2 mediates tamoxifen-stimulated endometrial carcinogenesis. Nature. **438**(7070), 981–987 (2005)
71. Feng, X., Macak, J.M., Schmuki, P.: Robust self-organization of oxide nanotubes over a wide pH range. Chem. Mater. **19**(7), 1534–1536 (2007)
72. Muniyappan, S., Shibu, J., Manovah, D., Sagayaraj, P., Murugakoothan, P.: Effect of pH on synthesis of TiO_2 nanotubes by electrochemical anodization. Indo-Asian J. Multi. Res. (IAJMR). **1**, 418–421 (2015)
73. Arif, H., Nur, A.Y., Azrina, A.A., Wu, L.L.: Progress on the influencing factors affecting the formation of TiO_2 nanotubes (TNTS) in electrochemical anodization. Acta Chemica Malaysia (ACMY) (2023)
74. Tesler, A.B., Altomare, M., Schmuki, P.: Morphology and optical properties of highly ordered TiO_2 nanotubes grown in NH_4F/o-H_3PO_4 electrolytes in view of light-harvesting and catalytic applications. ACS Appl. Nano Mater. **3**(11), 10646–10658 (2020)
75. Ismail, S., Khairul, K.A., Hisham, N., Mamat, M.S.: Effect of voltage on TiO_2 nanotubes formation in ethylene glycol solution. Jurnal Teknologi. **79**(5–2), 117–120 (2017)
76. Macak, J.M., Hildebrand, H., Marten, U., Jahns, P., Schmuki, P.: Mechanistic aspects and growth of large diameter self-organized TiO_2 nanotubes. J. Electroanal. Chem. **621**(2), 254–266 (2008)

77. Varghese, O., Gong, K.D., Paulose, M., Grimes, C.A., Dickey, E.C.: Crystallization and high-temperature structural stability of titanium oxide nanotube arrays. J. Mater. Res. **18**(1), 156–165 (2003)
78. Hoseinzadeh, T., Hoseinzadeh, T., Ghorannevis, Z.: Effects of various applied voltaGES on physical properties of TiO_2 nanotubes by anodization method. J. Theor. Appl. Phys. **11**, 243–248 (2017)
79. Pasikhani, J.V., Gilani, N., Pirbazari, A.E.: The effect of the anodization voltage on the geometrical characteristics and photocatalytic activity of TiO_2 nanotube arrays. Nano-Structures Nano-Objects. **8**, 7–14 (2016)
80. Yu, J., Wang, B.: Effect of calcination temperature on morphology and photoelectrochemical properties of anodized titanium dioxide nanotube arrays. Appl. Catal. B: Enviro. **94**(3–4), 295–302 (2010)
81. Macak, J.M., Zlamal, M., Krysa, J., Schmuki, P.: Self-organized TiO_2 nanotube layers as highly efficient photocatalysts. Small. **3**(2), 300–304 (2007)
82. Pang, Y.L., Lim, S., Ong, H.C., Chong, W.T.: A critical review on the recent progress of synthesizing techniques and fabrication of TiO_2-based nanotubes photocatalysts. Appl. Catal. A: General. **481**, 127–142 (2014)
83. Fleischer, C., Chatzitakis, A., Norby, T.: Intrinsic photoelectrocatalytic activity in oriented, photonic TiO_2 nanotubes. Mater. Sci. Semicond. Process. **88**, 186–191 (2018)
84. Wang, X., Zhao, J., Wang, X., Zhou, J.: Causes for the formation of titania nanotubes during anodization. IEEE Trans. Nanotechnol. **14**(1), 113–117 (2014)
85. Zheng, J., Zhou, H., Zou, Y., Wang, R., Lyu, Y., Wang, S.: Efficiency and stability of narrow-gap semiconductor-based photoelectrodes. Energy Environ. Sci. **12**(8), 2345–2374 (2019)
86. Isik, E., Tasyurek, L.B., Isik, I., Kilinc, N.: Synthesis and analysis of TiO_2 nanotubes by electrochemical anodization and machine learning method for hydrogen sensors. Microelectron. Eng. **262**, 111834 (2022)
87. Wang, D. Liu, Y., Yu, B., Zhou, F., Liu, W.: TiO_2 nanotubes with tunable morphology, diameter, and length: synthesis and photo-electrical/catalytic performance. Chem. Mater. **21**(7), 1198–1206 (2009)
88. Chen, J., Lin, J., Chen, X.: Self-assembled TiO_2 nanotube arrays with U-shaped profile by controlling anodization temperature. J. Nanomater. 1–4 (2010)
89. Mohan, L. Dennis, C., Padmapriya, N., Anandan, C.: Effect of electrolyte temperature and anodization time on formation of TiO_2 nanotubes for biomedical applications. Mater. Today Commun. **23**, 101103 (2020)
90. Cedrik, M., Marie, L., Ahmad, D., Patric, D., Thomas, C., Didier, R.: Upscalling anodic synthesis of TiO_2 nanotubes film as potential material for photoelectrocatalytic applications: influence of electrolyte overheating and aging on nanotube morphology and stability. J. Photocatal. 1, 43–49 (2020)
91. Sul, Y.T., Johansson, C.B., Jeong, Y.: The electrochemical oxide growth behaviour on titanium in acid and alkaline electrolytes. Med. Eng. Phy. **23**(5), 329–346 (2001)
92. Ayal, A.K.: Effect of anodization duration in the TiO_2 nanotubes formation on Ti foil and photoelectrochemical properties of TiO_2 nanotubes. Al-Mustansiriyah J. Sci. **29**(3), 77–81 (2019)
93. Regonini, D., Clemens, F.J.: Anodized TiO_2 nanotubes: effect of anodizing time on film length, morphology and photoelectrochemical properties. Mater. Letts. **142**, 97–101 (2015). https://doi.org/10.1016/j.matlet.2014.11.145
94. Mohamed, A.R., Rohani, S.: Modified TiO_2 nanotube arrays (TNTAs): progressive strategies towards visible light responsive photoanode, a review. Energy Environ. Sci. **4**(4), 1065–1086 (2011)
95. Ratnawati, R., Gunlazuardi, J., Slamet, S.: Effect of anodization time and temperature on the morphology of TiO_2 nanotube arrays for photocatalytic hydrogen production from glycerol solution. In: Proceeding the Regional Conference on Chemical Engineering (2014)
96. Huang, L., Peng, F., Yu, H., Wang, H., Yang, J., Li, Z.: The influence of ultrasound on the formation of TiO_2 nanotube arrays. Mater. Res. Bull. **45**(2), 200–204 (2010)

97. Ratnawati, R.: Similarity-effect of anodization time and temperature on the morphology of TiO_2 nanotube arrays for photocatalytic hydrogen production from glycerol solution (2023)
98. Li, Y., Ma, Q., Han, J., Ji, L., Wang, J., Chen, J., Wang, Y.: Controllable preparation, growth mechanism and the properties research of TiO_2 nanotube arrays. Appl. Surf. Sci. **297**, 103–108 (2014)
99. Cortes, F.J.Q., Arias, P.J.M., Phillips, J., Zea, H.: Empirical kinetics for the growth of titania nanotube arrays by potentiostatic anodization in ethylene glycol. Mater. Des. **96**, 80–89 (2016)
100. Nischk, M., Mazierski, P., Gazda, M., Zaleska, A.: Ordered TiO_2 nanotubes: the effect of preparation parameters on the photocatalytic activity in air purification process. Appl. Catal. B: Environ. **144**, 674–685 (2014)
101. Sul, Y.T., Johansson, C.B., Jeong, Y., Albrektsson, T.: The electrochemical oxide growth behaviour on titanium in acid and alkaline electrolytes. Med. Eng. Phys. **23**(5), 329–346 (2001)
102. Tsui, L., Homma, T., Zangari, G.: Photocurrent conversion in anodized TiO_2 nanotube arrays: effect of the water content in anodizing solutions. J. Phys. Chem. C. **117**(14), 6979–6989 (2013)

TiO_2-Based Heterostructure Coatings for Photocatalytic Water Purification

Liudmila Khoroshko, Aleksey Baglov, Maryia Yauseichyk, Suvankul Nurmanov, and Olim Ruzimuradov

Abstract Promising thin-film heterostructure photocatalysts based on copper and titanium oxides (TiO_2), and graphite-like carbon nitride (g-C_3N_4) could be obtained on silicon, metal, and glass substrates. The photocatalytic activity of the TiO_2/(CuO + Cu_2O) and TiO_2/g-C_3N_4 systems against a test organic pollutant upon activation by ultraviolet radiation with a wavelength of 365 nm was established. The possibility of functionalizing the surface of a polymer material with TiO_2 xerogel to obtain photocatalytic and antibacterial properties has been demonstrated. The prospects for using thin film coatings and heterostructures based on metal oxides in flow systems for water purification and disinfection are discussed.

Keywords Titania · Sol–gel synthesis · Photocatalysis · Heterostructures · Photocatalytic water purification

1 Introduction

Intensive studies are required to understand the critical dependencies of physical and chemical properties of oxide, semiconductor, and metallic structures on their size, structure, and chemical surface modification [1, 2]. The main focus is to obtain materials with specified and precisely controlled properties. The development of low-cost

L. Khoroshko (✉) · A. Baglov · M. Yauseichyk
R&D Department, Belarusian State University of Informatics and Radioelectronics, P. Browki Str. 6, 220013 Minsk, Belarus
e-mail: l.s.khoroshko@gmail.com

L. Khoroshko · A. Baglov
Faculty of Physics, Belarusian State University, 4 Nezavisimosti Av., 220030 Minsk, Belarus

S. Nurmanov · O. Ruzimuradov
Faculty of Chemistry, National University of Uzbekistan, University Str. 4, Tashkent, Uzbekistan 100174

O. Ruzimuradov
Department of Natural and Mathematic Sciences, Turin Polytechnic University in Tashkent, Kichik Khalqa Yoli Str. 17, Tashkent, Uzbekistan 100095

J. Prakash et al. (eds.), *Titanium Dioxide-Based Multifunctional Hybrid Nanomaterials*, Engineering Materials, https://doi.org/10.1007/978-3-031-80625-4_5

and effective methods for water purification retains its importance. The photocatalytic method for deep water purification from organic impurities is gaining attention due to its high efficiency and ability to destroy (mineralize) certain chlorine, nitro, and polyaromatic compounds without the need for consumable reagents [3]. Titanium dioxide (TiO_2), in the form of pure anatase or its mixture with rutile, is commonly used as a photocatalytic component [4, 5]. It can be found in dispersed particles or as coatings on various carriers. To enhance photocatalytic activity and improve charge separation efficiency, TiO_2-based materials can be augmented with layers or inclusions of oxides, nitrides, and other semiconductor materials to form heterojunctions [6–8]. Efficient thin-film catalysts development is especially important due to the increasing demand for portable water treatment devices and the design of drinking water final purification systems [9–11]. This chapter presents the summary of several practical results on the development of thin-film photocatalytic oxide and heterostructure coatings based on TiO_2.

2 Analysis of Features and Applications of Micro- and Nanostructured Photocatalytic Active Materials for Water Purification

The study of possible mechanisms of interaction between micro- and nanostructures for photocatalytic water purification receives significant attention. Efficient photocatalysts must be created and existing materials improved through composition modification and manufacturing technology enhancements [12, 13]. Traditionally, measures to increase the probability of capturing and destroying organic pollutants include developing a photocatalyst surface with a maximum specific area, mixing the reaction mixture to ensure pollutant delivery to the photocatalyst surface, and possibly aerating contaminated solutions [14, 15]. The using of active micro- and nanostructured materials in photocatalytic methods can effectively reduce organic contaminants in water. At the same time, the application of this method significantly reduces atmospheric pollution by various oxidizing agents. This ultimately avoids additional costs for neutralizing treatment products. The photocatalytic method of water purification does not require the utilization of accumulated pollutants, which positively affects the economic component of this method. The efficiency of the photocatalytic method for water purification from organic pollutants increases when nanostructured materials are used as the photocatalyst [14]. Oxygen adsorption on the surfaces of materials based on oxides of titanium, zinc, manganese, chromium, and other metals promotes the development of the process of charged particle formation during the transition of electrons from the metal cation to oxygen [16–18]. The combination of photocatalytic properties of various metal oxides and graphite-like carbon nitride is of considerable interest. These materials have shown effective water purification from organic contaminants through the synthesis of hydrogen with the help of photo-assisted destruction of water [19–21].

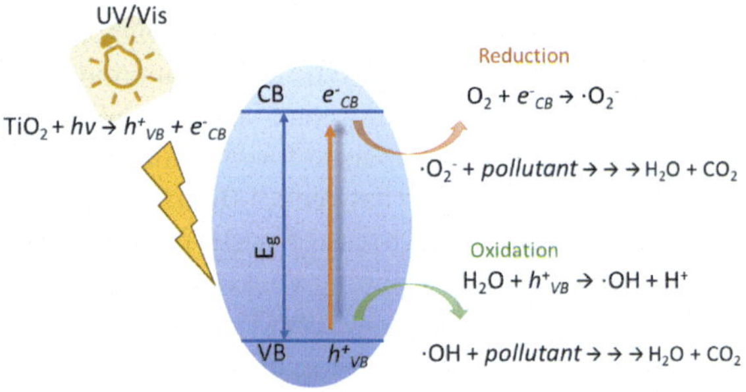

Fig. 1 Schematic representation of photocatalysis process (adapted from [22], MDPI, licensed under Creative Common License 4.0 (Creative Commons—Attribution 4.0 International—CC BY 4.0))

Photocatalysis is a complex phenomenon of chemical-physical origin that involves the acceleration of heterogeneous catalytic reactions by exciting the photocatalyst through to absorption of light with consequence of its interaction with the chemical reaction participants forming reaction intermediates and regenerates itself after each cycle of such interactions and closure of the catalytic cycle [3]. This process leads to the photochemical decomposition of compounds on the surface of a heterogeneous photocatalyst under the influence of radiation (Fig. 1).

These reactions occur only in the region where the photocatalyst can absorb light, which explains the absence of reaction in darkness or when illuminated with light of a certain wavelength but without the photocatalyst. The evaluation of illumination and catalyst enables the differentiation between catalytic or photochemical reactions.

Photocatalysts are semiconductor materials that use light to produce reaction products and do not change during the reaction, e.g. oxides materials TiO_2, ZnO, SnO_2, CdO, etc. [23–32]. The activation of photoadsorption, photodesorption, and photocatalytic reaction processes is connected to the excitation of the photocatalyst electron subsystem upon the absorption of light quanta. All reactions share common stages associated with energy migration from the semiconductor volume to the surface, dissipation of electron excitation energy, and the formation and destruction of active centers on the photocatalyst surface. The concentration of electrons near the surface promotes the emergence of one-electron bonding and the formation of surface compounds. The possibility of electron transition from the volume to the surface in the conduction zone is associated with the electronic structure of the semiconductor [33].

The high chemical inertness and low toxicity of the considered oxides make water purification using photocatalytic methods a promising solution for various industries, including ecology, chemical, oil, and gas, etc. The use of photocatalysis is expected to increase due to these advantages. At the same time, it is important to consider the comparatively low quantum efficiency of the process because weak separation of electron–hole pairs, as well as the limited absorption spectrum in the UV region. This significantly restricts the use of solar energy for industrial photocatalytic purification.

To minimize the influence of these factors, doping sensitization of nanoparticles of other materials is used to create heterostructures or non-stoichiometric compounds [34, 35]. The quantum size effects can be exploited by the synthesis of structured metal oxides in the form of nanosheets, nanorods, and other nanoscale objects. Currently, the most photocatalytic active nanomaterials are nanoparticles [36]. Therefore, the optimal composition and structure of materials for nanoparticle powders require them to be in a dispersed form. Additionally, the perfect thickness for photocatalytic active coatings is between 300 and 400 nm with a crystallite size of 10–70 nm [36]. Furthermore, the process of purifying with nanoparticles requires specialized equipment for its chemical removing or filtration and limits the practical application of nanosized metal oxide powders [37, 38]. To address this issue, nanosized powders can be embedded in a supporting matrix to form photocatalytic active coatings with a nanostructured surface on substrates. In the second case, it is possible to enhance the photocatalytic activity of the coating by using a substrate that has the desired semiconductor properties, such as TiO_2, and a structured coating for the formation of an effective surface.

The analysis of possible mechanisms of interaction of photocatalytic active micro- and nanostructured materials shows significant potential for water purification from organic contaminants. It is worth noting that the photocatalytic activity of oxide materials can be increased due to the formation of heterostructures.

The analysis of possible mechanisms of interaction of photocatalytic active micro- and nanostructured materials with organic pollutants shows significant potential for water purification from these contaminants [39–42]. The use of immobilized photocatalysts eliminates the need for additional filtration of the purified solution after photocatalysis, resulting in a more efficient water purification process (Fig. 2), and can provide new combined materials with photocatalytic and self-cleaning properties, for example [22, 43].

3 One-Step Synthesis of Combined Photocatalytic Active Structures Based on Titania and Copper Oxides

The sol–gel method allows the synthesis of a wide range of photocatalytic active materials due to its flexibility and compositional variability [44–46]. The ability to produce free-standing powders and immobilized catalysts covering various substrates is an important advantage of the sol–gel method. One of the most popular photocatalytic materials obtained by the sol–gel method is TiO_2 i.e. titania. The general disadvantage of titania is a low sensibility to optical spectra because wide gap of about 3.2 eV. But high stability, low cost, human and environmental friendliness, and high electronic conductivity make titania a very popular photocatalyst. There are many ways to sensitize TiO_2 for visible irradiation, such as doping with other metals and semiconductors, formation of heterostructures, etc.

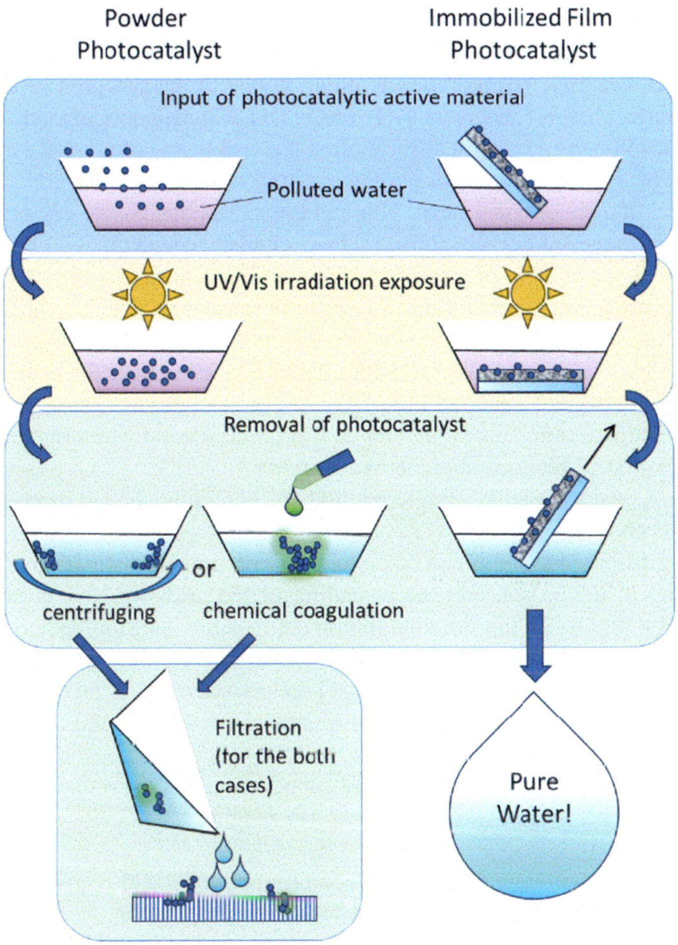

Fig. 2 Schematic representation of photocatalytic water purification using powder (left panel) and immobilized film (right panel) photocatalyst

3.1 Preparation and Study of Copper oxides/TiO_2 Heterostructures

Semiconductor materials with a smaller bandgap, such as Cu_2O and CuO (2.0 and 1.2 eV, respectively), can sensitize TiO_2-based photocatalysts to the visible wavelength range. These semiconductors enable the absorption of visible radiation and are also promising for use in optoelectronics, photovoltaics, and sensors. Additionally, materials with a high specific surface area are used to increase photocatalytic activity. Various methods, such as chemical etching, lithography, and template synthesis, are used to obtain materials with a developed surface. In [47], it was demonstrated that oxide layers containing arrays of nanosized CuO filaments can be easily formed by the thermal oxidation of copper. These structures are of interest for use in composite photocatalysts due to their high specific surface area. Composites containing CuO and/or Cu_2O and TiO_2 require special attention due to the potential separation of photogenerated charge carriers at TiO_2/CuO, TiO_2/Cu_2O, and CuO/Cu_2O heterojunctions, accompanied by light absorption in the visible and ultraviolet ranges. The fabrication of a composite photocatalytic coating using a sol–gel process is typically a series of technological steps. This can involve adding previously synthesized nanoparticles to the sol [24, 48] followed by a sophisticated synthesis [49], or using substrates with chemically or lithographically modified surfaces [50].

One-step produced materials are of interest because they simplify the scaling of the technology and improve prospects for commercialization. In the case of sol–gel technology, combined sols (including nanoparticles doped, for example, with commercial micro- and nanoparticles TiO_2, [41]), or combined thermal oxidation of substrate with xerogel annealing can be used. The second variant is represented by the formation of photocatalytic active systems TiO_2/(CuO + Cu_2O) by sol–gel technology. Consider a fairly simple method for obtaining such systems based on copper grids and foils.

Copper grid-based systems can be obtained by preliminary oxidation of the grid, deep coating in titania sol, and the annealing of the grid with titania xerogel on air. Paper [39] reports on the structure, phase composition, and antibacterial properties of these composite nanostructures. The nanostructures were produced through a combination of thermal oxidation and sol–gel synthesis. The preparation of the copper surface involved several chemical treatment steps. After the chemical treatment step, the samples were washed with distilled water and dried in flowing warm air at approximately 90 °C. Nanostructured copper oxides were produced via thermal oxidation of copper grids in air at temperatures ranging from 350 to 500 °C for 20 to 180 min.

TiO_2 films were produced on the surface of unoxidized and preoxidized copper grids using a sol–gel method. The sol used for TiO_2 preparation contained titanium tetraisopropoxide (Ti(i-OC_3H_7)$_4$) as a precursor and ethylene glycol monomethyl ether ($HOCH_2CH_2OCH_3$) as a solvent. The sol was adjusted to pH 1 by adding nitric acid. The grids were then coated with the sol by immersion to form 5 layers and dried at 200 °C for 5 min in air with each layer. The resulting TiO_2 xerogel films were air-annealed at 450–500 °C for 20–60 min to remove residual solvent and crystallize TiO_2 anatase.

Nanostructured copper oxides with a granular morphology and high porosity were formed on the surface of copper grids by thermal oxidation at temperatures ranging from 350 to 500 °C. Additionally, the formation of nanowires, with a thickness of 50 to 200 nm and a length of up to 10 μm, growing perpendicular to the wire surface was observed (Fig. 3a). The diameter and length of the nanowires were found to be dependent on temperature [39]. These nanowires are composed of copper (II) oxide (CuO) and have a mono- or bi-crystalline structure [39, 51]. Typically, these structures grow due to copper ion diffusion to the surface along grain boundaries during thermal oxidation. Based on the X-ray diffraction (XRD) data of the copper grid immersed in the TiO_2 sol, dried, and annealed at 500 °C for 20 min, it was found that CuO is the primary product of copper oxidation at this temperature, prevailing over Cu_2O in the grown structures. TiO_2 is also present, as an anatase according to interpretation of statistically significant peaks collected over several repeated measurements (Fig. 4, and see [25, 39, 52]). The results obtained from the foils were qualitatively similar, but they are not sufficiently illustrative due to the higher amount of metallic copper in the foil.

For the heterostructures based on copper foils, there is no necessity for peroxidation was found [25]. From the scanning electron microscopy (SEM, Fig. 3b) and XRD studies, it was established that the whiskers are composed of CuO, while the surface-grained layers consist of CuO and Cu_2O phases. Based on the calculation results from XRD, the phase relation of Cu:CuO:Cu_2O is approximately 81:15:4%. The conclusion regarding CuO whiskers is supported by the fact that they form at

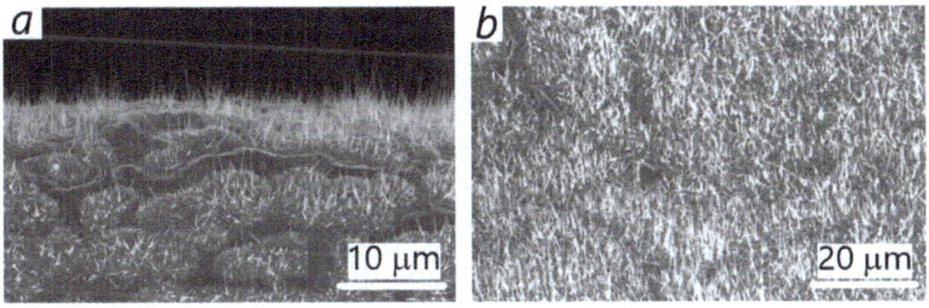

Fig. 3 SEM images of TiO_2/(CuO + Cu_2O) composites formed on the copper grid (**a**) and foil (**b**) after annealing at 450 °C

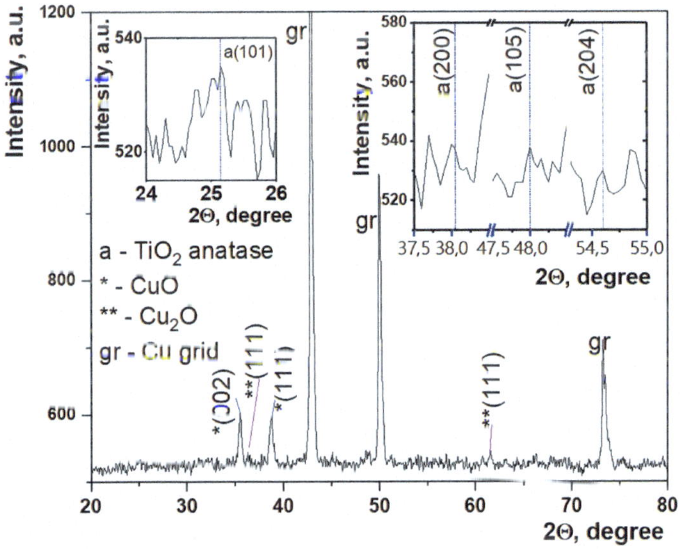

Fig. 4 XRD patterns of the copper grid with TiO_2 xerogel after annealing at 450 °C on air. On inserts, the interpretation of anatase peaks is provided

temperatures above 400 ºC, at which Cu_2O is not thermodynamically stable and transforms into CuO with sufficient oxygen concentration in the annealing environment [25].

3.2 *Photocatalytic Activity of the $TiO_2/(CuO + Cu_2O)$ Composites*

The photocatalytic activity of the composites was evaluated by measuring the degree of decomposition of the model contaminant Rhodamine B in an aqueous solution (10 mg/L) under UV irradiation (365 nm) for 1–60 min. To examine the photocatalytic activity, 3 g of the solution was poured into 30 mm diameter Petri dishes with the experimental samples at the bottom. Light exposure was provided by a 9 W mercury lamp placed 10 cm away from the surface of the solution. This experimental configuration allows to neglect of the attenuation of the radiation before reaching the photocatalyst surface. To ensure a uniform supply of contaminants to the catalyst, the solution was stirred by vibration. The concentration change of the analyte in the solution was determined by measuring the peak intensity of optical absorption of Rhodamine B at 553 nm. Before UV irradiation, the experimental samples were kept in the test solution under dark conditions for 10 min to allow for the chemical adsorption of the analyte onto the photocatalyst surface. Photocatalytic activity of obtained structures with TiO_2 and a combination of copper oxides ($CuO + Cu_2O$) expressed as a percentage of dye decomposition is provided in Fig. 5. Relative concentration of the dye was calculated using the ratio of the initial concentration assumed to be 100% in this and following cases, initial concentrations will be noted for every experiment. To obtain the degree of the dye decomposition, the relative concentration was subtracted from 100%, respectively.

The composites of $TiO_2/(CuO + Cu_2O)$ on grids exhibited antibacterial activity against *Escherichia coli* and *Staphylococcus aureus* when exposed to visible light, as previously reported in [39]. This suggests that such composites synthesized on copper foils may also have antibacterial properties.

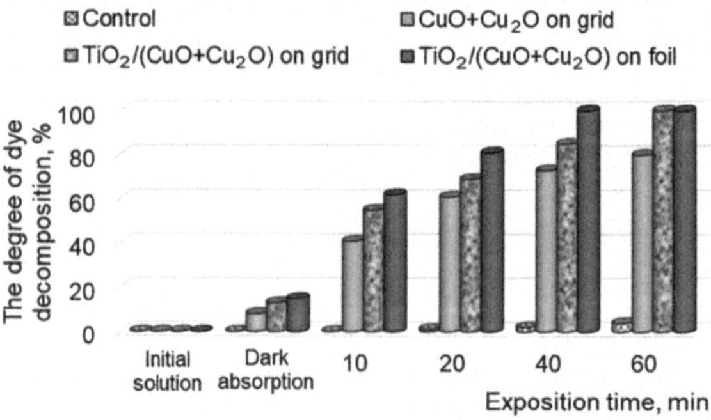

Fig. 5 Summary of photocatalytic activity investigation of titania and copper oxides composites, annealed at 450 °C

So, the sol–gel synthesis of $TiO_2/(CuO + Cu_2O)$ composites was used to efficiently fabricate immobilized photocatalytic active covers on copper foils and pre-oxidized copper grids in a one-step process. The resulting immobilized photocatalysts have large, efficient surfaces and can be fabricated in a variety of shapes. The grids are suitable for use as transparent elements for liquids and air applications. The one-step production of catalytically active materials shows promise for improving current water and air purification systems.

4 Coatings Based on TiO_2/g-C_3N_4 Heterostructures for Photocatalytic Water Purification

To obtain photocatalytic active structures based on micro- and nanostructured metal oxides for photocatalytic water purification, the TiO_2/g-C_3N_4 system was synthesized due to the possibility of obtaining by thermal deposition a developed surface of graphite-like carbon nitride [53], as well as the formation of the heterostructure to improve the efficiency of photocatalysts based on TiO_2.

4.1 Thin-Film Heterostructures

The starting material for TiO_2 production was titanium foil (metal grade VT1-0, GOST 19807–91). TiO_2 on the metal surface was formed by thermal oxidation at 550 °C for 30 min in a muffle furnace in air. Preliminary chemical treatment of the substrates in boiling hydrogen peroxide solution or concentrated sodium hydroxide was carried out to form a developed surface. The g-C_3N_4 was thermally deposited on oxidized substrates by pyrolytic decomposition of precursor melamine in a hermetically sealed crucible at 550 °C for 5 min, as described in [54]. The obtained heterostructures were investigated by SEM, XRD methods, and energy dispersive analysis (EDA).

Figure 6 displays SEM images of the surface of titanium foil following thermal oxidation. The foil surface exhibits titanium oxide islands, and the EDA results indicate the formation of TiO_2 for all samples. Chemical pretreatment enhances surface evolution, resulting in numerous greed-like fragments (Fig. 6b) after treatment with hydrogen peroxide or plate-like fragments (Fig. 6c) after treatment with sodium hydroxide. The presence of a more pronounced relief on the surface and a layer of TiO_2 has a positive effect on the deposition process of g-C_3N_4, resulting in thicker coatings with additional surface relief (Fig. 7). SEM results show that the coating thickness on samples pre-oxidized in hydrogen peroxide is three times greater than that obtained on the original titanium foil. The film heterostructures obtained exhibit a developed surface with clustered spherical g-C_3N_4 agglomerates, with individual element diameters reaching up to 500 nm. Additionally, even preliminary thermal oxidation of titanium contributes to a more than 60% increase in g-C_3N_4 coating thickness (from 303 nm for the initial foil to 999 nm for the foil after chemical oxidation in H_2O_2, thermal oxidation at 550 °C). Based on the EDA results, the concentration of sodium and its compounds in heterostructures obtained on foils pre-oxidized in sodium hydroxide does not exceed 0.45 at.%.

For all bilayers the results of XRD investigation are similar. In Fig. 8 we provide XRD spectra obtained with monochromatized CoK_α radiation. The thickest film (999 nm thickness) has sufficient adhesion and its photocatalytic activity was investigated with respect to Rhodamine B. For other samples, fragmentary film peeling was observed and it was not possible to evaluate photocatalytic activity clearly. To analyze photocatalytic activity, solutions of the test dye Rhodamine B were exposed to ultraviolet radiation of a mercury lamp with a maximum of 365 nm in the presence of the obtained structures. The initial concentration of the dye was 2.5 mg/L. Before exposure, the samples were kept in the dark to allow for complete dark adsorption of the dye. The study found that the adsorption equilibrium is established within 10 min for the tested structures. The degree of photocatalytic destruction of the test dye was determined by analyzing the absorption spectra of the solutions before and after exposure. Control measurements were taken every 10 min, and were stopped after the 30 min because of the low process speed.

The percentage of dye decomposition within 30 min is about 10%, at that pronounced adsorption–desorption processes on the photocatalyst surface are observed. The relatively low activity observed may be attributed to the underdeveloped relief of the outer layer and the limited interaction between the two semiconductors, owing to its considerable thickness. However, the technological outcomes of this

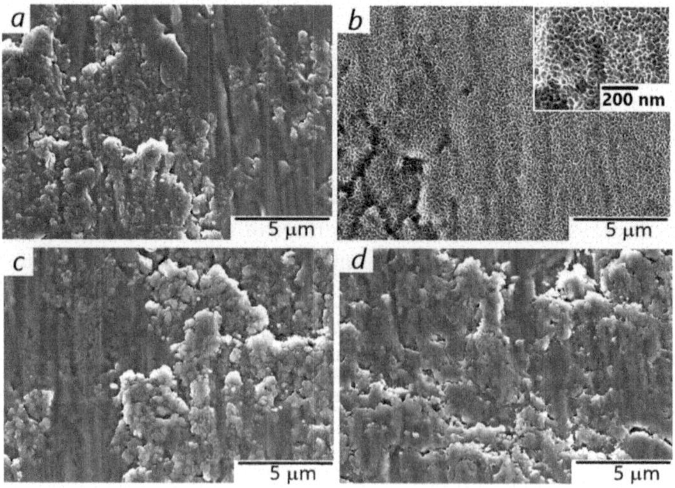

Fig. 6 SEM images of the surface of titanium foil with TiO_2 obtained by thermal oxidation at 550 °C (**a**); chemical oxidation in H_2O_2, thermal oxidation at 550 °C (**b**); chemical oxidation in NaOH, thermal oxidation at 550 °C (**c**); original foil (**d**)

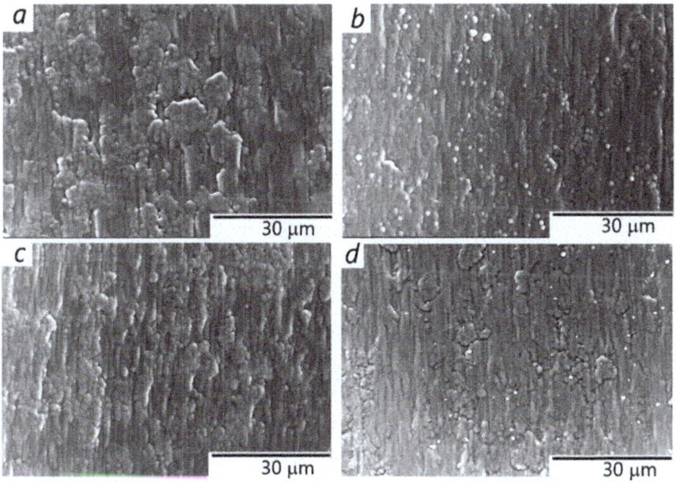

Fig. 7 SEM images of the surface of g-C_3N_4 layer formed on titanium foil with TiO_2 obtained by thermal oxidation at 550 °C (**a**); chemical oxidation in H_2O_2, thermal oxidation at 550 °C (**b**); chemical oxidation in NaOH, thermal oxidation at 550 °C (**c**); original foil (**d**)

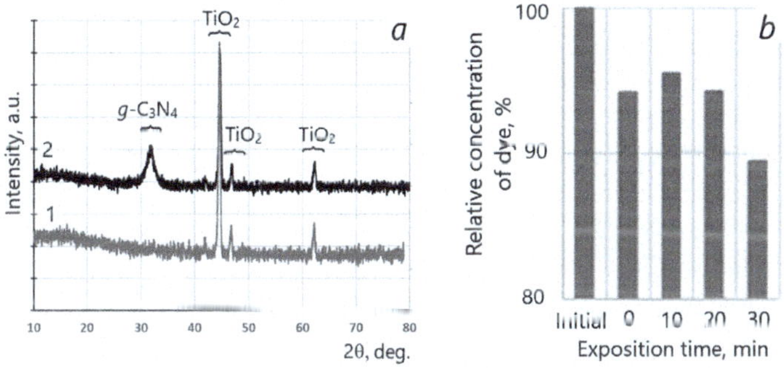

Fig. 8 **a** X-ray diffraction spectra of initial titanium foil (1) and bilayer with 999 nm g-C_3N_4 (2), **b** results of photocatalytic decomposition of Rhodamine B in a water solution with the presence of the bilayer sample

study are noteworthy in terms of semiconductor layer production. The methods used in this study enable the formation of TiO_2/g-C_3N_4 heterostructures. The increase in the thickness of the deposited g-C_3N_4 layer can be provided by the surface preparation of the titanium foil beforehand.

4.2 TiO_2/g-C_3N_4 Heterostructures Obtained by Combined and Template Synthesis

The advantages of using sol–gel synthesis of titania are the relative simplicity of laboratory technology with the possibility of scaling, control of the composition of the final material at the stage of synthesis of sols, and the possibility of forming both continuous and island coatings on various substrates. In addition, there is the possibility of one-stage synthesis of both heterostructures and materials with high values of specific surface due to the addition of a templating agent (polymer, organic compound), which due to lower temperatures of combustion and vaporization than the temperature of the main phase formation, forms voids, cavities and macroscopic grain structure in the finished films [53, 55].

The first variant is adding of prepared powders of g-C_3N_4 to titania sols and formation films from suspensions. We investigated these types of films. The combined coatings were synthesized using the sol–gel method. The g-C_3N_4 powders obtained from different precursors, melamine, and thiocarbamide, were dispersed in acid-free TiO_2 sol according to the recipe described in [23]. The proportion of g-C_3N_4 to TiO_2 was 1:2. The suspensions were homogenized by stirring on a laboratory vortex at 850 rpm for 60 min. The substrates used were glass slides that had been prewashed in a soap and soda solution and degreased in hot isopropyl alcohol. The coatings were formed by dipping the substrates in 1 mL of the suspension, keeping them in a horizontal position for 3 min, and then drying them in air at 60 °C. The coatings were annealed in air at 450 °C.

Thin TiO_2/g-C_3N_4 films were formed on glass in the anatase phase while preserving the graphite-like carbon nitride phase (Fig. 9). The peaks corresponding to TiO_2 were visible on the Raman spectra ($\lambda_{exc} = 532$ nm, using a confocal spectrometer Nanofinder HE). The peaks corresponding to g-C_3N_4 exhibit lower intensity due to the film's low material content. The Gaussian function approximation of this part of the spectrum reveals significant broadening, and the peak positions correspond to a higher degree of structurally distorted graphite-like carbon nitride [56]. We do not provide an approximation part, but notice the positions of the g-C_3N_4 peaks (Fig. 9a).

The study on photocatalytic activity indicates that both TiO_2/g-C_3N_4 heterostructures, derived from thiocarbamide powder and melamine, exhibit comparable efficiency results. The initial concentration of the dye in this case was 2.5 mg/L. For both samples, the photocatalytic degradation efficiency in the presence of the heterostructure is 5% higher than that of the pure TiO_2 film derived from the initial sol at the same annealing temperature (Fig. 9b).

The higher photocatalytic activity of the TiO_2/g-C_3N_4 films compared to the bilayers may be due to the more effective contact of both materials with the contaminated solution, as well as the higher UV-excitation efficiency of this configuration of semiconductors. The primary challenge in utilizing the created heterostructures is associated with the limited quantity of photocatalyst that comes into contact with the polluted solution. This is due to the thinness of the resulting coating. To solve this problem, one can increase the viscosity of the initial sols. This is because increasing

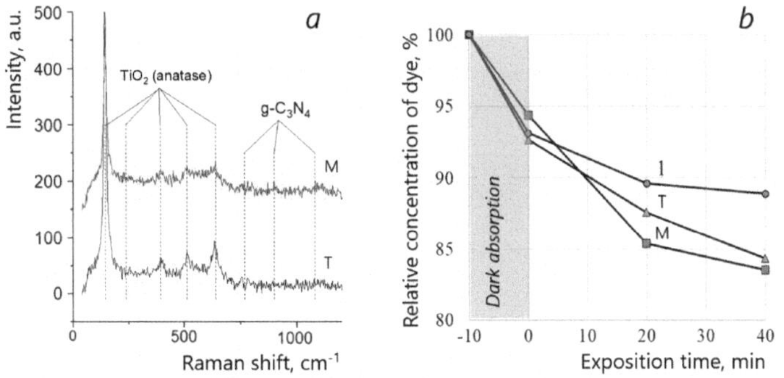

Fig. 9 **a** Raman spectra of combined films on glass containing g-C_3N_4 powder of thiocarbamide (T) and melamine (M), **b** kinetics of degradation of Rhodamine B in aqueous solution under UV-irradiation (365 nm) in the presence of film photocatalysts on glass: TiO_2 (1), TiO_2/g-C_3N_4 heterostructures based on thiocarbamide (T) and melamine (M)

the proportion of g-C_3N_4 powders in the TiO_2 sols-based suspension leads to agglomeration and rapid sedimentation, even with increased time and intensity of stirring, which makes it impossible to form films.

To obtain combined photocatalytic active TiO_2 films, we also used a modification of the classical sol–gel synthesis—template sol–gel synthesis. This variant assumes the possibility of complete thermal decomposition of the so-called "sacrificial" template. Thiocarbamide CS $(NH_2)_2$ (97.5%, GOST 6344–73), which was dissolved in acid-free sol, a precursor of TiO_2, was used as a template material. The concentration of template thiocarbamide varied from 0.012 to 0.096 g per 1 mL of sol and was limited by the wettability of the amount of the powder. Due to the moderate solubility of thiocarbamide in water and alcohols, which were used as the dispersion medium of the sol-base, recrystallization occurred during dissolution, resulting in the formation of a suspension. Homogeneity was achieved by stirring the mixture for 150 min on a laboratory shaker with an orbital shaking path (4.5 mm) IKA VORTEX 4 digital (IKA-Werke GmbH and Co. KG, Germany) at a speed of 850 rpm. Films were formed on 20 × 20 mm substrates cut from wafers of monocrystalline silicon by dipping them in 1 mL of the suspension. After that, the sample was air-dried at 60 °C. The procedure was repeated once more, resulting in a two-layer xerogel coating that includes the template material. Annealing time was 1 h under the different temperatures—550 and 850 °C.

The TiO_2 films formed on single-crystalline silicon are characterized by a developed surface texture with quasi-uniform macrogranularity and characteristic size of homogeneous areas up to 50 μm (Fig. 10). This island coating is a distinctive feature of template synthesis and is particularly relevant for photocatalytic coatings, as it significantly affects liquid microcirculation and enhances purification efficiency [57]. The morphology of the film is formed due to the low solubility and stability of the precursor, thiocarbamide, in the solvents used in the ashes. However, this does not hinder good wettability and the production of visually homogeneous film-forming dispersions.

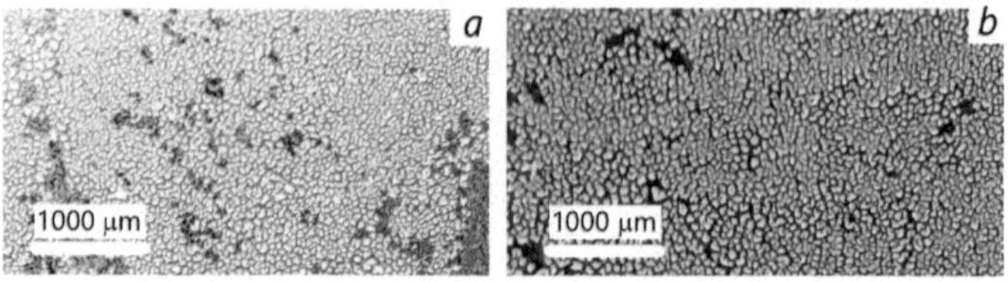

Fig. 10 Appearance of the surface of the samples formed from suspensions with 0.096 g/mL of thiocarbamide after annealing at 550 (**a**) and 850 °C (**b**) for 1 h

At an annealing temperature of 550 °C, a coating consisting mainly of anatase with a small admixture of rutile is formed, regardless of the template concentration. Increasing the annealing temperature up to 850 °C leads to a significant increase in the concentration of rutile in the film, see more details in [58].

The efficiency of photocatalytic degradation of the organic dye Rhodamine B in aqueous solution upon exposure to UV radiation in the presence of TiO_2/Si samples is comparable to that of thin-film samples obtained from the same sols without modification, as well as TiO_2/g-C_3N_4 heterostructures (Fig. 11).

The TiO_2 films exhibit a good level of dark adsorption. For TiO_2 films synthesized at 850 °C, decomposition is better due to the formation of a mixture of phases, including a higher level of UV radiation absorption by rutile. The efficiency of dye decomposition can be increased by the hydrophobicity of rutile, which gives the material's surface self-cleaning properties. This results in an increase in the percentage of dye decomposition for aqueous solutions and a change in the character of the curve of photocatalytic degradation. The obtained results demonstrate the promising development of template synthesis technologies for producing efficient thin-film photocatalysts.

Fig. 11 Kinetic of photocatalytic destruction of Rhodamine B in the aqueous solution under the UV-irradiation in the presence of the TiO_2/Si samples fabricated by annealing at 550 °C (1) and 850 °C (2)

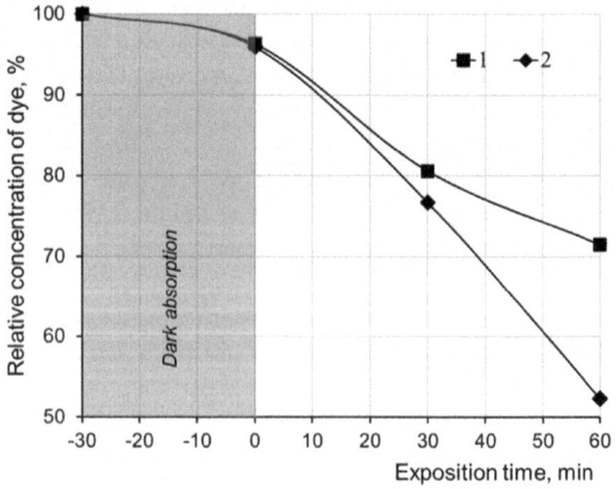

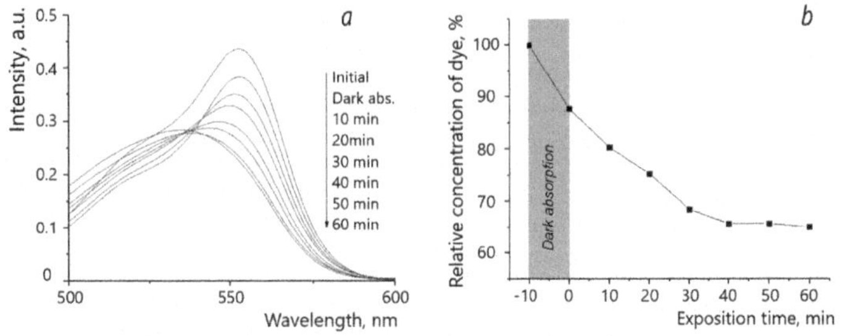

Fig. 12 Results of the study of photocatalytic activity of nanostructured TiO_2: absorption spectra of the test solutions during UV exposition (**a**); change of relative concentration of the test dye solution during the experiment (**b**)

5 Prospects for Further Development of Thin-Film Photocatalytic Active Structures

5.1 *Mono-TiO_2 Ultrathin Films*

Nanostructured TiO_2 films were obtained by chemical (using H_2O_2) and thermal oxidation methods onto titanium foil as previously described. To analyze photocatalytic activity, solutions of the test dye Rhodamine B were exposed to ultraviolet radiation of a mercury lamp with a maximum of 365 nm in the presence of the obtained structures. The initial concentration of the dye was 2.5 mg/L. The samples were kept in the dark before exposure to ensure complete dark adsorption of the dye. Based on the test results, it was determined that the studied structures reached adsorption equilibrium within 10 min.

In the dark mode, the dye is actively adsorbed by the surface of TiO_2, which is obtained through a combination of chemical and thermal oxidation (Fig. 12).

Under ultraviolet radiation, the adsorbed dye undertakes complete degradation through interaction with the photocatalyst surface, as demonstrated by a decrease in dye concentration in solution compared to the initial value. The shift in the position of the main peak also indicates the complete decomposition of Rhodamine B molecules during photocatalysis. Thus, preliminary structurization of the titanium surface by chemical and thermal oxidation allows to obtaining of thin film structures having a developed surface with registerable photocatalytic activity.

5.2 *Hybrid Polymer Materials Functionalized with TiO_2*

TiO_2 is an affordable and photocatalytic active material that shows promise in purifying water and air from organic compounds. One practical application involves using TiO_2 and other photocatalytic active materials as a coating on a structured or porous matrix [50, 59]. This thin-layer coating significantly increases the working surface of the photocatalyst and promotes process intensification. Roughness and porous structure increase the effective working surface of thin film coatings, providing efficient contact with contaminated solutions. Another important task is the development of hydrophobic coatings on which organic molecules in the water or airflow will not be easily adsorbed on the functionalized TiO_2 surface of the filter membrane. However, decomposed contaminants will be effectively removed from the surface of the photocatalytic active material due to its hydrophobic properties. In addition, the free radicals formed during photocatalysis can destroy harmful microorganisms that are resistant to UV light under normal conditions.

The thin film coatings developed in this study have potential applications in various fields that require water and air purification. Additionally, the photocatalyst can be used multiple times without the need for a recovery procedure. The use of film coatings offers several advantages over traditional fine powder photocatalysts, including the elimination of additional steps such as filtration and removal of photocatalysts, as well as the effective organization of flow-through solution purification. The use of film coatings offers several advantages over traditional fine powder photocatalysts, including the elimination of additional steps such as filtration and removal of photocatalysts, as well as the effective organization of flow-through solution purification. Sol–gel coatings have good adhesion, making them suitable for various substrate bases and expanding their application scope.

Specifically, they can produce air-permeable films with self-cleaning and antibacterial surfaces on PET ion-track membranes (PET-TM) coated with TiO_2 nanoparticles [43]. To fabricate photocatalytic active PET-TM systems, ion-track membranes were immersed in a hydrosol containing TiO_2 nanoparticles. The synthesis process is described in detail in [43]. After applying the sol and thermal treatment, a film coating of uniformly distributed sintered TiO_2 nanoparticles formed on the membrane surface. TiO_2 is present not only on the surface of PET-TM but also in the pores. The initial PET-TM has a wetting angle of about 65°. The surface morphology of PET-TM + TiO_2 systems was altered by the application of a nanostructured TiO_2 coating, resulting in increased hydrophobic properties as evidenced by a wetting angle of approximately 95°. Under UV irradiation, the concentration of Rhodamine B in solution decreased by approximately 34% in the presence of PET-TM + TiO_2 systems from 2.5 mg/L (Fig. 13a) [43]. Additionally, these systems exhibited antibacterial activity against standard laboratory strains. Both *E. coli* and *St. aureus* strains showed a decrease in colony numbers when exposed to UV irradiation (Fig. 13b [43]).

Based on the results of this experiment, we recommend using film photocatalysts made from metal oxides for various applications. These materials have inherent antibacterial properties, which make them promising for creating active functional materials. One advantage of the developed photocatalysts is their ability to be reused multiple times, increasing their economic efficiency.

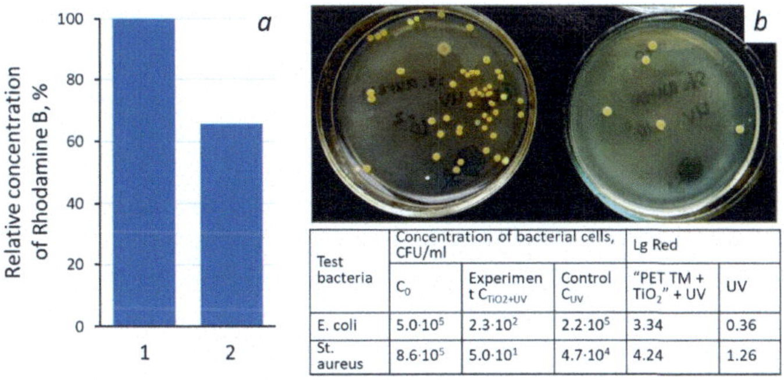

Test bacteria	Concentration of bacterial cells, CFU/ml			Lg Red	
	C_0	Experiment $C_{TiO2+UV}$	Control C_{UV}	"PET TM + TiO_2" + UV	UV
E. coli	$5.0 \cdot 10^5$	$2.3 \cdot 10^2$	$2.2 \cdot 10^5$	3.34	0.36
St. aureus	$8.6 \cdot 10^5$	$5.0 \cdot 10^1$	$4.7 \cdot 10^4$	4.24	1.26

Fig. 13 **a** Relative concentration of Rhodamine B in water solution before (1) and after (2) UV exposure in presence of; **b** antibacterial properties of PET-TM + TiO_2 systems in relation to test bacteria *St. aureus* ATCC 6538 without (left) and with (right) UV irradiation and the results of the antibacterial activity tests (adapted from [43], MDPI, licensed under Creative Common License 4.0 (Creative Commons –Attribution 4.0 International—CC BY 4.0))

6 Conclusion

The development of micro- and nanostructured materials for photocatalytic water purification is an important task due to the high efficiency of such materials, the possibility of destruction of several organic compounds, the absence of residual reaction products to be utilized, as well as, when using immobilized photocatalysts in the form of film coatings, the possibility of organizing a flowing purification process, including in closed-loop systems, and the absence of the need for filtration of the reaction mixture for removal of the water purification process.

The described method for obtaining photocatalytic active TiO_2/(CuO + Cu_2O) structures on copper foil substrates involves a one-step sol–gel synthesis combined with thermal treatment. This results in TiO_2 coatings perforated with copper oxide nanorods (whiskers), allowing both oxide systems to be in contact with contaminated solutions. The analysis of the photocatalytic activity of TiO_2/(CuO + Cu_2O) structures demonstrates the advantages of copper foils as a starting material over grids, as well as the enhancement of photocatalytic activity when using a combination of titanium and copper oxides compared to the results obtained for similar structures on copper grids.

Graphite-like carbon nitride g-C_3N_4 is a promising material for photocatalytic applications. Its layered structure, combined with a band gap of approximately 2.5 eV, makes it suitable for use in coatings for immobilized photocatalysts and the formation of heterostructures on TiO_2. The study of photocatalytic activity revealed that although the thin layers of TiO_2 exhibited photocatalytic activity at small thicknesses (up to 500 nm), the properties of the heterostructure were not improved by the formation of thick g-C_3N_4 films on the surface. To enhance the photocatalytic activity of heterostructures, it is necessary to increase the thickness of the structured layer of TiO_2 and form thinner films of graphite-like carbon nitride with a developed surface.

The photocatalytic activity of combined TiO_2/g-C_3N_4 films obtained from dispersions of prepared powders of graphite-like carbon nitride in TiO_2 sols is higher compared to bilayers. This may be due to the more effective contact of both materials with the contaminated solution and greater UV-excitation efficiency for this configuration of semiconductors.

The study observed the expressed photocatalytic activity in combined coatings made from dispersions based on TiO_2 sol-containing precursors for g-C_3N_4 production, namely melamine and thiocarbamide. The second case demonstrated the possibility of forming continuous coatings with a developed surface on smooth substrates such as glass and monocrystalline silicon. The film obtained from the suspension containing 0.096 g/mL thiocarbamide after annealing at 850 °C resulted in a maximum dye degradation percentage of 48%. This can be attributed to the formation of a developed relief due to the combustion of the template and the presumed optimal ratio of anatase and rutile phases in the TiO_2 film under these conditions. The results indicate promising developments in template synthesis technologies for producing efficient thin-film photocatalysts. Due to their high photocatalytic activity at small thicknesses, TiO_2/g-C_3N_4 heterostructures in the form of combined films are recommended for the development of photocatalytic water purification systems for organic pollutants. They can also be recommended for the additional purification of drinking water in flowing systems.

Acknowledgements The authors express their sincere gratitude to Prof. V. Borisenko, Mr. S. Maksimov, Dr. D. Yakimchuk, Dr. A. Shumskaya, Dr. O. Alisiyonak, Mr. N. Denisov, Dr. D. Tishkevich, Mr. D. Zhigulin, and Dr. O. Korolik for research assistance and stimulating discussion. The authors express gratitude for the State Program of Scientific Research "Photonics and Electronics for Innovations" (project 2.2.3) and the State Program of Scientific Research "Material Science, New Materials and Technologies" (project 2.24–4) of the Republic of Belarus for the partial supporting of the investigation. Authors are grateful for the financial support of the investigation to the Belarusian Foundation of Fundamental Research (Projects T23UZBG-111, F21UZBG-002) and the Ministry of Innovative Development of Uzbekistan (Projects FL-7923051904, MRB-2021-538).

References

1. Ali, S., Nasir, J.A., Dara, N.R., Rehman, Z.: Modification strategies of metal oxide photocatalysts for clean energy and environmental applications: a review. Inorg. Chem. Commun. **145**, 110011 (2022). https://doi.org/10.1016/j.inoche.2022.110011
2. Feng, C., Wu, Z.-P., Huang, K.-W., Ye, J., Zhang, H.: Surface modification of 2D photocatalysts for solar energy conversion. Adv. Mater. **34**(23), 2200180 (2022). https://doi.org/10.1002/adma.20220018
3. Braslavsky, S.E.: Glossary of terms used in photochemistry, 3rd edition (IUPAC Recommendations 2006). Pure Appl. Chem. **79**(3), 293 (2007). https://doi.org/10.1351/pac200779030293
4. Katal, R., Masudy-Panah, S., Tanhaei, M., Farahani, M.H.D.A., Jiangyong, H.: A review on the synthesis of the various types of anatase TiO_2 facets and their applications for photocatalysis. Chem. Eng. J. **384**, 123384 (2020). https://doi.org/10.1016/j.cej.2019.123384

5. Žerjav, G., Žižek, K., Zavašnik, J., Pintar, A.: Brookite vs. rutile vs. anatase: what's behind their various photocatalytic activities? J. Environ. Chem. Eng. **10**(3), 107722 (2022). https://doi.org/10.1016/j.jece.2022.107722
6. Kanakaraju, D., Chandrasekaran, A.: Recent advances in TiO_2/ZnS-based binary and ternary photocatalysts for the degradation of organic pollutants. Sci. Total Environ. **868**, 161525 (2023). https://doi.org/10.1016/j.scitotenv.2023.161525
7. Gopalan, A.-I., Lee, J.-C., Saianand, G., Lee, K.-P., Sonar, P., Dharmarajan, R., Hou, Y.-L., Ann, K.-Y., Kannan, V., Kim, W.-J.: Recent progress in the abatement of hazardous pollutants using photocatalytic TiO_2-based building materials. Nanomaterials **10**, 1854 (2020). https://doi.org/10.3390/nano10091854
8. Khan, H., Shah, M.U.H.: Modification strategies of TiO_2 based photocatalysts for enhanced visible light activity and energy storage ability: a review. J. Environ. Chem. Eng. **11**(6), 111532 (2023). https://doi.org/10.1016/j.jece.2023.111532
9. Horváth, E., Gabathuler, J., Bourdiec, G., Vidal-Revel, E., Benthem Muñiz, M., Gaal, M., Grandjean, D., Breider, F., Rossi, L., Sienkiewicz, A., Forró, L.: Solar water purification with photocatalytic nanocomposite filter based on TiO_2 nanowires and carbon nanotubes. npj Clean Water **5**, 10 (2022). https://doi.org/10.1038/s41545-022-00157-2
10. Dell'Edera, M., Lo Porto, C., De Pasquale, I., Petronella, F., Lucia, C.M., Agostiano, A., Comparelli, R.: Photocatalytic TiO_2-based coatings for environmental applications. Catal. Today **380**, 62 (2021). https://doi.org/10.1016/j.cattod.2021.04.023
11. Baglov, A.V., Radionov, A.A., Chubenko, E.B., Zaycev, V.A., Borisenko, V.E.: Installation for photocatalytic water purification from organic pollution in the flow reactor. Dokl. BGUIR **4**, 45 (2018). (In Russ.) https://doklady.bsuir.by/jour/article/view/995
12. Mai, H., Chen, D., Tachibana, Y., Suzuki, H., Abe, R., Caruso, R.A.: Developing sustainable, high-performance perovskites in photocatalysis: design strategies and applications. Chem. Soc. Rev. **50**, 13692 (2021). https://pubs.rsc.org/en/content/articlelanding/2021/cs/d1cs00684c/unauth
13. Wang, H., Li, X., Zhao, X., Li, C., Song, X., Zhang, P., Huo, P., Li, X.: A review on heterogeneous photocatalysis for environmental remediation: from semiconductors to modification strategies. Chin. J. Catal. **43**(2), 178 (2022). https://doi.org/10.1016/S1872-2067(21)63910-4
14. Samriti, Gupta, R., Ruzimuradov, O., Prakash, J.: Synthesis of multifunctional hybrid semiconductor nanomaterials. In: Prakash, J., Cho, J., Campos Janegitz, B., Sun, S. (eds.) Multifunctional Hybrid Semiconductor Photocatalyst Nanomaterials. Advances in Material Research and Technology. Springer, Cham (2023). https://doi.org/10.1007/978-3-031-39481-2_2
15. Thakur, S., Samriti, Ojha, A., Prakash, J.: Introduction to semiconductor photocatalyst nanomaterials: properties, modifications, and multifunctional applications. In: Prakash, J., Cho, J., Campos Janegitz, B., Sun, S. (eds.) Multifunctional Hybrid Semiconductor Photocatalyst Nanomaterials. Advances in Material Research and Technology. Springer, Cham (2023). https://doi.org/10.1007/978-3-031-39481-2_1
16. Arun, J., Nachiappan, S., Rangarajan, G., Alagappan, R.P., Gopinath, K.P., Lichtfouse, E.: Synthesis and application of titanium dioxide photocatalysis for energy, decontamination and viral disinfection: a review. Environ. Chem. Lett. **21**, 339 (2023). https://doi.org/10.1007/s10311-022-01503-z
17. Velempini, T., Prabakaran, E., Pillay, K.: Recent developments in the use of metal oxides for photocatalytic degradation of pharmaceutical pollutants in water—a review. Mater. Today Chem. **19**, 100380 (2021). https://doi.org/10.1016/j.mtchem.2020.100380
18. Danish, M.S.S., Estrella, L.L., Alemaida, I.M.A., Lisin, A., Moiseev, N., Ahmadi, M., Nazari, M., Wali, M., Zaheb, H., Senjyu, T.: Photocatalytic applications of metal oxides for sustainable environmental remediation. Metals **11**, 80 (2021). https://doi.org/10.3390/met11010080
19. Xia, C., Wang, H., Kim, J.K., Wang, J.: Rational design of metal oxide-based heterostructure for efficient photocatalytic and photoelectrochemical systems. Adv. Func. Mater. **31**(12), 2008247 (2021). https://doi.org/10.1002/adfm.202008247
20. Yuan, Y., Guo, R., Hong, L., Ji, X., Lin, Z., Li, Z., Pan, W.: A review of metal oxide-based Z-scheme heterojunction photocatalysts: actualities and developments. Mater. Today Energy **21**, 100829 (2021). https://doi.org/10.1016/j.mtener.2021.100829

21. Bai, L., Huang, H., Yu, S., Zhang, D., Huang, H., Zhang, Y.: Role of transition metal oxides in g-C_3N_4-based heterojunctions for photocatalysis and supercapacitors. J. Energy Chem. **64**, 214 (2022). https://doi.org/10.1016/j.jechem.2021.04.057
22. Wu, P., Xue, Z., Yu, T., Penkov, O.V.: Transparent self-cleaning coatings: a review. Coatings **13**, 1270 (2023). https://doi.org/10.3390/coatings13071270
23. Baglov, A.V., Denisov, N.M., Borisenko, V.E., Uglov, V.V., Malashevich, A.A.: Photocatalytic activity of nanostructured titania coatings on aluminum substrates. Inorg. Mater. **53**, 1180 (2017). https://doi.org/10.1134/S0020168517110036
24. Denisov, N.M., Baglov, A.V., Borisenko, V.E.: Role of iron and chromium in the photocatalytic activity of titanium dioxide films on stainless steel. Inorg. Mater. **53**, 176 (2017). https://doi.org/10.1134/S0020168517020030
25. Khoroshko, L., Borisenko, V., Baltrukovich, P., Nurmonov, S.: One-step sol-gel fabrication of TiO_2/(CuO+Cu_2O) photocatalysts. J. Sol Gel. Sci. Technol. **108**, 320 (2023). https://doi.org/10.1007/s10971-022-05906-w
26. Ruzimuradov, O., Musaev, K., Mamatkulov, S., Butanov, K., Gonzalo-Juan, I., Khoroshko, L., Turapov, N., Nurmanov, S., Razzokov, J., Borisenko, V., Riedel, R.: Structural and optical properties of sol-gel synthesized TiO_2 nanocrystals: effect of Ni and Cr (co)doping. Opt. Mater. **143**, 114203 (2023). https://doi.org/10.1016/j.optmat.2023.114203
27. Denisov, N.M., Chubenko, E.B., Bondarenko, V.P., Borisenko, V.E.: Black ZnO/C nanocomposite photocatalytic films formed by one-step sol–gel technique. J. Sol-Gel. Sci. Technol. **85**, 413 (2018). https://doi.org/10.1007/s10971-017-4554-1
28. Baglov, A.V, Khoroshko, L.S., Denisov, N.M.: Photocatalytic activity of nanostructured semiconducting metal oxides fabricated on microstructured aluminum foil. In: Borisenko, V.E., Gaponenko, S.V., Gurin, V.S., Kam, C.H. (eds.) Physics, Chemistry and Application of Nanostructures. World Scientific Publishing Co Pte Ltd (2017). https://doi.org/10.1142/9789813224537_0064
29. Deka, R.Ch., Deka, A., Deka, P., Saikia, S., Baruah, J., Sarma, P.: Recent advances in nanoarchitectonics of SnO_2 clusters and their applications in catalysis. J. Nanosci. Nanotechnol. **20**(8), 5153 (2020). https://doi.org/10.1166/jnn.2020.18539
30. Kulak, A.I., Sohrabi Anaraki, H., Gaponenko, N.V., Khoroshko, L.S., Kholov, P.A., Raichyonok, T.F.: Optical characteristics of strontium titanate films formed by sol-gel method on quartz substrates. J. Appl. Spectrosc. **84**, 132 (2017). https://doi.org/10.1007/s10812-017-0439-x
31. Chubenko, E.B., Baglov, A.V., Gnit'ko, A.A., Maksimov, S.E., Borisenko, V.E., Kulak, A.I., Zlotskii, S.V.: Structural and phase transformations in the thiourea/zinc acetate system. Inorg. Mater. **58**(2), 117 (2022). https://doi.org/10.1134/S0020168522020054
32. Singh, G., Kapoor, I.P.S., Dubey, R., Srivastava, P.: Synthesis, characterization and catalytic activity of CdO nanocrystals. Mater. Sci. Eng. B **176**(2), 121 (2011). https://doi.org/10.1016/j.mseb.2010.10.009
33. Morrison, S.R.: The chemical physics of surfaces: overview. Springer New York, NY (2013). https://doi.org/10.1007/978-1-4899-2498-8
34. Ribeiro, G., Ferreira, G., Menda, U.D., Alexandre, M., Brites, M.J., Barreiros, M.A., Jana, S., Águas, H., Martins, R., Fernandes, P.A., Salomé, P., Mendes, M.J., et al.: Sub-Bandgap sensitization of perovskite semiconductors via colloidal quantum dots incorporation. Nanomaterials **13**, 2447 (2023). https://doi.org/10.3390/nano13172447
35. Wang, Z., Chen, C., Ma, W., Zhao, J.: Sensitization of titania semiconductor: a promising strategy to utilize visible light. In: Pichat, P. (ed.) Photocatalysis and Water Purification: From Fundamentals to Recent Applications. Wiley (2013). https://doi.org/10.1002/9783527645404.ch8
36. Punia, P., Bharti, M.K., Chalia, S., Dhar, R., Ravelo, B., Thakur, P., Thakur, A.: Recent advances in synthesis, characterization, and applications of nanoparticles for contaminated water treatment-a review. Ceram. Int. **47**(2), 1526 (2021). https://doi.org/10.1016/j.ceramint.2020.09.050
37. Sousa, V.S., Corniciuc, C., Teixeira, M.R.: The effect of TiO_2 nanoparticles removal on drinking water quality produced by conventional treatment C/F/S. Water Res. **109**, 1 (2017). https://doi.org/10.1016/j.watres.2016.11.030

38. Honda, R.J., Keene, V., Daniels, L., Walker, S.L.: Removal of TiO_2 nanoparticles during primary water treatment: role of coagulant type, dose, and nanoparticle concentration. Environ. Eng. Sci. **31**(3), 127 (2014). https://doi.org/10.1089/ees.2013.0269
39. Denisov, N.M., Baglov, A.V., Borisenko, V.E., Drozdova, E.V.: Preparation and antibacterial properties of composite nanostructures from titanium and copper oxides. Inorg. Mater. **52**, 523 (2016). https://doi.org/10.1134/S0020168516050034
40. Chubenko, E.B., Baglov, A.V., Dudchik, N.V., Drozdova, E.V., Yemelyanova, O.A., Borisenko, V.E.: Estimation of the integral toxicity of photocatalysts based on graphitic carbon nitride in a luminescent test. Kinet. Catal. **63**(2), 166 (2022). https://doi.org/10.1134/S002315842202001X
41. Baglov, A.V., Shevtsova, T.A., Khoroshko, L.S., Dudchik, N.V., Drozdova, A.V., Borisenko, V.E.: Forming immobilized nanostructured titanium dioxide having antibacterial properties. Doklady of the National Academy of Sciences of Belarus **63**(2), 157 (2019). (In Russ.) https://doi.org/10.29235/1561-8323-2019-63-2-157-163
42. Chubenko, E.B., Baglov, A.V., Emeliyanova, O.A., Dudchik, N.V., Drazdova, A.V., Borisenko, V.E.: Formation and antibacterial properties of graphitic carbon nitride. Doklady of the National Academy of Sciences of Belarus **66**(4), 454 (2022). (In Russ.) https://doi.org/10.29235/1561-8323-2022-66-4-454-459
43. Alisiyonak, O., Lavitskaya, A., Khoroshko, L., Kozlovskiy, A.L., Zdorovets, M., Korolkov, I., Yauseichuk, M., Kaniukov, E., Shumskaya, A.: Breathable films with self-cleaning and antibacterial surfaces based on TiO_2-functionalized PET membranes. Membranes **13**, 733 (2023). https://doi.org/10.3390/membranes13080733
44. Zarkov, A.: Sol–gel technology applied to materials science: synthesis, characterization and applications. Materials **17**, 462 (2024). https://doi.org/10.3390/ma17020462
45. Gartner, M., Szekeres, A., Stroescu, H., Mitrea, D., Covei, M.: Advanced nanostructured coatings based on doped tio_2 for various applications. Molecules **28**, 7828 (2023). https://doi.org/10.3390/molecules28237828
46. Nevřivová, L., Zemánek, D.: Study of the mineralogical composition of an alumina–silica binder system formed by the sol–gel method. Materials **16**, 5466 (2023). https://doi.org/10.3390/ma16155466
47. Jiang, X., Herricks, T., Xia, Y.: CuO nanowires can be synthesized by heating copper substrates in air. Nano Lett. **2**, 1333 (2002). https://doi.org/10.3390/ma14051075
48. Linnik, O.P., Zhukovskiy, M.A., Starukh, G.N., Smirnova, N.P., Gaponenko, N.V., Asharif, A.M., Khoroshko, L.S., Borisenko, V.E.: Photocatalytic destruction of tetracycline hydrochloride on the surface of titanium dioxide films modified by gold nanoparticles. J. Appl. Spectrosc. **81**, 990 (2015). https://doi.org/10.1007/s10812-015-0040-0
49. Linnik, O., Khoroshko, L.: Non-porous nitrogen and ruthenium co-doped titania films for photocatalysis. Int. J. Nanosci. **18**, 194043 (2019). https://doi.org/10.1142/S0219581X1940043X
50. Khoroshko, L.S.: Two-dimensional porous anodic alumina for optoelectronics and photocatalytic application. J. Phys. Conf. Ser. **643**, 012110 (2015). https://doi.org/10.1088/1742-6596/643/1/012110
51. Filipič, G., Cvelbar, U.: Copper oxide nanowires: a review of growth. Nanotechnology **23**(19), 194001 (2012). https://doi.org/10.1088/0957-4484/23/19/194001
52. Jalali, E., Maghsoudi, S., Noroozian, E.: A novel method for biosynthesis of different polymorphs of TiO_2 nanoparticles as a protector for Bacillus thuringiensis from Ultra Violet. Sci. Rep. **10**, 426 (2020). https://doi.org/10.1038/s41598-019-57407-6
53. Chubenko, E.B., Denisov, N.M., Baglov, A.V., Bondarenko, V.P., Uglov, V.V., Borisenko, V.E.: Recovery behavior of the luminescence peak from graphitic carbon nitride as a function of the synthesis temperature. Cryst. Res. Technol. **55**, 190063 (2020). https://doi.org/10.1002/crat.201900163
54. Chubenko, E.B., Maximov, S.E., Bui, C.D., Pham, V.T., Borisenko, V.E.: Rapid chemical vapor deposition of graphitic carbon nitride films. Materialia **28**, 101724 (2023). https://doi.org/10.1016/j.mtla.2023.101724

55. Chubenko, E.B., Baglov, A.V., Fedotova, Y.A., Borisenko, V.E.: Synthesis of graphitic carbon nitride on the surface of fe_3o_4 nanoparticles. Inorg. Mater. **57**, 136 (2021). https://doi.org/10.1134/S0020168521020059
56. Wang, H., Zhang. X., Xie, J., Zhang, J., Ma, P., Pana, B., Xie, Y.: Structural distortion in graphitic-C_3N_4 realizing an efficient photoreactivity. Nanoscale **7**, 5152–5156 (2015). https://doi.org/10.1039/C4NR07645A
57. Baglov, A.V., Khoroshko, L.S., Yatskevich, P.A.: Optimization of two-scale oxide structure configuration for photocatalytic applications. Dokl. BGUIR **18**(3), 42 (2020). (In Russ.) https://doi.org/10.35596/1729-7648-2020-18-3-42-48
58. Yauseichyk, M.A., Maksimov, S.E., Khoroshko, L.S., Baglov, A.V., Korolik, O.V., Yakimchuk, D.V., Ruzimuradov, O.N., Mamatkulov, Sh.I.: One-step template sol-gel synthesis of TiO_2-based thin-film photocatalysts. J. BSU Phys. **2**, 58 (2023). (In Russ.) https://doi.org/10.33581/2520-2243-2023-2-58-65
59. Kutuzau, M., Shumskaya, A., Kaniukov, E., Alisienok, O., Shidlouskaya, V., Melnikova, G., Shemukhin, A., Nazarov, A., Kozlovskiy, A., Zdorovets, M.: Photocatalytically active filtration systems based on modified with titanium dioxide PET-membranes. Nucl. Instrum. Methods Phys. Res. Sect. B **460**, 212 (2019). https://doi.org/10.1016/j.nimb.2019.01.028

SERS Application of Titanium Dioxide Based Nanomaterials in Health and Environmental Monitoring

Samriti and Jai Prakash

Abstract This chapter explores the innovative combination of titanium dioxide (TiO_2) nanomaterials with surface enhanced Raman scattering (SERS) technology, highlighting their significant impact on enhancing detection sensitivity, specificity, and versatility. TiO_2, is a versatile semiconductor known for its stability and photocatalytic properties. TiO_2-based SERS substrates are highly effective and capable of detecting a wide range of analytes at ultra-low concentrations. TiO_2 alone and when combined with various plasmonic metals show enhanced SERS sensing due to charge transfer and electromagnetic enhancement mechanisms. In the context of health monitoring, TiO_2-based SERS substrates are enabling rapid and accurate detection of viruses and bacteria, paving the way for early diagnosis. Their ability to provide detailed molecular fingerprints and perform real-time, non-destructive analysis is revolutionizing clinical diagnostics and research. In environmental monitoring, TiO_2-based SERS platforms are enhancing the detection and quantification of volatile organic compounds, dye molecules, and pharmaceuticals in diverse environmental matrices. This advancement supports more effective pollution control and environmental protection efforts.

Keywords TiO_2 · SERS applications · Detection · Charge transfer · Health and environmental monitoring

1 Introduction

Surface Enhanced Raman Spectroscopy (SERS) has emerged as a powerful analytical tool that significantly enhances the Raman scattering signals of molecules adsorbed on nanostructured surfaces [1]. This enhancement, often by several orders of magnitude, enables the detection of analytes at extremely low concentrations, providing detailed molecular fingerprints crucial for various applications in health

Samriti · J. Prakash (✉)
Department of Chemistry, National Institute of Technology Hamirpur, Himachal Pradesh, Hamirpur 177005, India
e-mail: jaip@nith.ac.in

J. Prakash et al. (eds.), *Titanium Dioxide-Based Multifunctional Hybrid Nanomaterials*, Engineering Materials, https://doi.org/10.1007/978-3-031-80625-4_6

and environmental monitoring [2, 3]. The efficacy of SERS relies heavily on the properties of the substrate used, traditionally dominated by plasmonic metals and more recently, by semiconductor metal oxides.

Plasmonic metals, such as gold (Au) and silver (Ag), are well-known for their exceptional ability to enhance electromagnetic fields at their surfaces through localized surface plasmon resonances (LSPR). These materials have been extensively used to create SERS substrates due to their strong and tunable plasmonic properties, which contribute to significant signal enhancement. However, the high cost, potential toxicity, and chemical instability of these metals in certain environments have prompted the exploration of alternative materials [4–6]. In recent years, semiconductor metal oxides as novel SERS substrates had gained focus in nanomaterials research, due to their tunable properties, such as light trapping, band gap modulation, stoichiometry and morphological control, size-dependent plasmon/exciton resonances, and charge transfer (CT) between semiconductors metal oxide and analyte molecules. Another notable feature of metal oxide is the presence of internal vacancies or defects, which play a crucial role in enhancing SERS and photocatalytic activity [7]. The ability to easily modify and tailor these properties has made semiconductor nanomaterials highly attractive for various technological applications, including photocatalysis, field-effect transistors, electrical circuit components, solar cells, and sensing devices, particularly in SERS [7, 8]. Semiconductor metal oxides, particularly titanium dioxide (TiO_2), have gained considerable attention in SERS research as promising alternatives to traditional plasmonic materials. TiO_2 stands out due to its excellent photocatalytic properties, chemical stability, and ability to undergo band gap modulation and morphological control [9]. The presence of internal vacancies or defects in TiO_2 plays a crucial role in enhancing its SERS activity by facilitating CT processes between the semiconductor and the analyte molecules. This CT mechanism, combined with TiO_2 inherent properties, enables the creation of highly effective SERS substrates [9, 10].

Furthermore, TiO_2 ability to be functionalized with plasmonic nanoparticles (NPs), such as Au or Ag, enhances the electromagnetic field effects, leading to greater SERS signal enhancement [11–13]. The tunability of TiO_2 properties through methods such as doping, surface modification, and morphological control allows for the optimization of SERS substrates tailored for specific applications [10, 13]. This versatility has enabled TiO_2-based SERS substrates to achieve remarkable sensitivity and specificity in detecting a wide range of analytes, from environmental pollutants to biological molecules. By leveraging the unique properties of TiO_2, researchers can develop advanced SERS platforms that are not only highly effective but also stable and reproducible, addressing many of the current challenges in the field. These hybrid systems benefit from the enhanced electromagnetic field effects provided by the plasmonic NPs and the efficient CT facilitated by the TiO_2 matrix. The resulting synergetic effects lead to superior SERS performance, characterized by high sensitivity and specificity.

This chapter provides a comprehensive overview of the role of TiO_2-based nanomaterials in advancing SERS technology. It covers the fundamental principles of TiO_2 and SERS, explores recent innovations and applications, and addresses current challenges and future directions in the field. By delving into the transformative impact of TiO_2-based SERS in both health and environmental monitoring, this chapter aims to highlight the significant advancements and potential future developments in this dynamic area of research.

2 Fundamental Principles of TiO_2

TiO_2 is a versatile semiconductor metal oxide with significant applications across various fields due to its unique physical and chemical properties as shown in Fig. 1a. It exists primarily in three crystalline forms—rutile, anatase, and brookite—with rutile being the most stable and widely used in industry, while anatase is favored for its higher surface area and photocatalytic activity [14]. TiO_2 has a wide band gap of approximately 3.0–3.2 eV, which enables it to absorb ultraviolet (UV) light effectively and makes it suitable for photocatalytic applications, such as environmental remediation and water purification [13]. Its high refractive index enhances its optical properties, making it useful in coatings and pigments. TiO_2 photocatalytic activity is driven by its ability to generate electron–hole pairs under UV light, facilitating redox reactions that break down organic pollutants. Although TiO_2 primarily absorbs UV light due to its wide band gap, modifications such as doping with other elements can extend its light absorption into the visible spectrum [15]. Additionally, TiO_2 surface area and reactivity can be enhanced by controlling its morphology or introducing defects like oxygen vacancies, improving its performance in catalytic and sensing applications [16]. TiO_2 is also a key component in dye-sensitized solar cells (DSSCs) and various sensing devices due to its stability and reactivity. The ability to modify TiO_2 through doping or combining it with other materials further enhances its functionality, making it a valuable material in technology and environmental applications (Fig. 1b) [14].

3 Basic Principles of SERS

SERS is a powerful analytical technique that significantly enhances the Raman scattering of molecules adsorbed on rough metal surfaces or NPs. At its core, the Raman effect occurs when light interacts with a molecule, causing most photons to scatter elastically (Rayleigh scattering) while a small fraction scatters inelastically (Raman scattering), leading to a shift in the wavelength of the scattered light. This shift provides a unique molecular fingerprint, revealing information about the molecule's vibrational modes and chemical structure [1, 7, 17]. SERS enhances this Raman signal through two main mechanisms: electromagnetic enhancement and chemical enhancement. The dominant mechanism, electromagnetic enhancement, involves the excitation of LSPR in metal NPs such as Au, Ag, or Cu. This LSPR amplifies the local electromagnetic field at the nanoparticle surface, dramatically increasing the

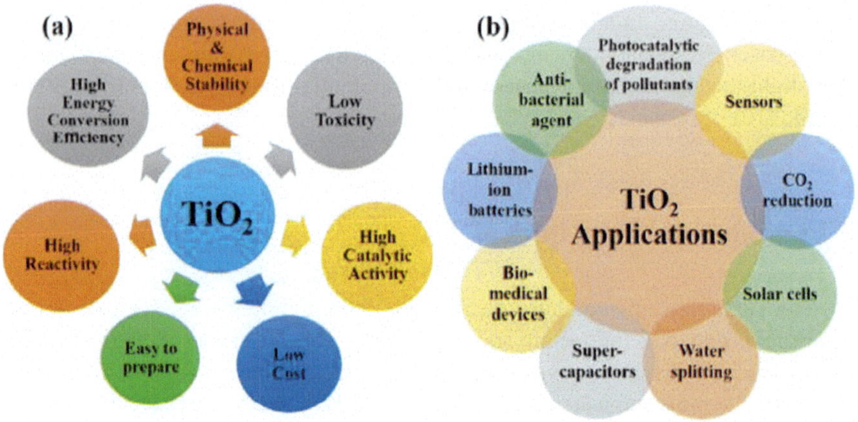

Fig. 1 **a** Diverse properties and **b** various applications of TiO_2 nanostructures. Reprinted with permission from [15]. Copyright (2021), Elsevier

Raman signal of nearby molecules. Chemical enhancement, although typically less significant, involves CT interactions between the adsorbed molecules and the metal surface, further contributing to the signal enhancement. Together, these mechanisms enable SERS to detect and analyze molecules at very low concentrations, making it a valuable tool in various fields, including health diagnostics and environmental monitoring [7].

The development of semiconductor-based SERS substrates—more especially, noble metal-free SERS substrates—has attracted a lot of interest lately. Utilizing the characteristics of Raman spectroscopy, research on metal oxide semiconductors produces excellent semiconductor-based SERS substrates. As mentioned in the introduction section, the favorable properties of metal oxide semiconductors have made them particularly promising noble metal-free substrates. When compared to noble metal substrates, semiconductor-based SERS substrates achieve SERS enhancement factors (EF) above 10^6 by facilitating CT between the adsorbed analyte molecules and the substrate [18, 19]. Noble metals show morphology-dependent LSPR and SERS amplification, as well as generally higher SERS EFs. But as the introduction explained, due to the difficulties in producing SERS substrates using noble metals, scientists are now looking into semiconductor-based SERS-active materials as viable substitutes.

3.1 *Mechanism of SERS in Noble Metal/ TiO_2 Nanohybrid*

The mechanism of SERS in noble metal/TiO_2 nanohybrids combines the unique properties of both materials to achieve significant Raman signal enhancement. Noble metals, such as Ag or Au, are renowned for their strong LSPRs, which occur when

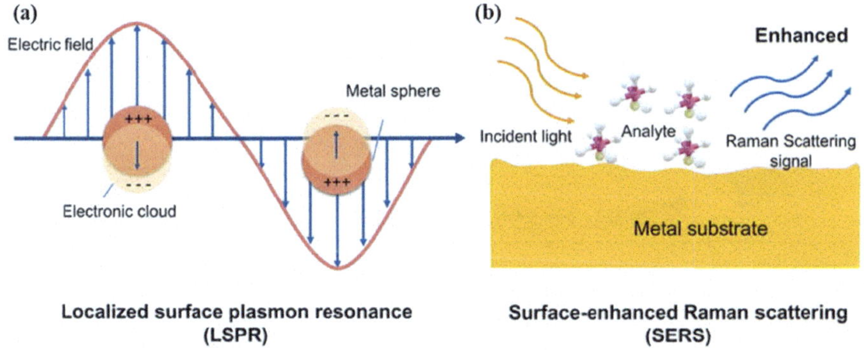

Fig. 2 **a, b** The schematic illustrating LSPR and SERS scattering. Reprinted with permission from [22]. Copyright (2022), IMR

light induces collective oscillations of free electrons on their surfaces as shown in Fig. 2. These plasmonic resonances create highly localized electromagnetic fields that substantially amplify the Raman scattering of molecules adsorbed on or near the metal surface [20]. The intensity and distribution of these plasmonic fields are influenced by the shape, size, and arrangement of the noble metal NPs, with features like sharp tips or narrow gaps creating "hot spots" where the fields are highly concentrated. TiO_2 enhances this effect by facilitating CT between the noble metal and the adsorbed analyte molecules [21]. Although TiO_2 does not exhibit strong plasmonic behavior itself, it can modify the electronic environment of the noble metal and improve the overall local electromagnetic field. This synergy between the noble metal's plasmonic enhancement and TiO_2 CT interactions leads to even greater Raman signal amplification, making noble metal/TiO_2 nanohybrids highly effective for various sensing applications [20]

3.2 *Mechanism of SERS in Noble Metal Free/ TiO_2 Nanohybrid*

SERS in noble metal-free/TiO_2 nanohybrids rely on the unique properties of TiO_2 combined with other materials to achieve Raman signal enhancement. In these nanohybrids, TiO_2 plays a crucial role in amplifying the Raman signal through several mechanisms. TiO_2 can facilitate CT between the analyte molecules and the semiconductor surface, enhancing the interaction and boosting the Raman signal [7]. This approach offers an alternative to noble metal-based SERS substrates, providing a cost-effective and versatile solution for various sensing applications while benefiting from the stability and functional advantages of TiO_2.

Surface Plasmon Resonance (SPR) or Mie Resonance, CT resonance, and molecular or excitonic resonance are the three forms of resonances that contribute to the

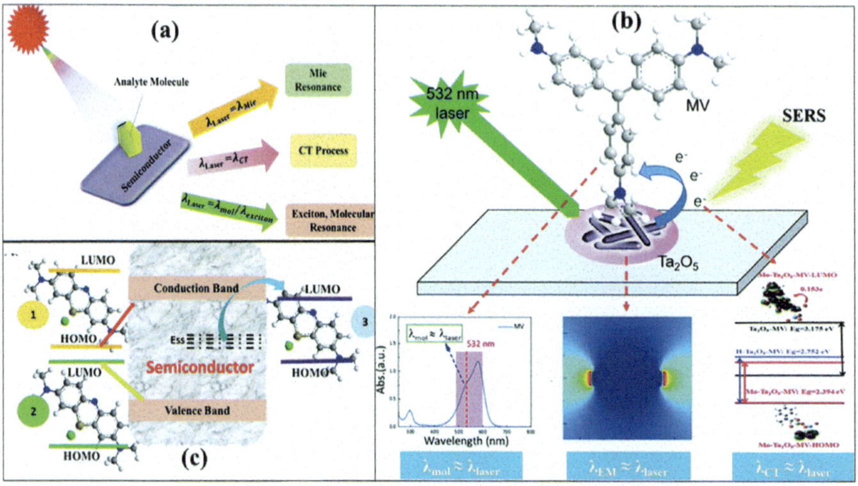

Fig. 3 **a** SERS mechanism in semiconductors. Reprinted with permission from [7]. Copyright (2022), RSC **b** diagram showing the mechanism of coupled resonance on the Ta_2O_5 SERS substrate. Reprinted with permission from [23]. Copyright (2019), Wiley **c** the CT pathways in semiconductor–molecule systems. Reprinted with permission from [7]. Copyright (2022), RSC

Herzberg-Teller vibronic coupling mechanism that drives SERS, rather than a single resonance process (Fig. 3a, b). The metal and the molecule are seen as one cohesive system in this context. Among these resonances, the molecule itself produces molecular resonance, the metal itself causes LSPR resonance, and the combined interactions of the molecule and metal provide CT resonance.

In the CT mechanism, the analyte molecule is chemically adsorbed onto the substrate surface, leading to a change in the molecule's polarizability (α) due to the generation of new electronic states. This chemisorption modifies the Raman spectra of the adsorbed species compared to the free molecules. The primary enhancement in SERS with semiconductor substrates arises from the direct CT process between the analyte and the substrate. Effective CT requires that the energy levels of the analyte and the substrate are properly aligned. Figure 3c illustrates the various ways in which the CT process can take place: (i) electrons from the adsorbed molecule's HOMO can be excited to the metal oxide semiconductor's conduction band (CB); (ii) electrons from the semiconductor's valence band (VB) can be photoexcited to the adsorbed molecule's LUMO; (iii) electrons from the semiconductor's VB can be excited to sub-band gap energy levels of the semiconductor, which can then be transferred to the LUMO of the adsorbed molecule. Furthermore, there are two less well-known CT pathways that involve electron transfer from a CT complex to the semiconductor's CB, where a strong link between the molecule and substrate creates a CT complex, and electron transfer from the semiconductor's CB to the HOMO of the adsorbed molecule [7].

4 TiO_2-Based Nanomaterials in Health and Environmental SERS Sensing

TiO_2-based nanomaterials have emerged as a pivotal component in the advancement of SERS for health and environmental sensing. For environmental monitoring, TiO_2-based SERS platforms significantly improve the detection and quantification of volatile organic compounds (VOCs), dye molecules, and pharmaceuticals in various environmental matrices. This progress aids in more effective pollution control and environmental protection efforts. In health monitoring, TiO_2-based SERS substrates facilitate the rapid and precise detection of viruses and bacteria, paving the way for early diagnosis.

4.1 *SERS Sensing of VOCs*

The detection of VOCs is crucial due to their significant impact on human health, the environment, and various industrial processes. TiO_2 has gained attention in SERS for sensing VOCs due to its advantageous properties discussed in the above sections. Recently, Xu et al. [24] developed an advanced SERS-based device for the sensitive and selective on-site screening of volatile aldehyde biomarkers for preclinical lung cancer diagnosis (Fig. 4a). Utilizing an Au-coated TiO_2 nanochannel membrane (Au/TiO_2) covered with ZIF-8, the design enhances Raman signal amplification and gaseous analyte capture through interfacial nanocavities. Open membrane ends facilitate efficient gas diffusion. Benzaldehyde detection, a lung cancer marker, is achieved using a Schiff base reaction with a Raman-active probe (4-aminothiophenol) on Au NPs, enabling trace detection. Combining machine learning with Raman spectroscopy further ensures accurate identification, promising a robust desktop SERS analysis system for point-of-care testing.

Furthermore, Wang al. [25] reported TiO_2 Nanotubes (NTs) decorated with AuNPs (Fig. 4b, c) and encapsulated in a metal–organic framework (MOF) for enhanced SERS detection of gaseous VOCs, specifically acetophenone. The combination of the high surface area of TiO_2 NTs, the plasmonic properties of Au NPs, and the porous structure of the MOF facilitates the adsorption of VOCs. This setup not only increases the number of active sites for Raman scattering but also enhances the overall sensitivity upto10^{-11} M (Fig. 4d, e) and selectivity of the SERS detection for gaseous VOCs like acetophenone. Li et al. [26] reported a recyclable PDMS/TiO_2/Ag hybrid substrate. The substrate demonstrated a limit of detection (LOD) as low as 5.8×10^{-9} M for 4-aminothiophenol after eight recycling cycles. The intrinsic Raman signal from the PDMS matrix improved detection linearity, enhancing the coefficient of linear correlation (R^2) from 0.967 to 0.997. The SERS substrate exhibited high sensitivity with a moderate EF of approximately 4.5×10^6 and good reproducibility, achieving a relative standard deviation (RSD) of 6.66% during tests.

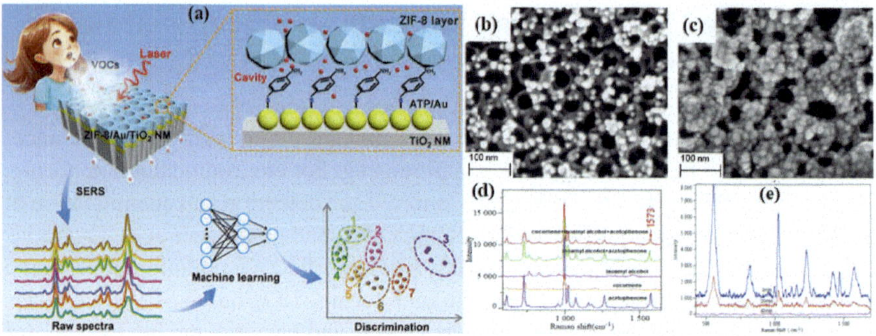

Fig. 4 **a** Au/TiO$_2$ SERS device fabricated for sensing of VOCs. Reprinted with permission from [24]. Copyright (2023), ACS **b, c** SEM image of TiO$_2$NTs/AuNPs and TiO$_2$NTs/AuNPs@ZIF−8 nanocomposites. **d** The SERS spectra of common exhaled VOCs and their mixtures. **e** SERS spectra of 10^{-7} M acetophenone on the TiO$_2$NTs/AuNPs@ZIF-8 surface under UV light irradiation at various times. Reprinted with permission from [25]. Copyright (2023), MDPI

Long-term stability was confirmed, with only an 8.02% decrease in detection efficiency over 70 days. The substrate facilitated practical quantitative detection of gaseous 2-nitroaniline (2-NAT), achieving a notable ultra-low limit of 1.2×10^{-7} M, highlighting its potential for real-time monitoring of hazardous gas molecules.

The advancements in TiO$_2$-based SERS substrates for VOCs detection have shown remarkable progress, offering enhanced sensitivity, selectivity, and stability. These developments promise robust, accurate, and reliable point-of-care testing systems and real-time monitoring solutions, underscoring the critical role of SERS technology in health diagnostics, environmental monitoring, and industrial applications.

4.2 SERS Sensing of Dyes

SERS is highly significant for the sensing of dyes due to its exceptional sensitivity, molecular specificity, and ability to perform rapid, non-destructive analysis. By enabling detection at extremely low concentrations, SERS is crucial for applications such as environmental monitoring and quality control, where trace amounts of dyes need to be identified.

Recent our group reported TiO$_2$ nanorods (NRs) and Ta-doped TiO$_2$ NRs, for SERS sensing of methylene blue dye molecules. The TiO$_2$ NRs with a band gap of 2.95 eV were obtained using TiO$_2$ nano powder as a precursor, optimized by soaking in HCl solution for a specific time. Similarly, highly uniform Ta-doped TiO$_2$ NRs with a band gap of 2.58 eV were synthesized as shown in Fig. 5a. Ta doping significantly improved the optical and CT properties of TiO$_2$ NRs, resulting in increased SERS activity due to additional defect/energy levels below the conduction band of TiO$_2$ as shown in Fig. 5b. The overall enhancement can be attributed to the synergistic effect

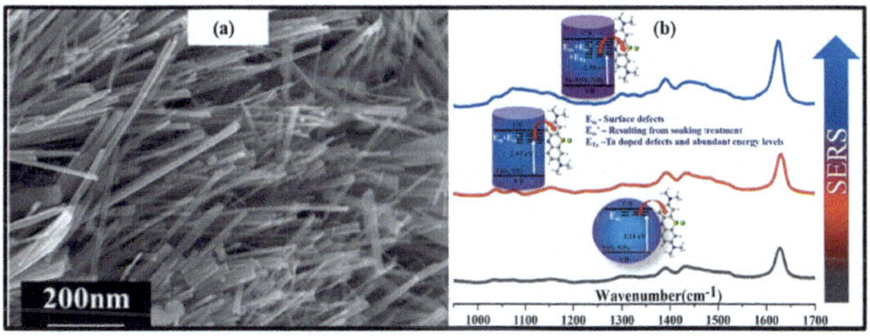

Fig. 5 **a** SEM image of Ta doped TiO_2 NRs. **b** Schematic showing SERS in TiO_2 NPs, TiO_2 NRs and in Ta-doped TiO_2 NRs. Reprinted with permission from [10]. Copyright (2022), Elsevier

of surface defects created by acid soaking treatment and the doped defects/energy levels introduced by Ta doping [10].

Ziad et al. [27] reported the fabrication of TiO_2/silver nanoprisms (SNP) hybrid arrays for sensing and catalytic applications of organic pollutants. TiO_2/SNP hybrid arrays were prepared with different weight ratios for sensing and degrading organic pollutants. The calculated EF for the TiO_2/SNP nanoarrays was found to be 1.58 $\times$ 10^9, indicating a significant enhancement in Raman signal intensity of methylene blue. The spot-to-spot reproducibility analysis of the TiO_2/SNP arrays showed excellent repeatability, with relative standard deviation (RSD) values of 11.7% for Rhodamine B and 10.9% for Methylene Blue dyes (Fig. 6a–d). The TiO_2/SNP substrates demonstrated a picomolar detection limit for organic dyes, showcasing their high sensitivity and efficiency in SERS applications. The high SERS performance of the TiO_2/SNP hybrid arrays is attributed to the electromagnetic enhancement provided by the silver nanoprisms. The sharp edges of the silver nanoprisms create strong light-trapping properties, enhancing the surface plasmon resonance effect and generating SERS active hotspots. Additionally, CT mechanisms can occur at the interface between the TiO_2 particles and the silver nanoprisms, further contributing to the SERS enhancement shown in Fig. 6e.

A highly ordered TiO_2 NT modified with plasmonic silver NPs (Ag) through a wet chemical method were reported by Ramachandran et al. [28]. The TiO_2/Ag nanostructures leverage both CT processes and electromagnetic field enhancement due to the synergy between the semiconductor TiO_2 and the plasmonic Ag NPs, creating a powerful SERS bi-component system. To optimize the density of Ag NPs on TiO_2 NTs, the photoinduction irradiation time was varied between 10 and 40 min. Structural and morphological properties of the TiO_2/Ag nanostructures were thoroughly validated. SERS performance was tested using rhodamine 6G at an initial concentration of 10^{-6} M, with the highest enhancement observed for samples irradiated for 20 min. This bicomponent SERS system demonstrated excellent reproducibility, with Raman mapping studies showing a low RSD of 9.8%. Additionally, it achieved pico-molar level detection of rhodamine 6G (10^{-12} M), showcasing the effective combination of CT and electromagnetic enhancement in detecting trace analytes.

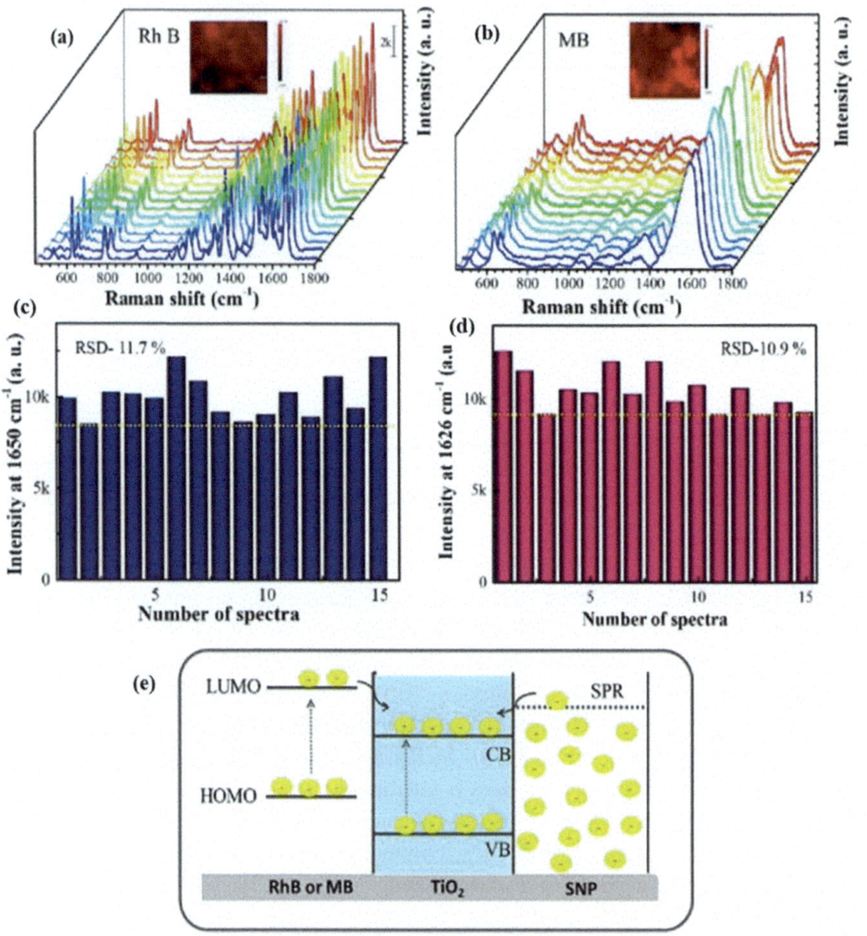

Fig. 6 **a–d** SERS spectra of methylene blue and rhodamine 6G dye molecules over TiO_2/SNP hybrid. **e** Mechanism of SERS in TiO_2/SNP hybrid. Reprinted with permission from [27]. Copyright (2023), Elsevier

The advancements in SERS sensing of dyes through various TiO_2-based nanostructures have showcased exceptional sensitivity and performance. These developments highlight the effective synergy of electromagnetic and CT mechanisms, offering highly reproducible and efficient SERS substrates. Such innovations promise substantial improvements in applications requiring precise and sensitive detection, including environmental monitoring and quality control.

4.3 SERS Sensing of Pharmaceutical Drugs

SERS plays a pivotal role in pharmaceutical drug sensing due to its high sensitivity, specificity, and rapid detection capabilities. It ensures drug quality and safety by providing precise molecular identification and detecting trace impurities, which is crucial for maintaining the integrity of pharmaceutical products. A semiconductor SERS substrate developed by Li et al. [29] having densely growing anatase TiO_2 NPs on the surface of cotton fabric using a filtration-hydrothermal method. This substrate is designed for in situ detection of trace analytes, such as antibiotic residues (quinolone antibiotic). The substrate exhibits high SERS activity, enabling the detection of analytes at concentrations as low as 10^{-7} M, which is significant for trace analysis. Additionally, the Cot/TiO_2 substrate demonstrates exceptional stability, maintaining its SERS activity even after 40 days of storage and withstanding corrosive environments, including strong acids and alkalis, as well as extreme temperatures ranging from − 20 to 100 °C. The substrate also shows excellent spectral repeatability, with low relative standard deviation values for SERS intensities, indicating reliable performance across multiple measurements. Furthermore, it exhibits good recyclability, retaining its effectiveness after several uses, which is crucial for practical applications in real-world settings.

Furthermore, Cao et al. [30] developed TiO_2 spheres coated with an ultrathin nitrogen-doped carbon (NC) shell and loaded with Ag NPs. This design successfully narrows the bandgap to 2.34 eV and enhances fluorescence quenching ability. SEM results revealed that the TiO_2@NC composites have a three-dimensional spherical structure. SEM images also confirmed the presence of Ag NPs on the TiO_2@NC/Ag composites, showing that the spherical morphology of the TiO_2@NC was maintained while a significant number of nanocrystals appeared on the surface (Fig. 7a). Additionally, HRTEM images revealed lattice fringes corresponding to the (101) plane of the anatase phase TiO_2, indicating the crystalline nature of the material. The presence of small Ag NPs attached to the surface of the TiO_2@NC spheres demonstrates a strong self-assembly effect (Fig. 7b). TiO_2@NC achieved a high enhancement factor of 8.0×10^6. The TiO_2@NC/Ag composites exhibited excellent sensitivity for detecting phenformin hydrochloride (PHE) in human urine, with a low detection limit of 5 nM and a linear range from 10^{-5} to 5×10^{-8} M (Fig. 7c, d). The SERS mechanism is detailed in Fig. 7e, where photogenerated electrons from the TiO_2@NC composite are transferred to the LUMO of the analyte molecules. The presence of Ag NPs facilitates this charge transfer, resulting in a stronger interaction between the analyte and the substrate, which further enhances the SERS signal. The combination of these mechanisms results in significantly improved sensitivity and detection capabilities for the TiO_2@NC/Ag composites.

SERS substrates for pharmaceutical drug sensing demonstrate significant progress in enhancing sensitivity, stability, and practical application. These developments highlight the growing potential of SERS technology in ensuring drug quality and safety, offering robust solutions for precise molecular identification and trace impurity detection in real-world settings.

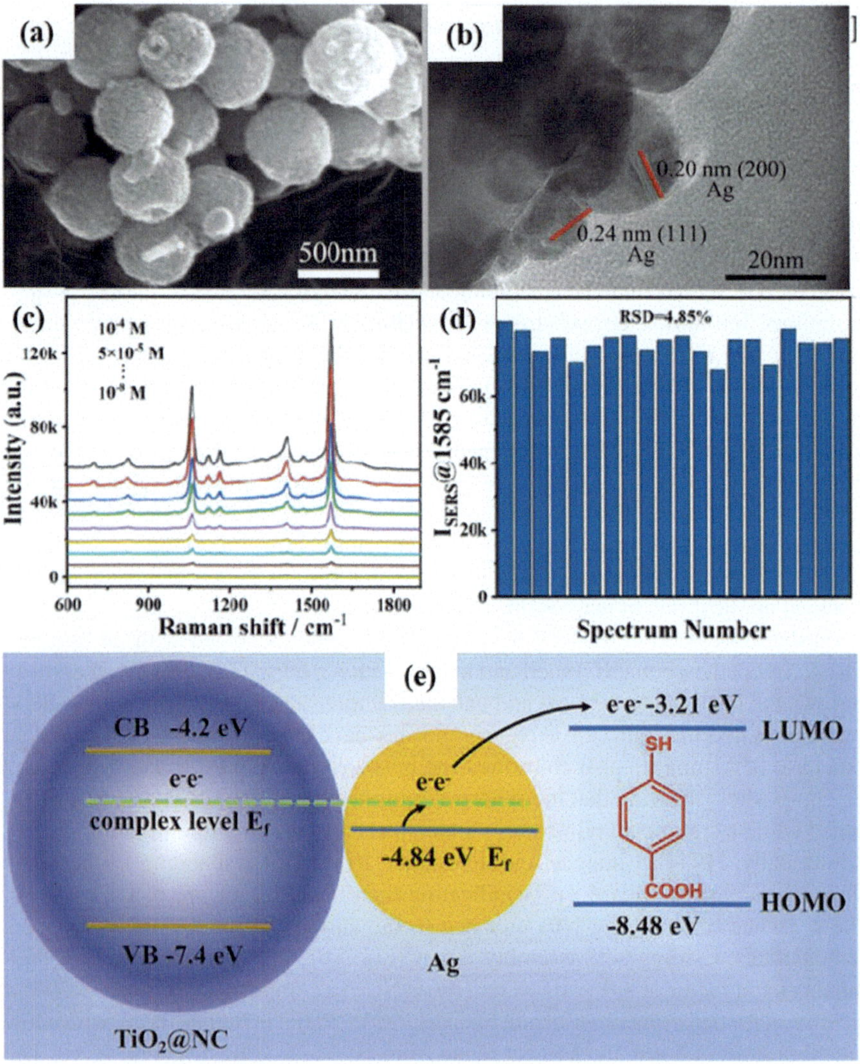

Fig. 7 **a** SEM and **b** HRTEM images of TiO_2@NC **c** SERS spectra of 4-MBA at various concentrations. **d** Uniformity in the SERS detection of 4-MBA. **e** Schematic showing SERS mechanism of TiO_2@NC. Reprinted with permission from [30]. Copyright (2022), Elsevier

4.4 SERS Sensing of Viruses/Bacteria

SERS is highly important in the sensing of viruses and bacteria due to its exceptional sensitivity, specificity, and rapid detection capabilities. SERS can detect and identify pathogens at extremely low concentrations, which is crucial for early diagnosis and prevention of disease spread. Its ability to provide distinct molecular fingerprints allows for the precise differentiation between various types of viruses and bacteria. The rapid detection offered by SERS enables timely responses in clinical settings, improving patient outcomes and helping control outbreaks. Additionally, the non-destructive nature of SERS preserves samples for further analysis, enhancing its utility in research and diagnostic applications. Hwang et al. [31] presents a novel method for the direct and label-free detection of SARS-CoV-2 in respiratory aerosols using a highly adsorptive Au-TiO_2 nanocomposite SERS face mask coupled with an ablation-assisted autoencoder (Fig. 8a). The Au–TiO_2 mask preconcentrates and captures oronasal aerosols, enhancing SERS signal intensities by 47% compared to Au nano islands. It detects SARS-CoV-2 spike proteins at 100 pM in artificial aerosols and quantifies SARS-CoV-2 lysates in aerosols with over 98% accuracy using deep learning, making it a promising tool for breath biopsy and early disease diagnosis.

Bacterial contamination in diets poses significant health risks, necessitating the development of rapid and sensitive methods for differentiating benign from pathogenic bacteria. Li et al. [32] introduces a SERS platform using self-standing Au/TiO_2 NTs, which discriminates bacteria through hydrophobic interactions between volatile metabolites from pathogenic bacteria and the SERS substrate shown in Fig. 8b. Modified with hydrocarbon moieties, the substrate enhances binding and enrichment of these metabolites, causing a notable decrease in Raman signals in the silent region (1800–2800 cm^{-1}). Using Escherichia coli (*E. coli*) as a model, the platform distinguishes benign from pathogenic bacteria within 4 min and detects E. coli at concentrations as low as 3×100 cells per mL. This technique, based on bacterial volatile metabolites, offers potential for guiding antibiotic treatments and ensuring rapid and reliable bacterial therapy.

SERS technology significantly advances the detection and differentiation of viruses and bacteria, offering unparalleled sensitivity and specificity. These innovations highlight SERS's role in early disease detection, outbreak control, and ensuring food safety, emphasizing its value in both clinical and research settings.

5 Future Challenges and Scope of TiO_2-based Nanomaterials in SERS Sensing

TiO_2 based materials are significantly advancing the field of SERS, particularly in health and environmental monitoring. Despite their promising capabilities, several challenges and areas for future development remain

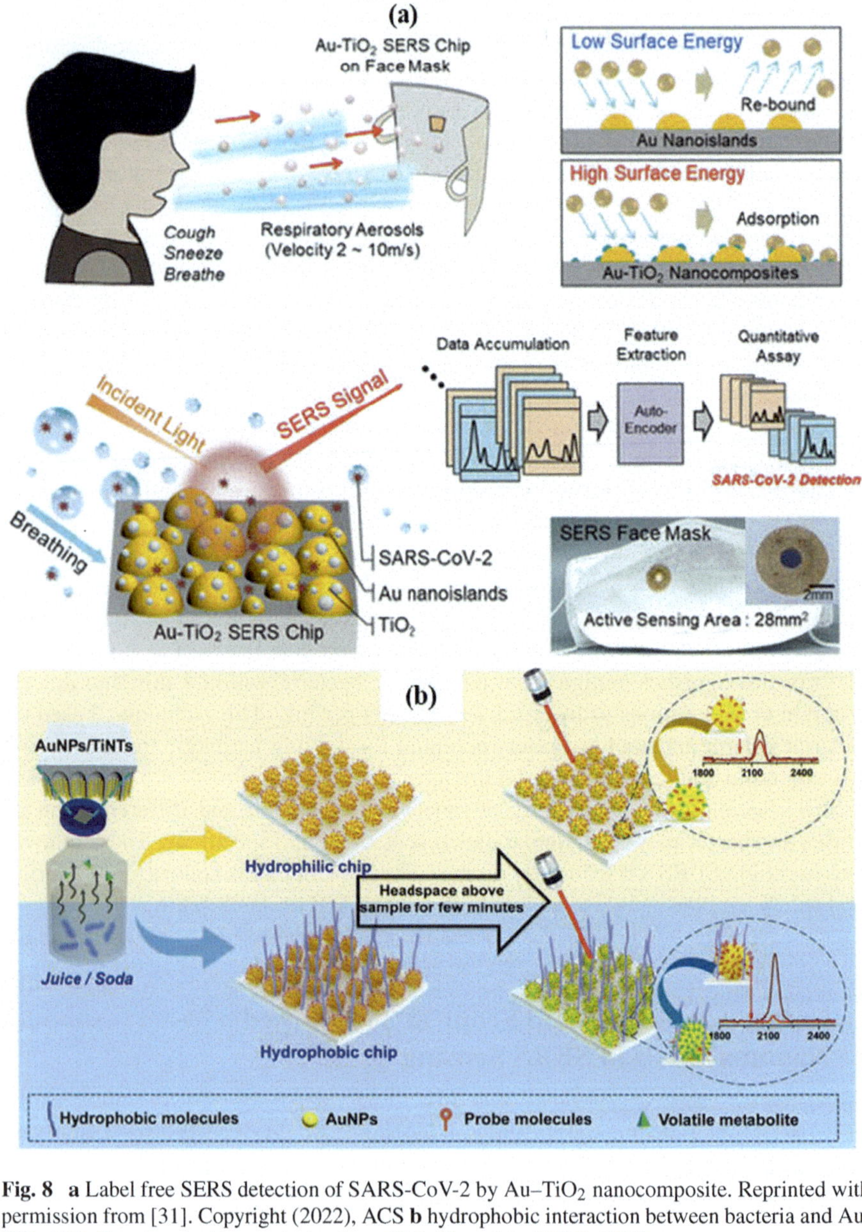

Fig. 8 **a** Label free SERS detection of SARS-CoV-2 by Au–TiO_2 nanocomposite. Reprinted with permission from [31]. Copyright (2022), ACS **b** hydrophobic interaction between bacteria and Au/TiO_2 NTs. Reprinted with permission from [32]. Copyright (2023), RSC

5.1 Future Challenges

Reproducibility and Consistency: Ensuring uniformity in the production of TiO_2-based SERS substrates is crucial. Variations in nanoparticle size, shape, and distribution can lead to inconsistent signal enhancements, affecting the reliability of the results.

Long-term Stability: The durability of TiO_2 substrates under various environmental conditions, such as humidity, temperature changes, and exposure to chemicals, needs improvement. Ensuring stable performance over time is essential for practical applications.

Scalability and Cost-Effectiveness: Developing scalable manufacturing processes that are also cost-effective is a major challenge. Current methods for fabricating TiO_2-based SERS substrates can be complex and expensive, limiting their widespread use.

Surface Functionalization: Optimizing the surface properties of TiO_2 to enhance the adsorption of target molecules while maintaining high SERS activity is challenging. Effective functionalization strategies tailored for specific applications are needed.

Detection in Complex Matrices: Discriminating target analytes in complex biological or environmental samples, which contain various interfering substances, remains difficult. Advanced techniques are required to isolate and identify specific signals amidst background noise.

Integration with Portable Devices: Combining TiO_2-based SERS substrates with portable and user-friendly devices is essential for real-world applications. Developing compact, robust, and easy-to-use SERS devices suitable for fieldwork is a key challenge.

Regulatory and Standardization Issues: Establishing standardized protocols and gaining regulatory approval for TiO_2-based SERS applications, especially in medical diagnostics and environmental monitoring, can be a lengthy and complex process.

5.2 Scope and Potential

Medical Diagnostics: TiO_2-based SERS substrates can revolutionize medical diagnostics by enabling the rapid and sensitive detection of biomarkers, pathogens, and disease-specific molecules. This can lead to early diagnosis and better patient outcomes.

Environmental Monitoring: These materials can enhance the detection of pollutants, contaminants, and pathogens in air, water, and soil. Their high sensitivity and specificity make them ideal for monitoring environmental health and safety.

Food Safety: TiO_2-based SERS can be used to detect contaminants, pesticides, and pathogens in food products, ensuring food safety and quality.

Pharmaceutical Analysis: In the pharmaceutical industry, TiO_2-based SERS can help in the identification and quantification of active pharmaceutical ingredients and impurities, ensuring product quality and compliance with regulations.

Biological Research: These materials can aid in studying complex biological systems and processes at the molecular level, providing insights into cellular mechanisms and interactions.

Forensic Science: TiO_2-based SERS can be applied in forensic analysis for the detection of trace evidence, including drugs, explosives, and biological materials.

By addressing these challenges and leveraging the full potential of TiO_2-based SERS substrates, significant advancements in health and environmental monitoring can be achieved, leading to safer and more efficient diagnostic and analytical methods.

6 Conclusion

The integration of TiO_2-based nanomaterials into SERS represents a significant advancement in analytical science, with profound implications for health and environmental monitoring. Throughout this chapter, we have explored how TiO_2, a robust and versatile semiconductor, enhances the capabilities of SERS by providing remarkable sensitivity, specificity, and versatility in detecting a wide range of analytes. In health monitoring, TiO_2-based SERS substrates offer unprecedented precision and speed in detecting biomarkers, pathogens, and pharmaceuticals. Their ability to deliver real-time, non-destructive analysis allows for early disease diagnosis and personalized treatment, ultimately improving patient outcomes and advancing clinical research. In environmental monitoring, TiO_2-based SERS has proven invaluable for identifying and quantifying pollutants, contaminants, and hazardous substances in complex environmental matrices. This capability is crucial for effective pollution control, environmental protection, and ensuring public safety. Despite these advancements, challenges remain, including issues related to reproducibility, substrate stability, and scalability. Addressing these challenges through ongoing research and innovation is essential for the continued development and widespread adoption of TiO_2-based SERS technologies. Looking ahead, the future of TiO_2-based SERS is promising, with potential advancements expected to overcome current limitations and expand the range of applications. Continued exploration of novel TiO_2 nanomaterials, improved fabrication techniques, and integration with emerging technologies will further enhance the impact of SERS in both health and environmental monitoring. In conclusion, TiO_2-based nanomaterials are revolutionizing SERS by providing powerful tools for sensitive, specific, and rapid detection. Their transformative potential is poised to drive significant progress in analytical science, with broad implications for improving health diagnostics and environmental safety.

Acknowledgements Author (Samriti) gratefully acknowledge the MHRD and Department of Chemistry, National Institute of Technology, Hamirpur for financial support throughout their Ph.D. research work.

References

1. Sharma, B., et al.: SERS: materials, applications, and the future. Mater. Today **15**(1–2), 16–25 (2012)
2. Gong, T., et al.: Development of SERS tags for human diseases screening and detection. Coord. Chem. Rev. **470**, 214711 (2022)
3. Zhou, X., et al.: Bacteria detection: from powerful SERS to its advanced compatible techniques. Adv. Sci. **7**(23), 2001739 (2020)
4. Pham, T.T.H., et al.: The structural transition of bimetallic Ag–Au from core/shell to alloy and SERS application. RSC Adv. **10**(41), 24577–24594 (2020)
5. Awiaz, G., Lin, J., Wu, A.: Recent advances of Au@ Ag core–shell SERS-based biosensors. In: Exploration. Wiley Online Library (2023)
6. Li, C., et al.: Combining surface-accessible Ag and Au colloidal nanomaterials with SERS for in situ analysis of molecule–metal interactions in complex solution environments. Nat. Protoc. **18**(9), 2717–2744 (2023)
7. Samriti, R., Vishal, R.K., Gupta, Prakash, J.: Engineering metal oxide semiconductor nanostructures for enhanced charge transfer: fundamentals and emerging SERS applications. J. Mater. Chem. C **10**(1), 73–95 (2022)
8. Yaipimai, W., et al.: Multifunctional metal and metal oxide hybrid nanomaterials for solar light photocatalyst and antibacterial applications. J. Mater. Sci. **50**, 7681–7697 (2015)
9. Samriti, et al.: Highly efficient visible light active doped metal oxide photocatalyst and SERS substrate for water treatment. Environ. Sci. Pollut. Res. **30**(12), 34054–34068 (2023)
10. Samriti, J., Chandra, M., Gupta, R.K., Prakash, J.: Hydrothermal synthesis and Ta doping of TiO2 nanorods: effect of soaking time and doping on optical and charge transfer properties for enhanced SERS activity. Mater. Chem. Phys. **278**, 125642 (2022)
11. Li, Z., et al.: Honeycomb-like AgNPs@ TiO2 array SERS sensor for the quantification of micro/nanoplastics in the environmental water samples. Talanta **266**, 125070 (2024)
12. Li, X., et al.: A bioinspired Ag nanoparticle/PPy nanobowl/TiO2 micropyramid SERS substrate. Nanomaterials **12**(22), 4104 (2022)
13. Xie, Z., et al.: TiO2 nanorod arrays decorated with Au NPs as sensitive and recyclable SERS substrates. J. Alloy. Compd. **861**, 157999 (2021)
14. Prakash, J., et al.: Novel rare earth metal–doped one-dimensional TiO2 nanostructures: fundamentals and multifunctional applications. Mater. Today Sustain. **13**, 100066 (2021)
15. Gupta, T., Cho, J., Prakash, J.: Hydrothermal synthesis of TiO2 nanorods: formation chemistry, growth mechanism, and tailoring of surface properties for photocatalytic activities. Mater. Today Chem. **20**, 100428 (2021)
16. Chakraborty, A., et al.: TiO2 nanoflower photocatalysts: synthesis, modifications and applications in wastewater treatment for removal of emerging organic pollutants. Environ. Res. **212**, 113550 (2022)
17. Cialla, D., et al.: Surface-enhanced Raman spectroscopy (SERS): progress and trends. Anal. Bioanal. Chem. **403**, 27–54 (2012)
18. Wang, L., et al.: SERS-based test strips: principles, designs and applications. Biosens. Bioelectron. **189**, 113360 (2021)
19. Li, C., et al.: Towards practical and sustainable SERS: a review of recent developments in the construction of multifunctional enhancing substrates. J. Mater. Chem. C **9**(35), 11517–11552 (2021)
20. Prakash, J., et al.: Noble metals-TiO2 nanocomposites: from fundamental mechanisms to photocatalysis, surface enhanced Raman scattering and antibacterial applications. Appl. Mater. Today **11**, 82–135 (2018)
21. Chen, Q., et al.: EM and CT synergistic SERS enhancement in Si/TiO2/Ag heterostructures via local interfacial effect. Chem. Eng. J. **493**, 152316 (2024)
22. Huang, X., et al.: Noble metal nanomaterials for the diagnosis and treatment of hematological malignancies. Front. Biosci. Landmark **27**(2), 40 (2022)

23. Yang, L., et al.: A novel ultra-sensitive semiconductor SERS substrate boosted by the coupled resonance effect. Adv. Sci. **6**(12), 1900310 (2019)
24. Xu, J., et al.: Construction of high-active SERS cavities in a TiO2 nanochannels-based membrane: a selective device for identifying volatile aldehyde biomarkers. ACS Sens. **8**(9), 3487–3497 (2023)
25. Wang, C., et al.: Facile preparation of TiO2NTs/Au@ MOF nanocomposites for high-sensitivity SERS sensing of gaseous VOC. Sensors **24**(14), 4447 (2024)
26. Li, X., et al.: PDMS/TiO2/Ag hybrid substrate with intrinsic signal and clean surface for recyclable and quantitative SERS sensing. Sens. Actuators B Chem. **351**, 130886 (2022)
27. Ziad, R., et al.: Multi-functional silver nanoprism-titanium dioxide hybrid nanoarrays for trace-level SERS sensing and photocatalytic removal of hazardous organic pollutants. Spectrochim. Acta Part A Mol. Biomol. Spectrosc. **297**, 122701 (2023)
28. Ramachandran, K., et al.: Highly ordered TiO2 NTs decorated with plasmonic silver NPs (TiO2/Ag) for ultra-sensitive SERS detection of textile dye contaminant Rhodamine 6 G. J. Phys. Conf. Ser. (2024) (IOP Publishing)
29. Li, K., et al.: A flexible semiconductor SERS substrate by in situ growth of tightly aligned TiO2 for in situ detection of antibiotic residues. Microchim. Acta **191**(2), 113 (2024)
30. Cao, Y., et al.: Killing two birds with one stone: silver NPs loaded on ultrathin N-doped carbon-coated TiO2 porous spheres with narrow bandgap for efficient SERS sensing and photoinduced antibacterial applications. Appl. Surf. Sci. **583**, 152512 (2022)
31. Hwang, C.S., et al.: Highly adsorptive Au-TiO2 nanocomposites for the SERS face mask allow the machine-learning-based quantitative assay of SARS-CoV-2 in artificial breath aerosols. ACS Appl. Mater. Interfaces **14**(49), 54550–54557 (2022)
32. Li, J., et al.: Hydrophobic interaction enables rapid enrichment of volatile metabolites on Au/TiO 2 based SERS substrates for ultrasensitive bacteria detection. J. Mater. Chem. B **11**(17), 3877–3884 (2023)

One-Dimensional TiO_2 Nanostructures for the Degradation of Emerging Contaminants

Chinthalapudi Naga Lakshmi, Rahul Sinha, and Narendra Singh

Abstract Water ecosystems are under constant threat from human activities, which release pollutants resistant to removal in conventional wastewater treatment plants, exacerbating concerns for water quality. Emerging contaminants can impact both the environment and human health with prolonged exposure. Current wastewater treatment research focuses on developing advanced technologies to remove emerging contaminants at a full scale efficiently. Titanium dioxide (TiO_2) has attracted considerable attention from environmental scientists as a potential solution for tackling emerging contaminants. One-dimensional TiO_2 (1D TiO_2) nanostructures distinguished by their structural and functional characteristics, serving as essential components in numerous applications. Recent advancements have enabled the tailored synthesis of 1D TiO_2 nanostructures with precise size and morphology control. Compared to zero-dimensional (0D) and two-dimensional (2D) counterparts, the focus on 1D TiO_2 nanostructures underscores their expansive specific surface area, elevated aspect ratio, and exceptional charge transport capabilities. In this chapter, we elucidate the TiO_2 crystal structure and examine the current trends in the synthesis of 1D TiO_2 nanostructures. Additionally, we analyze essential engineering strategies contributing to the exceptional characteristics of these 1D TiO_2 nanostructures, including optimized light absorption, augmented surface area, and effective electron/hole pair separation, all of which augment their potential applications for the emerging contaminants degradation through photocatalysis (PC), electrocatalysis (EC), photoelectrocatalysis (PEC).

Keywords Emerging contaminants · One-dimensional TiO_2 · Photocatalysis · Electrocatalysis · Photoelectrocatalysis

---

C. N. Lakshmi · R. Sinha · N. Singh (✉)
Department of Chemical Engineering, Indian Institute of Technology Tirupati, Tirupati, Andhra Pradesh 517619, India
e-mail: narendra@iittp.ac.in

J. Prakash et al. (eds.), *Titanium Dioxide-Based Multifunctional Hybrid Nanomaterials*, Engineering Materials, https://doi.org/10.1007/978-3-031-80625-4_7

1 Introduction

The rise in production, utilization, and disposal of various chemicals prevalent in industry, agriculture, medicine, and daily life has led to the uncontrolled release of these substances into aquatic ecosystems, often at concentrations ranging from nanograms per liter (ng/L) to milligrams per liter (mg/L) [1]. These minute compounds, primarily organic and termed as emerging contaminants, cause potential risks to aquatic ecosystems and human health [2]. These substances are commonly identified in effluents, drinking water, and surface water sources because they persist despite undergoing treatment in conventional wastewater treatment plants (WWTPs), posing a threat to water quality standards [3].

Various techniques are available to treat emerging contaminants. The inherent low biodegradability and high stability of most emerging contaminants make conventional treatment methods (oxidation, flocculation/coagulation, sedimentation, chemical precipitation and activated sludge) ineffective in their removal. Thus, alternative technologies like adsorption, advanced oxidation processes, and membrane technology have been devised to enhance emerging contaminants removal efficiency. Nevertheless, advanced oxidation processes (AOPs) like sono-catalysis, ozonolysis, Fenton oxidation, electrochemical oxidation, and photocatalysis (PC) have attracted considerable interest for their effectiveness in eradicating emerging contaminants [4]. These AOPs produce reactive oxygen species like hydroperoxyl radicals ($HO_2^{\cdot}$) and singlet oxygen ($^1O_2$), which can degrade and decompose ECs found in wastewater. Moreover, limitations exist; for instance, ferrous (iron) salts, obtaining in the formation of substantial quantities of iron-based sludge. Additionally, commercial ozonation is linked with elevated operating expenses and the potential generation of hazardous secondary by-products like bromate and dichlorobromomethane. Among advanced oxidation processes (AOPs), photocatalysis (PC) stands out as a sustainable technology employed for water treatment, air purification, and renewable energy generation. [5–7].

With the increasing demand for renewable energy-driven solutions, electrochemical (EC) and photoelectrochemical (PEC) systems are poised to revolutionize water treatment methodologies [8]. Electrochemical treatment systems facilitate electron transfer between dissolved pollutants and the electrode surface under an applied bias potential. Advancing beyond electrochemcial systems, PEC systems combine photochemistry with electrochemical reactions using light energy. Unlike traditional PC systems, PEC systems ensure closer contact between the semiconductor (anode) and a charge-collecting substrate (cathode), facilitating efficient carrier separation and transport under an external bias. Moreover, through careful selection of (photo) cathode and (photo)anode materials, the entire process can be driven by renewable energy sources, resulting in improved efficiencies compared to single-semiconductor PC systems [9].

Titanium dioxide (TiO_2) stands out as a preferred option due to its chemical stability and lack of harmful effects, prompting widespread interest from researchers globally. TiO_2 manifests in three distinct polymorphs: anatase, rutile, and brookite

[10]. Rutile, serving as the principal source and the most stable form of TiO_2, holds significant prominence. Despite the relative ease of creating all three polymorphs in laboratory setups, the less stable anatase and brookite frequently transition into rutile, the thermodynamically favoured form, when heated to calcination temperatures above approximately 600 °C. Notably, Fujishima and Honda's groundbreaking discovery of water splitting on a single-crystalline TiO_2 (rutile) anode under UV light has spurred extensive exploration into TiO_2-based photocatalysts for their high activity in degrading pollutants. In each of the three forms, titanium (Ti^{4+}) ions bond with six oxygen (O^{2-}) ions to create TiO_6 octahedra. Anatase features octahedra that share corners (vertices), forming (001) planes and exhibiting a tetragonal structure. On the other hand, rutile demonstrates octahedra sharing edges at (001) planes, contributing to its tetragonal structure. Conversely, brookite showcases edge and corner-sharing, leading to an orthorhombic structure [11]. TiO_2 typically behaves as an n-type semiconductor owing to oxygen deficiency. The band gap values differ among its polymorphs: anatase exhibits a 3.2 eV band gap, rutile displays a 3.0 eV band gap, and brookite demonstrates a band gap of approximately 3.2 eV [12].

In recent years, significant advancements have been achieved in synthesizing, modifying, and utilizing one-dimensional TiO_2 (1D TiO_2) nanostructures with modifications like doping with metal, forming heterojunction with other semiconductors like Fe_2O_3, WO_3, ZnO, Al_2O_3, etc. One-dimensional nanostructures, encompassing nanofibers (NFs), nanotubes (NTs), nanowires (NWs) and nanorods [NRs], have emerged as focal points in scientific inquiry owing to their unique properties including expansive surface areas, high aspect ratios, and facile functionalization capabilities. Employing prevalent fabrication methodologies like hydrothermal synthesis and electrospinning, scientists have achieved the production of diverse nanostructures, ranging from rings and wires to rods and belts [13]. The initial three categories exhibit solid structures, while the nanotube features a hollow core. Owing to the notable differences in geometry, properties, and fabrication methods between the nanotubes (NTs) and other structures, researchers focused exclusively on 1D nanos tructures falling within the NWs, NRs, and Nanobelts (NBs) classifications [14]. NRs can be thought of as shorter NWs, often presenting a stiff, linear appearance. NBs represent a specialized form of NWs with a rectangular cross-section, typically composed of single crystals featuring distinct side crystal facets. In certain instances, the term "nanofibers" is utilized to describe the structure of NWs, especially when these NWs are markedly elongated and devoid of single-crystalline attributes [15].

This chapter includes the latest developments in creating 1D TiO_2 nanostructures. It explores essential structure engineering strategies that improve the properties of these materials using various synthesis techniques. Finally, this chapter provides an outline of the 1D TiO_2 nanostructures for the emerging contaminants degradation through PC, EC and PEC treatment techniques.

2 Synthesis Techniques

The preparation of 1D TiO_2 nanostructures boasts a range of synthesis methodologies, each offering unique advantages and outcomes. The main methods include hydrothermal synthesis, chemical vapor deposition (CVD), sol–gel processing, template-assisted fabrication, electrochemical anodization, electrospinning, and microwave assisted synthesis etc. Each approach provides unique control over the morphology and characteristics of the resultant nanostructures, addressing diverse application requirements.

2.1 Solvothermal/Hydrothermal Methods

The solvothermal/hydrothermal approach is a prevalent method in the ceramics various applications for forming small particles, utilizing either aqueous or nonaqueous solutions [16, 17]. Utilizing a closed vessel under high-temperature, high-pressure conditions, this technique offers a straightforward process and minimal production expenses, thereby attracting considerable attention within the scientific community [18]. Under these specific circumstances, the supercritical fluid phenomenon manifests, leading to a marked increase in the solubility of solid precursors. As a result, the process aids in the formation of nanostructures sourced from a broad spectrum of inorganic materials. The simplicity of the setup, operational ease, and positive growth results have prompted numerous researchers to adopt this technique specifically for synthesizing 1D TiO_2 nanostructures. Initially, the focus was on TiO_2 nanotube growth, but more recently, researchers have adapted these techniques for solid 1D TiO_2 nanostructures like NWs and NRs. Moreover, both methodologies typically rely on TiO_2 nanoparticles, $C_6H_{22}N_2O_8Ti$ [19], $Ti(OC_3H_7)_4$ [20], and $TiCl_4$ [21] as precursor materials. Nonetheless, the solvothermal technique is generally carried out in organic solvents (like ethylene glycol, ethanol, n-hexane, etc.), whereas the hydrothermal technique predominantly operates in aqueous solutions media [21]. Nian et al. prepared suspensions of titanate NTs in an acidic solution, leading to the creation of single-crystalline anatase NRs characterized by a unique elongation direction of crystal growth. The synthesis process initiated with the preparation of nanotube suspensions by treating TiO_2 with NaOH, followed by pH adjustment through HNO_3 addition. Notably, the size of the resultant anatase titanate NTs escalated with increasing pH levels of the suspensions. The mechanism proposed for the transition from tubes to rods, based on the comparison of titanate and anatase crystalline structures, suggests that the nanotubes undergo localized shrinkage. This shrinkage is accomplished by decreasing the interlayer distance between the tube walls. Transmission electron microscopy (TEM) analysis confirmed the development of consistent nanorods sized approximately 20×120 nm [22].

In other paper, TiO_2 NTs with an impressive surface area of 469 m^2/g have been successfully fabricated via a straightforward hydrothermal approach. Following this synthesis, the TiO_2 NTs were subject to surface modification with silver nanoparticles (Ag NPs), leading to the creation of Ag-decorated TiO_2 nanotubes (Ag–TiO_2 NTs). The resultant material exhibits a broadened optical response, spanning from UV to visible wavelengths, alongside the emergence of a distinct surface plasmon resonance (SPR) band. These distinctive optical properties render Ag–TiO_2 NTs promising candidates for applications in plasmonic photocatalysis [25]. The fabrication of multiwalled TiO_2 nanotubes (MWTNTs) using solvothermal techniques, with the reaction between TiO_2 nanoparticles and a NaOH aqueous solution under medium temperatures, is described in their study. Following the synthesis, MWTNTs are obtained with lengths ranging from 70 to 200 nm and an average diameter between 4 to 10 nm [26]. Zhang et al., employed the synthesis of anatase TiO_2 single crystalline NWs using a facile hydrothermal technique from TiO_2 nanoparticles. The XRD analysis (Fig. 1a) suggests that the NWs are predominantly made up of anatase TiO_2, with crystalline cell constants comparable to those reported in literature. Additionally, there is a minor presence of brookite. The SEM image of the TiO_2 NWs indicated in Fig. 1b, indicates a notable abundance of NWs, appearing clean and uncontaminated on their surfaces. However, aggregation into bundles is observed, likely attributed to solution dynamics or SEM sample preparation procedures, contributing to variations in nanowire dimensions [23]. In another paper, the growth of TiO_2 NWs membranes on titanium (Ti) substrates is accomplished through hydrothermal technique. This process involves the use of diverse organic solvents abundant in carbonyl groups to initiate the oxidation of titanium. Initially, within a solution containing 10 M sodium hydroxide (NaOH), sodium titanate nanosheets are formed and then converted into hydrogen titanate nanowires via ion exchange. The phase of TiO_2 NWs changes from Type-B to anatase at the calcination temperature of 700 °C, despite their diameter ranging from 10 to 20 nm and length up to 100 μm. The resulting TiO_2 nanowire membranes exhibit attributes essential for water treatment applications, allowing for the utilization of their photocatalytic capabilities in environmental pollutant remediation [27]. Vazquez et al., the attention was on synthesizing TiO_2 NWs through hydrothermal treatment at 180 °C for 2 h, utilizing $TiCl_4$ as the precursor. Observations revealed that these NWs initiated growth as individual entities after 30 min of the reaction, gradually organizing into a branched structure. Characterized by elongated structures with a rutile crystalline phase, the individual NWs underwent further organization into a branched structure after approximately 45 min, as depicted in Fig. 1c. The TEM image (Fig. 1d) displayed NWs with diameters varying in the range of 10 and 20 nm and lengths of several hundred nanometers [24].

2.2 Electrospinning Method

Electrospinning represents an accessible method for creating ultrafine fibers, characterized by diameters in the range of tens of nanometers (nm) and sub micrometers (μm). In a standard electrospinning arrangement, there are three essential components: a spinneret with a small metallic needle, a high-voltage power source, and a

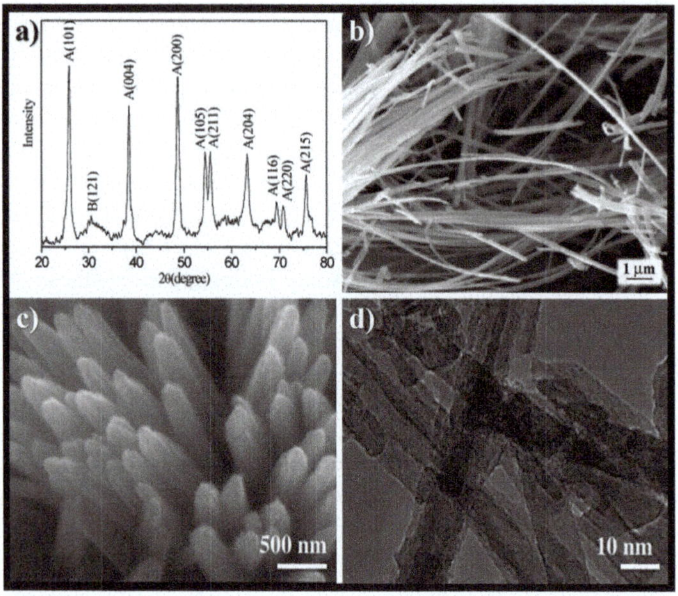

Fig. 1 **a** XRD patterns, **b** SEM image of TiO_2 NRs. Reprinted with permission from Ref. [23]. Copyright 2002, Elsevier **c** SEM image, **d** TEM of the as-synthesized TiO_2 NWs. Reprinted with permission from Ref. [24]. Copyright 2019, Elsevier

grounded conductive collector, as illustrated in Fig. 2. By charging the surface of a polymer solution droplet using high voltage, this process initiates with the ejection of a liquid jet that solidifies upon landing on the collector. As electrostatic repulsion dominates the droplet's surface tension, it generates a charged liquid jet, continuously elongating and directed towards the collector [28]. The usual precursor solution comprises three components: a titanium precursor, a binder, and a stabilizer [29].

Typical precursors for the production of titanium dioxide (TiO_2) fibers include various compounds such as Titanium tetra isopropoxide (TTiP) [30, 31], titanium tetraisopropoxide ($Ti(OiPr)_4$) [32], and titanium tetrabutoxide ($Ti(OBu)_4$) [33]. In

Fig. 2 Schematic representation of electrospinning technique

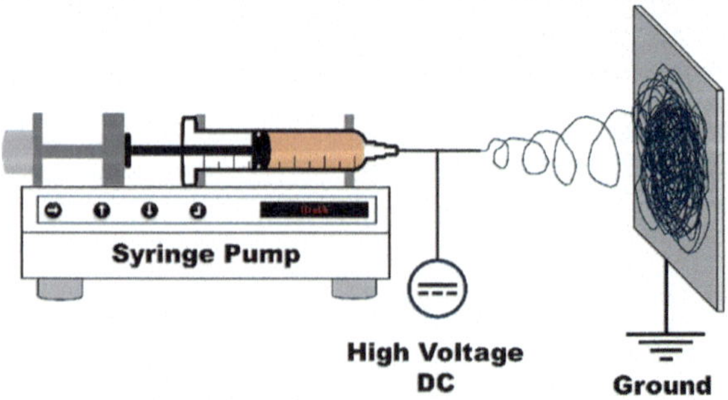

the context of manufacturing TiO_2 fibers, a binder is a crucial polymer constituent responsible for bonding TiO_2 fine powders into a fibrous shape following spinning. It's worth noting that the binder can also act as a solvent and is commonly removed through combustion at the time of calcination stage. Acetic acid is commonly employed as a stabilizing agent in the solution, regulating the hydrolysis reaction [34]. Jin et al., investigated the water steam modified mesoporous TiO_2 NFs using polyethylene (PEO) as a binder. This phenomenon involves the rearrangement of $TiO_6{}^{2-}$ octahedral units, induced by steam for the formation anatase basic units. Additionally, the increased reactivity of water vapor induces the protonation of base functionalities, promoting phase transformation. Continuously, in the presence of water vapor, dissolution–precipitation processes proceed, generating numerous mesoscopic pores, which substantially enhances the surface area. The changes in microstructure observed in TiO_2 NFs after thermal treatment at various temperatures (150 °C, 350 °C, 550 °C). At steam temperature 150 °C (Fig. 3a), pores are visible on both the surface and cross-section of the NFs, with an uneven distribution across the surface. Higher-resolution SEM images (Fig. 3a2, b2, c2) reveal individual NFs. As temperature rises, pore distribution becomes more uniform, and pore diameter increases. These NFs exhibit a porous structure in water steam, potentially enhancing their properties. These mesoporous structures, known for their enhancive surface area and crystallinity, can significantly boost the photocatalytic efficiency of TiO_2 nanofibers [35].

Lee et al., synthesized the porous TiO_2 NFs using biopolymer system of PVP and PVDF. The elimination of polyvinylpyrrolidone (PVP) in a composite material featuring titanium dioxide (TiO_2) can create fiber pores, thereby boosting surface area and facilitating better approach to TiO_2. The introduction of internal pores within the TiO_2 nanofibers resulted in a remarkable increase in the BET surface area, elevating it from 30.6 to 117.2 m^2/g [36]. In another report by Pascariu et al., La-doped TiO_2 nanofibers were synthesized and then subjected to calcination within a temperature

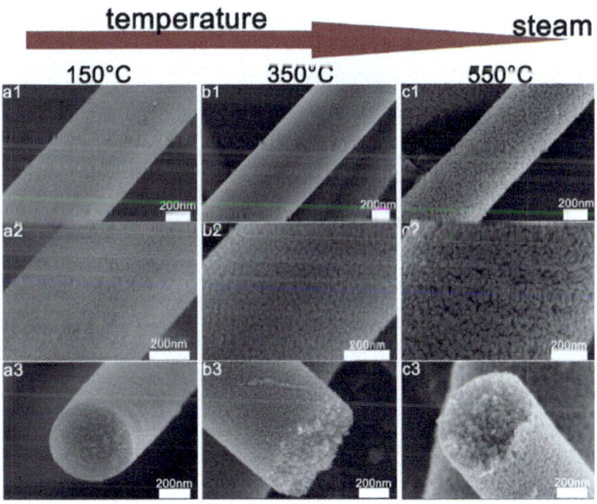

Fig. 3 SEM images of TiO_2 NFs (a1, b1, c1); high magnification surface SEM images of the TiO_2 NFs (a2, b2, c2); cross section of the TiO_2 NFs (a3, b3, c3) heat-treated at different temperatures in water steam Reprinted with permission from Ref. [35]. Copyright 2018, Elsevier

range of 400–700 °C after the electrospinning process. Analysis indicated that pure TiO_2 nanofibers had diameters of 129 nm, whereas La-doped TiO_2 materials exhibited diameters measuring 101 nm. The study aimed to employ the impact of varying La^{3+}-dopant concentrations, calcination temperatures, and photocatalyst's dosage on photodegradation efficiency [37].

2.3 Sonochemical Method

Acoustic cavitation refers to the process where bubbles are formed and then collapse within a liquid medium, induced by sonic or ultrasound waves (Fig. 4). Additionally, sonication offers the benefit of cavitation bubbles generating shock waves, deagglomeration, high-speed microjets, and powerful hydrodynamic shear forces [38]. This phenomenon results in the development of intense localized conditions, including ultrahigh temperatures (~5000 K), ultrahigh pressures (up to 1000 atm), and extremely rapid heating and cooling rates exceeding 10^{10} K/s. Within these areas, solvents and volatile solutes undergo homolytic bond cleavage, producing hydroxyl radical ($^{\cdot}OH$) through the sonolysis of water. These radicals catalyze various chemical reactions [39]. Applying high energy "shock waves" to a chemical system is referred to as sonochemistry, a technique that has shown success in the synthesis of 1-D nanostructures of TiO_2. The advantages of sonochemical techniques encompass rapid reaction kinetics, precise modulation of reaction conditions, straightforward and safe execution, and the production of nanomaterials featuring uniform spherical shapes, symmetrical distributions, and high purity [40].

Zhu et al. introduced a novel approach utilizing sonication to generate titania whiskers and nanotubes. Initially, the process yields titanate whiskers, taking the form

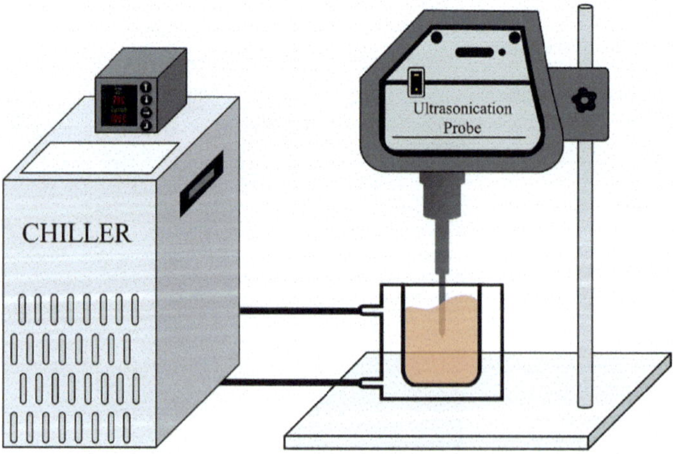

Fig. 4 Schematic representation of sonochemical technique

of slender sheets measuring approximately 1 mm in length and 60 nm in width. These titanate whiskers serve as precursors for producing arrays of titania whiskers with a diameter of 5 nm. Furthermore, the method results in the synthesis of titania NTs, featuring a diameter of around 5 nm and a length ranging from 200 to 300 nm. This method highlights the efficacy of sonication in precisely controlling the morphology and dimensions of nanostructures [41]. Guo et al. introduced a gentle, template-free sonochemical technique for synthesizing mesoporous TiO_2 NRs from bulk Ti powder sourced from industry. The synthesis was conducted under standard atmospheric pressure and relatively low temperatures. As shown in Fig. 5a, the XRD spectra of TiO_2 indicates the diffraction peaks synonymous with anatase TiO_2, confirming their crystalline nature. The XPS spectra showcased in Fig. 5b provides insights into the composition of TiO_2, with distinctive peaks indicating the chemical states present. XPS analysis post-annealing confirms TiO_2 crystal formation (Ti^{4+} peak at 458.4 eV, 5.7 eV splitting). O/Ti ratio $\approx$2 supports TiO_2 composition. The SEM image in Fig. 5c illustrates the presence of olive-like nanorod aggregates in TiO_2, characterized by a layered structure. Subsequent TEM analysis, as depicted in Fig. 5d, confirms the prevalence of irregular NRs as the dominant morphology of TiO_2. These NRs exhibit diameters ranging from 8 to 12 nm and lengths varying from 15 to 100 nm, with a significant portion varying from 70 to 100 nm. Furthermore, Fig. 5e presents a high-resolution TEM image providing detailed insights into the crystalline structure of TiO_2 [42].

2.4 Chemical Vapor Deposition Method

Chemical vapor deposition (CVD) is widely adopted for its efficacy in producing high-quality thin films and coatings. This method involves introducing volatile precursors into a reaction chamber, where they decompose upon encountering a heated substrate surface, causing to the fabrication of the desired material as shown in Fig. 6. The fabrication of high-quality 1D TiO_2 nanostructures like nanowires or nanorods has been achieved through CVD. These nanostructures are grown on a silicon (Si) substrate that has been coated with titanium (Ti), $Ti(OC_3H_7)_4$, or $TiCl_4$ inside a closed chamber maintained at elevated temperatures and reliance on pure chemicals, often sourced from virgin and non-renewable reservoirs, emphasize the need for sustainable practices to minimize environmental impact while maintaining CVD's efficiency [43].

Shi et al. investigated the growth of TiO_2 nanostructures via surface reaction-limited CVD at elevated temperatures, coupled with the presence of gold (Au), revealed a shift in TiO_2 phase from anatase to rutile. Elevated deposition temperatures prompted a transition of TiO_2 phase from anatase to rutile. The morphology of NWs-containing strips, characterized by SEM shows a uniform orientation, was distinct (Fig. 7a). TEM image provided insights into the structure of NWs on flakes, revealing their likely formation through bifurcations of the flake matrix as shown in Fig. 7b [44].

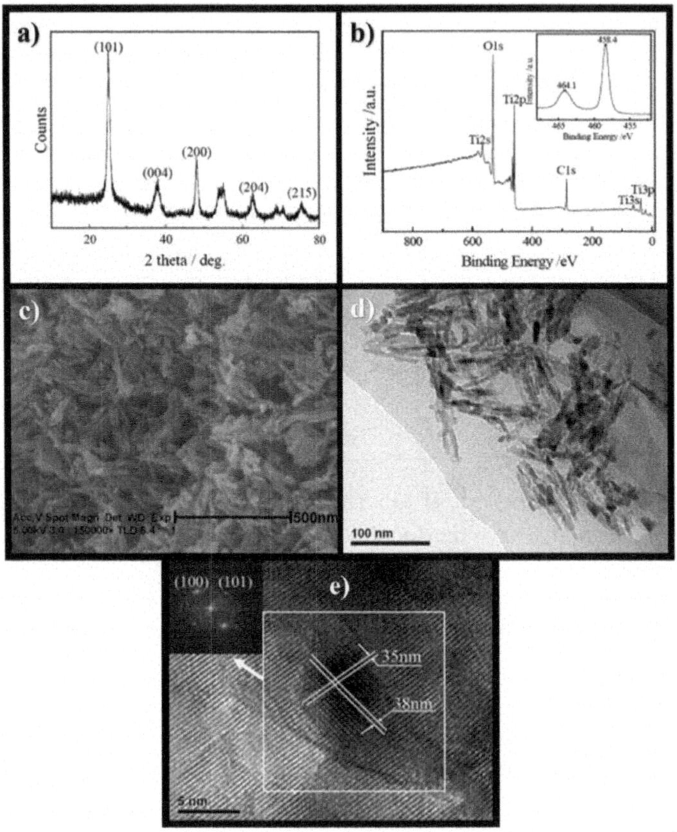

Fig. 5 **a** X-ray diffraction (XRD) pattern, **b** X-ray photoelectron spectroscopy (XPS) of TiO_2 NRs; **c** SEM image of TiO_2 NRs, **d** TEM image of TiO_2 NRs; and **e** high-resolution image of TiO_2 NRs. Reprinted with permission from Ref. [42]. Copyright 2009, Elsevier

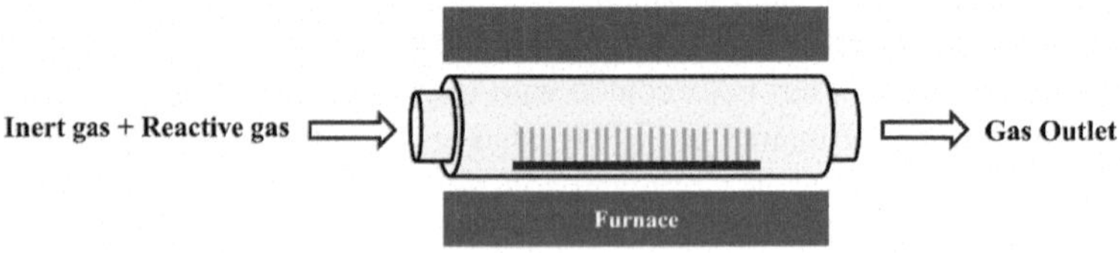

Fig. 6 Schematic representation of CVD technique

Later, Liu et al. synthesized the TiO_2@C core–shell NWs and nanowalls were fabricated using a straightforward CVD method, offering the capability to adjust structural attributes through modifications in experimental parameters [45]. Figure 7c, d present TEM images showing a standard nanorod and nanowall, respectively, distinguished by lighter shading on their sides. The carbon-enveloped nanowire exhibits a diameter below 100 nm, while the nanowall's width extends up to 1 μm. Later, high-density TiO_2 NWs have been successfully synthesized on a $TiSi_2$ layer via atmospheric pressure chemical vapor deposition. Operating at 720 °C, the process results in nanowires with diameters in the range of 20 and 50 nm and lengths approximately 3 μm. Formation of TiO_2 NWs involves the gradual utilization of $TiSi_2$. These nanowires align with the (001) orientation of the tetragonal rutile TiO_2 crystal, initiating growth from the lower end and extending upwards as the tip propels forward [46].

The high-temperature surface reaction-limited growth of TiO_2 nanostructures via CVD prompted a anatase to rutile phase transition in the presence of Au. Au coating facilitated crystal growth, leading to particle formation, while anisotropic nanowire growth displayed a preference for no Au coating. The emergence of 1D TiO_2 nanostructures during shorter purging intervals was attributed to growth bifurcation, which may result from the surface reaction-limited CVD process [28].

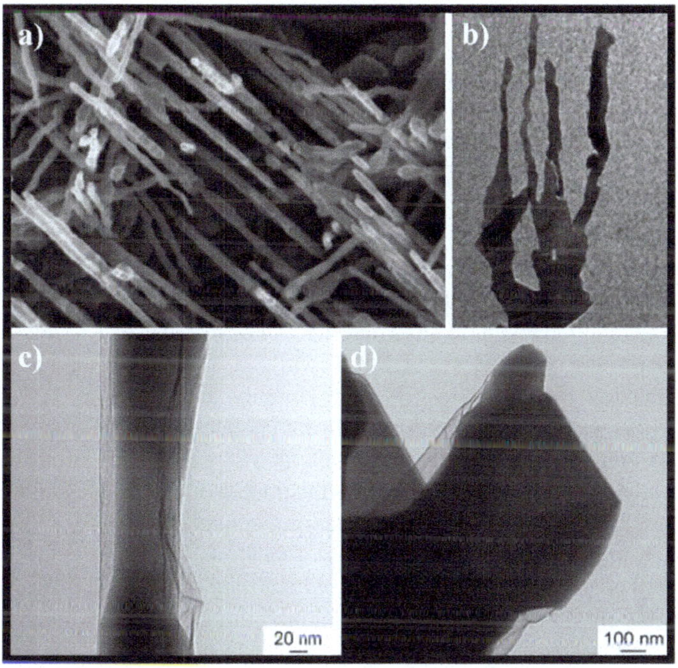

Fig. 7 **a** SEM top view of the array of the TiO_2 NWs-on-flake structure showing a parallel arrangement. **b** TEM analysis of the TiO_2 NWs-on-flake structure. Reprinted with permission from Ref. [44]. Copyright 2011, American Chemical Society. TEM images of TiO_2@C nanostructures: **c** nanorod and **d** nanowall. Reprinted with permission from Ref. [45]. Copyright 2012, Elsevier

2.5 Electrochemical Anodization Method

The electrochemical anodization process is an affordable and straightforward technique for generating TiO_2 nanotubes (NTs), having been extensively investigated over the last few decades. Their extensive use is attributed to their inherent stability, nontoxicity, environmental compatibility, low cost, recyclability, and ease of synthesis. In the electrochemical anodization of titanium and its alloys, practice involves the repeated anodization of non-target titanium in freshly prepared electrolyte as a means of 'aging' or 'conditioning' it prior to anodizing the target titanium to generate a porous oxide layer, showcasing its straightforward nature and broad applicability (Fig. 8). However, this technique remains inadequately specified or explored. Aging phenomena have been observed in electrolytes containing organic solvents like ethylene glycol, dimethyl sulfoxide (DMSO), and glycerol. Mohammadpour et al. explored the successful synthesis of uniform and precisely structured TiO_2 nanotubes (NTs) through anodization of Ti foil in an electrolyte solution comprising ethylene glycol with 0.15 M NH_4F and 3 vol% deionized water. Their research aimed to explore with changes in water content, electrolyte temperature, and fluoride concentration affect the transition from nanoporous materials to NTs [47]. In the same way, employing the electrochemical method, TiO_2 NTs were synthesized by adjusting the levels of water content and hydrofluoric acid (HF) percentages. These conditions included altering water concentrations (5% and 10%) and introducing a minute amount of HF (0.25% by weight) into the anodizing solution for 60 min at a constant 40 V potential. The resulting nanotube lengths varied based on these modifications. HR-SEM analysis indicates that higher levels of acidic electrolyte result in augmented nanotube length and diameter as shown in Fig. 9. For instance, NTs achieve a length of 2.6 mm at a water content level of 5 wt%, while at 10 wt% water content and 0.25 wt% HF, their length extends to 4.71 mm. Notably, adding HF at 0.25% concentration and 10% water content led to significant changes, with nanotubes exhibiting maximum height and diameter. Minimal alterations were observed with differing water content alone. The study underscores the impact of electrolyte composition on nanotube morphology and potential cell viability enhancement [30].

In another report, a self-assembled array of nanotubular titanium dioxide (TiO_2) were synthesized through the anodization of pure titanium in a solution comprising distilled water (8% vol.), glycerol, and ammonium fluoride, using a dual electrode setup. They controlled the size and distribution of nanopores by adjusting the DC voltage between 30 and 60 V. The diameter of the TiO_2 nanopores ranged from 59 to 128 nm, with variations dependent on the applied anodizing voltage [48]. Nanotubes exhibiting notably different diameters were successfully synthesized across the voltage span of 30–60 V. Remarkably, TiO_2 maintained its chemical composition regardless of the changes in anodizing voltage.

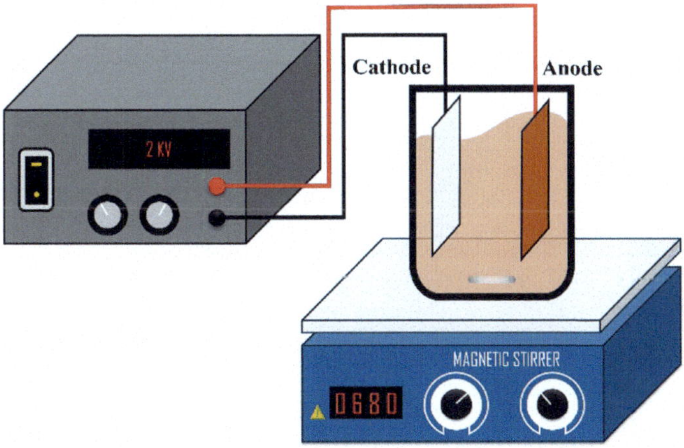

Fig. 8 Schematic representation of electrochemical anodization technique

2.6 *Other Techniques*

Transparent arrays of 1D β-TiO_2 nanoneedles were successfully grown on conducting glass substrates using a hot-filament metal vapor deposition method. These nanoneedles, with a conical shape and orthorhombic crystalline structure resembling brookite, exhibit diameters in the range of 24 and 26 nm and lengths of approximately 650 nm. The distinct conical morphology of these nanoneedles is pivotal, influencing their optical and electronic features and potentially enabling applications like light trapping in solar cells or utilization as components in optoelectronic devices. The conical morphology of the nanoneedles is particularly noteworthy, as it influences their optical and electronic properties, making them potentially useful for applications such as light trapping in solar cells or as components in optoelectronic devices. The expansive 1D β-TiO_2 nanoneedle array expedites Na^+ ion diffusion and charge transfer by virtue of its generous surface area and well-defined textural boundaries [49]. Using microwave radiation, Wu et al. synthesized TiO_2 NTs by subjecting TiO_2 crystals (anatase, rutile, or mixed phase) to NaOH aqueous solution under precise microwave power parameters. These NTs exhibited a multiwall structure with an open-ended central hollow morphology, demonstrating diameters ranging from 8 to 12 nm and lengths spanning 200 to 1000 nm [50]. Hui Li et al. conducted a study in 2015 employing a microwave-assisted chemical reduction technique utilizing $NaBH_4$ to synthesize bulk Ti^{3+} self-doped TiO_2 NTAs. This innovative method facilitated the incorporation of abundant Ti^{3+} species into the bulk structure of TNTs, resulting in bulk Ti^{3+} self-doped TiO_2 NTAs photoelectrodes displaying remarkable and consistent performance in photoelectrochemical (PEC) water splitting, attributed to their effective absorption of visible light and rapid charge separation [51]. Table 1 provides a summary of various synthesis techniques used for producing 1D TiO_2 nanostructures.

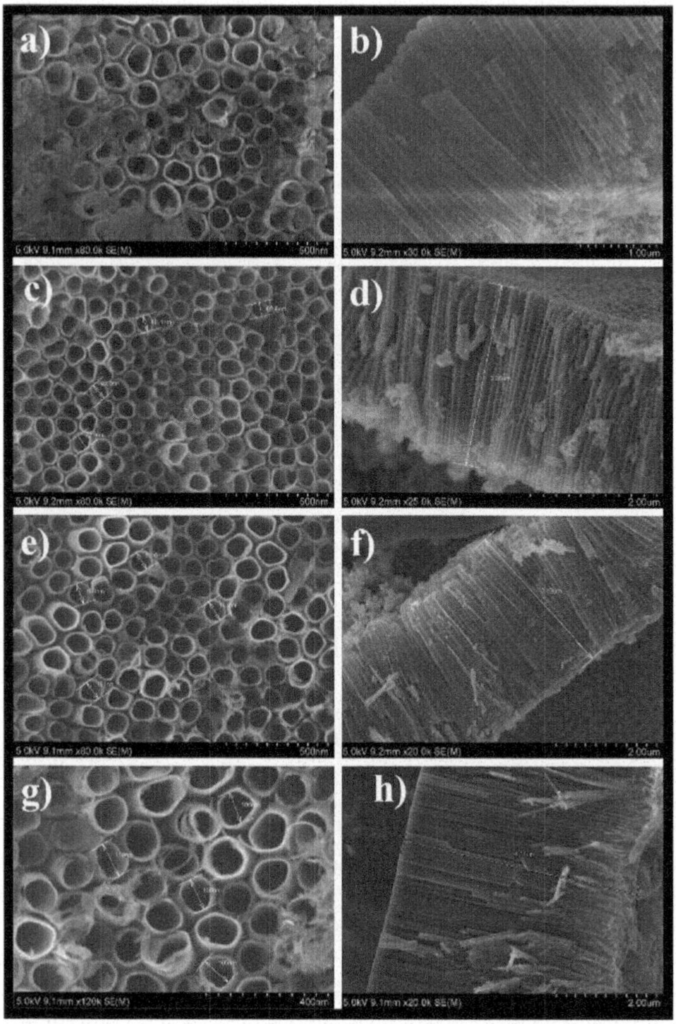

Fig. 9 HR-SEM images of **a** 5 vol.% of water content; **b** cross sectional view; **c** 10 vol.% of water content; **d** cross sectional view; **e** 2 vol.% of HF; **f** cross sectional view; **g** 5 vo.l% of HF; **h** cross sectional view NTs. Reprinted with permission from Ref. [30]. Copyright 2021, Elsevier

Attar et al. investigated the TiO_2 nanorod arrays of varying diameters through the sol–gel template-assisted technique. They used three alumina-based templates with distinct pore sizes, approximately 50–80 nm, 100–150 nm, and 200–300 nm, were utilized in the study. As depicted in Fig. 10a, the FE-SEM image displays TiO_2 nanorod arrays with diameters in the range of 40 and 70 nm and lengths extending several μm. Figure 10b provides a magnified view of these arrays from Fig. 10a, highlighting the densely packed nanorods arranged in a structured manner. High-resolution FE-SEM images in Fig. 10c, d reveal TiO_2 nanorod arrays with

Table 1 Various synthesis techniques for the synthesis of 1D TiO_2 nanostructures

S. No	Material	Synthesis parameters	Properties	Ref.
1	Anatase TiO_2 rods	Hydrothermal method Temperature: 200 °C, 180 °C Time: 24 h	Diameter (D): 20–200 nm	[52]
2	TiO_2/WO_3 composite nanofibers	Electrospinning Voltage: 20 kV Flowrate: 1 ml/h Needle tip to collector distance: 15 cm	D: 630–800 nm, 720–1050 nm, 920–1170 nm, 980–1200 nm, 1930–2950 nm	[19]
3	TiO_2 nanorods	Hydrothermal method Temperature: 180 °C Time: 24 h	D: 70–100 nm	[20]
4	TiO_2 nanotubes	Electrochemical anodization Electrode: Ti foil (working electrode), Ag/AgCl (3 M KCl) (reference electrode) Potential: 55 V Time: 30 min Calcination: 450 °C for 1 h	Length (L): 4.7–5.6 μm	[53]
5	Titania nanorods	Sonochemical Power: 24 or 56 W Time: Temperature: 60 °C Drying: 90 °C for 16 h	L: 30 nm	[54]
6	TiO_2 nanorods	CVD Substrate: WC–Co Temperature: 400 °C, 500 °C, 600 °C Pressure: 50 Torr	D: 50–100 nm L: 0.5–2 μm	[55]
7	Ta doped TiO_2 nanofibers	Electrospinning Voltage: 13 kV Flowrate: 20 μL/min Distance: 10 cm	D: 40–60 nm	[56]
8	N doped TiO_2 nanorods	Hydrothermal Temperature: 200 °C Pressure: 48 h	D: 12–60 nm L: 70–500 nm	[57]
9	TiO_2 nanotubes	Electrochemical anodization Electrode: Ti foil (both anode and cathode) Distance between electrode: 3 cm Potential: 40 V Time: 1.5 h, 6 h Calcination: 550 °C for 2 h	L: 1800 nm D: 70 nm	[58]

(continued)

Table 1 (continued)

S. No	Material	Synthesis parameters	Properties	Ref.
10	TiO_2@Ag@Cu_2O nanofibers	Electrospinning Spinneret diameter: 0.4 mm Voltage: 18 kV Flowrate: 1.5 mL/min Needle tip to collector distance: 15 cm Calcination temperature: 250 °C, 550 °C Calcination time: 3 h	D: 200 nm (at 250 °C)	[59]
11	TiO_2 nanowires	CVD Substrate: borosilicate glass substrates Temperature: 660 °C 740 °C Time: 0–150 min	D: 20–50 nm L: 3 μm	[46]
12	TiO_2 nanotubes	Electrochemical anodization Electrode: Ti foil Distance between electrode: Potential: 60 V Time: 1 h Calcination: 400 °C for 2 h	L:32.76 μm	[60]
13	$TiO_{2-\delta}$ nanobelts	CVD Substrate: Au-coated Si Temperature: 510 °C Total flow rate: 300 cm^3/min Pressure: 0.7–100 Torr	L: 2–4 μm Width (0.3–0.4 μm), thickness (100–200 nm)	[61]
14	Fe_2O_3/TiO_2 nanorods	Hydrothermal method Temperature: 150 °C Time: 6 h, 12 h, 24 h, 48 h	L: 3.29 μm	[62]
15	TiO_2 nanorods	Microwave: 900 W, at 2.45 GHz Time: 2 min	D: 400 nm	[63]

diameters approximately ranging from 80–130 nm and 160–250 nm, respectively. Differences in size may occur due to volume reduction caused by the elimination of residual organic components and the subsequent compaction process during annealing. Furthermore, selecting a suitable template enables the fabrication of nanorod arrays with desired diameters and lengths [64].

Chiang et al., 2014 investigated the modification TiO_2 nanorods using Cu dopant. They synthesized the Cu–TiO_2 nanorod composites using a combination of microwave-assisted sol–gel synthesis and chemical reduction methodologies. The TEM image displayed in Fig. 10e exhibits the TiO_2 NRs characterized by a particle diameter of 21 nm. Figure 10f illustrates the consistent attachment of copper ions onto the TiO_2 NRs surface subsequent to the chemical reduction of Cu ions with $NaBH_4$. The copper ions on the Cu–TiO_2 NRs consisted primarily of Cu_2O and Cu^0 [65].

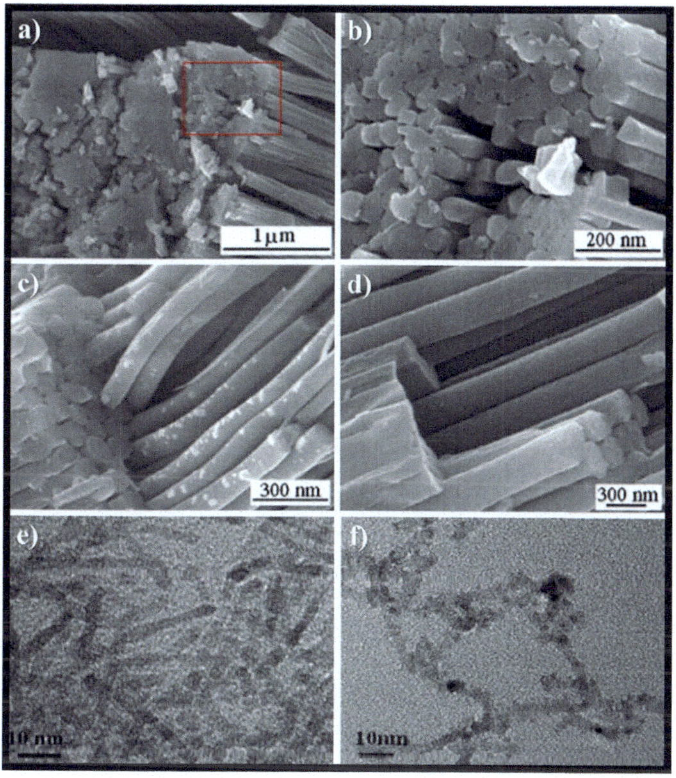

Fig. 10 FE-SEM images of TiO_2 nanorod arrays with diameter of about **a** 40–70 nm; **b** local high magnification image shown in (**a**); **c**, **d** 80–130 and 160–250 nm, respectively. Reprinted with permission from Ref. [64]. Copyright 2009, Elsevier. TEM images of **e** TiO_2 nanorods; **f** Cu–TiO_2 nanorods. Reprinted with permission from Ref. [65]. Copyright 2014, Elsevier

3 Applications of TiO_2 Nanostructures

Through various efficient fabrication techniques, different types of 1D TiO_2 nanostructures have been synthesized. Their remarkable ion-exchange/intercalation activities and adsorption/photocatalytic properties have considerable interest for potential applications. These applications encompass photocatalytic pollutant degradation, CO_2 conversion, water splitting, energy fuels and their roles in supercapacitors, lithium-ion batteries and solar cells. In this section, we will provide utilization of 1-D TiO_2 nanostructures for the pollutants degradation via photocatalysis, electrocatalysis or photoelectrocatalysis.

3.1 Photocatalysis

TiO_2 nanostructured materials stand out as effective photocatalysts due to their highest surface area, bio-compatibility, non-toxicity, and enhansive reaction rates. They exhibit exceptional performance in addressing both aqueous and atmospheric pollutants, showcasing good photocatalytic activity. Moreover, TiO_2 displays significant potential in the photo-reduction of CO_2 into hydrocarbon energy fuels. With the rapid growth of economies, environmental issues, notably clean water scarcity and contamination, have become increasingly severe. Addressing these challenges requires immediate and effective solutions. TiO_2 stands out as a remarkable photocatalyst owing to its cost-effectiveness, nontoxic nature, enduring stability, and potent oxidizing capabilities. These qualities make it a valuable tool for breaking down unwanted organic compounds, offering hope for mitigating environmental degradation. When subjected to UV light, electrons within the valence band undergo a transition to the conduction band due to photon energy absorption. In the conduction band, these electrons can initiate reactions with oxygen molecules, yielding peroxide anions. Meanwhile, the vacancies left in the valence band, or "holes," react with water molecules, producing hydroxyl radicals. This simultaneous mechanism facilitates the degradation of pollutants by actively breaking down undesired compounds.

TiO_2 NBs synthesized through the hydrothermal technique using concentrated alkaline NaOH, effectively degrading pharmaceuticals under UV light. TiO_2 NBs decomposes carbamazepine, naproxen, and theophylline in water. Elevated temperatures, alkaline conditions, and higher concentrations expedite degradation primarily by generating hydroxyl radicals. Notably, the photocatalytic mechanism exhibits non-selective removal of persistent organic compounds, showcasing its potential for water treatment. These findings offer a promising approach for developing eco-friendly methods to combat water contamination, aiding in the removal of persistent pollutants from water effluents [67]. Naraginti et al. synthesized the ternary ZrO_2/Ag@TNR photocatalyst using hydrothermal and photo deposition techniques. TEM imaging unveiled a distinct evolution from TiO_2 nanoparticles to nanorods, as depicted in Fig. 11a. Further treatment involving ZrO_2 and Ag decoration led to the formation of nanoparticles, all under 10 nm in size, evenly distributed on the TNR surface. These observations strongly support the fabrication of ZrO_2/Ag@TNR nanocomposites, as demonstrated in Fig. 11b. The photocatalytic degradation of azithromycin (AZY) degradation using ZrO_2/Ag@TNR in the presence of visible light was shown in Fig. 11c. While TiO_2 and ZrO_2@TNR showed minimal degradation (<5%) over 8 h, Ag@TNR and ZrO_2/Ag@TNR exhibited higher degradation rates, with the latter achieving over 90% efficiency. This superior performance is attributed to the augmented surface area and effective charge transfer between metals and metal oxides. The probable mechanism for the degradation of AZY using ZrO_2/Ag@TNR was shown in Fig. 11d. When subjected to visible light, adjacent Ag becomes excited, leading to the formation of electron–hole pairs. These electrons swiftly migrate to the conduction band (CB) of both titanium dioxide (TiO_2)

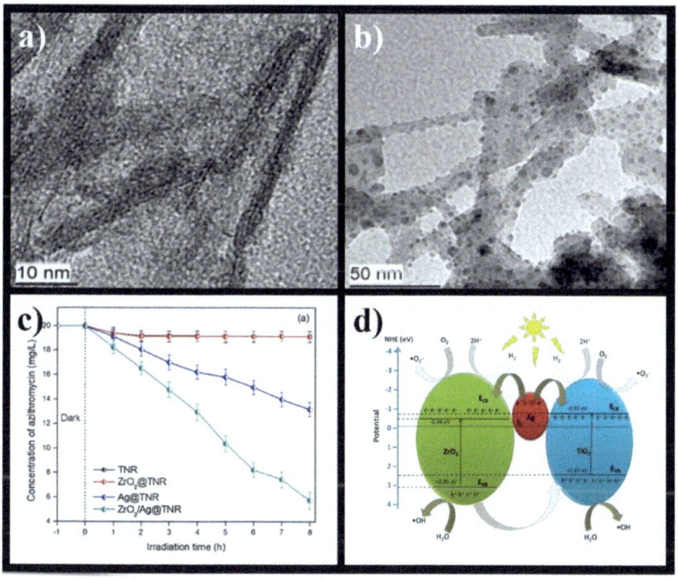

Fig. 11 TEM images of **a** TNR, **b** ZrO_2/Ag@TNR; **c** photocatalytic degradation of AZY by different synthesized nanorods under visible light; **d** proposed photocatalytic mechanism of ZrO_2@Ag/TNR nanocomposites. Reprinted with permission from Ref. [66]. Copyright 2019, Elsevier

and zirconium dioxide (ZrO_2), intensifying the photocatalytic capacity. Additionally, photogenerated electrons have the ability to transfer from silver to defect levels within ZrO_2 and TiO_2, prolonging the persistence of charge carriers. Consequently, this ternary composite facilitates efficient separation of photo-induced carriers, minimizing recombination and promoting exceptional photocatalytic performance [66]. Table 2 presents a diversity of 1D TiO_2 nanocomposites with detailed insights into their photocatalytic performance.

Emerging contaminants is a contributors to environmental pollution, raising considerable global concern. Utilization of 1-D TiO_2 and modified 1-D TiO_2 photocatalysts offers a feasible solution for the removal of emerging contaminants from wastewater.

3.2 *Electrocatalysis*

Electrocatalysis focuses on electrochemical reactions occurring at the interface between an electrode and an electrolyte. In electrocatalysis, redox reactions proceed within an electrochemical cell in the presence of an electrocatalyst. This catalyst

Table 2 Photocatalytic degradation of various emerging contaminants using 1D TiO_2 nanostructures

S. No	Catalyst	Targeted pollutants	Reaction conditions	Degradation efficiency	Ref.
1	TNTs TNWs TNRs	Amicarbazone	Light source: blacklight lamps (23.3 W/cm^2) Pollutant concentration: 10 mg/L Time: 120 min	TNTs = 23%, TNRs = 5%, TNWs = 53%	[68]
2	TNTs	Caffeine (CAF)	Light source: UV lamp 0.5 (0.5 mW/cm^2) Pollutant concentration: 15 mg/L Time: 4 h	51%	[69]
3	TiO_2/TiNTs	Phenanthrene	Light source: 150 W UV lamp (2.5 mW/cm^2) Pollutant concentration: 200 μg/L Time: 4 h	> 95%	[70]
4	ZnO/TiO_2-nanorod MXene	Antibiotic ceftriaxone sodium	Light source: 150 W halogenate high pressure lamp (100 mW/cm^2) Pollutant concentration: 5 mg/L Time: 180 min	Rate constant (k): k_1 = 0.01597 min^{-1} $k_2 = 0.00263$ g mg^{-1} min^{-1}	[71]
5	$K_2Ti_6O_{13}/TiO_2$ (HKT-5)	Tetracycline	Light source: 300 W xenon lamp (260 mW/cm^2) Pollutant concentration: 20 mg/L Time: 60 min	> 90%	[72]
6	Fe/SneTiO_2	Tetracycline	Light source: 300 W Xenon arc lamp (100 mW/cm^2) Pollutant concentration: 30 mg/L Time: 60 min	96.96%	[72]
7	g-C_3N_4/TiO_2	Ibuprofen	Light source: visible light (50 mW/cm^2) Pollutant concentration: 10 ppm Time: 90 min	95%	[73]

(continued)

Table 2 (continued)

S. No	Catalyst	Targeted pollutants	Reaction conditions	Degradation efficiency	Ref.
8	g-C_3N_4/TNTs	Sulfamethazine	Light source: 450 W xenon-ozone free short arc lamp (85 ± 0.5) Pollutant concentration: 5 mg/L Time: 5 h	100%	[74]
9	TiO_2 nanotube	2,3-dichlorophenol	Light source: 8-W medium pressure mercury lamp Pollutant concentration: 20 mg/L Time: 300 min	93%	[75]
10	Rutile TiO_2	Cinnamic acid (CA) Ibuprofen (IBP)	Light Source: 60 W UV–Vis lamp Pollutant concentration: 10 ppm Time: 5 h	CA = 98%, IBP = 80%	[76]
11	Ag/TiO_2 nanotube arrays	17α-Ethinylestradiol	Light source: 8W low-pressure UV lamp (2.36 × 10^{-7} einstein/s) Pollutant concentration: 5 mg/L Time: 120 min	92.1%	[77]
12	N-doped TiO_2 nanotube arrays	Acephate	Light source: sunlight Pollutant concentration: 8 × 10^{-4} mol/L Time: 2 h	84%	[78]
13	Cu–TiO_2 nanorod	Bisphenol A	Light source: 8 W UV or visible lamps Pollutant concentration: 10 mg/L Time: 120 min	k: 0.116 ± 0.022 min^{-1}	[65]
14	N-TiO_2 NTs	Acetaminophen	Light source: 400 W linear halogen lamp Pollutant concentration: 5 mg/L Time: 210 min	95.7%	[79]

(continued)

Table 2 (continued)

S. No	Catalyst	Targeted pollutants	Reaction conditions	Degradation efficiency	Ref.
15	TiO_2 nanowires	Atrazine	Light source:: photocatalysis Light source: two 15 W Philips UV light lamps (2.47 ± 0.16 mW cm^{-2}) Pollutant concentration: 5 mg/L Time: 60 min	k: 0.019 min^{-1}	[80]
16	graphene-TiO_2 nanowires	Metronidazole	Light source: 20 W, tungsten-halogen Pollutant concentration: 35 mg/L Time: 80 min	99.26%	[81]
17	N, S-co doped TiO_2 nanowires	Atrazine	Light source: two 15 W fluorescent lamps (3.99×10^{-4} Wcm^{-2}) Pollutant concentration: 5 mg/L Time: 6 h	60%	[82]
18	Graphene encapsulated TiO_2 nanowires	Nitrophenol	Light source: LED lamp Pollutant concentration: 0.5 mM in 50 ml Time: 100 min	> 95%	[83]
19	TiO_2/CdS/Au	2-nitrophenol (2-np), 4-nitrophenol (4-np), CIP	Light source: solar light Pollutant concentration: 2-np = 100 ppm, 4-np = 50 ppm CIP = 20 mg/L Time: 60 min	2-np = 99% 4-np = 67.68%, CIP = 99%	[84]
20	CdS QDs loaded hollow TiO_2 nanofibers	4-NP	Light source: UV light (1.07 mW/cm^2) Pollutant concentration: 10^{-4} M, Time: 90 min	k: 0.1078 ± 0.0054 min^{-1}	[85]
21	FeOOH@TiO_2 and Fe_2O_3@TiO_2	1H-benzotriazole (BT)	Light source: 15 W fluorescent bulb (1.35 mW/cm^2) Pollutant concentration: 5 mg dm^{-3} Time: 60 min	FeOOH@TiNT, k = 0.00123 0.8 mM_Fe_2O_3@TiNT = 0.00111	[86]

reduces the overpotential of the electrochemical reactions, facilitating specific electrochemical reactions. The efficiency and selectivity of electrocatalytic reactions are determined by the electrocatalyst utilized.

A study was conducted to explore the electrochemical oxidation efficiency of carbamazepine utilizing a distinctive blue-colored TiO_2 NTA membrane filter anode. Featuring a porous structure, the synthesized materials displayed a specific surface area of 2.01 $m^2\ g^{-1}$ and an electroactive surface area of 9.32 $cm^2\ cm^{-2}$. They exhibited an oxygen evolution potential of 2.61 V versus SCE. When operated in a flow-through mode, these materials efficiently decomposed aqueous carbamazepine according to a first-order kinetic model, owing to the presence of sulfate radicals ($SO_4^{\cdot -}$), oxygen radicals ($O_2^{\cdot -}$), and hydroxyl radicals ($^{\cdot}OH$) generated during the process. At the optimized current density of 3 mA cm^{-2}, the degradation rate constant was determined as 0.403 min^{-1}, with an Electrical Energy per Order of 0.086 kW h m^{-3}. Subsequently, after 180 min, mineralization efficiency reached 50.8%, alongside a mineralization current efficiency of 9.5% [87]. Zeng et al. examined the electrochemical degradation of sulfamethoxazole utilized a titanium membrane coated with Sb-SnO_2 coated reactive electrochemical membrane (REM) over reduced TiO_2 nanotube arrays (RNTA), presenting an eco-friendly and highly efficient approach. The use of antimony (Sb) as a dopant for SnO_2 exhibited greater efficacy in enhancing electrocatalytic activity compared to bismuth (Bi). Moreover, the study revealed that Sb-SnO_2-coated REM (TM/RTNA/ATO) exhibited lower energy consumption compared to Bi-SnO_2-coated REM (TM/RTNA/BTO). Moving from batch mode to flow-through mode resulted in a remarkable enhancement, increasing the efficiency of sulfamethoxazole removal by up to 8.4 times. Factors like fluid velocity, initial sulfamethoxazole concentration, current density, and electrolyte concentration played a role in determining sulfamethoxazole removal efficiency and energy consumption [88].

By employing a combination of seed-assisted growth and nitridation transformation, scientists have engineered a specialized configuration of hollow nanorods, incorporating TiN/TiO_2 heterostructures within. Described as TiN/TiO_2-n, where 'n' (1, 2, 3) indicated the concentration of dicyanamide (400, 600, 800 mg) added. As shown in Fig. 12a. the surface of the TiO_2 nanorod appears smooth surface. After undergoing nitridation, the surface morphology of TiN/TiO_2-1 transitions to a roughened state, characterized by the presence of numerous black nanoparticles dispersed on its surface, as illustrated in Fig. 12b, indicating the surface interaction of TiO_2 nanorods with HCN and NH_3. Furthermore, the augmentation of dicyanamide dosage results in the emergence of a hollowed structure in TiN/TiO_2-2 (Fig. 12c), indicating continuous surface reactions and the inward migration of titanium species within the nanorods, ultimately leading to the progressive formation of the hollowed structure (Fig. 12d). As depicted in Fig. 12e, the composite of TiN/TiO_2-2 exhibits the lowest values for both charge transfer resistance (R_{ct}) and internal resistance (R_{Ω}), indicating enhanced charge transfer efficiency at the NRs surface and interface. Figure 12f illustrates the cyclic voltammetry (CV) of TiO_2/CC, TiN/TiO_2-1, TiN/TiO_2-2, and TN/TiO_2-3, displaying distinguishable oxidation as well as reduction peaks suggestive of Ti species undergoing redox reactions. As

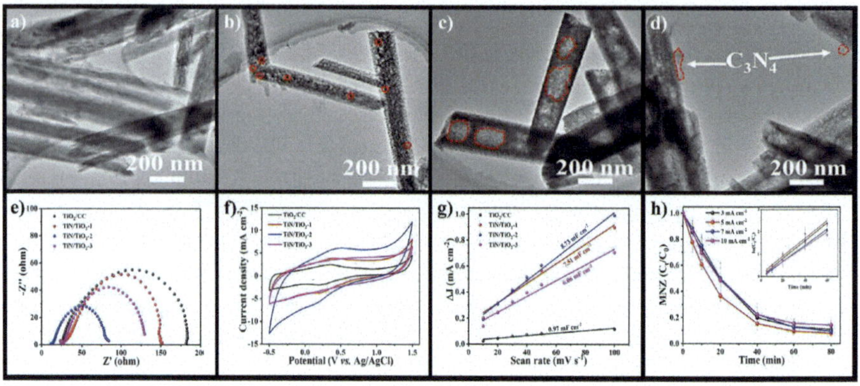

Fig. 12 TEM images of the **a** TiO_2/CC, **b** TiN/TiO_2-1, **c** TiN/TiO_2-2, **d** TiN/TiO_2-3, **e** EIS, **f** CV curves and **g** ESCA of the TiO_2/CC, TiN/TiO_2-1, TiN/TiO_2-2 and TiN/TiO_2-3 composites, effects of the reaction environment to the electrochemical oxidation performance of the TiN/TiO_2-2 electrode: **h** current densities. Reprinted with permission from Ref. [89]. Copyright 2023, Elsevier

the dicyanamide quantity increases, notable changes are observed in the TiN/TiO_2-n CV curve, suggesting enhanced electrochemical activity. The TiN/TiO_2-2 composite demonstrates an electrochemical active surface area (ESCA) approximately nine times greater as compared to TiO_2/CC (Fig. 12g), reflecting its substantial accessible surface. The efficiency of metronidazole removal displays an initial increase followed by a decrease with varying current densities (Fig. 12h). Notably, the highest removal efficiency of 92.51% is observed at a current density of 5 mA cm^{-2}, primarily due to the occurrence of the oxygen evolution side reaction at high current densities [89]. Table 3 summarizes the EC degradation of various emerging contaminants using 1D TiO_2 nanostructures.

Utilizing advanced electrocatalytic approaches and materials provides a promising strategy for efficiently managing these emerging contaminants in wastewater, thus aiding environmental restoration efforts.

3.3 Photoelectrocatalysis (PEC)

PEC, a technology combining heterogeneous photocatalysis and electrochemical oxidation, shows potential in the degradation of emerging pollutants. By uniting these methods, PEC provides a viable solution, that can tackles the challenge of electron–hole pair recombination resulting from the semiconductor anode's exposure to irradiation, leading to enhanced hole lifetimes and improved efficiency in contaminant oxidation.

Cheng et al. introduced a simple technique for synthesizing reduced TiO_2 NTs as photoelectrodes. In the FESEM images, TiO_2 NTs were depicted with 120 nm

Table 3 EC treatment of various emerging contaminants using 1D TiO_2 nanostructures

S. No	Catalyst	Targeted pollutants	Reaction conditions	Degradation efficiency	Ref
1	TiN/TiO_2	Metronidazole	Electrode: TiN/TiO_2 (anode), stainless-steel mesh (cathode) Current density: 5 mA cm^{-2} Pollutant concentration: 30 ppm Time: 80 min	97.97%	[89]
2	TiO_2 nanotube	Tetracycline (TC) and metronidazole (MNZ)	Electrode: TiO_2 nanotube (anode), stainless-steel plate (cathode) Current density: 10 mA cm^{-2} Pollutant concentration: TC = 50 mg/L, MNZ = 30 mg/L Time: 240 min	99%	[90]
3	TiO_2 nanotube	Nitrobenzene	Electrode: TiO_2 NTA (working), Pt mesh (counter), Hg/HgO (reference) Applied potential: –1.20 V (vs. SCE) Pollutant concentration: 300 mg/L Time: 8 h	95%	[91]
4	TM/RTNA/ATO	Sulfamethoxazole	Electrode: TM/RTNA/ATO (anode), porous platinum plate (cathode) Applied potential: 3.3–3.5 V Pollutant concentration: 10 μM SMX in 80 ml Time: 18 min	91%	[88]
5	Co/Bi/TiO_2 nanotube	Meropenem	Electrode: Co/Bi/TiO_2 nanotube (working) Current density: 4–12 mA cm^{-2} Pollutant concentration: 500 μg L^{-1} Time: 15 min	94%	[92]

(continued)

Table 3 (continued)

S. No	Catalyst	Targeted pollutants	Reaction conditions	Degradation efficiency	Ref
6	TiO_2 nanotubes	Salicylic acid	Electrode: TiO_2 nanotubes (working), Pt coil (counter), Ag/AgCl (reference) Current density: 2–5 mA cm^{-2} Pollutant concentration: 30 ppm Time: 3 h	90%	[93]
7	TiO_2 nanotube	Sulfamethoxazole	Electrode: TiO_2 nanotube (anode), Ti plate (cathod) Current density: 2 mA/cm^2 Pollutant concentration: 80 mL SMX, 10 μM Time: 50 min	86.1%	[94]
8	TNTs/3D-SnO_2	Benzoic acid	Electrode: TNTs/3D-SnO_2 (anode), Stainless steel (cathod) Current density: 25 mA cm^{-2} Pollutant concentration: 120 mg/L Time: 4 h	95.6%	[95]
9	TiO_2 nanotube array	Chloramphenicol	Electrode: TiO_2 nanotube array (working), Ti plate (cathode) Current density: 2–10 mA cm^{-2} Pollutant concentration: 5–30 mg/L Time: 60 min	95%	[96]

average diameter and a wall thickness of 20 nm, as illustrated in Fig. 13a, b. Moreover, as shown in the inset of Fig. 13a, b, while the bare TiO_2 NTs had a tube length of about 5 μm, the reduced TiO_2 NTs showed a decreased length of 3.5 μm, thus resulting the breakdown of surface nanotubes during the microwave reduction technique. The reduced TiO_2 NTs photoanode achieved an enhancive photocurrent of 0.147 mA cm^{-2}, compared to 0.067 mA cm^{-2} for the bare TiO_2 NTs, indicating enhanced separation of the of photogenerated electron–hole charge carriers, as observed in Fig. 13c. However, under Xenon lamp irradiation for 10 h, reduced TiO_2 NTs exhibited superior photocatalytic (PC) performance (99.5%) compared to bare TiO_2 as shown in Fig. 13d. Enhanced visible light absorption and charge separation efficiency contributed to this improvement. The proposed mechanism of the diclofenac degradation using TiO_2 NTs was shown in Fig. 13e. Weak light absorption and rapid recombination of charge carriers posed challenges for the TiO_2 NTs photoelectrode [97].

Kmentova et al., investigated the fabrication of TiO_2 nanorods on the surface of transparent FTO substrate [98]. TiO_2 NRAs were fabricated via hydrothermal methods, while nanotube arrays were created through electrochemical anodization. Despite identical annealing conditions, nanorods exhibited rutile crystalline structure, while nanotubes showed anatase phase. However, both nanotube photoanodes displayed inferior photoelectrochemical (PEC) performance compared to nanorod-based photoanode, attributed to prolonged higher purity, electron lifetime, and enhanced light absorption into visible spectrum. The lowest PEC activity was obtained in hydrothermally synthesized nanotubes utilizing ZnO NRs as sacrificial templates, likely due to limited development of nanotube-FTO substrate interface.

Teng et al. investigated the MoS_2 nanobelts loaded TiO_2 NTAs using photo electrodeposition method for the degradation of sulfadiazinmu. The degradation rates achieved by MoS_2/TiO_2 NTs are more than 3 times greater than those observed with pure TiO_2 NTs. The higher effectiveness of MoS_2/TiO_2 relative to pure TiO_2 is attributed to the visible light absorption of MoS_2, its band structure compatibility with TiO_2, and the enlarged surface area resulting from MoS_2 nanoparticle integration, leading to enhanced visible-light photoelectrocatalytic activity in pollutant degradation [99].

A significant focus in the future should be directed towards improving the photoelectrocatalytic degradation of pollutants by modifying 1D TiO_2 nanostructured materials. Table 4 indicates the PEC degradation of various emerging contaminants using 1D TiO_2 nanostructures.

4 Conclusions

The exploration of 1D TiO_2 nanostructured materials has been a focal point among many researchers. This chapter represents the advancements in fabricating, modifying, and employing 1D TiO_2-based materials, underscoring the significance of meticulous size and morphology regulation. In accordance with the fundamental principles governing bulk material formation through solid-state reactions, it has employed various techniques to synthesize one-dimensional TiO_2 nanostructures. These techniques include hydrothermal synthesis, chemical vapor deposition, electrochemical anodization, template-assisted fabrication, electrospinning, and sol–gel processes, each offering unique advantages in terms of control and scalability of 1-D nanostructures. Enhancements in the performance of 1D TiO_2 nanomaterials often involve strategies like expanding the surface area and active sites by incorporating metals, nonmetals, and semiconductors. In various sectors, 1D TiO_2 nanostructured materials are extensively employed due to their large specific surface area, high aspect

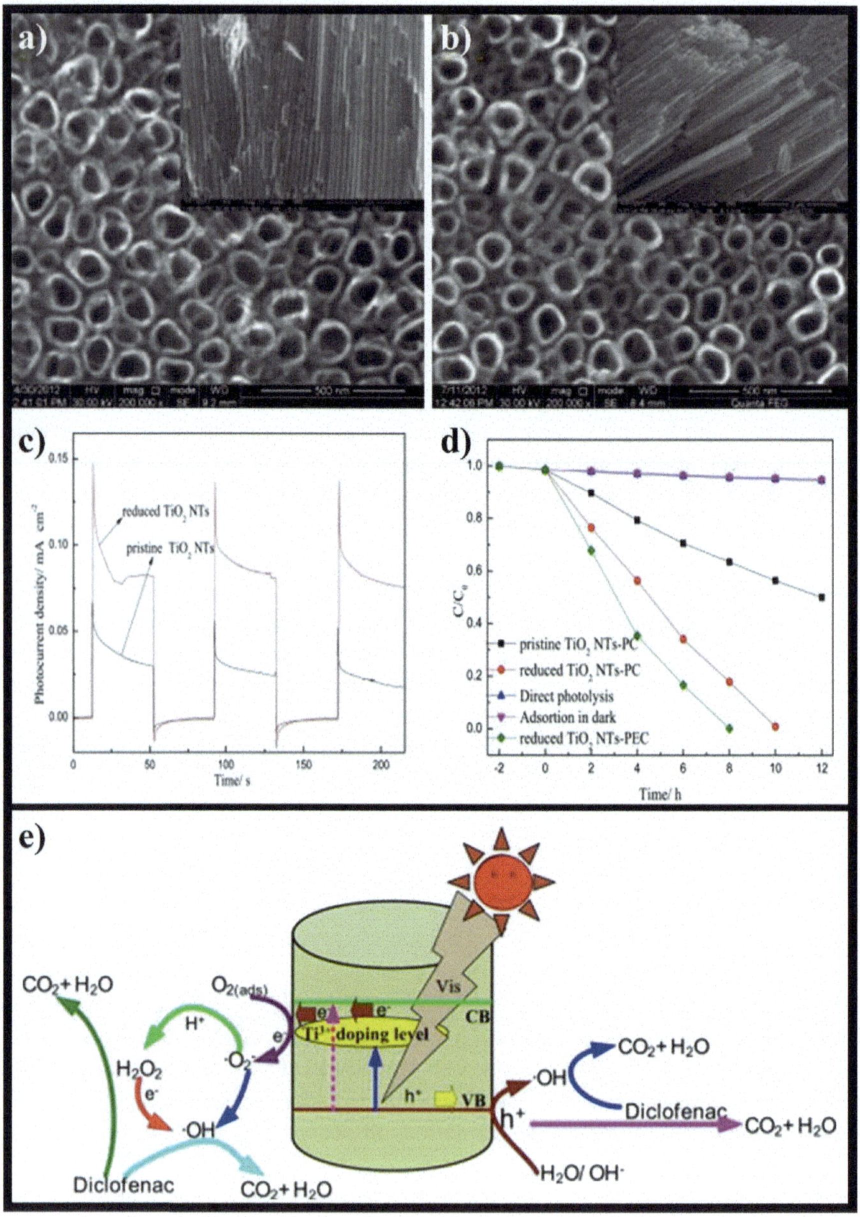

Fig. 13 Typical top-view and side view (inset) FE-SEM images (**a**: pristine, **b**: reduced); **c** photocurrent density (vs. SCE); **d** PC degradation of diclofenac; **e** the probable mechanism for degradation of diclofenac using pristine and reduced TiO_2 NTs photoelectrodes. Reprinted with permission from Ref. [97]. Copyright 2016, Elsevier

Table 4 PEC treatment of several emerging contaminants using 1D TiO_2 nanostructures

S. No	Catalyst	Targeted pollutants	Reaction conditions	Degradation efficiency	Ref.
1	MoS_2/TiO_2	Levofloxacin	Electrode: MoS_2/ TiO_2 nanotube (photoanode), Pt foil (counter), SCE (reference) Applied potential: 0.3 V Light source: 500 W Xenon lamp (33 W/cm2) Pollutant concentration: 10 mg/L Time: 180 min	100%	[100]
2	NCDs/TiO2 NA-Vo OXYGEN vacancies (Vo)	Tetracycline (TC)	Electrode: Ti foam (photoanode), Pt foil (counter), SCE (reference) Applied potential: 0.6 V Light source: 300 W Xe lamp Pollutant concentration: 50 mg/L Time: 3.5 h	70%	[101]
3	$MoSe_2/TiO_2$ nanotubes	Cefotaxime sodium	Electrode: $MoSe_2$/ TiO_2 nanotubes (photoanode), Pt foil (counter) Applied potential: 0–0.6 V Light source: 300 W Xe lamp Pollutant concentration: 20 mg/L Time: 180 min	95.2%	[102]

(continued)

Table 4 (continued)

S. No	Catalyst	Targeted pollutants	Reaction conditions	Degradation efficiency	Ref.
4	N, S–TiO_2/TiO_2 NTs	Diclofenac	Electrode: N, S–TiO_2/TiO_2 (photoanode), SCE (reference) Applied potential: 0.4 V Light source: 35 W xenon lamp Pollutant concentration: 5 mg/L Time: 12 h	73.3%	[103]
5	$Bi_2Sn_2O_7$/TiO_2 NTAs	Sulfamethazine	Electrode: $Bi_2Sn_2O_7$/TiO_2 NTAs (photoanode), Pt foil (counter) Applied potential: 1 V Light source: 50 W LED lamp Pollutant concentration: 10 mg/L Time: 120 min	90.3%	[104]
6	TiO_2 NNs/Ti, NCDs/Co_3O_4/Ti	Sulfadiazine	Electrode: TiO_2 NNs/Ti mesh (photoanode), NCDs/Co_3O_4/Ti mesh (photocathode) Applied potential: Light source: 30 W LED lamp Pollutant concentration: 10 mg/L Time: 75–360 min	98.54%	[105]

(continued)

Table 4 (continued)

S. No	Catalyst	Targeted pollutants	Reaction conditions	Degradation efficiency	Ref.
7	BiOBr/TNAs	p-Chloronitrobenzene	Electrode: BiOBr/TNAs (photoanode), Pt foil (counter) Applied potential: 0–1.5 V Light source: 300 W Xenon lamp Pollutant concentration: 50 mL p-CNB (0.1 mM) Time: 360	71.96%	[106]
8	Au/g-C_3N_4/TNAs	o-chloronitrobenzene	Electrode: Au/g-C_3N_4/TNAs Applied potential: 1.5 V Light source: Xenon lamp Pollutant concentration: 50 mL o-CNB (0.18 mM) Time: 300 min	80.8%	[107]
9	Mn0.5Zn0.5Fe_2O_4/TiO_2 NTAs	2,4-dichlorophenol	Electrode: Mn0.5Zn0.5Fe_2O_4/TiO_2 NTAs (photoanode), Pt foli (counter), SCE (reference) Applied potential: 0.6 V Light source: Sunlight (36 mW cm^{-2}) Pollutant concentration: 20 mg/L Time: 120 min	100%	[108]

(continued)

Table 4 (continued)

S. No	Catalyst	Targeted pollutants	Reaction conditions	Degradation efficiency	Ref.
10	N doped CQDs/ TiO_2 NW array	Tetracycline	Electrode: N doped CQDs/TiO_2 NW array on Ti foam (2 cm × 4 cm) Applied potential: 0.6 V Light source: 300 W xenon lamp	96%	[101]
11	$BiVO_4/TiO_2$ NRs	4-NP	Electrode: $BiVO_4/$ TiO_2 NRs Pollutant concentration: 20 mg/L Light source: 500 W Xe lamp	95.7%	[109]

ratio, and impressive ionic or electronic charge transfer properties. The versatile nature of 1D TiO_2 nanostructured materials, attributed to their elongated aspect ratio, substantial specific surface area, and impressive charge transfer properties, enables their widespread adoption and integration into diverse industrial processes and applications. These include the degradation of emerging contaminants including phenols, pharmaceuticals and personal care products via PC, EC, and PEC techniques.

References

1. Lakshmi, C.N., Singh, N.: Removal of pharmaceutical compounds: overview of treatment methods. In: Singh Swatantra, P., Agarwal, A.K., GT and MSM (eds.) New Trends in Emerging Environmental Contaminants, pp. 161–180. Springer Singapore, Singapore (2022)
2. Nawaz, T., Sengupta, S.: Chapter 4—contaminants of emerging concern: occurrence, fate, and remediation. In: Ahuja, S. (ed.) Advances in Water Purification Techniques, pp. 67–114. Elsevier (2019)
3. Murgolo, S., De Ceglie, C., Di Iaconi, C., Mascolo, G.: Novel TiO2-based catalysts employed in photocatalysis and photoelectrocatalysis for effective degradation of pharmaceuticals (PhACs) in water: a short review. Curr. Opin. Green Sustain. Chem. **30** (2021)
4. Deegan, A.M., Shaik, B., Nolan, K., et al.: Treatment options for wastewater effluents from pharmaceutical companies. Int. J. Environ. Sci. Tech. **8**, 649–666 (2011)
5. Naga Lakshmi, C., Irfan, M., Sinha, R., Singh, N.: Magnetically recoverable Ni-doped iron oxide/graphitic carbon nitride nanocomposites for the improved photocatalytic degradation of ciprofloxacin: investigation of degradation pathways. Environ. Res. **242** (2024). https://doi.org/10.1016/j.envres.2023.117812
6. Mishra, S., Sundaram, B.: A review of the photocatalysis process used for wastewater treatment. Mater. Today Proc. (2023). https://doi.org/10.1016/j.matpr.2023.07.147
7. Mohammad, I., Jeshurun, A., Ponnusamy, P., Reddy, B.M.: Mesoporous graphitic carbon nitride/hydroxyapatite (g-C3N4/HAp) nanocomposites for highly efficient photocatalytic

degradation of rhodamine B dye. Mater. Today Commun. **33** (2022). https://doi.org/10.1016/j.mtcomm.2022.104788
8. Wu, S., Hu, Y.H.: A comprehensive review on catalysts for electrocatalytic and photoelectrocatalytic degradation of antibiotics. Chem. Eng. J. **409** (2021)
9. Divyapriya, G., Singh, S., Martínez-Huitle, C.A., et al.: Treatment of real wastewater by photoelectrochemical methods: an overview. Chemosphere **276** (2021). https://doi.org/10.1016/j.chemosphere.2021.130188
10. Ge, M., Cao, C., Huang, J., et al.: A review of one-dimensional TiO2 nanostructured materials for environmental and energy applications. J. Mater. Chem. A Mater. **4**, 6772–6801 (2016)
11. Pelaez, M., Nolan, N.T., Pillai, S.C., et al.: A review on the visible light active titanium dioxide photocatalysts for environmental applications. Appl. Catal. B **125**, 331–349 (2012)
12. Asahi, R., Taga, Y., Mannstadt, W., Freeman, A.J.: Electronic and optical properties of anatase TiO 2
13. Weng, B., Liu, S., Tang, Z.R., Xu, Y.J.: One-dimensional nanostructure based materials for versatile photocatalytic applications. RSC Adv. **4**, 12685–12700 (2014)
14. Li, Y., Yang, X.Y., Feng, Y., et al.: One-dimensional metal oxide nanotubes, nanowires, nanoribbons, and nanorods: synthesis, characterizations, properties and applications. Crit. Rev. Solid State Mater. Sci. **37**, 1–74 (2012)
15. Soto-Vázquez, L., Cotto, M., Ducongé, J., et al.: Synthesis and photocatalytic activity of TiO2 nanowires in the degradation of p-aminobenzoic acid: a comparative study with a commercial catalyst. J. Environ. Manage. **167**, 23–28 (2016). https://doi.org/10.1016/j.jenvman.2015.11.006
16. Aravind, M., Amalanathan, M., Aslam, S., et al.: Hydrothermally synthesized Ag-TiO2 nanofibers (NFs) for photocatalytic dye degradation and antibacterial activity. Chemosphere **321** (2023). https://doi.org/10.1016/j.chemosphere.2023.138077
17. Gupta, T., Samriti, Cho, J., Prakash, J.: Hydrothermal synthesis of TiO2 nanorods: formation chemistry, growth mechanism, and tailoring of surface properties for photocatalytic activities. Mater. Today Chem. **20** (2021)
18. Yang, Z., Wang, B., Cui, H., et al.: Synthesis of crystal-controlled TiO2 nanorods by a hydrothermal method: rutile and brookite as highly active photocatalysts. J. Phys. Chem. C **119**, 16905–16912 (2015). https://doi.org/10.1021/acs.jpcc.5b02485
19. Odhiambo, V.O., Ongarbayeva, A., Kéri, O., et al.: Synthesis of TiO2/WO3 composite nanofibers by a water-based electrospinning process and their application in photocatalysis. Nanomaterials **10** (2020). https://doi.org/10.3390/nano10050882
20. Santhi, K., Navaneethan, M., Harish, S., et al.: Synthesis and characterization of TiO2 nanorods by hydrothermal method with different pH conditions and their photocatalytic activity. Appl. Surf. Sci. **500** (2020). https://doi.org/10.1016/j.apsusc.2019.144058
21. Akilavasan, J., Wijeratne, K., Moutinho, H., et al.: Hydrothermally synthesized titania nanotubes as a promising electron transport medium in dye sensitized solar cells exhibiting a record efficiency of 7.6% for 1-D based devices. J. Mater. Chem. A Mater. **1**, 5377–5385 (2013). https://doi.org/10.1039/c3ta01576a
22. Nian, J.N., Teng, H.: Hydrothermal synthesis of single-crystalline anatase TiO2 nanorods with nanotubes as the precursor. J. Phys. Chem. B **110**, 4193–4198 (2006). https://doi.org/10.1021/jp0567321
23. Zhang, Y.X., Li, G.H., Jin, Y.X., et al.: Hydrothermal synthesis and photoluminescence of TiO 2 nanowires
24. Soto-Vázquez, L., Rolón-Delgado, F., Rivera, K., et al.: Catalytic use of TiO2 nanowires in the photodegradation of Benzophenone-4 as an active ingredient in sunscreens. J. Environ. Manage. **247**, 822–828 (2019). https://doi.org/10.1016/j.jenvman.2019.07.005
25. Pugazhenthiran, N., Murugesan, S., Anandan, S.: High surface area Ag-TiO2 nanotubes for solar/visible-light photocatalytic degradation of ceftiofur sodium. J. Hazard. Mater. **263**, 541–549 (2013). https://doi.org/10.1016/j.jhazmat.2013.10.011
26. Zulfiqar, M., Chowdhury, S., Omar, A.A.: Hydrothermal synthesis of multiwalled TiO2 nanotubes and its photocatalytic activities for Orange II removal. Sep. Sci. Technol. (Philadelphia) **53**, 1412–1422 (2018). https://doi.org/10.1080/01496395.2018.1444050

27. Hu, A., Zhang, X., Oakes, K.D., et al.: Hydrothermal growth of free standing TiO2 nanowire membranes for photocatalytic degradation of pharmaceuticals. J. Hazard. Mater. **189**, 278–285 (2011). https://doi.org/10.1016/j.jhazmat.2011.02.033
28. Doh, S.J., Kim, C., Lee, S.G., et al.: Development of photocatalytic TiO2 nanofibers by electrospinning and its application to degradation of dye pollutants. J. Hazard. Mater. **154**, 118–127 (2008). https://doi.org/10.1016/j.jhazmat.2007.09.118
29. Sarkodie, B., Amesimeku, J., Frimpong, C., et al.: Photocatalytic degradation of dyes by novel electrospun nanofibers: a review. Chemosphere **313** (2023)
30. Sivaprakash, V., Narayanan, R.: Synthesis of TiO2 nanotubes via electrochemical anodization with different water content. In: Materials Today: Proceedings, pp. 142–146. Elsevier Ltd. (2020)
31. Misra, M., Singh, N., Gupta, R.K.: Enhanced visible-light-driven photocatalytic activity of Au@Ag core-shell bimetallic nanoparticles immobilized on electrospun TiO2 nanofibers for degradation of organic compounds. Catal. Sci. Technol. **7**, 570–580 (2017). https://doi.org/10.1039/c6cy02085b
32. Kiennork, S., Nakhowong, R., Chueachot, R., Tipparach, U.: Preparation and characterization of electrospun TiO2 nanofibers via electrospinning. In: Integrated Ferroelectrics, pp. 131–137. Taylor and Francis Ltd. (2015)
33. Lee, S.S., Bai, H., Liu, Z., Sun, D.D.: Novel-structured electrospun TiO2/CuO composite nanofibers for high efficient photocatalytic cogeneration of clean water and energy from dye wastewater. Water Res. **47**, 4059–4073 (2013). https://doi.org/10.1016/j.watres.2012.12.044
34. Song, J., Guan, R., Xie, M., et al.: Advances in electrospun TiO2 nanofibers: design, construction, and applications. Chem. Eng. J. **431** (2022)
35. Jin, X., Yuan, K., Xu, C., et al.: Water steam modified crystallization and microstructure of mesoporous TiO2 nanofibers. Ceram. Int. **44**, 2158–2164 (2018). https://doi.org/10.1016/j.ceramint.2017.10.168
36. Lee, C.G., Javed, H., Zhang, D., et al.: Porous electrospun fibers embedding TiO2 for adsorption and photocatalytic degradation of water pollutants. Environ. Sci. Technol. **52**, 4285–4293 (2018). https://doi.org/10.1021/acs.est.7b06508
37. Pascariu, P., Cojocaru, C., Homocianu, M., et al.: New La3+ doped TiO2 nanofibers for photocatalytic degradation of organic pollutants: effects of thermal treatment and doping loadings. Ceram. Int. **48**, 4953–4964 (2022). https://doi.org/10.1016/j.ceramint.2021.11.033
38. Suslick, K.S., Hammerton, D.A., Cline, R.E.: Sonochemical hot spot. J. Am. Chem. Soc. **108**, 5641–5642 (1986). https://doi.org/10.1021/ja00278a055
39. Wang, X., Li, Z., Shi, J., Yu, Y.: One-dimensional titanium dioxide nanomaterials: nanowires, nanorods, and nanobelts. Chem. Rev. **114**, 9346–9384 (2014)
40. Calderón-Jiménez, B., Montoro Bustos, A.R., Pereira Reyes, R., et al.: Novel pathway for the sonochemical synthesis of silver nanoparticles with near-spherical shape and high stability in aqueous media. Sci. Rep. **12** (2022). https://doi.org/10.1038/s41598-022-04921-9
41. Zhu, Y., Li, H., Koltypin, Y., et al.: Sonochemical synthesis of titania whiskers and nanotubes. Chem. Commun. **24**, 2616–2617 (2001). https://doi.org/10.1039/b108968b
42. Guo, S., Wu, Z., Wang, H., Dong, F.: Synthesis of mesoporous TiO2 nanorods via a mild template-free sonochemical route and their photocatalytic performances. Catal. Commun. **10**, 1766–1770 (2009). https://doi.org/10.1016/j.catcom.2009.05.027
43. You, Y., Mayyas, M., Xu, S., et al.: Growth of NiO nanorods, SiC nanowires and monolayer graphene: via a CVD method. Green Chem. **19**, 5599–5607 (2017). https://doi.org/10.1039/c7gc02523h
44. Shi, J., Wang, X.: Growth of rutile titanium dioxide nanowires by pulsed chemical vapor deposition. Cryst. Growth Des. **11**, 949–954 (2011). https://doi.org/10.1021/cg200140k
45. Liu, H., Zhang, Y., Li, R., et al.: Synthesis and characterization of TiO 2@C core-shell nanowires and nanowalls via chemical vapor deposition for potential large-scale production. J. Colloid Interface Sci. **367**, 115–119 (2012). https://doi.org/10.1016/j.jcis.2011.09.022
46. Du, J., Gu, X., Guo, H., et al.: Self-induced preparation of TiO2 nanowires by chemical vapor deposition. J. Cryst. Growth **427**, 54–59 (2015). https://doi.org/10.1016/j.jcrysgro.2015.07.004

47. Mohammadpour, F., Moradi, M.: Double-layer TiO2 nanotube arrays by two-step anodization: used in back and front-side illuminated dye-sensitized solar cells. Mater. Sci. Semicond. Process. **39**, 255–264 (2015). https://doi.org/10.1016/j.mssp.2015.04.048
48. Zakir, O., Idouhli, R., Elyaagoubi, M., et al.: Fabrication of TiO2 nanotube by electrochemical anodization: toward photocatalytic application. J. Nanomater **2020** (2020). https://doi.org/10.1155/2020/4745726
49. Devan, R.S., Ma, Y.R., Patil, R.A., Lukas, S.M.: Highly stable supercapacitive performance of one-dimensional (1D) brookite TiO2 nanoneedles. RSC Adv. **6**, 6218–6225 (2016). https://doi.org/10.1039/c6ra11348f
50. Wu, X., Jiang, Q.Z., Ma, Z.F., et al.: Synthesis of titania nanotubes by microwave irradiation. Solid State Commun. **136**, 513–517 (2005). https://doi.org/10.1016/j.ssc.2005.09.023
51. Li, H., Chen, J., Xia, Z., Xing, J.: Microwave-assisted preparation of self-doped TiO2 nanotube arrays for enhanced photoelectrochemical water splitting. J. Mater. Chem. A Mater. **3**, 699–705 (2015). https://doi.org/10.1039/c4ta05021e
52. Yu, Y., Xu, D.: Single-crystalline TiO2 nanorods: highly active and easily recycled photocatalysts. Appl. Catal. B **73**, 166–171 (2007). https://doi.org/10.1016/j.apcatb.2006.07.017
53. Sánchez-García, G., Da Silva, E., Fernández-Domene, R.M., et al.: TiO2 nanostructures synthesized by electrochemical anodization in green protic ionic liquids for photoelectrochemical applications. Ceram. Int. **49**, 26900–26909 (2023). https://doi.org/10.1016/j.ceramint.2023.05.227
54. Qayyum, A., Giannakoudakis, D.A., LaGrow, A.P., et al.: High-frequency sonication for the synthesis of nanocluster-decorated titania nanorods: making a better photocatalyst for the selective oxidation of monoaromatic alcohol. Catal. Commun. **163** (2022). https://doi.org/10.1016/j.catcom.2022.106406
55. Pradhan, S.K., Reucroft, P.J., Yang, F., Dozier, A.: Growth of TiO2 nanorods by metalorganic chemical vapor deposition. J. Cryst. Growth **256**, 83–88 (2003). https://doi.org/10.1016/S0022-0248(03)01339-3
56. Singh, N., Prakash, J., Misra, M., et al.: Dual functional ta-doped electrospun TiO2 nanofibers with enhanced photocatalysis and SERS detection for organic compounds. ACS Appl. Mater. Interfaces **9**, 28495–28507 (2017). https://doi.org/10.1021/acsami.7b07571
57. Gai, L., Duan, X., Jiang, H., et al.: One-pot synthesis of nitrogen-doped TiO 2 nanorods with anatase/brookite structures and enhanced photocatalytic activity. CrystEngComm **14**, 7662–7671 (2012). https://doi.org/10.1039/c2ce25563d
58. Janekbary, K.K., Gilani, N., Pirbazari, A.E.: One-step fabrication of Ag/RGO doped TiO2 nanotubes during anodization process with high photocatalytic performance. J. Porous Mater. **27**, 1809–1822 (2020). https://doi.org/10.1007/s10934-020-00954-5
59. Li, X., Raza, S., Liu, C.: Directly electrospinning synthesized Z-scheme heterojunction TiO2@Ag@Cu2O nanofibers with enhanced photocatalytic degradation activity under solar light irradiation. J. Environ. Chem. Eng. **9** (2021). https://doi.org/10.1016/j.jece.2021.106133
60. Tong, M.H., Chen, Y.X., Lin, S.W., et al.: Synchronous electrochemical anodization: a novel strategy for preparing cerium doped TiO2 nanotube arrays toward visible light PEC water splitting. Electrochim. Acta **463** (2023). https://doi.org/10.1016/j.electacta.2023.142793
61. Hoa, N.T.Q., Kim, E.T.: Self-catalytic growth of TiO2-δ nanobelts and nanosheets using metallorganic chemical vapor deposition. Electrochem. Solid State Lett. **11** (2008). https://doi.org/10.1149/1.2795838
62. Bashiri, R., Samsudin, M.F.R., Mohamed, N.M., et al.: Influence of growth time on photoelectrical characteristics and photocatalytic hydrogen production of decorated Fe2O3 on TiO2 nanorod in photoelectrochemical cell. Appl. Surf. Sci. **510** (2020). https://doi.org/10.1016/j.apsusc.2020.145482
63. Pol, V.G., Langzam, Y., Zaban, A.: Application of microwave superheating for the synthesis of TiO2 rods. Langmuir **23**, 11211–11216 (2007). https://doi.org/10.1021/la7020116
64. Attar, A.S., Ghamsari, M.S., Hajiesmaeilbaigi, F., et al.: Sol-gel template synthesis and characterization of aligned anatase-TiO2 nanorod arrays with different diameter. Mater. Chem. Phys. **113**, 856–860 (2009). https://doi.org/10.1016/j.matchemphys.2008.08.040

65. Chiang, L.F., Doong, R.: Cu-TiO2 nanorods with enhanced ultraviolet- and visible-light photoactivity for bisphenol A degradation. J. Hazard Mater. **277**, 84–92 (2014). https://doi.org/10.1016/j.jhazmat.2014.01.047
66. Naraginti, S., Yu, Y.Y., Fang, Z., Yong, Y.C.: Visible light degradation of macrolide antibiotic azithromycin by novel ZrO2/Ag@TiO2 nanorod composite: transformation pathways and toxicity evaluation. Process Saf. Environ. Prot. **125**, 39–49 (2019). https://doi.org/10.1016/j.psep.2019.02.031
67. Liang, R., Hu, A., Li, W., Zhou, Y.N.: Enhanced degradation of persistent pharmaceuticals found in wastewater treatment effluents using TiO2 nanobelt photocatalysts. J. Nanopart. Res. **15** (2013). https://doi.org/10.1007/s11051-013-1990-x
68. Conceição, D.S., Ferreira, D.P., Graça, C.A.L., et al.: Photochemical and photocatalytic evaluation of 1D titanate/TiO 2 based nanomaterials. Appl. Surf. Sci. **392**, 418–429 (2017). https://doi.org/10.1016/j.apsusc.2016.09.067
69. Arfanis, M.K., Adamou, P., Moustakas, N.G., et al.: Photocatalytic degradation of salicylic acid and caffeine emerging contaminants using titania nanotubes. Chem. Eng. J. **310**, 525–536 (2017). https://doi.org/10.1016/j.cej.2016.06.098
70. Cheng, K., Cai, Z., Fu, J., et al.: Synergistic adsorption of Cu(II) and photocatalytic degradation of phenanthrene by a jaboticaba-like TiO2/titanate nanotube composite: an experimental and theoretical study. Chem. Eng. J. **358**, 1155–1165 (2019). https://doi.org/10.1016/j.cej.2018.10.114
71. Abbas, K.K., Abdulkadhimal-Ghaban, A.M.H., Rdewi, E.H.: Synthesis of a novel ZnO/TiO2-nanorod MXene heterostructured nanophotocatalyst for the removal pharmaceutical ceftriaxone sodium from aqueous solution under simulated sunlight. J. Environ. Chem. Eng. **10** (2022). https://doi.org/10.1016/j.jece.2022.108111
72. Ghoreishian, S.M., Ranjith, K.S., Lee, H., et al.: Tuning the phase composition of 1D TiO2 by Fe/Sn co-doping strategy for enhanced visible-light-driven photocatalytic and photoelectrochemical performances. J. Alloys Compd. **851** (2021). https://doi.org/10.1016/j.jallcom.2020.156826
73. Liu, R., Sun, L., Qiao, Y., et al.: Efficient photocatalytic degradation of pharmaceutical pollutants using plasma-treated g-C3N4/TiO2. Energy Technol. **8** (2020). https://doi.org/10.1002/ente.202000095
74. Ji, H., Du, P., Zhao, D., et al.: 2D/1D graphitic carbon nitride/titanate nanotubes heterostructure for efficient photocatalysis of sulfamethazine under solar light: catalytic "hot spots" at the rutile–anatase–titanate interfaces. Appl. Catal. B **263** (2020). https://doi.org/10.1016/j.apcatb.2019.118357
75. Liang HCh, Li XZ.: Effects of structure of anodic TiO2 nanotube arrays on photocatalytic activity for the degradation of 2,3-dichlorophenol in aqueous solution. J. Hazard Mater. **162**, 1415–1422 (2009). https://doi.org/10.1016/j.jhazmat.2008.06.033
76. Huyen, T.T.T., Chi, T.T.K., Dung, N.D., et al.: Enhanced photocatalytic activity of {110}-faceted TiO2 rutile nanorods in the photodegradation of hazardous pharmaceuticals. Nanomaterials **8** (2018). https://doi.org/10.3390/nano8050276
77. Mao, K., Li, Y., Zhang, H., et al.: Photocatalytic degradation of 17α-ethinylestradiol and inactivation of *Escherichia coli* using Ag-modified TiO2 nanotube arrays. Clean (Weinh) **41**, 455–462 (2013). https://doi.org/10.1002/clen.201100698
78. Zhang, X., Zhou, J., Gu, Y., Fan, D.: Visible-light photocatalytic activity of N-doped TiO2 nanotube arrays on acephate degradation. J. Nanomater. **2015** (2015). https://doi.org/10.1155/2015/527070
79. Sayegh, S., Abid, M., Tanos, F., et al.: N-doped TiO2 nanotubes synthesized by atomic layer deposition for acetaminophen degradation. Colloids Surf. A Physicochem. Eng. Asp. **655** (2022). https://doi.org/10.1016/j.colsurfa.2022.130213
80. Zhang, Y., Han, C., Zhang, G., et al.: PEG-assisted synthesis of crystal TiO2 nanowires with high specific surface area for enhanced photocatalytic degradation of atrazine. Chem. Eng. J. **268**, 170–179 (2015). https://doi.org/10.1016/j.cej.2015.01.006

81. Wang, X., Wang, A., Lu, M., Ma, J.: Synthesis of magnetically recoverable Fe0/graphene-TiO2 nanowires composite for both reduction and photocatalytic oxidation of metronidazole. Chem. Eng. J. **337**, 372–384 (2018). https://doi.org/10.1016/j.cej.2017.12.090
82. Zhang, Y., Liu, P., Wu, H.: Development of high efficient visible light-driven N, S-codoped TiO 2 nanowires photocatalysts. Appl. Surf. Sci. **328**, 335–343 (2015). https://doi.org/10.1016/j.apsusc.2014.12.043
83. Lee, H.G., Sai-Anand, G., Komathi, S., et al.: Efficient visible-light-driven photocatalytic degradation of nitrophenol by using graphene-encapsulated TiO2 nanowires. J. Hazard. Mater. **283**, 400–409 (2015). https://doi.org/10.1016/j.jhazmat.2014.09.014
84. Kandi, D., Behera, A,. Martha, S., et al.: Quantum confinement chemistry of CdS QDs plus hot electron of Au over TiO2 nanowire protruding to be encouraging photocatalyst towards nitrophenol conversion and ciprofloxacin degradation. J. Environ. Chem. Eng. 7 (2019). https://doi.org/10.1016/j.jece.2018.102821
85. Singh, N., Mondal, K., Misra, M., et al.: Quantum dot sensitized electrospun mesoporous titanium dioxide hollow nanofibers for photocatalytic applications. RSC Adv. **6**, 48109–48119 (2016). https://doi.org/10.1039/c6ra04305d
86. Čižmar, T., Kojić, V., Rukavina, M., et al.: Hydrothermal synthesis of feooh and Fe2 O3 modified self-organizing immobilized TiO2 nanotubes for photocatalytic degradation of 1h-benzotriazole. Catalysts **10**, 1–19 (2020). https://doi.org/10.3390/catal10121371
87. Xu, L., Niu, J., Xie, H., et al.: Effective degradation of aqueous carbamazepine on a novel blue-colored TiO2 nanotube arrays membrane filter anode. J. Hazard Mater. **402** (2021). https://doi.org/10.1016/j.jhazmat.2020.123530
88. Zeng, W., Zhang, H., Wu, R., et al.: Environment-friendly and efficient electrochemical degradation of sulfamethoxazole using reduced TiO2 nanotube arrays-based Ti membrane coated with Sb-SnO2. J. Hazard Mater. **446** (2023). https://doi.org/10.1016/j.jhazmat.2022.130642
89. Liu, P., Deng, K., Huang, Y., et al.: In-situ generation of hollowed nanorods array with TiN/TiO2 heterostructures for electrocatalytic degradation of metronidazole. Chem. Eng. Sci. **281** (2023). https://doi.org/10.1016/j.ces.2023.119148
90. Pang, D., Liu, Y., Song, H., et al.: Trace Ti3+- and N-codoped TiO2 nanotube array anode for significantly enhanced electrocatalytic degradation of tetracycline and metronidazole. Chem. Eng. J. **405** (2021). https://doi.org/10.1016/j.cej.2020.126982
91. Ahmadi, A., Wu, T.: Electrocatalytic reduction of nitrobenzene using TiO2 nanotube electrodes with different morphologies: kinetics, mechanism, and degradation pathways. Chem. Eng. J. **374**, 1241–1252 (2019). https://doi.org/10.1016/j.cej.2019.06.020
92. Ahmadi, A., Vogler, B., Deng, Y., Wu, T.: Removal of meropenem from environmental matrices by electrochemical oxidation using Co/Bi/TiO2 nanotube electrodes. Environ. Sci. (Camb) **6**, 2197–2208 (2020). https://doi.org/10.1039/d0ew00184h
93. Chang, X., Thind, S.S., Chen, A.: Electrocatalytic enhancement of salicylic acid oxidation at electrochemically reduced TiO2 nanotubes. ACS Catal. **4**, 2616–2622 (2014). https://doi.org/10.1021/cs500487a
94. Zeng, W., Liang, H., Zhang, H., et al.: Efficient electrochemical oxidation of sulfamethoxazole by a novel reduced TiO2 nanotube arrays-based flow-through electrocatalytic membrane. Sep. Purif. Technol. **289** (2022). https://doi.org/10.1016/j.seppur.2022.120720
95. Chen, Y., Tu, Y., Zhang, Y., et al.: Fabrication and enhanced electrocatalytic activity of TiO2nanotubes based three-dimensionally macroporous SnO2with mesoporous walls. Chem. Eng. J. **311**, 100–110 (2017). https://doi.org/10.1016/j.cej.2016.11.071
96. Wang, P., Chu, G., Gao, G., et al.: Efficient electrochemical oxidation of chloramphenicol by novel reduced TiO2 nanotube array anodes: kinetics, reaction parameters, degradation pathway and biotoxicity forecast. Materials **16** (2023). https://doi.org/10.3390/ma16113971
97. Cheng, X., Cheng, Q., Deng, X., et al.: A facile and novel strategy to synthesize reduced TiO2 nanotubes photoelectrode for photoelectrocatalytic degradation of diclofenac. Chemosphere **144**, 888–894 (2016). https://doi.org/10.1016/j.chemosphere.2015.09.070

98. Kmentova, H., Kment, S., Wang, L., et al.: Photoelectrochemical and structural properties of TiO2 nanotubes and nanorods grown on FTO substrate: comparative study between electrochemical anodization and hydrothermal method used for the nanostructures fabrication. Catal. Today **287**, 130–136 (2017). https://doi.org/10.1016/j.cattod.2016.10.022
99. Teng, W., Wang, Y., Huang, H.H., et al.: Enhanced photoelectrochemical performance of MoS 2 nanobelts-loaded TiO 2 nanotube arrays by photo-assisted electrodeposition. Appl. Surf. Sci. **425**, 507–517 (2017). https://doi.org/10.1016/j.apsusc.2017.06.297
100. Zeng, L., Li, X., Fan, S., et al.: Enhancing interfacial charge transfer on novel 3D/1D multidimensional MoS2/TiO2 heterojunction toward efficient photoelectrocatalytic removal of levofloxacin. Electrochim. Acta **295**, 810–821 (2019). https://doi.org/10.1016/j.electacta.2018.10.153
101. Li, S., Liu, C., Chen, P., et al.: In-situ stabilizing surface oxygen vacancies of TiO2 nanowire array photoelectrode by N-doped carbon dots for enhanced photoelectrocatalytic activities under visible light. J. Catal. **382**, 212–227 (2020). https://doi.org/10.1016/j.jcat.2019.12.030
102. Li, T., Zhang, P., He, H., et al.: Highly efficient photoelectrocatalytic degradation of cefotaxime sodium on the MoSe2/TiO2 nanotubes photoanode with abundant oxygen vacancies. J. Solid State Chem. **303** (2021). https://doi.org/10.1016/j.jssc.2021.122455
103. Cheng, X., Wang, P., Liu, H.: Visible-light-driven photoelectrocatalytic degradation of diclofenac by N, S-TiO2/TiO2 NTs photoelectrode: performance and mechanism study. J. Environ. Chem. Eng. **3**, 1713–1719 (2015). https://doi.org/10.1016/j.jece.2015.06.015
104. Wu, H., Hu, Z., Liang, R., et al.: Novel Bi2Sn2O7 quantum dots/TiO2 nanotube arrays S-scheme heterojunction for enhanced photoelectrocatalytic degradation of sulfamethazine. Appl. Catal. B **321** (2023). https://doi.org/10.1016/j.apcatb.2022.122053
105. Hu, Z., Zhou, M., Maitlo, H.A., et al.: Novel dual-photoelectrode photoelectrocatalytic system based on TiO2 nanoneedle arrays photoanode and nitrogen-doped carbon dots/Co3O4 photocathode for efficient water purification at low/no applied voltage. Appl. Catal. B **331** (2023). https://doi.org/10.1016/j.apcatb.2023.122676
106. Ma, B., Xin, S., Xin, Y., et al.: Visible-light-driven photoelectrocatalytic degradation of p-chloronitrobenzene by BiOBr/TiO2 nanotube arrays photoelectrodes: mechanisms, degradation pathway and DFT calculation. Sep. Purif. Technol. **268** (2021). https://doi.org/10.1016/j.seppur.2021.118699
107. Xin, S., Ma, X., Lu, J., et al.: Enhanced visible light photoelectrocatalytic degradation of o-chloronitrobenzene through surface plasmonic Au nanoparticles and g-C3N4 co-modified TiO2 nanotube arrays photoanode. Appl. Catal. B **323** (2023). https://doi.org/10.1016/j.apcatb.2022.122174
108. Pan, J., Li, X., Zhao, Q., et al.: Construction of Mn0.5Zn0.5Fe2O4 modified TiO2 nanotube array nanocomposite electrodes and their photoelectrocatalytic performance in the degradation of 2,4-DCP. J Mater Chem C Mater **3**, 6025–6034 (2015). https://doi.org/10.1039/c5tc01008j
109. Ma, C., Wei, M.: BiVO4-nanorod-decorated rutile/anatase TiO2 nanofibers with enhanced photoelectrochemical performance. Mater. Lett. **259** (2020). https://doi.org/10.1016/j.matlet.2019.126849

Advancing Sunscreen Technology with Multifunctional Hybrid Titanium Dioxide Nanoparticles

Richa Arora, Areeba Suhail, Garima Arora, and Upasana Issar

Abstract Exposure to UV radiation leads to skin damage, encompass aging, sunburn, and even skin cancer. Crystalline TiO_2 has been widely used as a UV blocker but poses certain challenges like skin damage due to its photocatalytic activity. Nanoparticles of titanium dioxide (TiO_2) is extensively utilized and favoured in a variety of cosmetic products, including sunscreens as alternatives to crystalline counterpart as well as organic UV filters. Hybridization of TiO_2 with biopolymer such as lignin or any other heterogeneous polymer such as melanoidin have shown to efficiently enhance the UV protection ability. Moreover, the use of hybrid nanocomposites can mitigate the white cast effect caused by TiO_2, indicating a more promising application of TiO_2 in cosmetic products. This chapter intends to present a comprehensive picture on the UV protection by TiO_2 nanoparticles. It also proposes to cover the recent developments in the synthesis of multi-functionalised hybrid TiO_2 nanomaterials as an alternative and viable option for sunscreens.

Keywords Multifunctional · Hybrid-TiO_2 nanoparticles · Sunscreen · UV protectors

Abbreviations

APTES	Aminopropyltriethoxysilane
BNC	Bionanocomposite
BNNSs	Boron nitride nanosheets
CF	Cellulosic fabrics
CNF	Cellulose nanofibers
IR	Infrared

R. Arora
Department of Chemistry, Shivaji College, University of Delhi, Delhi 110027, India

A. Suhail · G. Arora · U. Issar (✉)
Department of Chemistry, Kalindi College, University of Delhi, Delhi 110008, India
e-mail: upasana@kalindi.du.ac.in

J. Prakash et al. (eds.), *Titanium Dioxide-Based Multifunctional Hybrid Nanomaterials*, Engineering Materials, https://doi.org/10.1007/978-3-031-80625-4_8

LS	Lignosulfonate
MPN	Metal phenolic network
NPs	Nanoparticles
PAN NFs	Polyacrylonitrile nanofibers
PMMA	Poly(methyl methacrylate)
PPCH	Polyphosphonium chitosanhydrazone
PVDF	Poly(vinylidene fluoride)
ROS	Reactive oxygen species
SiQAC	3-(Trimethoxysilyl)propyl]ammonium chloride
SPF	Sun protection factor
TA	Tannic acid
TAML	TiO_2/arginine-modified lignin
TPMP	3-(Trihydroxysilyl)propyl methylphosphonate
UV	Ultraviolet
UVA	Ultraviolet A
UVAPF	Ultraviolet-A protection factor
UPF	UV protection factor
UVB	Ultraviolet B
UVC	Ultraviolet C
UV-vis	Ultraviolet–visible
WHO	World Health Organization

1 Introduction

The Sun emits a broad spectrum of radiation, comprising infrared (IR), visible light, and ultraviolet (UV) radiation, covering the electromagnetic spectrum from 100 nm to 1 mm. In terms of wavelength, IR radiation is the longest, ranging from 0.7 to 1000 μm, followed by visible radiation, which covers 380–700 nm, and UV radiation with the shortest wavelength, spanning 10–400 nm. Among these, UV radiation is the most hazardous owing to its high energy. UV radiation is classified into three types: (1) ultraviolet A (UVA) rays, (2) ultraviolet B (UVB) rays, and (3) ultraviolet C (UVC) rays. UVA has the longest wavelength (315–400 nm/3.94–3.10 eV) and the lowest energy, causing relatively minimal damage. UVB falls within the intermediate wavelength range (280–320 nm/4.43–3.94 eV) and has moderate energy. UVC, with the shortest wavelength (100–280 nm/12.4–4.43 eV) and the highest energy, poses the greatest risk of harm [1–3]. The boundary between visible and ultraviolet spectra is not sharply defined, as there is a gradual transition between these regions (Fig. 1a).

The earth's ozone layer completely absorbs UVC radiation and partially absorbs UVB, resulting in most of the UV radiation reaching the earth's surface being UVA (~95%) and a smaller portion being UVB (~5%). Both UVA and UVB can cause DNA damage. UVA primarily contributes to tanning and skin aging, while UVB penetrates deeper layers of the skin, leading to sunburn (Fig. 1b). UV radiation

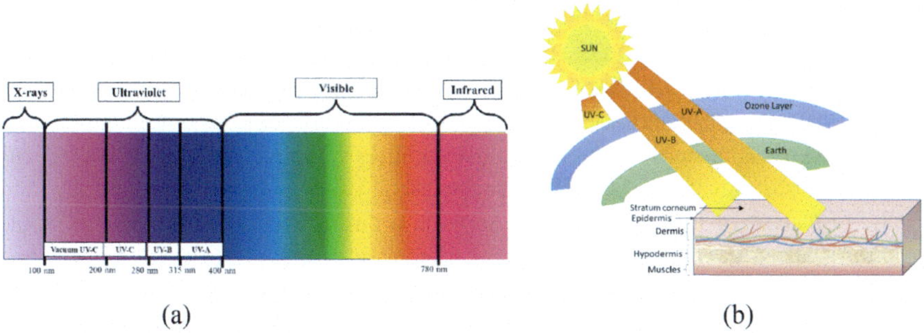

Fig. 1 **a** Electromagnetic spectrum from X-rays to infrared radiations focusing on the UV region. **b** Penetration of UV radiation into the skin

generates reactive oxygen species (ROS) in the skin, initiating photochemical reactions and skin sensitization. Melanin, a natural antioxidant, alleviates UV damage by neutralizing ROS or absorbing and dispersing energy. Prolonged UV exposure can lead to severe complications such as eye disorders and skin cancers, including malignant melanomas and non-melanoma keratinocyte cancers (e.g., basal cell carcinomas and squamous cell carcinomas) [4]. Despite these harmful effects, sunlight has benefits such as improving cardiovascular health through the production of nitric oxide, antimicrobial properties, and stimulating vitamin D synthesis. Preventive measures can decline sun damage. The World Health Organization (WHO) recommends wearing protective clothing, sunglasses, and hats, and using sunscreen on exposed skin. Sunscreens play a crucial role in protecting the skin from the harmful effects of solar radiation, including UV rays. Regular sunscreen use reduces the risk of skin cancer and premature aging. Skin tone influences UV sensitivity, with darker skin tones showing greater resistance to sunburn but still being susceptible to UV-induced damage. Regular sunscreen application is advised for all skin types to prevent cancer, as sunburn and tanning may not be readily apparent in darker skin tones [5].

Sunscreens are generally classified into two categories based on their mechanisms of protection: physical sunscreens and chemical sunscreens. Physical sunscreens act by reflecting UVA and UVB rays, offering immediate protection upon application. However, they often leave a noticeable white residue on the skin. Chemical/Organic sunscreens, on the other hand, function by absorbing UV light and converting it into heat, which is subsequently dissipated from the skin. These sunscreens need to be applied 20 min before sun exposure to ensure proper absorption. Common UV filters used in chemical sunscreens include titanium dioxide (TiO_2) and zinc oxide (ZnO). These compounds can generate ROS, which may lead to skin irritation, itching, and redness [6].

The development of sunscreen technology has shown a great advancement after the onset of the nanotechnology. The nanotechnology has gained global recognition for its wide range of applications, including improving the properties of conventional topical formulations in the cosmetics and drug industries. It addresses challenges such as low skin permeation and retention, instability of active ingredients, poor water resistance, and the production of reactive ROS by inorganic filters [7]. Studies show that incorporating UV filters into nanosystems, known as nanosunscreens, offers numerous advantages over traditional sunscreen formulations with non-encapsulated UV filters. Utilizing nanosystems requires careful evaluation of their toxicological effects, exposure levels, and chemical properties.

The objective of this chapter is to provide a comprehensive overview of solar radiation, focusing on UV radiation, its effects on human skin, and the technological advancements in sunscreens, including the use of nanohybrids for effective protection. It underlines the significance of preventive measures, including the use of sunscreens. Furthermore, the chapter highlights the role of nanotechnology in advancing sunscreen formulations, specifically the development of nanohybrids and the integration of TiO_2 nanoparticles. TiO_2, known for its high refractive index and UV-blocking capabilities, is widely used in sunscreens to enhance sun protection factor (SPF) and provide broad-spectrum UV protection. In all, the chapter discusses the benefits and challenges associated with TiO_2, including its photocatalytic activity and the potential generation of ROS, as well as strategies to mitigate these effects through surface modifications and multifunctional hybrid formulations.

2 TiO_2 Nanoparticles as Powerful UV-Protector

TiO_2 naturally exists in three crystalline forms: rutile, anatase, and brookite. Figure 2a–c represents unit lattices of these three crystalline forms. Among these, rutile is the most prevalent and stable form. A key optical property of this crystal is its refractive index, which is particularly high in the UV and visible wavelength ranges. This high refractive index contributes to the brightness and whiteness of TiO_2 pigments [8].

TiO_2 is a semiconductor with an electronic structure characterized by a series of orbital bands separated by an energy band gap. Light absorption that equals or exceeds the band gap energy between the valence band and the conduction band leads to the excitation of a valence band electron (e^-) to the conduction band, creating a hole (h^+) in the valence band. This electron–hole pair can travel to the crystal's surface and engage in redox reactions with absorbed substrates [8]. Natural TiO_2 exhibits optical band gaps just above 3 eV, with specific values for its polymorphs as: rutile at approximately 3.0 eV [9, 10], anatase at around 3.4 eV [11], and brookite at about 3.3 eV [12]. This restricts its photoactivity to the UV region of the electromagnetic spectrum, making it a suboptimal choice for efficient sunscreen application.

Nanosized TiO_2 is commonly used in sunscreens as inorganic physical blocker and exhibits high SPF owing to its transparency and large surface area-to-volume ratio. Figure 2d–f depicts the spherical nanoclusters (size 5 Å) of various forms of TiO_2. Absorption happens when the incident radiation's energy matches the band gap of semiconductors like TiO_2 [13]. Starting at the sizes of 20 nm to < 100 nm, these nanoparticles (NPs) provide better scattering and enhanced protective effects.

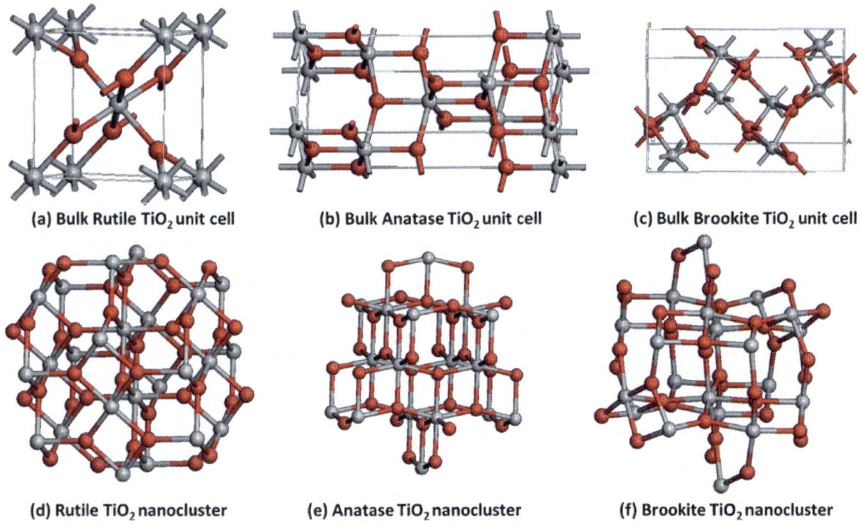

Fig. 2 Representation of bulk and nanoclusters of various forms of TiO_2 (red balls = O atoms, grey balls = Ti atoms)

TiO_2 NPs are more efficient at filtering UVB rays than UVA. As the size of the NPs increases, the peak point is shifted to longer wavelengths (red shift) indicating that annealing increases the NPs size (Fig. 3). Moreover, it gives the flexibility to use these NPs in the broad range of spectra [14].

With every advantage, there also come some repercussions. It has been observed that exposure to nanosunscreens leads to the incorporation of TiO_2 NPs into the deepest layers of the stratum corneum (Fig. 1b) and hair follicles, which may act as long term reservoirs. Within the skin, nanoparticle aggregation and physicochemical interactions between particles, skin, and light influence the overall efficacy of UV reduction. Both in the presence and absence of light, ROS can be generated, leading to cyto- and genotoxicity. Anatase TiO_2 exhibits the highest phototoxicity compared to rutile TiO_2 NPs [15].

Although TiO_2 NPs are widely used in the commercial sunscreens, they have high photocatalytic reactivity due to its wide bandgap. Upon UV light absorption, TiO_2 generates high-energy electron–hole pairs that produce ROS, such as hydroxyl radicals and superoxide anions. These ROS can cause DNA damage, inflammation, skin aging, carcinogenesis, and the degradation of chemicals and synthetic polymers, presenting a significant limitation for TiO_2-based sunscreens. To overcome this issue, scientists have tried various alternatives to enhance the UV protection of these NPs keeping in mind their cytotoxic behaviour. These alternatives include coating of TiO_2 NPs with other inorganic/organic based substrates such as silica or creating a multifunctional TiO_2 hybrid structure. In the following sections, emphasis is laid on the multifunctional hybrids of TiO_2 that are synthesised to enhance the optical activity of TiO_2 NPs.

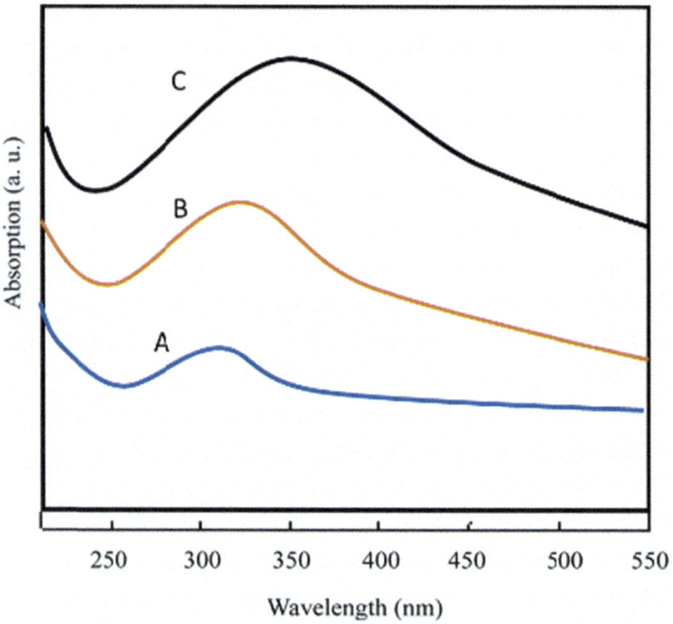

Fig. 3 Ultraviolet–visible (UV–vis) spectrum of TiO_2 NPs samples: as synthesized TiO_2 NPs (**a**); TiO_2 NPs annealed at 600 °C (**b**) and TiO_2 NPs annealed at 800 °C (**c**) (as adapted from [14])

3 Multifunctional Nanocomposites of TiO_2

Surface modification of TiO_2 NPs is crucial in order to enhance their dispersibility, UV-blocking efficiency and reducing the photocatalytic activity. This can be accomplished through inorganic coatings with metal oxides or organic coatings by employing coupling agents or by reacting with the surface hydroxyl groups.

With the aim to minimize skin irritation and providing antibacterial protection, a chitosan-TiO_2 nanocomposite was synthesised using wet impregnation method [16]. Chitosan is a deacetylated derivative of chitin, which is one of the most abundant natural biopolymers and is primarily found in the exoskeletons of shellfishes, the cuticles of insects, certain algae, and the cell walls of fungi. Chitosan is a linear polysaccharides composed of varying ratios of N-acetyl-2-amino-2-deoxy-D-glucose (*N*-acetyl-glucosamine, GlcNAc) and 2-amino-2-deoxy-D-glucose (glucosamine, GlcN) units, linked via β(1→4) glycosidic bonds [17]. It can mitigate the harmful effects of ROS by converting ROS into glucosamine and chito-oligosaccharides, which do not irritate the skin [18]. Chitosan also possesses antibacterial properties [19]. Fourier-transform IR spectroscopy analysis reveals substantial interactions between the amino groups of chitosan and the oxygen atoms in TiO_2. UV–vis spectroscopy demonstrates that chitosan effectively reduces the band gap energy of TiO_2, thereby, increasing its activity under light exposure. The Chitosan/TiO_2 nanocomposite exhibits significant antibacterial properties, which remain effective with chitosan

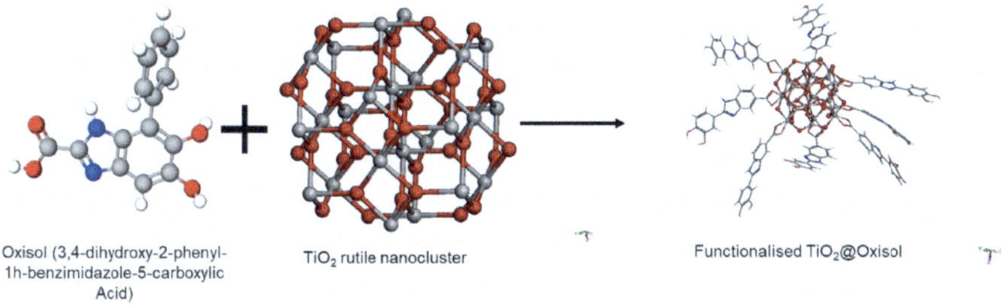

Fig. 4 Representation of Oxisol functionalised TiO_2 nanocluster

loading levels up to 3%. Additionally, the creams exhibited moderate UV protection, with SPF values ranging from 21.30 to 22.77 [16].

Another new class of sunscreen UV filters were synthesised by combining physical filters with Oxisol, an antioxidant molecule with a booster effect. Organic UV filters are compounds that absorb UV radiation and dissipate the absorbed energy through various mechanisms, including fluorescence, internal conversion, and molecular vibrations. Boosters, like dihydroxyphenyl benzimidazole carboxylic acid (Oxisol), improve UV protection by affecting the formulation's flow properties and working together with UV filters [20]. The synthesis involves the direct surface functionalization of inorganic filters, specifically TiO_2, with Oxisol generating a nanohybrid (Fig. 4).

The results show that Oxisol/TiO_2 hybrid improves the SPF values and demonstrates a high antioxidant activity. The formulations became more stable and had lower photocatalytic activity. Cytotoxicity tests showed minimal harmful effects, suggesting these next-generation sunscreens are safer and more effective [21].

Lignin is a complex organic polymer found in the cell walls of plants, particularly in the wood and the bark. Chemically, lignin is composed of phenolic compounds, making it highly aromatic and resistant to the degradation. It is often a byproduct of the paper and pulp industry, where it is separated from the cellulose fibres. It has natural UV-absorbing and antioxidant properties, making it a promising material for applications in food, cosmetics, and pharmaceutical industries [22]. Amalgamation of lignin with TiO_2 can modify physical sunscreens forming stable complexes that reduce photocatalytic activity and enhance UV protection [23, 24]. The sulfonate derivative of lignin called lignosulfonate (LS), also exhibit good dispersibility and have been used in various applications, including as UV protectants in the polymers. Coating of LS on TiO_2 surface via hydrothermal esterification technique utilises the carboxyl groups of LS and the hydroxyl groups on TiO_2. The synthesis of LS@TiO2 nanocomposites offers an efficient and eco-friendly approach for the valorisation of lignin biomass, while simultaneously enhancing the UV protection properties of nano-TiO_2 materials. The esterification reaction between LS and TiO_2 improves the UV-blocking ability as well as the dispersibility of TiO_2. The modified multifunctional TiO_2 hybrid NPs display a rougher surface, which helps in the effective scattering and reflection of UV rays. Moreover, the synthesis process also creates

Ti^{3+} defects in TiO_2, which results in the broadening of the absorption spectrum, thereby, boosting its UV absorption ability. Additionally, oxidation of LS during the process increased its UV-absorbing capacity by transforming into hydroquinone and quinone structures. Therefore, creams with 10% LS@TiO_2 nanocomposites display SPF of 50+, a prominent property for skincare products [25].

Recently, TiO_2/arginine-modified lignin (TAML) NPs were synthesized to enhance the UV protection and the antioxidant properties. Antioxidant assays demonstrated that TAML NPs effectively neutralize the hydroxyl radicals (·OH), exhibiting a scavenging efficiency 17 times greater than that of the pure TiO_2. UV protection tests indicated that TAML achieved SPF value of 149.5, the highest reported for lignin-based sunscreen particles. These findings offer significant insights into the development of advanced and biocompatible emulsifiers for various high-value applications [26].

Silver–TiO_2 NPs hybrid have gathered considerable attention in the cosmetics and hygiene industries owing to their enhanced antibacterial properties, which surpass those of the pure TiO_2. Moreover, the incorporation of ZnO NPs into sunscreen formulations provides robust protection against harmful UV radiation and improved cosmetic aesthetics. Keeping all this in mind, Chegeni and co-workers utilised sol–gel technique to synthesise a nanocomposite composed of TiO_2, ZnO, and Ag NPs ($TiO_2/Zn_2TiO_4/Ag$) in order to fully utilise the properties of each component. The $TiO_2/Zn_2TiO_4/Ag$ nanocomposite displays unique antibacterial and UV-blocking properties promising for sunscreens. This makes it a useful ingredient in skincare products for better protection and antibacterial benefits [27].

Another characteristic of UV filters is that of triggering a redox reaction between the oxygen and water molecule in the presence of sunlight, thereby, forming ROS which degrade products and cause adverse biological effects. To address this issue, an alternative photochemical pathway was proposed that couples UV filters with the redox-active molecules. Tannic acid (TA) (Fig. 5) is a polyphenolic compound with great antioxidant activity. In order to develop a new type of anti-UV coating that suppresses ROS via redox chemistry, the layer-by-layer self-assembly of TiO_2 NPs with TA is introduced. TA and TiO_2 NPs were deposited onto porous poly(methyl methacrylate) (PMMA) microspheres, resulting in strong UV absorptivity and a precise control over the absorption intensity.

The TA-TiO_2 hybrid PMMA microspheres exhibited higher SPF and better UV protection compared to TiO_2 NPs alone. The hybrid structure generated lower amounts of the superoxide anions and the hydroxyl radicals under UV irradiation, indicating a reduction in ROS generation [28].

In another study, the tannic acid-metal complex known as Metal Phenolic Network (MNP) was loaded on TiO_2 NPs. The resultant composite TiO_2-MPN NPs is proposed to be a sole active ingredient of the sunscreen. These hybrids are formed by complexing the tannic acid with multivalent metal ions, creating a stable coating shell. Figure 6a depicts a pictorial representation of the hybrid composite formed between tannic acid, metal and TiO_2 NPs. The hybrid material expands TiO_2's ability to absorb light, including high-energy visible light, increasing the SPF and ultraviolet-A protection factor (UVAPF) by about four times (Fig. 6b). It also suppresses photoinduced

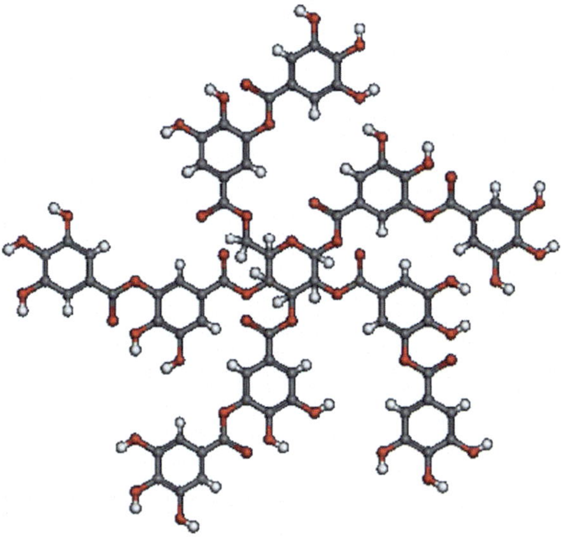

Fig. 5 Tannic acid (grey balls: C atoms, red balls: O atoms; white balls: H atoms)

ROS by 99.9% for up to six hours under the simulated solar irradiation, thereby, significantly dropping the cytotoxicity and demonstrating the hybrid NPs protective effect on skin cells.

This is an eco-friendly and biocompatible approach for designing sunscreens that are free from synthetic polymers and organic UV filters, providing broad-spectrum protection and ROS suppression. This new method solves environmental and health problems, leading to more eco-friendly sunscreens [29].

Adding organic and inorganic compounds to protein-based materials is a major research focus because it improves their properties and expands their range of uses. Functionalization of protein-based materials through the incorporation of TiO_2 NPs modifies both technological properties viz. mechanical, thermal, adsorptive, etc. and functional properties viz. antimicrobial activity, photodegradation, UV protection, etc. These hybrids are biodegradable and show great strength, elasticity, thermal stability, and resistance to oxygen and water, all of which are dependent on the concentration of TiO_2.

Recent research has explored the potential of protein–TiO_2 hybrid composites. For example, soy protein isolated films combined with TiO_2 demonstrated antimicrobial properties against *E. coli* and *S. aureus* [30]. Similarly, marine algae protein based films functionalized with TiO_2 showed promising results [31]. TiO_2-enhanced protein films, such as those made from gelatin or sesame protein, exhibited improved mechanical and UV-protective properties [32, 33]. Protein–TiO_2 hybrid composites are promising for developing eco-friendly packaging materials with antimicrobial and UV-protective effects. The preparation method and the protein-TiO_2 mixing ratio can also influence the material's properties [34].

Recently Alalaiwe and co-workers attempted to create a multifunctional nanocomposite comprising of TiO_2/avobenzone-loaded mesoporous silica which cater a wide

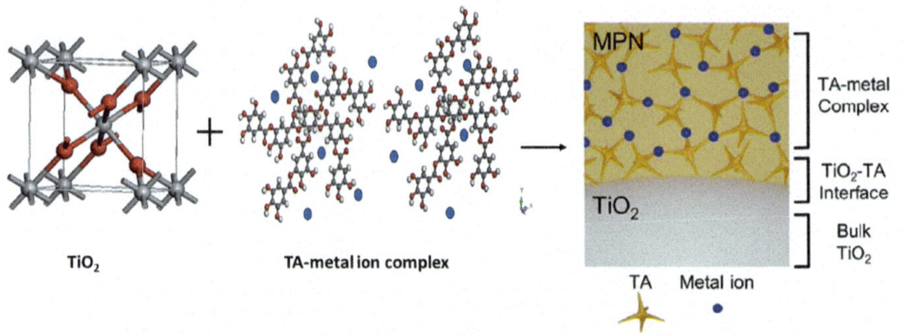

(a)

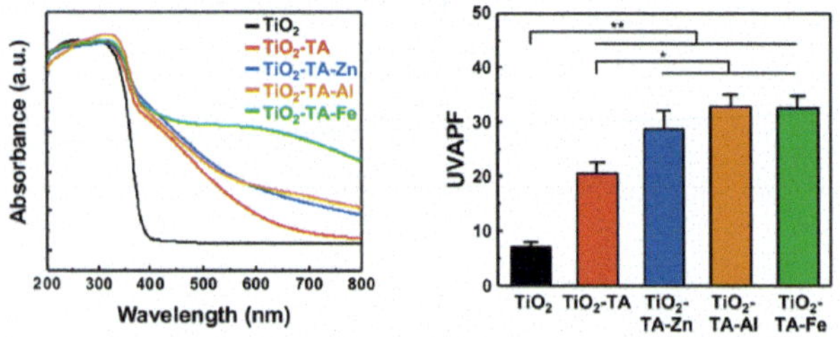

(b)

Fig. 6 **a** Schematic description of the TiO_2–MPN interface; **b** UV–vis absorption spectra of TiO_2, TiO_2-TA, TiO_2-TA-Zn, TiO_2-TA-Al, and TiO_2-TA-Fe NPs and UVAPF of 20 wt % of TiO_2, TiO_2-TA, TiO_2-TA-Zn, TiO_2-TA-Al, and TiO_2-TA-Fe NPs. Reprinted (adapted) with permission from [29], Copyright (2024), American Chemical Society

degree of spectrum as far as UV protection is concerned. Mesoporous silica is effective for pollutant adsorption due to its porous structure, large surface area, and stability [35]. TiO_2 is an excellent photocatalyst, however its use becomes limited owing to its aggregation and low quantum efficiency. Combination of mesoporous silica with TiO_2 addresses these concerns by increasing NPs surface area, thus enhancing its photocatalytic efficiency [36]. Additionally, Avobenzone, a common UVA filter, used in many of the commercially available sunscreens, has issues with skin penetration and photostability [37]. The objective was to develop a multifunctional mesoporous silica material embedded with TiO_2 NPs and avobenzone for effective pollutant adsorption and UVA protection. TiO_2/SBA-S, characterized by its low porosity, demonstrated superior pollutant adsorption and synergistic UVA protection when combined with avobenzone. This formulation effectively reduced the avobenzone penetration compared to the formulations with higher porosity. Animal studies indicated that the TiO_2/SBA-S hydrogel significantly mitigated signs of photoaging and enhanced skin barrier function [38].

Efforts to decrease TiO_2 NPs photocatalytic activity include creating multifunctional hybrids with other materials or by coating it with silica shells. However, excessive inorganic coatings can reduce light absorption and transmittance. Alternatives like di- and triblock copolymers have been used to reduce the photocatalytic activity of NPs without losing its UV protection ability. Nevertheless, these materials are complex and incur huge costs, and concerns about their biodegradability and biocompatibility persist [39]. Plant-based natural polymers, such as cellulose nanofibers (CNF) derived from most abundant natural polymer cellulose, offers a renewable, hydrophilic material that can disperse in water and act as a moisturizer. Despite its potential, not much emphasis has been laid on the use of CNF in the sunscreen industries. It has been found that incorporating CNF into commercial sunscreens with SPF ratings of 35+ and 50+ enhanced their transmittance performance, indicating synergistic interactions between CNF and other active ingredients. Considering the properties of CNFs, a nanocomposite of CNF and TiO_2 is made using the sol gel method. The resultant TiO_2-CNF hybrid shows enhanced UV protection even at low concentrations of TiO_2. Moreover, combining this hybrid with avobenzone displayed remarkable improvement in UV photoprotection [40].

Melanin is a natural pigment found in many living organisms and plays a vital role in their biology. It has impressive properties such as protection against UV light, binding to metals, neutralizing free radicals, and acting as an antioxidant. One of melanin's key features is its ability to protect from UV radiation by converting over 90% of absorbed UV energy into heat within a few billionths of a second [41, 42]. This rapid dissipation of energy helps prevent DNA damage, thereby, preventing from malignant melanoma. Melanin absorbs a wide range of light, from visible to UV. It is naturally produced in the body through the oxidation of L-tyrosine by an enzyme called tyrosinase. Synthetic melanin, often called polydopamine, is made from dopamine in a basic solution. This synthetic form has similar chemical and physical properties as that of the natural melanin. Polydopamine NPs can mimic natural melanin's behaviour in the body, being taken up by skin cells (keratinocytes) and offering protection by forming protective caps around cell nuclei, much like natural melanin does [43]. Apart from this, melanin also has a strong ability to stick to a variety of materials, both inorganic and organic. Therefore, a hybrid of synthetic Melanin and TiO_2 NPs was prepared using oxidation and self-polymerization of dopamine. UV–vis absorption spectra show that a thinner melanin layer coating on TiO_2 NPs provides better UV absorption. Optimal performance is achieved by carefully controlling the thickness of the melanin layer. A very thin melanin layer, about 1 nm thick, gives the best SPF [44].

On similar lines, very recently a study reported the synthesis of Natural Melanin/TiO_2 hybrid nanocomposite using ultrasonication under UV radiation. Natural Melanin was extracted from *Nigella sativa* (black seeds). The incorporation of natural melanin into TiO_2 NPs displayed thermal stability and enhanced properties making it a promising candidate for various applications, including solar cells and optoelectronics [45].

Porous 2D nanomaterials have been in extensive use owing to their ability to interact with a wide range of organic and inorganic species. This interaction occurs not only on their surfaces but also at the pore edges due to the abundance of exposed and active radicals. Among the various 2D nanomaterials, boron nitride nanosheets (BNNSs), often referred to as "white graphene" due to their structural resemblance to graphene, consist of a honeycomb lattice of covalently bonded boron and nitrogen

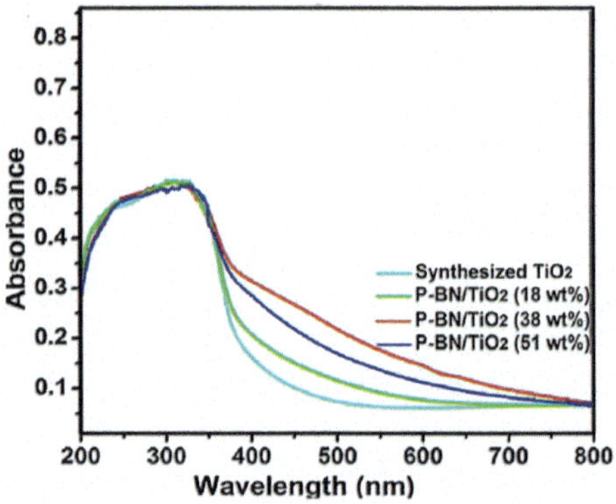

Fig. 7 UV–vis absorption spectra of synthesized TiO_2 and different porous BN/TiO_2 hybrid nanosheets (for comparison). Reprinted (adapted) with permission from [50]. Copyright (2017), Elsevier

atoms [46]. BNNSs have unique electronic properties, including a wide bandgap of up to 5.5 eV, making them ideal for deep-ultraviolet light emitters and substrates for graphene-based devices [47]. While these are not commonly used as catalytic materials for photocatalysis and photovoltaic applications due to strong optical absorption in the ultraviolet–visible range, BNNSs are promising as support materials for catalysts [48, 49]. Liu and coworkers produced functionalised BN/TiO_2 nanohybrids. TiO_2 NPs are uniformly grown on porous BNNSs to create a novel photocatalyst, resulting in the formation of B–O–Ti bonds within the porous BN/TiO_2 hybrid nanosheets. The UV–vis absorption spectra of synthesized TiO_2 and porous BN/TiO_2 hybrid nanosheets with varying TiO_2 loadings are displayed in Fig. 7. The porous BN/TiO_2 hybrid nanosheets show enhanced absorption in the visible light region compared to the synthesized TiO_2. Notably, the porous BN/TiO_2 hybrid nanosheets with 38 wt% TiO_2, showed the best photocatalytic performance, displayed a much stronger absorption tail across the entire visible light spectrum, highlighting their excellent ability to capture visible light.

Moreover, the porous BN/TiO_2 hybrid nanosheets show excellent durability, maintaining 97% of their high photocatalytic activity under visible light after five cycles. These findings help in designing advanced hybrid photocatalysts with active chemical bonds, useful for environmental protection, water splitting, and photo-electrochemical conversion [50].

4 Applications of Multifunctional Hybrids of TiO_2 in Textile Industries

In order to provide the UV protection, many textile industries are fabricating the cloths that have a layer of nanomaterial coating on it which effectively protects the skin from UV radiations. There is a high market demand for fabrics with multiple functionalities, but current methods to achieve these properties are often complex, time-consuming, and costly.

There are reports involving studies on some unique particles which are characterised by having surfaces with two or more distinct physical properties. These microparticles or NPs are known as Janus particles. These particles are used to enhance superhydrophobicity of fabrics, are UV resistant, and display antimicrobial properties [51]. The Janus particles comprising of TiO_2 NPs and poly(vinylidene fluoride) (PVDF) a hydrophobic polymer, on one side, and an epoxy resin on the other, and which adhere to the fabric surface through the adhesion effect of the epoxy resin. Fabrics treated with these Janus particles achieved UV protection factor (UPF) values as high as 733, indicating strong UV protection. The optimal treated fabric is extremely hydrophobic and resistant to staining, with self-cleaning properties. Additionally, the finished fabric also displayed strong antibacterial properties against *E. coli*, attributed to the antimicrobial action of TiO_2 under UV pre-irradiation. This method provides a novel platform for nano-finishing fabrics, enabling the incorporation of new functionalities while preserving the fabric's inherent properties [52].

Another innovative method for developing multifunctional nanocomposite fabric surfaces that are UV-protective, photocatalytic, antimicrobial, and flame-retardant is proposed by Rashid and co-workers. Utilizing a sol–gel/hydrothermal process, a phosphorus-based flame retardant, 3-(trihydroxysilyl)propyl methylphosphonate (TPMP), was applied in combination with Ag-doped TiO_2 nanocluster to the surface of cotton fibers. Analysis of the fabric confirms successful incorporation of Ag-TiO_2 and TPMP on the cotton fabric surface. The results indicate the formation of a unique TPMP–polysiloxane nanocomposite matrix on the cotton fiber with well-distributed Ag-TiO_2 (Fig. 8). The synergy between all components of the nanocomposite resulted in an exceptional UV protection in the UVA and UVB regions, with a UPF above 50, self-sterilizing activity against both the tested bacteria and enhanced thermo-oxidative stability. This approach ensures further development of advanced technical textiles with multifunctional protective surfaces [53].

Extending the study, a multifunctional cotton fabric exhibiting superior UV protection, self-sterilization, photocatalytic self-cleaning, and flame-retardant properties is developed. For the first time, a sustainable sol–gel/hydrothermal approach was employed to functionalize Ag-loaded TiO_2 in situ with a combination of flame-retardant TPMP and two types of biobarrier-forming trialkoxysiloxanes, namely 3-(trimethoxysilyl)propyl]ammonium chloride (SiQAC) and aminopropyltriethoxysilane (APTES), on cotton fabric. Characterization studies showed the successful functionalization of TiO_2 by TPMP, SiQAC, and APTES through Si–O–Ti bond formation and the well-dispersed Ag/TiO2/TPMP/SiQAC, Ag/TiO2/TPMP/APTESx (where x can be 1 or 2 for different concentrations of APTES) coatings on the cotton fabric surface. Figure 9 illustrates the schematic process for preparing multifunctional cotton samples (a), along with the coating formation on Ag/TiO_2/TPMP/SiQAC samples (b) and Ag/TiO_2/TPMP/APTESx samples (c).

The combination of TPMP/SiQAC led to the formation of TiO_2/Ag/AgCl, which provided enhanced UV protection and superior photocatalytic self-cleaning capabilities. This can be attributed to the creation of a semiconductor–metal–semiconductor heterojunction, which generated highly active oxygen species (OH·) and reactive

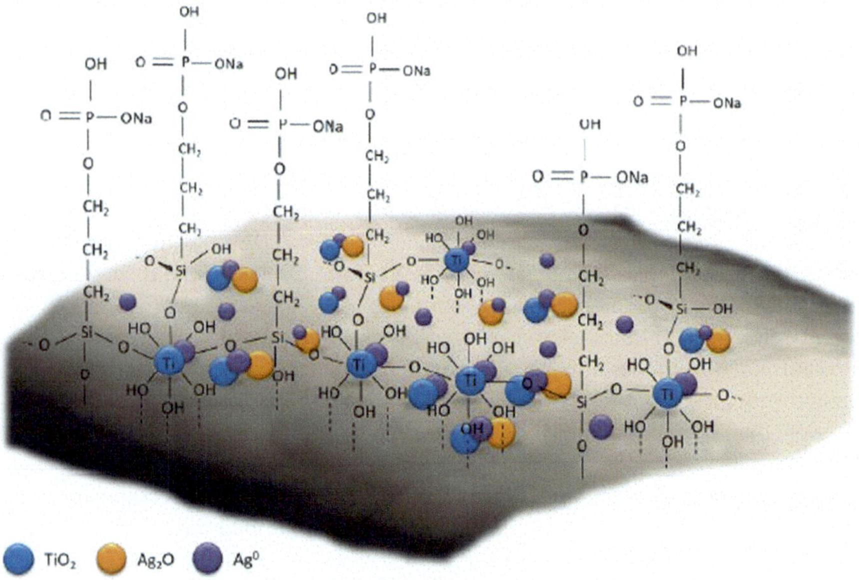

Fig. 8 Proposed representation of the formation of the TiO_2-functionalized hybrid nanocomposite film on the surface of cotton fibres. Reprinted with permission from [53]. Copyright (2023), Elsevier

chlorine species (Cl·) on the AgCl surface. Conversely, the TiO_2/Ag combined with TPMP/APTES offered excellent UV protection, as the Ag NPs effectively complemented TiO_2 in blocking UVA and UVB rays. However, this combination exhibited slightly lower photocatalytic self-cleaning activity, as it only formed OH· species, resulting in a slower rate of Rhodamine B dye discoloration. Nonetheless, both coatings demonstrated high photocatalytic self-cleaning efficiency, and complete self-sterilization activity [54].

A new eco-friendly bio-reductant called polyphosphonium chitosanhydrazone (PPCH) is created from low molecular weight shrimp chitosan. Bionanocomposites (BNC) comprising of binary (PPCH-Ag) and ternary (PPCH-TiO_2-Ag) systems have been synthesized and applied to cellulosic fabrics (CFs) through one-step in situ process. PPCH serves as both a reducing and a stabilizing agent when exposed to microwave radiation, facilitating the in situ reduction of Ag^+ ions to AgNPs on the surface of the cotton fabric, forming a coating known as PPCH-Ag (BNC1). Antimicrobial tests revealed that fabrics treated with BNCs exhibited significantly enhanced antibacterial properties compared to the untreated fabrics or those coated only with PPCH. This enhancement is due to the antimicrobial properties of both PPCH and the firmly bound Ag and TiO_2 NPs. Although the ternary BNC (PPCH-TNPs-Ag) showed slightly lower antimicrobial activity than the binary BNC (PPCH-Ag), likely because of a lower proportion of AgNPs, it still provided excellent UV protection. The BNC2-CF fabric demonstrated a huge reduction in UV transmission and achieved the highest UPF (49.2 ± 3.3), indicating strong UV-protective capabilities [55].

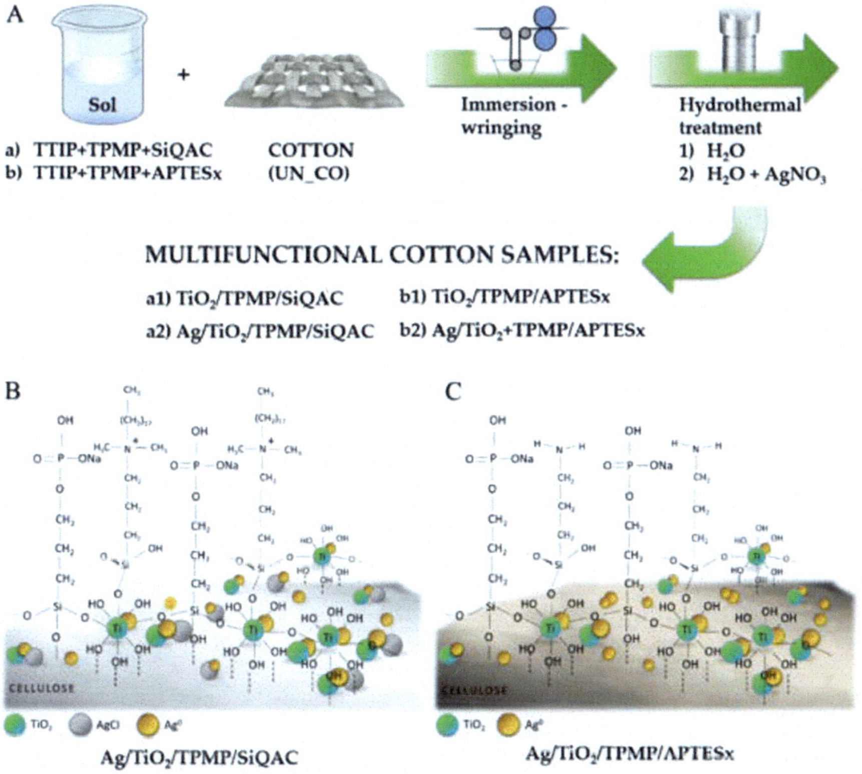

Fig. 9 Schematic representation of the preparation of multifunctional cotton samples (**a**) and nanohybrid samples Ag/TiO_2/TPMP/SiQAC (**b**) and Ag/TiO_2/TPMP/APTESx (**c**). Reprinted with permission from [54]. Copyright (2024), Elsevier

A straightforward and cost-efficient method is developed for fabricating highly non-wetting nanocomposite polyacrylonitrile nanofibers (PAN NFs) modified with varying concentrations of TiO_2 and Ag NPs called Ag@PAN-TiO_2 NFs. These nanohybrids exhibit antibacterial effects against *E. coli* and *S. aureus*, as well as antiviral activity against bovine coronavirus, self-cleaning abilities under light due to their photocatalytic feature, and pollutant detection via surface-enhanced Raman scattering. It is believed that when TiO_2 NPs are exposed to UV light, they release electrons that reduce the fibre diameter and make the surface rougher. These electrons gather especially in areas where the TiO_2 NPs are concentrated near the fibre surface, causing the thin PAN layers in those spots to be etched away. Overall, the study supports the potential of chemically and physically adjustable self-cleaning platforms representing significant advancements in sensing and catalysis [56].

5 Concluding Remarks

The development of multifunctional hybrid TiO_2 NPs represents a significant advancement in sunscreen technology. This chapter highlights the potential of hybrid TiO_2 nanomaterials to enhance UV protection while addressing common limitations such as the white cast effect and photocatalytic activity. By hybridizing TiO_2 with biopolymers like lignin and melanoidin, or incorporating additional molecules like chitosan and tannic acid, the UV-blocking efficacy is improved while mitigating the generation of ROS. These multifunctional hybrids exhibit enhanced stability, increased SPF, and additional benefits such as antibacterial properties and reduced skin irritation.

This chapter demonstrates the feasibility of producing eco-friendly and biocompatible sunscreen formulations that provide broad-spectrum protection. The incorporation of natural and synthetic compounds into TiO_2 nanostructures not only boosts UV protection but also offers promising applications in various industries beyond cosmetics, such as textiles and pharmaceuticals.

6 Future Challenges

Despite the promising results, several challenges remain to be addressed in the future development and application of multifunctional hybrid TiO_2 NPs in sunscreen technology. Comprehensive studies on the long-term safety and toxicity of these hybrid NPs are essential. Understanding the chronic effects of prolonged exposure, especially the potential for nanoparticle penetration into deeper skin layers and systemic circulation, is critical. In addition, the environmental impact of these NPs, particularly their persistence and bioaccumulation in aquatic and terrestrial ecosystems, needs thorough investigation. The potential release of NPs during product use and disposal must be assessed to ensure environmental safety. While significant steps have been made to reduce the photocatalytic activity of TiO_2 NPs, completely mitigating this effect remains a challenge. Innovative coating techniques and the development of non-photocatalytic TiO_2 composites should be prioritized. The scalability of synthesizing these multifunctional hybrids on an industrial scale must be explored. Efficient, cost-effective, and environment friendly manufacturing processes need to be developed to facilitate widespread adoption.

By addressing these challenges, the potential of multifunctional hybrid TiO_2 NPs can be fully realized, paving the way for safer, more effective, and environmentally sustainable sunscreens and related products.

References

1. Kneissl, M., Seong, T.-Y., Han, J., Amano, H.: The emergence and prospects of deep-ultraviolet light-emitting diode technologies. Nat. Photonics **13**, 233–244 (2019). https://doi.org/10.1038/s41566-019-0359-9
2. Zou, Y., Zhang, Y., Hu, Y., Gu, H.: Ultraviolet detectors based on wide bandgap semiconductor nanowire: a review. Sensors **18** (2018). https://doi.org/10.3390/s18072072
3. Herndon, J.M., Hoisington, R.D., Whiteside, M.: Deadly ultraviolet UV-C and UV-B penetration to earth's surface: human and environmental health implications. J. Geo. Env. Earth Sci. Int. **14**, 1–11 (2018). https://doi.org/10.9734/JGEESI/2018/40245
4. D'Orazio, J., Jarrett, S., Amaro-Ortiz, A., Scott, T.: UV radiation and the skin. Int. J. Mol. Sci. **14**, 12222–12248 (2013). https://doi.org/10.3390/ijms140612222
5. Wacker, M., Holick, M.F.: Sunlight and vitamin D: a global perspective for health. Dermatoendocrinology **5**, 51–108 (2013). https://doi.org/10.4161/derm.24494
6. Geoffrey, K., Mwangi, A.N., Maru, S.M.: Sunscreen products: rationale for use, formulation development and regulatory considerations. Saudi Pharm J. **27**, 1009–1018 (2019). https://doi.org/10.1016/j.jsps.2019.08.003
7. Santos, A.C., Marto, J., Chá-Chá, R., Martins, A.M., Pereira-Silva, M., Ribeiro, H.M., Veiga, F.: Nanotechnology-based sunscreens—a review. Mater. Today Chem. **23**, 100709 (2022). https://doi.org/10.1016/j.mtchem.2021.100709
8. Tanemura, S., Miao, L., Wunderlich, W., Tanemura, M., Mori, Y., Toh, S., Kaneko, K.: Fabrication and characterization of anatase/rutile–TiO2 thin films by magnetron sputtering: a review. Sci. Technol. Adv. Mater. **6**, 11–17 (2005)
9. Amtout, A., Leonelli, R.: Optical properties of rutile near its fundamental band gap. Phys. Rev. B **51**, 6842–6851 (1995). https://doi.org/10.1103/PhysRevB.51.6842
10. Pascual, J., Camassel, J., Mathieu, H.: Fine structure in the intrinsic absorption edge of Ti\mathrmO_2. Phys. Rev. B **18**, 5606–5614 (1978). https://doi.org/10.1103/PhysRevB.18.5606
11. Tang, H., Lévy, F., Berger, H., Schmid, P.E.: Urbach tail of anatase \mathrmTiO_2. Phys. Rev. B **52**, 7771–7774 (1995). https://doi.org/10.1103/PhysRevB.52.7771
12. Mattsson, A., Österlund, L.: Adsorption and photoinduced decomposition of acetone and acetic acid on anatase, brookite, and rutile TiO2 nanoparticles. J. Phys. Chem. C **114**, 14121–14132 (2010). https://doi.org/10.1021/jp103263n
13. Lautenschlager, S., Wulf, H.C., Pittelkow, M.R.: Photoprotection. Lancet **370**, 528–537 (2007)
14. Kargan, N.H., Aliahmad, M., Azizi, S.: Detecting ultra-violet radiation by using titanium dioxide nanoparticles. www.scirp.org/journal/paperinformation.aspx?paperid=19809 (2012)
15. Smijs, T.G., Pavel, S.: Titanium dioxide and zinc oxide nanoparticles in sunscreens: focus on their safety and effectiveness. Nanotechnol. Sci. Appl. **4**, 95–112 (2011). https://doi.org/10.2147/NSA.S19419
16. Petrick, J., Ibadurrohman, M.: Slamet: development of chitosan/TiO2 nanocomposite for multifunctional sunscreen application. IOP Conf. Ser. Mater. Sci. Eng. **722**, 012030 (2020). https://doi.org/10.1088/1757-899X/722/1/012030
17. Aranaz, I., Alcántara, A.R., Civera, M.C., Arias, C., Elorza, B., Heras Caballero, A., Acosta, N.: Chitosan: an overview of its properties and applications. Polymers (Basel) **13**, (2021). https://doi.org/10.3390/polym13193256
18. Chang, K., Tai, M., Cheng, F.: Kinetics and products of the degradation of chitosan by hydrogen peroxide. J. Agric. Food Chem. **49**, 4845–4851 (2001). https://doi.org/10.1021/jf001469g
19. Goy, R., Britto, D., Assis, O.: A review of the antimicrobial activity of chitosan. Polímeros **19**, 241–247 (2009). https://doi.org/10.1590/S0104-14282009000300013
20. Bino, A., Baldisserotto, A., Scalambra, E., Dissette, V., Vedaldi, D.E., Salvador, A., Durini, E., Manfredini, S., Vertuani, S.: Design, synthesis and biological evaluation of novel hydroxyphenyl-1H-benzimidazoles as radical scavengers and UV-protective agents. J. Enzym. Inhib. Med. Chem. **32**, 527–537 (2017)

21. Battistin, M., Dissette, V., Bonetto, A., Durini, E., Manfredini, S., Marcomini, A., Casagrande, E., Brunetta, A., Ziosi, P., Molesini, S., Gavioli, R., Nicoli, F., Vertuani, S., Baldisserotto, A.: A new approach to UV protection by direct surface functionalization of TiO2 with the antioxidant polyphenol dihydroxyphenyl benzimidazole carboxylic acid. Nanomaterials **10** (2020). https://doi.org/10.3390/nano10020231
22. Kai, D., Tan, M.J., Chee, P.L., Chua, Y.K., Yap, Y.L., Loh, X.J.: Towards lignin-based functional materials in a sustainable world. Green Chem. **18**, 1175 (2016)
23. Chen, X.Y., Kuo, D.H., Lu, D.F., Hou, Y.X., Kuo, Y.R.: Synthesis and photocatalytic activity of mesoporous TiO2 nanoparticle using biological renewable resource of un-modified lignin as a template. Microporous Mesoporous Mater. **223**, 145 (2016)
24. Morsella, M., d'Alessandro, N., Lanterna, A.E., Scaiano, J.C.: Improving the sunscreen properties of TiO2 through an understanding of its catalytic properties. ACS Omega **1**, 464 (2016)
25. Yu, J., Li, L., Qian, Y., Lou, H., Yang, D., Qiu, X.: Facile and green preparation of high UV-blocking lignin/titanium dioxide nanocomposites for developing natural sunscreens. Ind. Eng. Chem. Res. **57**, 15740–15748 (2018). https://doi.org/10.1021/acs.iecr.8b04101
26. Zhou, X., Liu, E., Li, J., Chen, F., Zhang, Y., Chen, K., Qi, D.: Facile synthesis of rodlike TiO2/lignin nanostructures as multifunctional emulsifiers for high internal phase emulsions. ACS Appl. Nano Mater. **6**, 14820–14830 (2023). https://doi.org/10.1021/acsanm.3c02384
27. Chegeni, M., Pour, S.K., Faraji Dizaji, B.: Synthesis and characterization of novel antibacterial Sol-gel derived TiO2/Zn2TiO4/Ag nanocomposite as an active agent in sunscreens. Ceram. Int. **45**, 24413–24418 (2019). https://doi.org/10.1016/j.ceramint.2019.08.163
28. Son, H.Y., Koo, B.I., Lee, J.B., Kim, K.R., Kim, W., Jang, J., Yoon, M.S., Cho, J.-W., Nam, Y.S.: Tannin-titanium oxide multilayer as a photochemically suppressed ultraviolet filter. ACS Appl. Mater. Interfaces **10**, 27344–27354 (2018). https://doi.org/10.1021/acsami.8b09200
29. Choi, S., Rahman, R.T., Kim, B., Kang, J., Kim, J., Shim, J., Nam, Y.S.: Photochemically inert broad-spectrum sunscreen by metal-phenolic network coatings of titanium oxide nanoparticles. ACS Appl. Mater. Interfaces **16**, 16767–16777 (2024). https://doi.org/10.1021/acsami.4c00174
30. Wang, S.-Y., Zhu, B.-B., Li, D.-Z., Fu, X.-Z., Shi, L.: Preparation and characterization of TIO2/SPI composite film. Mater. Lett. **83**, 42–45 (2012)
31. He, Q., Huang, Y., Lin, B., Wang, S.: A nanocomposite film fabricated with simultaneously extracted protein-polysaccharide from a marine alga and TiO 2 nanoparticles. J. Appl. Phycol. **29**, 1541–1552 (2017)
32. Fan, X., Chen, K., He, X., Li, N., Huang, J., Tang, K., Li, Y., Wang, F.: Nano-TiO2/collagen-chitosan porous scaffold for wound repairing. Int. J. Biol. Macromol. **91**, 15–22 (2016)
33. He, Q., Zhang, Y., Cai, X., Wang, S.: Fabrication of gelatin–TiO2 nanocomposite film and its structural, antibacterial and physical properties. Int. J. Biol. Macromol. **84**, 153–160 (2016)
34. Anaya-Esparza, L.M., Villagrán-de la Mora, Z., Rodríguez-Barajas, N., Sandoval-Contreras, T., Nuño, K., López-de la Mora, D.A., Pérez-Larios, A., Montalvo-González, E.: Protein–TiO2: a functional hybrid composite with diversified applications. Coatings **10** (2020). https://doi.org/10.3390/coatings10121194
35. Cashin, V.B., Eldridge, D.S., Yu, A., Zhao, D.: Surface functionalization and manipulation of mesoporous silica adsorbents for improved removal of pollutants: a review. Environ. Sci. Water Res. Technol. **4**, 110–128 (2018)
36. Barroso-Martín, I., Moretti, E., Talon, A., Storaro, L., Rodríguez-Castellón, E., Infantes-Molina, A.: Au and AuCu nanoparticles supported on SBA-15 ordered mesoporous titania-silica as catalysts for methylene blue photodegradation. Materials **11**, 890 (2018)
37. Pavlou, P., Siamidi, A., Vlachou, M., Varvaresou, A.: UV filters and their distribution on the skin through safe, non-penetrating vehicles. J. Cosmet. Sci. **72** (2021)
38. Alalaiwe, A., Lin, Y.-C., Lin, C.-F., Huang, C.-C., Wang, P.-W., Fang, J.-Y.: TiO2-embedded mesoporous silica with lower porosity is beneficial to adsorb the pollutants and retard UV filter absorption: a possible application for outdoor skin protection. Eur. J. Pharm. Sci. **180**, 106344 (2023). https://doi.org/10.1016/j.ejps.2022.106344

39. Du, J., Sun, H.: Polymer/TiO2 hybrid vesicles for excellent UV screening and effective encapsulation of antioxidant agents. ACS Appl. Mater. Interfaces **6**, 13535–13541 (2014). https://doi.org/10.1021/am502663j
40. Rabani, I., Jang, H.-N., Park, Y.-J., Tahir, M.S., Lee, Y.-B., Moon, E.-Y., Song, J.W., Seo, Y.-S.: Titanium dioxide incorporated in cellulose nanofibers with enhanced UV blocking performance by eliminating ROS generation. RSC Adv. **12**, 33653–33665 (2022). https://doi.org/10.1039/D2RA06444H
41. Liu, Y., Ai, K., Liu, J., Deng, M., He, Y., Lu, L.: Dopamine-melanin colloidal nanospheres: an efficient near-infrared photothermal therapeutic agent for in vivo cancer therapy. Adv. Mater. **25**, 1353–1359 (2013)
42. Wang, J., Sun, S.J., Ding, H., Chen, W.T., Liang, Y.: Preparation of a composite photocatalyst with enhanced photocatalytic activity: smaller TiO2 carried on SiO2 microsphere. Appl. Surf. Sci. **493**, 146 (2019)
43. Huang, Y., Li, Y., Hu, Z., Yue, X., Proetto, M.T., Jones, Y., Gianneschi, N.C.: Mimicking melanosomes: polydopamine nanoparticles as artificial microparasols. ACS Cent. Sci. **3**, 564–569 (2017)
44. Li, S., Wang, Y., Li, T., Ma, P., Zhang, X., Xia, B., Chen, M., Du, M., Dong, W.: Skin bioinspired anti-ultraviolet melanin/TiO2 nanoparticles without penetration for efficient broad-spectrum sunscreen. Colloid Polym. Sci. **299**, 1797–1805 (2021). https://doi.org/10.1007/s00396-021-04905-7
45. Algessair, S., Madkhali, N.: Formation of natural melanin/TiO2 nanostructure hybrids with enhanced optical, thermal and magnetic properties as a soft material. J. Wuhan Univ. Technol. Mater. Sci. Ed. **39**, 613–620 (2024). https://doi.org/10.1007/s11595-024-2917-3
46. Lei, W., Liu, D., Chen, Y.: Highly crumpled boron nitride nanosheets as adsorbents: scalable solvent-less production. Adv. Mater. Interfaces **2** (2015). https://doi.org/10.1002/admi.201400529
47. Lei, W., Mochalin, V.N., Liu, D., Qin, S., Gogotsi, Y., Chen, Y.: Boron nitride colloidal solutions, ultralight aerogels and freestanding membranes through one-step exfoliation and functionalization. Nat. Commun. **6** (2015). https://doi.org/10.1038/ncomms9849
48. Wang, M., Li, M., Xu, L., Wang, L., Ju, Z., Li, G., Qian, Y.: High yield synthesis of novel boron nitride submicro-boxes and their photocatalytic application under visible light irradiation. Catal. Sci. Technol. **1**, 1159–1165 (2011). https://doi.org/10.1039/c1cy00111f
49. Ide, Y., Liu, F., Zhang, J., Kawamoto, N., Komaguchi, K., Bando, Y., Golberg, D.: Hybridization of Au nanoparticle-loaded TiO2 with BN nanosheets for efficient solar-driven photocatalysis. J. Mater. Chem. A. **2**, 4150–4156 (2014). https://doi.org/10.1039/C3TA13769D
50. Liu, D., Zhang, M., Xie, W., Sun, L., Chen, Y., Lei, W.: Porous BN/TiO2 hybrid nanosheets as highly efficient visible-light-driven photocatalysts. Appl. Catal. B **207**, 72–78 (2017). https://doi.org/10.1016/j.apcatb.2017.02.011
51. Walther, A., Müller, A.H.E.: Janus particles: synthesis, self-assembly, physical properties, and applications. Chem. Rev. **113**, 5194–5261 (2013). https://doi.org/10.1021/cr300089t
52. Hou, J., Yang, Y., Yu, D.-G., Chen, Z., Wang, K., Liu, Y., Williams, G.R.: Multifunctional fabrics finished using electrosprayed hybrid Janus particles containing nanocatalysts. Chem. Eng. J. **411**, 128474 (2021). https://doi.org/10.1016/j.cej.2021.128474
53. Rashid, M.M., Tomšič, B., Simončič, B., Jerman, I., Štular, D., Zorc, M., Korošin, N.Č: In situ tailoring of Ag-doped-TiO2/TPMP/cotton nanocomposite with UV-protective, self sterilizing and flame-retardant performance for advanced technical textiles. Polym. Degrad. Stab. **216**, 110504 (2023). https://doi.org/10.1016/j.polymdegradstab.2023.110504
54. Rashid, M.M., Simončič, B., Zorc, M., Čelan Korošin, N., Jerman, I., Tomšič, B.: Sustainable in situ synthesis of silver/titanium dioxide/organofunctional trialkoxysilane nanocomposite coatings on cotton fabric to tailor superior multifunctional protective properties. Surf. Interfaces **49**, 104447 (2024). https://doi.org/10.1016/j.surfin.2024.104447
55. Refaee, A.A., Mostafa, T.B., El-Naggar, M.E., Alfaifi, M.Y., Shati, A.A., Elbehairi, S.E.I., Elshaarawy, R.F.M., Ismail, L.A.: Cellulosic fabrics modified with polyphosphonium chitosan hydrazone-TiO2-Ag nanobiocomposites for multifunctional applications. Int. J. Biol. Macromol. **220**, 482–492 (2022). https://doi.org/10.1016/j.ijbiomac.2022.08.104

56. Ceylan, A., Karagoz, S., Tekin, S., Kayaci, N., Pekdemir, S., Salem, S., Gokduman, E.M., Demirci, E., Sahmetlioglu, E., Onses, M.S., Kiremitler, N.B., Yilmaz, E.: TiO2 and Ag NPs modified polyacrylonitrile NFs: antimicrobial, self-cleaning and SERS sensing capabilities for protective clothing applications. Colloids Surf. A Physicochemical Eng. Aspects **678**, 132432 (2023). https://doi.org/10.1016/j.colsurfa.2023.132432

Properties of Photocatalytically Active TiO_2 Nanoparticles and Prospects for Their Applications in Agriculture

Olga A. Shilova, Anastasiya S. Kovalenko, Tamara V. Khamova, Anton M. Nikolaev, Sergey V. Myakin, Gennady P. Kopitsa, Olga R. Udalova, Dmitry L. Kornyukhin, Anna M. Artemyeva, Anna S. Zhuravleva, and Gayane G. Panova

Abstract Nanopowders of titania (TiO_2) features a unique combination of UV-stimulated photocatalytic activity and biocompatibility making ones promising for various biochemical applications, including agrotechnology in respect to the promotion of plants germination and growth as well as their protection from different stressors and pathogens. However, the considered target properties strongly depend on such factors as phase composition (i.e. ratio between anatase and rutile polymorph modifications), particle size (especially in the case of nanoparticles possessing an overall enhanced reactivity and size-depending specific properties) and surface characteristics, particularly presence and content of different centers responsible for acid–base interactions with the environment, incl. aqueous medium. This chapter presents a review of recent researches on the analysis of relationships among these

---

O. A. Shilova (✉) · A. S. Kovalenko · T. V. Khamova · A. M. Nikolaev · G. P. Kopitsa
I.V. Grebenshchikov Institute of Silicate Chemistry, Saint-Petersburg 199034, Russia
e-mail: olgashilova@bk.ru

O. A. Shilova · A. M. Nikolaev
Saint-Petersburg State Electrotechnical University "LETI", Saint-Petersburg 197022, Russia

O. A. Shilova · S. V. Myakin
Saint-Petersburg State Institute of Technology (Technical University), Saint-Petersburg 190013, Russia

S. V. Myakin
Institute for Analytical Instrumentation, Russian Academy of Sciences, St. Petersburg 190103, Russia

G. P. Kopitsa
Petersburg Nuclear Physics Institute Named By B.P. Konstantinov of NRC «Kurchatov Institute», Leningradskaya Oblast, Gatchina 188300, Russia

O. R. Udalova · A. S. Zhuravleva · G. G. Panova
Agrophysical Research Institute, St. Petersburg 195220, Russia

D. L. Kornyukhin · A. M. Artemyeva
N.I. Vavilov All-Russian Institute of Plant Genetic Resources (VIR), St. Petersburg 190031, Russia

J. Prakash et al. (eds.), *Titanium Dioxide-Based Multifunctional Hybrid Nanomaterials*, Engineering Materials, https://doi.org/10.1007/978-3-031-80625-4_9

properties of TiO_2 nanoparticles differing in the preparation conditions in order to determine the optimum structural and surface features of titanium dioxide providing the highest photocatalytic activity and certain agricultural applications, incl. the effect of pre-sowing treatment with aqueous suspensions of TiO_2 nanoparticles on the seed germination and plant seedling growth, as well as phytoprotection activity against specific fungal pathogens.

Keywords Nonopowder TiO_2 · Preparation conditions · Surface properties · Photocatalytic activity · Agriculture

1 Introduction

Titanium oxide (TiO_2) powders are multi-purpose materials. Due to the combination of low cost, ease of preparation and variety of physical and chemical properties, researchers have already found numerous possible applications and continue to find new ones [1]. It should be noted that titanium oxide exists in several polymorphs, including anatase, rutile, brookite TiO_2 (B), TiO_2 (II, α-PbO_2-like form), TiO_2 (H, hollandite-like form) etc. [2–5], among which anatase and rutile are the most widely spread.

Titania powders are used as colorants in food, cosmetics and paints, ceramics and electronics, wastewater treatment, biomedicine and biology, agriculture. One of the main properties of titanium oxide providing such a wide range of applications is photocatalytic activity [6–8], which depends on phase composition, particle size and shape, degree of crystallinity, bandgap, and specific surface area [9, 10]. For a long time, it was considered that anatase phase of titania is the most photocatalytically active. However, numerous recent studies reveal that TiO_2 with mixed phase composition (particularly anatase and rutile in various proportions) exhibits the highest photocatalytic activity [11]. Our research [12, 13] showed that commercial TiO_2 nanopowder (Plasmotherm®) with a mixed phase composition (anatase:rutile ratio 1:4) and large nanoparticles (anatase CSR ~70 nm, rutile CSR ~90 nm) features a much higher photocatalytic activity compared to weakly crystallized anatase (CSR size ~7 nm) obtained by precipitation from $TiCl_4$.

So far, the effect of TiO_2 nanoparticles on living organisms is ambiguous. In some works, the authors note that photocatalytic activity can have a depressing effect on the human body [14], while anatase is considered more toxic than rutile due to its greater photocatalytic activity [15].

In recent years, special attention has been paid to the use of titanium oxide nanoparticles in agriculture [16–19]. Thus, in [20], the effect of introducing TiO_2 nanoparticles into the soil in various quantities (0, 20, 40, 60, 80, 100 mg/kg) on wheat growth was studied. It has been established that an amount of 60 mg/g leads to a maximum increase in the length of roots and shoots, and a further increase in concentration leads to a decrease in growth rates and biomass. In [21], the effect of treating radish crops with different concentrations of TiO_2 nanoparticles (50 and 100 μmol/L) in combination with irradiation with light of different intensities was studied. It has been established that, depending on the processing conditions, both redistribution of biomass towards the underground part and an increase in the efficiency of photosynthesis can be achieved. The example of contradictory influence of TiO_2 nanoparticles on the development of various agricultural crops was considered in [22]. On the one hand, acceleration of seed germination and growth was recorded due to increased efficiency of photosynthesis and absorption of micronutrients, formation of chlorophyll, carotenoids and anthocyanins when TiO_2 had been used in small quantities (specifically, on the order of 0.01–0.03%). On the other hand, a clear phytotoxic effect was also observed due to the formation of radicals and hydrogen peroxide, which destroy chlorophyll molecules, proteins and cell membranes. Usually, titania nanoparticles are introduced into the soil like fertilizer [23], or they are used to treat the roots of plant seedlings during planting [24], or they are used for foliar treatment of plants [21]. Also, pre-sowing seed treatment is considered a very effective method of increasing plant productivity [25, 26]. However, the most common methods of pre-sowing seed treatment include soaking the seeds in various solutions of useful micro- and macroelements as well as biocides or formation of polymer functional coatings containing useful substances [27–34].

Our studies have shown the effectiveness of the application of aqueous suspensions of TiO_2 nanoparticles for pre-sowing treatment of vegetable seeds [12, 13, 35]. In addition, for the first time we demonstrated the possibility of forming a moisture- and air-permeable functional shell on the surface of the seeds of main agricultural crops. This shell made of silica sol is enriched with micro- and macroelements useful for plants, including TiO_2 nanoparticles [36].

Thus, the use of TiO_2 nanoparticles in agricultural technologies is very promising. However, among the publications on this topic insufficient attention is paid to the influence of the phase composition of titania, their photocatalytic activity as well as peculiarities of interaction with the aquatic environment on the growth and development of plants, and on phytoprotective properties.

This chapter provides an overview of recent research, including the results obtained by our scientific group, on the relationship between methods and conditions for obtaining TiO_2 nanopowders, their degree of crystallinity, phase composition (especially the rutile:anatase ratio), particle size, surface properties (including specific surface area, acid–base characteristics and ζ-potential), interaction with water, photocatalytic activity (involving both overall values and specific mechanisms) and efficiency of agricultural applications, including germination and growth stimulation as well as phytoprotection properties.

2 Effect of TiO_2 Phase Composition, Particle Size and Specific Surface Area upon Photocatalytic Activity

TiO_2 is the most widely applied photocatalyst. In addition to high stability, biological and chemical inertness, this semiconductor retains a high catalytic activity during a longtime and multiple exploration [37]. Photocatalysis is the acceleration of chemical reactions upon activation with light. In this case, the absorption of sufficient energy upon UV radiation with the wavelength below 380 nm exceeding the semiconductor bandgap results in the transition of electrons from the valence band (VB) to the conduction band (CB) yielding holes in the valence band according to the reaction:

$$\mathrm{TiO_2} + \mathrm{h}\nu \rightarrow \mathrm{e}^-_{\mathrm{CB}} + \mathrm{h}^+_{\mathrm{VB}} \quad (1)$$

Photogenerated electrons and holes can behave in different ways. A part of them undergo recombination inside titania particles, while others reach the particle surface and joins red-ox reactions with adsorbed oxygen or water molecules in the case of their presence on the surface:

$$\begin{gathered} e^-_{\mathrm{CB}} + \mathrm{O_2} \rightarrow \mathrm{O_2^-} \\ \mathrm{O_2^-} + e^-_{\mathrm{CB}} \rightarrow \mathrm{O_2^{2-}} \rightarrow \mathrm{O^-} + \mathrm{O^-}, \\ \mathrm{O_2^{2-}} + 2\mathrm{H^+} \rightarrow \mathrm{H_2O_2}, \\ \mathrm{H_2O_2} + e^-_{\mathrm{CB}} \rightarrow \mathrm{OH^*} + \mathrm{OH^-}, \\ \mathrm{O^-} + e^-_{\mathrm{CB}} \rightarrow \mathrm{O^{2-}}, \\ \mathrm{O^-} + \mathrm{H^+} \rightarrow \mathrm{OH^*}, \\ e^-_{\mathrm{CB}} + \mathrm{H_2O} \rightarrow \mathrm{OH^-} + \mathrm{H^*}, \\ h^+_{\mathrm{VB}} + \mathrm{H_2O} \rightarrow \mathrm{OH^*} + \mathrm{H^+}. \end{gathered} \quad (2)$$

Some of the above processes are schematically illustrated in Fig. 1.

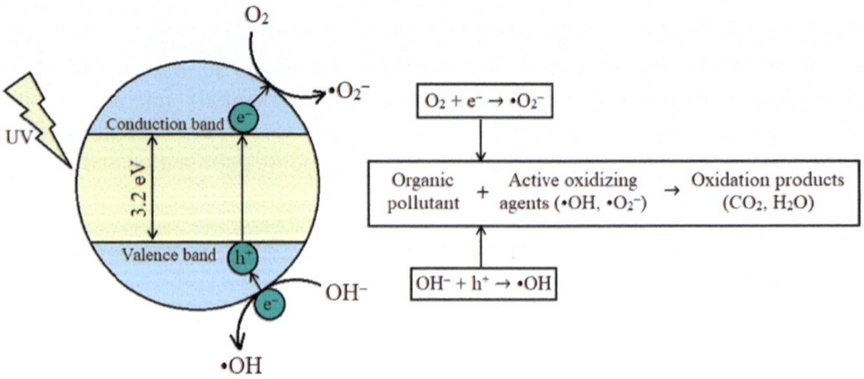

Fig. 1 Mechanism of photocatalysis using TiO_2 (anatase)

Photocatalytic reactions proceed due to the resulting O^- and $OH^{\cdot}$ species featuring strong oxidizing properties and facilitating the decomposition of organic impurities and bacteria to carbon dioxide and water:

$$O^- + \textit{organic compounds (pollutants)} \rightarrow H_2O + CO_2 \tag{3}$$

$$OH^* + \textit{organic compounds (pollutants)} \rightarrow H_2O + CO_2 \tag{4}$$

Properties of TiO_2 (including phase composition, particle size, crystallinity, surface functional composition) strongly affect the photocatalytic activity [9, 10]. A common approach used to the enhancement of photocatalytic processes involving TiO_2 is based on the adjustment of these parameters depending on titania synthesis methods and conditions [38–43]. Generally, phase composition is considered as the most important factor affecting the photocatalytic activity of TiO_2.

According to numerous references, anatase for a long time has been considered to possess the best photocatalytic properties among TiO_2 phases. Particularly, the enhanced activity of anatase compared with rutile and brookite was explained on the basis of the calculation of electronic structures and effective masses of charge carries in these phases using a plane wave pseudopotential method [44]. The differences in calculation results are determined by the peculiarities of band gap. Anatase features an indirect bandgap making impossible direct transitions of photogenerated electrons to the conduction band, compared to rutile with a direct bandgap. Furthermore, the average effective mass of photogenerated electrons and holes in anatase is the lowest among TiO_2 modifications providing the fastest migration of charge carriers to the surface of particles and consequently lowest rate of their recombination.

The above calculation results are confirmed by experimental data in [45] relating to the synthesis and comparative study of all the discussed titania modifications with the conclusion regarding the highest photocatalytic activity of anatase among them (due to indirect bandgap, small particle size and high specific surface area of the powder).

However, recent studies indicate that nanopowders with mixed anatase: rutile phase composition exhibit better photocatalytic properties compared with those comprising only rutile or anatase phase [46]. Particularly, a commercial nanopowder P-25 (Degussa) considered as a standard of photocatalytic activity has a mixed composition including 80% anatase and 20% rutile.

Our recent studies of TiO_2 nanopowders [12, 13] were aimed at the revelation and analysis of effects of phase composition, particle size, surface characteristics (including textural parameters and functional composition) determined by the synthesis method and conditions of titania upon its photocatalytic activity. The studied samples involved both commercial (produced by Plasmotherm® and Evonik-Degussa®) TiO_2 nanopowders and those synthesized in our laboratory via the precipitation from acidified TiCl4 solution with ammonia according to the procedure described in [12, 13], followed by annealing at different temperatures (300, 500,

800 °C). All the samples were characterized by the determination of phase composition (anatase:rutile ratio), particle size, textural parameters (specific surface area, porosity) and surface characteristics (pH change upon the immersion of samples in water and ζ-potential). The obtained data were compared with a photocatalytic activity of the materials determined by detection method [47].

The results summarized in Table 1 indicate that photocatalytic activity of the studied titania samples depend most strongly upon the phase composition, particle size and crystallinity [12, 13]. Nanopowders with a mixed anatase:rutile composition possess a higher activity compared with pure rutile and anatase samples (Fig. 2; Table 1). The highest activity is observed for the nanopowder with the ratio anatase:rutile = 20:80% (sample TiO_2-Pl-2 in Table 1) featuring the highest crystallinity (maximum CSR values for both phases, including 70 nm for anatase and 90 nm) and similar particle size of both phases.

Among the synthesized samples rutile phase (anatase:rutile ratio 90:10%) is present only upon annealing at 800 °C. This sample also provides an increased photocatalytic activity compared with pure anatase samples prepared with heating at 300 and 500 °C (Table 1; Fig. 2).

The increased photocatalytic activity of mixed-phase titania nanopowders can be determined by the migration of photogenerated electrons from rutile to anatase and holes from anatase to rutile [46, 49, 50] (or in opposite directions [49–51]) under UV excitation (Fig. 3). The resulting separation of electrons and holes prevents from their recombination thus facilitating a prolonged and more efficient photocatalysis on the material surface [46, 49–51].

Similar results were obtained in [11] at annealing of TiO_2 nanoparticles prepared by sol–gel method from $TiCl_4$. Heating at 500 and 900 °C results in the formation of anatase and rutile, respectively. The study of annealing temperature effect upon the photocatalytic activity determined by the decomposition of potassium permanganate ($KMnO_4$) solution revealed that anatase has a higher photocatalytic activity compared to rutile, probably due to a higher adsorption activity of anatase surface and lower rate of electron–hole recombination. Rutile produced at higher annealing temperatures (900 °C) features higher crystallite size due to aggregation under the direct sunlight. Electrons generated during photocatalysis are permanently trapped by enlarged aggregates of rutile crystallites, resulting in an increase of electron–hole recombination rate and a decrease in photocatalytic activity. In addition, an increase in crystallite size leads to a decrease in specific surface area. Nevertheless, the highest photocatalytic activity was exhibited by the sample, subjected to heat treatment at 800 °C and containing both anatase and rutile phases, likely due to the transfer of photoinitiated electrons from anatase to rutile leading to a decrease in the rate of electron–hole recombination.

Furthermore, as reported in [45], for better separation of photogenerated charges a direct contact of the anatase and rutile phases is necessary, as well as their optimal sizes, i.e. close sizes of rutile and anatase particles are desirable. In the considered case where anatase phase features larger particles than rutile, so-called "shading" of anatase by rutile occurred, resulting in a partial blocking of UV radiation by rutile and reduction of the photocatalysis efficiently. For this reason, anatase and rutile

Table 1 Comparison data for the studied TiO2 nanopowders [12, 13, 48]

TiO_2 samples	Synthesized by chemical precipitation					Commercial*		
	Using nitric acid with heating at different temperatures [12, 13, 48]			Using hydrochloric acid with heating at different temperatures [12, 13, 48]				
Heating temperature, °C	300	500	800	100	500			
Sample designation	TiO_2-HNO_3-300	TiO_2-HNO_3-500	TiO_2-HNO_3-800	TiO_2-HCl-100	TiO_2-HCl-500	TiO_2-Pl-1	TiO_2-Pl-2	TiO_2-Ev-D
Rutile:anatase ratio, %	0:100	0:100	10:90	0:100	0:100	100:0	80:20	20:80
CSR, nm	10	15	70:45	7	16	80	90:70	47:23
S_c, m^2/g	138	75	5	369	69	13	6	54
Singlet oxygen yield, a.u.**	128	122	308	160	208	100	2033	492

Notes *TiO_2 powders were produced by Plasmctherm® (TiO_2-Pl-1; TiO_2-Pl-2) and Evonik-Degussa® (TiO_2-E-D); **photocatalytic activity of TiO_2 nanopowders was characterized by the amount of singlet oxygen yielded upon their UV-irradiation ($P = 10$ W, $\lambda = 365$ nm) as a relative value (relating to the data for a material TiO_2-Pl-1 comprising a pure rutile phase and featuring the lowest activity in this series of samples) [49]

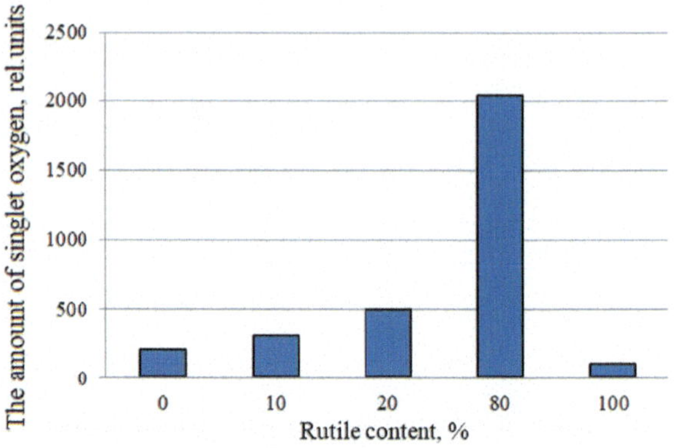

Fig. 2 Singlet oxygen yield upon UV-irradiation of mixed-phase TiO_2 samples as a function of rutile content [12, 13]

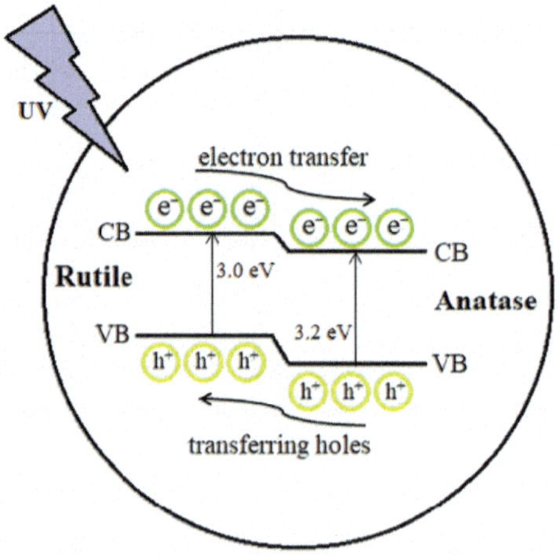

Fig. 3 Mechanism of photocatalysis using a double-phase (anatase:rutile) TiO_2

particles should be of approximately the same size and directly contact with each other in the course of photocatalysis.

This conclusion was confirmed in our studies [12, 13] indicating the highest photocatalytic activity for the sample TiO_2-Pl-2 (Table 1) featuring similar size (CSR) of anatase and rutile phases (70 and 90 nm, respectively) directly contacting with each other (Fig. 4).

In [51] it was stated that in their study photogenerated electrons in mixed TiO_2 (anatase:rutile) powders are transferred from rutile to anatase. This mechanism was just presented in the case of rutile and anatase particles of approximately the same size, which (as noted earlier) is an important factor ensuring the separation of photogenerated charge carriers for a longer and more efficient photocatalysis process.

The effects of specific surface area on the photocatalytic activity of TiO_2 nanopowders can be described as follows. As previously noted, photocatalysis is a process that occurs on the surface of nanoparticles. Therefore, it is expected that an increase in the specific surface area will contribute to an increase in photocatalytic activity. In

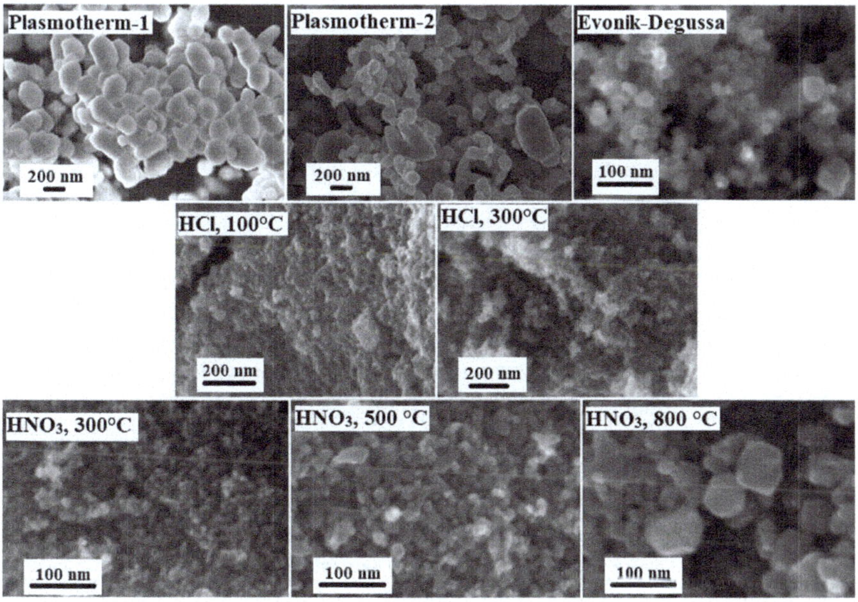

Fig. 4 SEM images of the studied titania nanopowders

our studies [12, 13], a decrease in the specific surface area of particles was observed from ~ 370 to 5 m^2/g with an increase in firing temperature from 100 to 800 °C (Table 1). However, an increase in the heat treatment temperature contributed to an increase in the crystallization of the resulting TiO_2 particles, which led to an increase in photocatalytic activity. Consequently, in our studies [12, 13] it was revealed that the specific surface area is not such a dominant factor influencing the photocatalytic activity of TiO_2, compared to the phase composition, particle size and degree of crystallization, as well as the state of the surface,—particularly when interacting with the aqueous environment.

According to [51], photogenerated electrons in mixed anatase:rutile titania powders migrate from rutile to anatase. This mechanism is particularly effective in the case of rutile and anatase particles of similar size providing the separation of photogenerated charge carriers and consequently resulting in a prolonged and more efficient photocatalysis, as considered above.

The effect of specific surface area on the photocatalytic activity of TiO_2 nanopowders can be described as follows. As discussed above, photocatalysis is a process taking place on the surface of nanoparticles. Therefore, the increase of the specific surface area can be expected to promote a growth of photocatalytic activity. However, in our studies [12, 13], the increase of annealing temperature from 100 to 800 °C resulted in the decrease in the specific surface area of titania particles from ~370 to 5 m^2/g (Table 1), however this temperature growth also caused an increase in crystallinity of TiO_2 particles and a certain enhancement of their photocatalytic activity.

3 Surface Chemistry and Interaction with Aqueous Media

Chemical features of the surface, including the predominance, concentration and spatial distribution of certain atoms and functional groups determined by the preparation and storage conditions of materials is one of the key factors responsible for their reactivity. Surface properties are particularly significant in respect to hydrophilic-hydrophobic characteristics, acid–base and donor–acceptor interactions with the environment and irradiation-stimulated phenomena such as photocatalysis. These effects are especially important for the behavior of insoluble solids in aqueous and other liquid media mediating the target reactions. For the considered titania nanoparticles, their photocatalytic activity is based on the interaction of their surface with UV-radiation and their agricultural applications essentially involve water-mediated effects of titania surface upon seeds and growing plants. Therefore, a detailed study of interaction with water is a promising approach to comparative characterization of surface properties of TiO_2 particles depending on their phase composition, synthesis and preparation conditions in order to understand the influence of these factors upon the efficiency of titania in photocatalysis and various specific applications.

In our research the surface properties of the considered titania samples and their interaction with water were studied using the following methods:

ζ-Potential measurements by dynamic and electrophoretic light scattering using a NanoBrook 90 Plus Zeta particle size analyzer (with the cell BI-SCP and electrode AQ-1327) in suspensions with 0.1 g/L concentration in 1 h after their preparation. The obtained results summarized in Table 2 indicate that ζ-potential depends both on the phase composition and preparation conditions. Particularly, positive ζ-potential values are observed for pure anatase samples prepared using nitric acid and treated at 300 and 500 °C, as well as predominantly rutile commercial samples TiO_2-Pl-1 and TiO_2-Pl-2. Other materials, including the sample treated with HNO_3 and annealed at 800 °C with a small part of rutile, as well anatase samples synthesized in the presence of HCl have negative ζ-potentials.

The comparison of these data with structural features of titania modifications suggests that negative ζ-potentials can be determined by either predominant occupation of the surface with electron-withdrawing oxygen atoms or acidic hydroxyls ≡Ti–OH dissociating into H^+ ions and negatively charged ≡Ti–O^- surface. Positive ζ-potential values of the considered samples can be determined by $=Ti(OH)_2$ and $-Ti(OH)_3$ groups dissociating yielding OH^- ions and forming a positively charged surface. Probably, such groups prevail on the surface of rutile as well as anatase oxidized in HNO_3 and heated at 300 and 500 °C, but their treatment at 800 °C results in dehydration with the transformation of these groups into oxygen bridges or ≡Ti–OH. A similar effect can be provided by hydrochloric acid in the case of both TiO_2-HCl samples.

Table 2 ζ-Potential values of the studied titania samples (mV)

TiO_2-HNO_3-300	TiO_2 HNO_3-500	TiO_2-HNO_3-800	TiO_2-HCl-100	TiO_2-HCl-500	TiO_2-Pl-1	TiO_2-Pl-2
Rutile:anatase ratio, %						
0:100	0:100	7:93	0:100	0:100	100:0	80:20
6 ± 3	6 ± 3	−24 ± 5	−30 ± 3	−21 ± 3	30 ± 3	6 ± 2

In respect to the properties of aqueous suspensions, their highest stability is expected for TiO_2-Pl-1 and TiO_2-HCl-100 samples featuring the highest absolute values of ζ-potential.

Dynamic pH-metry technique [52, 53] by measuring the changes in pH value of the suspensions obtained by immersing 30 mg of the studied powders in 30 mL of distilled water at permanent agitation with a magnetic stirrer. pH measurements were performed using Multitest IPL-301 pH-meter (NPP SEMICO, Russia) in 5, 10, 15, 30, 60 s after the sample immersion and subsequently in every 30 s up to 5 min from the powder immersion.

The results summarized in Fig. 5 show that TiO_2-Pl-2 sample featuring a mixed rutile:anatase = 80:20% phase composition and the highest photocatalytic activity (singlet oxygen yield) prominently differs from other samples in acidic properties of the surface demonstrated by an abrupt decrease in pH of the suspension in the first 15 s after immersion followed by a more gradual decrease up to 1 min.

The observed behavior can be accounted for the interaction of water with Lewis acidic centers $Ti^{4+} + nH_2O \rightarrow [Ti\text{–}(OH)n]^{(4-n)+} + nH^+$ and dissociation of acidic hydroxyls $\equiv Ti\text{–}OH \rightarrow \equiv Ti\text{–}O^- + H^+$ respectively.

Probably this ratio between rutile and anatase phases provides a certain distortion of Ti–O–Ti oxygen bridging bonds on the surface to form the optimal amounts of titanium cations generating holes and hydroxyl groups yielding $^{\cdot}OH$ radicals upon interaction with holes according to the photocatalysis mechanism described above.

For all other samples, except TiO_2-HCl-500, an almost neutral or weakly basic reaction of the surface is observed probably due to the predominance of either titanium (Lewis acidic) and oxygen (Lewis basic) atoms mutually neutralizing each other in respect to interaction with water or nearly neutral $=Ti(OH)_2$ and weakly basic $-Ti(OH)_3$ groups, in accordance with the above considerations regarding ζ-potentials. The most prominent pH growth is observed for TiO_2-HCl-100 and TiO_2-HNO_3-300

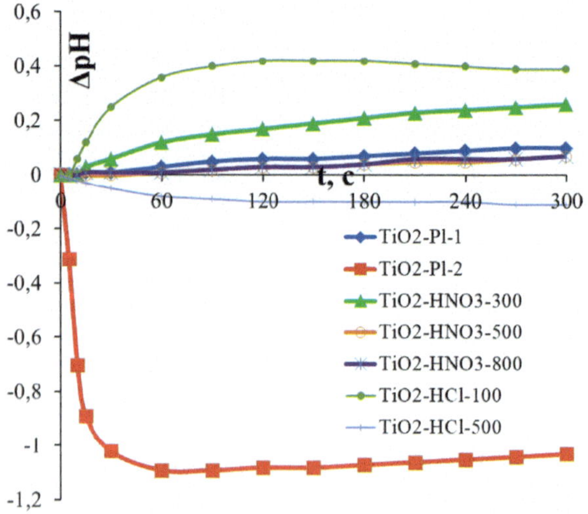

Fig. 5 pH changes in aqueous suspensions of the studied samples after their immersion in distilled water

samples probably containing $-Ti(OH)_3$ groups due to the lowest treatment temperatures (100 and 300 °C, respectively). On the contrary, TiO_2-HCl-500 sample annealed at 500 °C features a weakly acidic surface reaction reflecting the transformation of these groups into ≡Ti–OH. A nearly neutral behavior of TiO_2-HNO_3-500, TiO_2-HNO_3-800 and TiO_2-Pl-1 samples with a slight pH increase can be accounted for the presence of $=Ti(OH)_2$ groups.

The considered surface acid–base characteristics of titania samples can significantly affect their biological activity, since acidity of soil and aqueous media is one of the key factors responsible for the development of plants, with certain crops featuring individual optimal pH levels for germination and growth [54–56].

4 Aqueous Suspensions and Silica Sols with TiO_2 Nanoparticles in Agriculture

Nanotechnology, along with biotechnology, plays an important role in the development of agriculture. The main premise for the application of nanoparticles in agriculture is determined by the prospects for the reduction of fertilizers and relating adverse effects onto the environment.

TiO_2 nanoparticles are among the most widely used nanomaterials as microfertilizers to improve the uptake of nutrients by plants. For example, in [23] foliar application of TiO_2 nanoparticles barley with provides a similar effect compared to feeding with chelated microfertilizers. TiO_2 nanoparticles facilitate bacteria to interact (attach) to plant roots, increasing survival and competitiveness. Researchers from Swedish University of Agricultural Sciences (Uppsala) used a hydrogel (CaptiGel) synthesized from titanium tetraethoxide $Ti_4(OCH_2CH_3)_{16}$, followed by the modification of the surface of resulting TiO_2 nanoparticles with triethanolamine and phospholipids [24]. Thus, prepared that titania nanoparticles modified with phospholipids significantly increased the number of beneficial bacteria (*Bacillus amyloliquefaciens UCMB5113*) protecting rape seed (*Brassica napus*) from the fungal pathogen *Alternaria brassicae*.

Unfortunately, so far almost no research relating to the application of TiO_2 nanoparticles in agriculture involved the consideration of their phase composition and photocatalytic activity. Our recent studies revealed that high photocatalytic activity of nanoparticles actually contributes to the suppression of phytopathogens, but can also inhibit plant development [12]. The data on the effect of pre-sowing treatment of white cabbage seeds with aqueous suspensions of TiO_2 nanoparticles TiO_2-HCl-100 and TiO_2-Pl 2 nanoparticles differing in phase composition, crystallinity and, as a consequence, photocatalytic activity influence on seed germination and plant seedling growth are summarized in Table 3 [13].

Treatment of seeds with TiO_2 suspensions does not have a significant effect on their germination energy and germination process. However, a slight trend to the increase of seed germination is observed upon the introduction of both considered

Table 3 Influence of pre-sowing treatment with aqueous suspensions of TiO_2 nanoparticles on the seed germination and plant seedling growth of white cabbage cv. *Penca de Povoa*

TiO_2 NPs aqueous suspensions		Germination energy		Germination		Sprout length		Root length	
Nanoparticles*	***C***, g/L	%	Percent of control	%	Percent of control	cm	Percent of control	cm	Percent of control
TiO_2-HCl-100	0.3	88	105	90	113	$3.0 \pm 0.2^{**}$	136^{**}	$6.2 \pm 0.5^{**}$	155^{**}
	3.0	80	95	77	96	$2.9 \pm 0.2^{**}$	132^{**}	4.5 ± 0.4	113
TiO_2-Pl-2	0.3	75	89	86	108	$2.7 \pm 0.2^{**}$	123^{**}	$5.3 \pm 0.5^{**}$	133^{**}
	3.0	75	89	78	98	$2.8 \pm 0.2^{**}$	127^{**}	$5.2 \pm 0.4^{**}$	130^{**}
Control (H_2O)	0	84	100	80	100	2.2 ± 0.2	100	4.0 ± 0.4	100

Note *Photocatalytic activity of TiO_2-Pl-2, indirectly characterized by the amount of yielded singlet oxygen is 10 times higher compared to TiO_2-HCl-100—see Table 1 [13]; ***C***—concentration of the aqueous suspensions of TiO_2 nanoparticles; **the value differs from the control at a 5% significance level additives in the concentration 0.3 g/L, more prominent (by 13% relating to the control seeds) in the case of less photocatalytically active TiO_2-HCl-100. Furthermore, in all the considered cases treatment of cv. *Penca de Povoa* white cabbage seeds with suspensions of TiO_2 nanoparticles provided a significant stimulation of the growth of sprouts (by 23–36%) and roots (by 13–55%). The highest growth rates of both roots and sprouts is achieved upon the application of less photocatalytically active TiO_2-HCl-100 at a relatively low concentration of 0.3 g/L [13], while the increase of suspension concentration to 3 g/L resulted in a decrease of the stimulating effect.

The most significant effect of pre-sowing treatment was observed in respect to the survival of cabbage plants, grown in the peat-based soil substitute with blackleg root rot fungal pathogens (*Pyhtium debaryanum* Hesse, *Olpidium brassicae* Wor., *Phizoctonia aderholdii* Kolosh., *Leptosphaeria maculans (Sowerby)* P. Karst., *Fusarium* sp.) [12]. As shown in Table 4, TiO_2-Pl-2 sample provides an improvement of cabbage resistance to blackleg disease judging by the number of surviving plants at application in the amount of 0.3 g/L whereas the increase of its concentration to 3.0 g/L results in a negative effect. On the contrary, the use of TiO_2-HCl-100 improves the survival of these plants only in a relatively high concentration of 3.0 g/L while 0.3 g/L is insufficient and deteriorates the target performances.

Perhaps, the observed phenomena are determined by both acidity of the suspension medium and magnitude of photocatalytic activity. Generally, excessive concentration of TiO_2 nanoparticles with high photocatalytic activity inhibits the growth and development of plants.

The search for new methods for seed treatment in order to stimulate plant growth, increase the productivity of agricultural crops and increase their resistance to phytopathogens is a field of extensive research [25]. One of the most promising approaches to pre-sowing seed treatment is based on the formation of an air-permeable coating on the surface of seeds [30, 58]. In addition to facilitating the plant growth at the initial stage of development by supplying nutrients

Table 4 Survival of cv. Penca de Povoa white cabbage plants grown from seeds treated with TiO_2 suspensions and sown in a substrate with blackleg root rot fungal pathogens (*Pyhtium debaryanum* Hesse, *Olpidium brassicae* Wor., *Phizoctonia aderholdii* Kolosh., *Leptosphaeria maculans (Sowerby)* P. Karst., *Fusarium* sp.)

Seed treatment		Average part of survived plants, %	Weighted average lesion score, M (VIR scale [57])
Sample	Suspension concentration, g/L		
Reference sample (water)	0	38.9 ± 3,4	7.9
TiO_2-HCl-100	0.3	34.3 ± 2,6	6.1
	3.0	57.2 ± 4,1	5.3
TiO_2-Pl-2	0.3	45.7 ± 1,2	5.2
	3.0	34.3 ± 0,8	5.1
Sterilized soil	–	100	1

(macro- and microelements) directly to the seed and sprout, such coatings protect the seeds of cultivated plants from harmful influences, including abiotic and biotic stressors. Materials tested for the considered application include starch, gum, gelatin and polyvinyl alcohol, cellulose derivatives, chitosan, amphiphilic polymers, dolomite limestone and aluminum silicate, protein-based biomaterials, etc. [59]. Among them, silicon-containing additives are of particular interest for the use in agricultural technologies. Silicon compounds are easily absorbed by the root system of plants and reduce the impact of abiotic and biotic stresses [60, 61]. In some plant species, silica creates intracellular or extracellular silica bodies (phytoliths) necessary for growth, mechanical strength, rigidity, and protection from fungal phytopathogens [62]. At the same time, for seed treatment, aqueous suspensions of silica nanoparticles [63] or aqueous solutions of water-soluble silicates or low concentration organosilicon compounds [64] are mainly used.

The authors of this chapter were the first to apply and study the pre-sowing seed treatment using film-forming silica sols prepared by acid hydrolysis of tetraethoxysilane (TEOS–$Si(OEt)_4$) in the presence of macro- and microelements beneficial for plants in the form of aqueous solutions of salts and acids [65]. During the sol–gel transition process, such silica sols are capable of covering the surface of seeds with thin-film coatings comprising a polysiloxane matrix with uniformly distributed nutrients. The data of scanning electron microscopy (Fig. 6) confirm the presence of an almost continuous coating on the surface of agricultural seeds after their treatment with silica sol. In contrast to permeable polymer coatings, in this case the discontinuity and even the "island" morphology of the coating is beneficial due to providing the access of moisture and air to the surface of seeds.

Germination tests revealed that the treatment in silica sols does not have a depressing effect on the seeds. The coated seeds germinated simultaneously with the

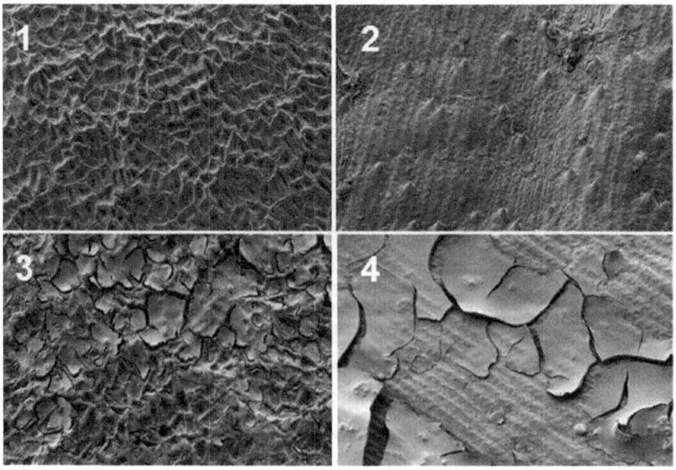

Fig. 6 SEM images of Chinese cabbage seeds (left) and spring barley (right) untreated (1, 2) and treated with silica sol with 20% TEOS (3, 4)

Fig. 7 SEM images of the surface of Chinese cabbage seeds after treatment in water (1), silica sol with 1% TEOS (2) and silica sol containing TiO_2-Ev-D (3)

control ones. Moreover, the deposited coatings positively affected the growth characteristics of spring barley and Chinese cabbage, as well as the resistance of barley plants to the causative agent of root rot [36, 65]. Similar results were obtained later in [59] regarding the effect of pre-sowing treatment of pea seeds with film-forming silica sol based on TEOS in the presence of the Azofoska fertilizer. Generally it can be concluded that the application of sol–gel technology is a promising approach in respect to the search for approaches to increase the yields of crops and protect them from adverse environmental impacts.

Our subsequent studies on application of sol–gel technologies for the treatment of seeds involved the introduction of small additives of nanoparticles, including TiO_2, into silica sols [36], in order to enhance the above positive effects.

The addition of nanoparticles, including TiO_2, to silica sol extends the possibilities for the improvement of growth and development of plants. The introduced nanoparticles are quite uniformly distributed in the prepared silica-based functional coatings (Fig. 7).

It should be particularly mentioned that the application of TiO_2 nanoparticles aqueous suspensions and TiO_2-containing sol–gel compositions based on silica sols significantly promotes the development of beneficial bacteria in the soil, which facilitate the assimilation of nutrients (especially nitrogen compounds) by plants (Table 5). As shown in Table 5, the considered pre-sowing treatment of seeds provides a prominent increase in the ratio between beneficial bacteria and micromycetes.

The obtained results indicate a positive effect of the synthesized aqueous suspensions and sol–gel compositions containing TiO_2 nanoparticles in low concentrations in development of plant-friendly bacteria, which, apparently, due to competitive interactions, contributes to reducing the share of micromycetes, including pathogenic ones.

Table 5 Ratio between the number of bacteria using organic forms of nitrogen and the number of micromycetes on the surface of Chinese cabbage seeds after pre-sowing treatment with TiO_2 aqueous suspension and TiO_2-containing silica sols

Seed treatment option	Ratio of the number of bacteria using organic forms of nitrogen[1] to the number of micromycetes[2]
Water (control)	0.27
Aqueous suspension of TiO_2 (0.003 g/ml)	1.06
Silica sol (1%TEOS) + S	1.01
Silica sol (1%TEOS) + S + TiO_2-Ev-D	0.68

Note the number of microorganisms was determined by inoculation on different solid or liquid nutrient media, including cabbage agar[1] or acidified wort agar[2]; S is a plant-friendly aqueous solution containing the required micro- and macro elements

5 Conclusion

TiO_2 nanoparticles in anatase or mixed anatase–rutile crystalline modifications exhibit photocatalytic activity, mainly affected by the degree of crystallization, phase composition and nanoparticle size, while the effect of the surface development is not so prominent. Our studies revealed that the highest photocatalytic activity is observed for 70–90 nm sized double-phase TiO_2 nanoparticles, preferably with a similar size of anatase and rutile particles directly contacting each other.

The phase composition of TiO_2 nanoparticles significantly affects their surface acid–base properties and interaction with aqueous environment.

TiO_2 nanoparticles are successfully used in agricultural technologies for root and foliar treatment of plants. One of the promising approaches in this field is the application of titania nanoparticles in water suspensions and silica sols for pre-sowing treatment of vegetable crop seeds. A particularly important factor responsible for the target positive effect is the optimal ratio between the concentration of nanoparticles and their photocatalytic activity, since excessive photocatalytic activity inhibits the development of plants. The phase composition and acid–base properties of the surface of TiO_2 nanoparticles also significantly affect the growth and development of plants from treated seeds.

The application of water suspension and silica sol-based preparations containing optimal amounts of TiO_2 nanoparticles facilitates the development of plant-friendly microflora, which, among other useful effects, increases the competitiveness of plants in the early stages of their development and enhances potential protection against phytopathogens and other stress factors.

References

1. Javed, R., Ain, N., Gul, A. et al.: Diverse biotechnological applications of multifunctional titanium dioxide nanoparticles: an up-to-date review. IET Nanobiotechnol. **16**(5), 171–189 (2022). https://doi.org/10.1049/nbt2.12085
2. Simons, P.Y., Dachille, F.: The structure of TiO_2(II), a high-pressure phase of TiO2. Acta Crystallogr. A **23**(2), 334–336 (1967)
3. Latroche, M., Brohan, L., Marchand, R. et al.: New hollandite oxides: TiO_2(H) and K0.06TiO2. J. Solid State Chem. (81), 78–82 (1989)
4. Hu, Y., Tsai, H.-L., Huang, C.-L.: Effect of brookite phase on the anatase–rutile transition in titania nanoparticles. J. Eur. Ceram. Soc. **23**(5), 691–696 (2003)
5. Nicholls, D.: Complexes and First-Row Transition Elements, p. 215. Macmillan, London (1974)
6. Parrino, F., Palmisano, L.: Titanium Dioxide (TiO_2) and its Application. Elsevier (2021). https://doi.org/10.1016/B978-0-12-819960-2.00024-9
7. Al-Mamun, M.R., Kader, S., Islam, M.S. et. al.: Photocatalytic activity improvement and application of UV-TiO2 photocatalysis in textile wastewater treatment: a review. J. Environ. Chem. Eng. **7**(5) (2019). https://doi.org/10.1016/j.jece.2019.103248
8. Vaskevich, V.V., Kovalenko, D.L., Semchenko, A.V. et al.: Nanostructured photocatalytic sol-gel coatings based on titania. In Problems of Physics, Mathematics and Technics, vol. 4, no. 25, pp. 7–10. F. Scorina Gomel State University (2015)
9. Lee, K.-Ya, Sato, K., Mohamed, A.R.: Facile synthesis of anatase-rutile TiO_2 composites with enhanced CO_2 photoreduction activity and the effect of Pt loading on product selectivity. Mater. Lett. **163**, 240–243 (2016). https://doi.org/10.1016/j.matlet.2015.10.094
10. Silva-Osuna, E.R., Vilchis-Nestor, A.R., Villarreal-Sanchez, R.C. et al.: Study of the optical properties of TiO_2 semiconductor nanoparticles synthesized using Salvia rosmarinus and its effect on photocatalytic activity. Optical Mater. **124**, 112039 (2022). https://doi.org/10.1016/j.optmat.2022.112039
11. Haider, A., Al-Anbari, R., Kadhim, G., et al.: Synthesis and photocatalytic activity for TiO_2 nanoparticles as air purification. MATEC Web Conf. **162**, 05006 (2018). https://doi.org/10.1051/matecconf/201816205006
12. Shilova, O.A., Panova, G.G., Mjakin, S.V., et al.: Structure, properties, and phytoprotective functions of titanium dioxide nanopowders and their aqueous suspensions. Russ. J. Inorg. Chem. **66**, 765–772 (2021). https://doi.org/10.1134/S0036023621050181
13. Khamova, T.V., Kopitsa, G.P., Nikolaev, A.M., et al.: The structure and properties of TiO_2 nanopowders for use in agricultural technologies. Biointerface Res. Appl. Chem. **1**(4), 12285–12300 (2021)
14. Khitab, A., Ahmad, S., Munir, M.J., et al.: Synthesis and applications of nano titania particles: a review. Rev. Adv. Mater. Sci. **53**, 90–105 (2018)
15. Sahare, P., Alvarez, P.G., Yanez, J.M.S., et al.: Engineered titania nanomaterials in advanced clinical applications. Beilstein J. Nanotechnol. **13**, 201–218 (2022). https://doi.org/10.3762/bjnano.13.15
16. Bellani, L., Siracusa, G., Giorgetti, L., et al.: TiO_2 nanoparticles in a biosolid-amended soil and their implication in soil nutrients, microorganisms and Pisum sativum nutrition. Ecotoxicol. Environ. Saf. **190**, 110095 (2020). https://doi.org/10.1016/j.ecoenv.2019
17. Wu, X., Hu, J., Wu, F., et al.: Application of TiO_2 nanoparticles to reduce bioaccumulation of arsenic in rice seedlings (*Oryza sativa* L.): a mechanistic study. J. Hazard. Mater. 124047 (2020). https://doi.org/10.1016/j.jhazmat.2020.124047
18. Ullah, S., Adeel, M., Jilani, M., et al.: Physiological and biochemical response of wheat (*Triticum aestivum*) to TiO_2 nanoparticles in phosphorous amended soil: a full life cycle study. J. Environ. Manag. **263**, 110365 (2020). https://doi.org/10.1016/j.jenvman.2020.110365
19. Lian, J., Zhao, L., Wu, J., et al.: Foliar spray of TiO_2 nanoparticles prevails over root application in reducing Cd accumulation and mitigating Cd-induced phytotoxicity in maize (*Zea mays* L.). Chemosphere **239**, 124794 (2020). https://doi.org/10.1016/j.chemosphere.2019.124794

20. Rafique, R., Arshad, M., Khokhar, M.F., et al.: Growth response of wheat to titania nanoparticles application. NUST J. Eng. Sci. **7**(1) 42–46 (2014)
21. Vatankhah, A., Aliniaeifard, S., Moosavi-Nezhad, M., et al.: Plants exposed to titanium dioxide nanoparticles acquired contrasting photosynthetic and morphological strategies depending on the growing light intensity: a case study in radish. Sci. Rep. **13**, 5873 (2023)
22. Chaudhary, I., Singh, V.: Titanium dioxide nanoparticles and its impact on growth, biomass and yield of agricultural crops under environmental stress: a review. Res. J. Nanosci. Nanotechnol. **10**(1) (2020)
23. Janmohammadi, M., Amanzadeh, T., Sabaghnia, N., et al.: Impact of foliar application of nano micronutrient fertilizers and titanium dioxide nanoparticles on the growth and yield components of barley under supplemental irrigation. Acta Agricult. Slovenica **107**, 265–276 (2016). https://doi.org/10.14720/aas.2016.107.2.01
24. Palmqvist, N.G.M., Bejai, S., Meijer, J., et al.: Nano titania aided clustering and adhesion of beneficial bacteria to plant roots to enhance crop growth and stress management. Sci. Rep. **13**(5), 10146 (2015). https://doi.org/10.1038/srep10146
25. Taylor, A.G., Amirkhani, M., Hill, H.: Modern seed technology. Agriculture **11**, 630 (2021)
26. Mayton, H., Amirkhani, M., Yang, D., et al.: Tomato seed coat permeability: optimal seed treatment chemical properties for targeting the embryo with implications for internal seed-borne pathogen control. Agriculture **11**, 199 (2021)
27. Janmohammadi, M., Mohamadzadeh-Alghoo, M., Sabaghnia, N., et al.: Seed pre-sowing treatments and essential trace elements application effects on wheat performance. Acta Agricult. Slovenica **119**(1), 1–10 (2023). https://doi.org/10.14720/aas.2023.119.1.2671
28. Mallick, S., Mohanty, S., Das, S., et al.: Effect of polymer coating on seed quality during storage of groundnut seeds. Pharma Innov. J. **11**, 1396–1401 (2022)
29. Reddy, B.P., Bara, B.M., Krishina, R.Y.: Effect of polymer seed coating and seed treatment on seed quality parameters and yield attributing characters of hybrid maize (*Zea mays* L.). Int. J. Curr. Microbiol. App. Sci. **8**, 1175–1182 (2019)
30. Sohail, M., Pirzada, T., Opperman, C.H., et al.: Recent advances in seed coating technologies: transitioning toward sustainable agriculture. Green Chem. **24**, 6052–6085 (2022)
31. Fu, Y., Bhunia, A.K., Yao, Y.: Alginate-based antimicrobial coating reduces pathogens on alfalfa seeds and sprouts. Food Microbiol. **103**, 103954 (2022)
32. Saberi Riseh, R., Skorik, Y.A., Thakur, V.K., et al.: Encapsulation of plant biocontrol bacteria with alginate as a main polymer material. Int. J. Mol. Sci. **22**, 11165 (2021)
33. Saberi Riseh, R., Moradi Pour, M., Ait Barka, E.: A Novel route for double-layered encapsulation of *Streptomyces fulvissimus* Uts22 by alginate-arabic gum for controlling of *Pythium aphanidermatum* in cucumber. Agronomy **12**, 655 (2022)
34. Aisvarya, S., Kalyanasundaram, M., Kannan, M., et al.: Dual activity of biogenic zinc oxide nanoparticlesas a protectant against seed weevil and growth promoter in maize seeds. In Book of extended abstracts, proceedings of the GPCP 2021, Coimbatore, India, pp. 137–138 (2021)
35. Khamova, T.V., Kopitsa, G.P., Nikolaev, A.M., et al.: The structure and properties of TiO_2 nanopowders for use in agricultural technologies. Biointerface Res. Appl. Chem. **11**(4), 12285–12300 (2021). https://doi.org/10.33263/BRIAC114.1228512300
36. Shilova, O.A., Khamova, T.V., Panova, G.G., et al.: Synthesis and research of functional layers based on titanium dioxide nanoparticles and silica sols formed on the surface of seeds of Chinese cabbage. Russ. J. Appl. Chem. **93**, 25–34 (2020). https://doi.org/10.1134/S1070427220010036
37. Ioannou, L.A. Hapeshi, E., Vasquez, M.I., et al.: Solar/TiO_2 photocatalytic decomposition of [beta]-blockers atenolol and propranolol in water and wastewater. Solar Energy **85**(9), 1915–1926 (2011)
38. Park, H., Park, Y., Kim, W., et al.: Surface modification of TiO_2 photocatalyst for environmental applications. J. Photochem. Photobiol. C **15**, 1–20 (2013)
39. Jesus, M.A.M.L., Ferreira, A.M., Lima, L.F.S., et al.: Micro-mesoporous TiO_2/SiO_2 nanocomposites: sol-gel synthesis, characterization, and enhanced photodegradation of quinolone. Ceramics Int. **47**(17), 23844–23850 (2021)

40. Pham, K., Pelisset, S., Kinnunen, N., et al.: Controlled photocatalytic activity of TiO_2 inverse opal structures with atomic layer deposited (ALD) metal oxide thin films. Mater. Chem. Phys. **277**, 125533 (2022)
41. Angela, S., Lunardi, V.B., Kusuma, K., et al.: Facile synthesis of hierarchical porous ZIF-8@TiO_2 for simultaneous adsorption and photocatalytic decomposition of crystal violet. Environ. Nanotechnol. Monit. Manag. **16**, 100598 (2021)
42. Musaev, K., Mirkhamitova, D., Yarbekov, A., et al.: Facile synthesis of SiO_2–TiO_2 photocatalyst nanoparticles for degradation of phenolic water pollutants. SN Appl. Sci. **1**(10) 1164 (2019)
43. Hou, C., Hu, B., Zhu, J.: Photocatalytic degradation of methylene blue over TiO_2 pretreated with varying concentrations of NaOH. Catalysts **8**(12), 575–587 (2018)
44. Zhang, J., Zhou, P., Liu, J., et al.: New understanding of the difference of photocatalytic activity among anatase, rutile and brookite TiO_2 physical chemistry. Chem. Phys. **16**(38), 20382–20386 (2014)
45. Žerjav, G., Žižek, Kr., Zavašnik, J., et al.: Brookite vs. rutile vs. anatase: what's behind their various photocatalytic activities? J. Environ. Chem. Eng. **10**(3), 107722 (2022)
46. Wang, K., Zhuo, Y., Chen, J., et al.: Crystalline phase regulation of anatase–rutile TiO_2 for the enhancement of photocatalytic activity. RSC Adv. **10**(71), 43592–43598 (2020)
47. Daimon, T., Nosaka, Y.: Formation and behavior of singlet molecular oxygen in TiO_2 photocatalysis studied by detection on near-infrared phosphorescence. J. Phys. Chem. C **111**(11), 4420–4424 (2007)
48. Shilova, O.A., Kovalenko, A.S., Nikolaev, A.M., et al.: Surface and photocatalytic properties of sol–gel derived TiO_2@SiO_2 core-shell nanoparticles. J. Sol-Gel Sci. Technol. **108**, 263–273 (2023). https://doi.org/10.1007/s10971-022-05943-5
49. Scanlon, D.O., Dunnill, Ch.W., Buckeridge, J., et al.: Band alignment of rutile and anatase TiO_2. Nat. Mater. **12**, 798 (2013)
50. Mi, Y., Wemg, Y.: Band alignment and controllable electron migration between rutile and anatase TiO2. Sci. Rep. **5**, 11482 (2015)
51. Li, G.H., Richter, C.P., Milot, R.L., et al.: Synergistic effect between anatase and rutile TiO_2 nanoparticles in dye-sensitized solar cells. Dalton Trans. **45**, 10078–10085 (2009)
52. Sychov, M.M., Zakharova, N.V., Mjakin, S.V.: Effect of milling on the surface functionality of $BaTiO_3$–$CaSnO_3$ ceramics. Ceram. Int. **39**, 6821–6826 (2013). https://doi.org/10.1016/j.ceramint.2013.02.013
53. Fedorenko, N.Yu., Mjakin, S.V., Khamova, T.V., et al.: Relationship among the composition, synthesis conditions, and surface acid-basic properties of xerogel particles based on zirconium dioxide. Ceram. Int. **48**, 6245–6249. https://doi.org/10.1016/j.ceramint.2021.11.165 (2022)
54. Deska, J., Jankowski, K., Bombik, A., et al.: Effect of growing medium pH on germination and initial development of some grassland plants. Acta Sci. Pol. Agricultura. **10**(4), 45–56 (2011)
55. Gentili, R., Ambrosini, R., Montagnani, C., et al.: Effect of soil pH on the growth, reproductive investment and pollen allergenicity of *Ambrosia artemisiifolia* L. Front. Plant Sci. **9** (2018). https://doi.org/10.3389/fpls.2018.01335
56. Rymuza, K., Radzka, E.: Effect of pH levels on soybean seed germination dynamics. Acta Sci. Pol. Agricultura **20**(2), 81–88 (2021). https://doi.org/10.37660/aspagr.2021.20.2.4
57. Guidelines for studying and maintaining a collection of vegetable plants: (carrots, celery, parsley, parsnips, radishes and radishes), VASKHNIL, All-Russian Research Institute of Plant Growing named after. N. I. Vavilova; [Compiled by L.V. Sazonova and others]. Leningrad: VIR, 1981, 192 p (in Russian)
58. Afzal, I., Javed, T., Amirkhan, M., et al.: Modern seed technology: seed coating delivery systems for enhancing seed and crop performance. Agriculture **10**, 526 (2020)
59. Borak, B.: Sol-gel coatings with Azofoska fertilizer deposited onto pea seeds. Polymers **14**, 4119 (2022). https://doi.org/10.3390/polym14194119
60. Zhu, Y.-X., Gong, H.-J., Yin, J.-L.: Role of silicon in mediating salt tolerance in plants: a review. Plants. **8**(6), 147 (2019). https://doi.org/10.3390/plants8060147

61. Ma, J.F.: Role of silicon in enhancing the resistance of plants to biotic and abiotic stresses. Soil Sci. Plant Nutrit. **50**(1), 11–18 (2004). https://doi.org/10.1080/00380768.2004.10408447
62. Neethirajan, S., Gordon, R., Wang, L.: Potential of silica bodies (phytoliths) for nanotechnology. Trends Biotechnol. **27**(8), 461 (2009). https://doi.org/10.1016/j.tibtech.2009.05.002
63. Goswami, P., Mathur, J., Srivastava, N.: Silica nanoparticles as novel sustainable approach for plant growth and crop protection. Heliyon **8**(7), E09908 (2022). https://doi.org/10.1016/j.heliyon.2022.e09908
64. Slastya, I.V.: Use of silicon compounds as a factor of raising spring barley cultivars productivity under water stress. Agricult. Biol. **2** (2013) (in Russian)
65. Shilova, O.A., Khamova, T.V., Panova, G.G., et al.: Using the sol–gel technology for the treatment of barley seeds. Glass Phys. Chem. **44**(1), 26 (2018). https://doi.org/10.1134/S108765961801011X

Recent Development and Future Prospects of Hierarchical Nanostructures of TiO_2 in Energy Applications

Sudhir Kumar Singh and Deepak Sharma

Abstract Surface modification through nanomaterials deposition is highly promising method for enhancing the pool boiling heat transfer across various energy applications such as wastewater and incinerator heat recovery, thermal management of battery and electronics device cooling. The diverse enhancement techniques have been studied to improve heat transfer, with nanoparticle coating emerging as the most effective method for various energy and environmental applications. TiO_2 stands out for its exceptional structure, electrochemical characteristics, chemical stability, non-toxic, and feasibility. The TiO_2 nanorods and nanotubes structures exhibit superior photocatalytic scheme because of enhanced delocalization and diminished charge recombination. The surface characteristics are influenced by several factors, including materials, nanoparticles size, thermal conductivity of the nanomaterial, nanomaterial deposition techniques, thickness of the coating layer, concentration of nanofluid. Among the various deposition techniques explored, electrochemical deposition stands out as particularly advantageous due to its simplicity, cost-effectiveness, excellent adhesion properties, and ability to control surface characteristics such as coating thickness, porosity, and wettability through modulation of current density. During the electrochemical deposition process, the emergence of hydrogen bubbles from the surface generated a significant quantity of cavities. This phenomenon increases the surface porosity and roughness as related to bare substrate. The surface wettability has been found to be an important surface parameter affecting the dry-out occurrence which significantly influenced by the surfaces' topological and chemical characteristics.

Keywords TiO_2 · Nanomaterial · Surface coating · Electrochemical deposition · Surface wettability

S. K. Singh · D. Sharma (✉)
Department of Mechanical Engineering, National Institute of Technology Hamirpur, Hamirpur, Himachal Pradesh 177005, India
e-mail: dsharma@nith.ac.in

J. Prakash et al. (eds.), *Titanium Dioxide-Based Multifunctional Hybrid Nanomaterials*, Engineering Materials, https://doi.org/10.1007/978-3-031-80625-4_10

1 Introduction

Over the past few decades, there has been extensive research into TiO_2 nanostructures with complex shapes, and their wide-ranging uses across various fields. By precisely regulating the size and composition of TiO_2, researchers have unlocked new and enhanced properties, leading to significant progress in materials chemistry, particularly in energy and environmental applications [1]. The enhanced efficiency and versatility of these materials can be attributed to their unique physical and chemical properties, like stability, effective transfer of ionic charges, and larger specific surface area [2]. The exceptional properties exhibited by zero, one, two, & three-dimensional (0-3D) TiO_2 nanostructures are the main reason behind its further study. Structural hierarchy, a fundamental principle observed in nature, manifests across various scales, from microscopic cells to vast forests. Nature often employs a repetitive growth rule to construct successive stages of these structures. Similarly, many man-made materials exhibit hierarchical features, which significantly influence their overall properties. When inspired by natural structures, materials designed at the nanoscale can undergo substantial modifications, presenting vast opportunities for enhancing their performance across diverse technological applications [3]. The hierarchical nanostructures encompass well-organized subunits at the nanoscale, with architectures ranging from (0-3D) [4]. Introducing hierarchical porosity at various length scales can simulate environmental approaches, offering a plethora of appropriate functionalities for numerous purposes. Recognizing the intricate association between these structural designs and their properties is crucial for the advancement of porous materials with sophisticated hierarchical structures. However natural assembly exhibit extremely organized hierarchical 3D structures, they often lack stability and resilience in harsh environments [5].

Metal oxides are highly valued in heterogeneous catalysis, particularly in photocatalysis, due to their efficient redox and acid–base properties [6]. Among these oxides, TiO_2 stands out for its exceptional structure, electrochemical characteristics, chemical stability and feasibility [7]. In recent years, TiO_2-based materials with hierarchical structures have garnered significant notice related to their corresponding spherical nanoparticle. Numerous research has focused on enhancing the mechanical and thermal properties of by increasing specific surface area, porosity, etc. TiO_2 has garnered immense interest and stands out as significantly researched metal oxides because of its photocatalytic capabilities [8]. TiO_2 displays outstanding efficacy in oxidatively breaking down organic pollutants regardless of non-toxicity, durability, and comparatively economical. It is used in applications like photocatalysis, sensors, lithium-ion batteries, dye-sensitized solar cells, and biomaterials. TiO_2 occurs in different polymorphs including anatase, rutile, brookite, and the amorphous form. It is widely acknowledged that anatase form demonstrates superior photocatalytic performance compared to another forms [9].

The significance of TiO_2 in photocatalysis has been widely acknowledged and their effectiveness relies on a range of properties including surface area, crystal structure, shape, size, and band gap energy [10]. TiO_2 nano/microstructured and hybrid-based nanocomposites have become focal points because of its varied physicochemical attributes. The TiO_2 nanorods and nanotubes structures exhibit superior photocatalytic scheme because of enhanced delocalization and diminished charge recombination. One-dimensional TiO_2 has garnered a significant recognition than their zero- and two-dimensional owing to their advanced aspect ratio, augmented surface area, and effective electronic charge properties [11]. Various fabrication techniques have been explored to enhance the light absorption, charge separation, and surface area characteristics. The distinct characteristics of TiO_2 are significantly affected by morphology and particle size, which have been extensively investigated in the past and applied in diverse applications. Numerous novel synthetic methods have emerged to control properties like structural design, compound size, morphology, and surface characteristics in the preparation of TiO_2 materials [12].

Today's world faces two major challenges: the escalating demand for energy and the imperative to produce it cleanly. Concurrently, there's a pressing need to conserve and safeguard the environment, closely intertwined with energy production and consumption. The hierarchical nanostructures utilized in photocatalysis offer a promising avenue to address both of these issues, albeit on a modest yet impactful scale [13]. TiO_2 is mainly categorized into four crystal structures: the stable rutile, the metastable anatase, brookite, and monoclinic forms. The monoclinic variant demonstrates notable thermal stability. In both powder and thin-film states, brookite exhibits good stability and superior photocatalytic activity compared to anatase. Anatase consists of octahedral units with shared corners, while rutile shares edges, and brookite shares both corners and edges. Rutile is usually regarded as the highest stable phase of TiO_2, whereas in nanoscale dimensions, brookite and anatase phases are esteemed as more stable [14].

This chapter deals with the recent development and future prospects of TiO_2 based nanomaterials and their applications in energy and environment. Initially, the basic concept of pool boiling heat transfer, their energy related applications and parameters affecting heat transfer performance have been discussed. In the next section, the effect of surface porosity and wettability on heat transfer has been reported. Finally, the different coating methods and their advantages over other methods have been studied.

2 Pool Boiling Heat Transfer

2.1 Overview

The pool boiling heat transfer has been the most effective method in recent decades due to the practical application in nuclear reactor cooling, evaporators in thermal desalination and refrigeration systems, the chemical industry, and, more recently, electronics cooling [15]. With very little temperature fluctuations, a heated surface may release a significant quantity of heat under one of the most effective heat transfer regimes. Figure 1 depicts the heat transport mechanisms in this boiling regime, which

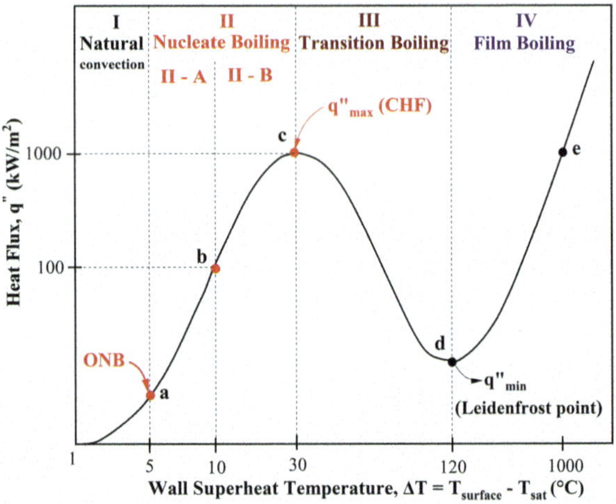

Fig. 1 Pool boiling curve

may be summed up as follows: There are various types of convection that occur in liquid–liquid systems: (i) heat transfer from a wall to a liquid occurs at low levels of heat flow, without bubbles merging or interacting; (ii) small-scale convection arises due to liquid movement caused by bubble expansion; (iii) the thin layer of liquid beneath a bubble evaporates; (iv) Evaporation happens from the heated layer of liquid surrounding the bubble; (v) Heat moves through the liquid temporarily as it re-establishes contact with the surface after a bubble leaves; (vi) Heat transfer is driven by external forces [16].

It is important to note that system pressure affects the behavior of surface microstructure. The range of active nucleation sites in high pressure systems, such as thermal desalination, power production, and refrigeration and air conditioning, includes nanometer-sized cavities. Pool boiling has been used to the cooling of electronics by immersion cooling, wherein the high-density data centre is submerged straight into a dielectric liquid. One drawback of using these liquids is having a high wetting point, little latent heat, and a smooth, i.e. surface with limited surface modification options. This led to poor heat transfer rates, critical heat flux (CHF), and substantial wall superheat at boiling incipience, meaning the chip may get overheated before boiling begins. As a result, other methods, thermosyphon, for instance that involve attaching the heat sink to the chip like on-chip cooling, became practical [17].

Based on current research, there are two types of surface alterations related to pool boiling improvement. The first kind involves passively creating artificial modifications or optimizations to the heating surface beforehand, such as raising the density of the nucleation sites, expanding the surface area, or introducing nanoparticles to the fluid. On the other hand, creating unique surfaces that can adaptably alter their geometrical forms or properties like wettability during the boiling process is the proactive approach. Another method to effectively enhance boiling heat transfer ability is surface coating [18].

2.2 *Parameters Affecting Pool Boiling Heat Transfer*

The parameters affecting the pool boiling heat transfer are shown in Fig. 2. Surface microgeometry (size and shape of cracks and pores), interactions between the solid surface, liquid, and vapor (wettability, adhesion, adsorption), and thermophysical properties of the surface material (thermal conductivity and thermal absorption) influences the pool boiling surface. These variables have an impact on the HTC. Research on the improvement of CHF by surface modification in saturated pool boiling has typically looked at two things: (1) uniform or modulated porous coatings; or (2) structures of different sizes that are manufactured or put on the heated surface. Based on the modified hydrodynamic stability model, the impacts of (a) increased surface area, (b) nucleation site density, (c) wettability, (d) capillary wicking, and (e) wavelength decrease generally result in CHF amplification in saturated pool boiling [19].

2.2.1 Surface Porosity

Porosity refers to the volume of empty or unoccupied space within a material. A porous surface is formed because of hydrogen bubbles growing from the surface of the coated heater. An increase in the number of cavities indicates greater surface irregularity or an increase in nucleation sites, which in turn leads to an increase in surface area. Because bubbles originate from these cavities, an increase in cavities accelerates bubble formation, resulting in enhanced nucleate boiling heat transfer.

2.2.2 Surface Wettability

Surface wettability is recognized to be significantly influenced by the surfaces' topological and chemical characteristics. It is the ability of liquid to spread on a surface. It

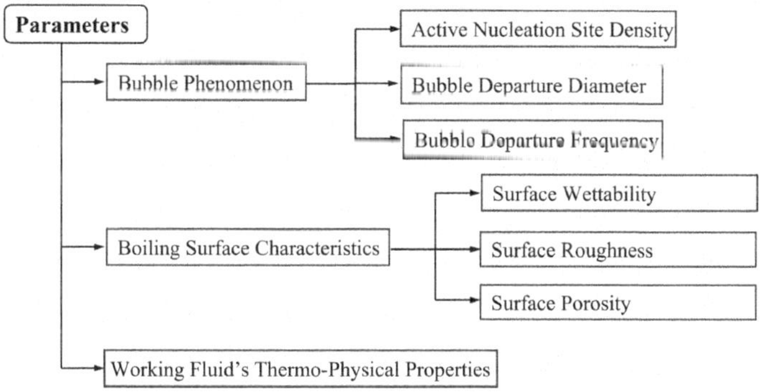

Fig. 2 Parameters affecting pool boiling heat transfer

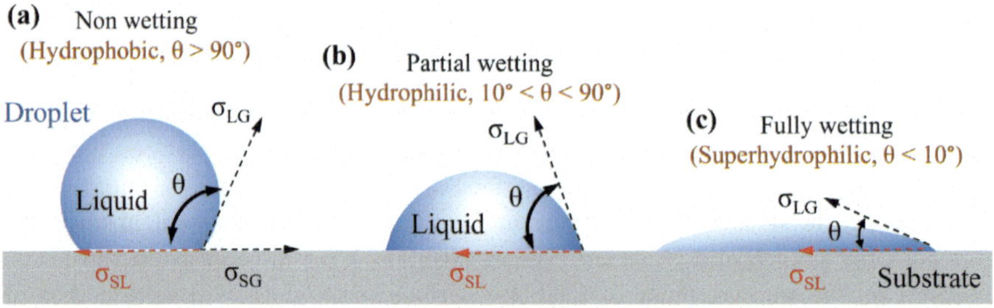

Fig. 3 Wettability of droplet **a** non wetting **b** partial wetting **c** fully wetting

measures the contact angle between liquid and heater surface. The contact angle (θ) is the angle between a tangential to the liquid surface at the line of meeting three phases and the plane of the solid surface on which liquid resides. Pool boiling heat transfer performance may be improved by manufacturing processes that create morphological nanostructured surfaces and regulate the surface wettability, as these factors are strongly correlated with surface wettability. The liquid thoroughly wets the whole solid surface by filling the spaces between the nanostructures. The perceived contact area (also known as the projection area) is less than the entire real solid–liquid contact area. The bubble's ebullition mechanism on boiling surface interfaces is regulated by nanostructures, which also alter surface contact angles and wettability [20]. This results in the onset of critical heat flux conditions. The significance of surface wettability and these nanostructures in modifying bubble nucleation and heat transmission rate cannot be overstated (Fig. 3).

Yim et al. [20] fabricated the hydrophobic TiO_2 coated and hydrophilic TiO_2 coated surfaces for PBHT enhancement. The TiO_2 coated surface significantly improved boiling HTC as compared the hydrophilic and bare surface. However, the PBHT performance of the uncoated surface surpassed that of the hydrophilic TiO_2 surface. The TiO_2 deposited surface exhibited a 64% increase in HTC compared to the uncoated surface. Kim et al. [21] used a wetting transition from a hydrophobic to a hydrophilic surface to study PBHT. They covered one side of a silicon surface with TiO_2 to create a hydrophobic surface, and then heated the TiO_2 coated hydrophobic surface to create a hydrophilic surface. One side of the silicon substrate with a heating surface area of 1 cm^2 was covered with TiO_2 for the wetting transition experiment, while the other side was insulated with SiO_2. They found that the length of the heat treatment and its temperature had an impact on the wetting transition. An inverse link between contact angle and time/temperature was indicated by the wetting contact angle increasing as both time and temperature decreased. Table 1 shows the effect of surface wettability on pool boiling heat transfer performance.

Table 1 Studies of surface modification using TiO_2 coating

Ref.	Nanomaterial	Coating method	Surface wettability	Fluid	CHF (%)	HTC (%)
[22]	TiO_2/water	Boiling method	Superhydrophilic	DI water	106	
[20]	TiO_2	Photolithography	Hydrophilic	HFE-7100	47	82
[21]	TiO_2	Heat treatment	Hydrophobic	DI water	86	95
[23]	TiO_2 nanowire	e-beam evaporator	Hydrophobic	R134a	76	81
[24]	TiO_2	Surface treatment	Hydrophilic	DI water	86	95

3 TiO_2 Based Nanomaterials, Surface Modification, and Their Applications in Pool Boiling Heat Transfer

3.1 Selection of Nanomaterials

The selection of materials for the nanostructures is crucial as well; these materials need to be robust mechanically and able to withstand the extreme heat and conditions that occur during pooling.

3.1.1 TiO_2 Nanomaterials

TiO_2 has an intermediate material cost per unit area based on cost, mechanical and thermal qualities, and manufacturability. Because of its outstanding mechanical strength and readily available micro/nanofabrication equipment, TiO_2 is frequently preferred [25]. Excellent mechanical and thermal qualities of metal materials also make them suitable for use in boiling. Conversely, materials made of ceramic and polymers may have less desirable mechanical qualities or have poor heat conductivity. To achieve nanostructured surfaces with controlled wettability, a suitable mix of materials and manufacturing techniques must be chosen in order to match our needs in boiling applications [26]. To maximize heat transmission, decisions must be taken regarding surface properties (such as wettability and structural shape), as well as material durability and interfacial strength from a reliability perspective. Table 2 shows the stability of various nanomaterial coated surface.

3.1.2 TiO_2 Nanotube

The study of TiO_2 nanotubes garnered significant interest shortly after the remarkable discovery of graphitic carbon microtubules [27]. TiO_2 nanotubes exhibit superior charge separation and a greater surface area compared to TiO_2 nanoparticles, their bandgaps, approximately 3.0 eV, show little disparity [28]. Further enhancements

Table 2 Stability of various coated nanomaterial on surface

S. no	Nanomaterial	Stability
1	Oxide layer coating (TiO_2)	Most stable
2	Metals porous coating	High stability
3	CNTs/MWCNTs coating	Low stability (at larger mass flux, the deformation of structure was greater)
4	Graphene based coating	Low stability
5	Metal nanowires coating	Low stability

of TiO_2 nanotube properties for photoelectrochemical water splitting remain necessary. This could involve reducing the bandgap through chemical doping, establishing heterojunctions with other materials to diminish charge recombination. Additionally, sensitizing the surface of TiO_2 nanotubes with plasmonic materials, quantum dots, and dye-sensitizer molecules significantly enhances light absorption capabilities [29].

The unique one-dimensional nanotube structure of titanium dioxide continues to hold appeal. However, its nanotubes possess a relatively wide bandgap of around 3.0 eV, presenting challenges such as limited light energy utilization and facile recombination of photogenerated electrons and holes [30]. To address these issues, various strategies have been devised: (i) enhancing surface morphology to augment the catalytic active area, (ii) modifying band structure to minimize charge recombination, and (iii) sensitizing the surface to enhance light absorption [9]. These approaches have yielded notable improvements in TiO_2 nanotubes, encompassing fundamental principles of photocatalytic water splitting, as well as methods for their preparation and polymorphs [31]. Hu et al. [32] investigated PBHT using TiO_2 nanotube coating on a titanium plate. The heptanol at different concentrations served as the wetting solution. The modified surface achieved a maximum CHF of 534 kW/m^2 and HTC of 21 kW/m^2 K, which depicted an increment of 62.9% and 121%, respectively over uncoated surface.

3.2 *Surface Coating Techniques*

The different surface coating method is shown in Fig. 4. The top-down and bottom-up approach are mainly two types of surface modification methods. In the top-down approach, molecules and atoms are disassembled from the bulk so as to obtain desired microstructured surface. Whereas in the bottom-down approach, the atoms and molecules are assembled to have anticipated nano/microstructured surface. The bottom-up approach is better as compared to top-down because it promises a

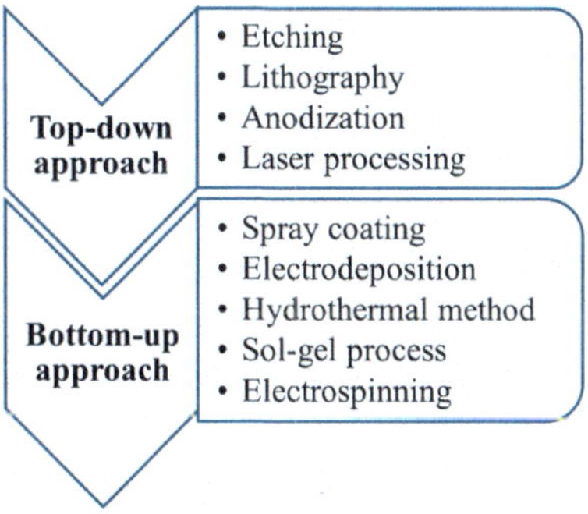

Fig. 4 Different surface modification method

better chance to obtain nanostructure with less defects, more homogeneous chemical composition [33].

From a manufacturing perspective, it is crucial to have tangible examples of the methods used to create microstructured surfaces with intended wettability. These fabrication techniques have advantages and disadvantages, even though they provide a great way to create structured surfaces and manage surface wettability. While some of them provide straightforward, scalable, and affordable production methods, others can produce nanostructures with high controllability and regularity; yet others call for costly instruments [34]. All of them are only able to handle a portion of the requirement to enhance the boiling applications' heat transfer performance. Compared to alternative high-temperature substrates, the manufacturing methods employing nanoporous materials appear attractive and beneficial because they may produce structures with density and dimensions that are comparatively controlled. They are scalable and reasonably priced [35]. Mostly based on chemical synthesis or self-assembly, these structured surfaces consisting of nanotubes (and nanowires) and nanoparticles are also easy, scalable, and affordable [36]. The methods may be used on various metal, oxide, and polymer materials. It is still difficult to deal with the wide range of forms, sizes, and spacing seen in nanotubes and nanoparticles [37]. Furthermore, in some situations, the worst condition created by bubbles at high temperature during boiling may be too much for the strength between the nanoparticles and boiling surface materials to withstand, leading to film delamination, particle damage, and detachment.

3.2.1 Electrochemical Deposition Method

The electrochemical deposition method has received attention because of its cost effectiveness, less complexity and operation in ambient conditions. The surface morphologies can be controlled by changing the electrochemical parameters such as electrolyte, current density and deposition time. Figure 5 shows the schematic diagram of electrochemical deposition process [38].

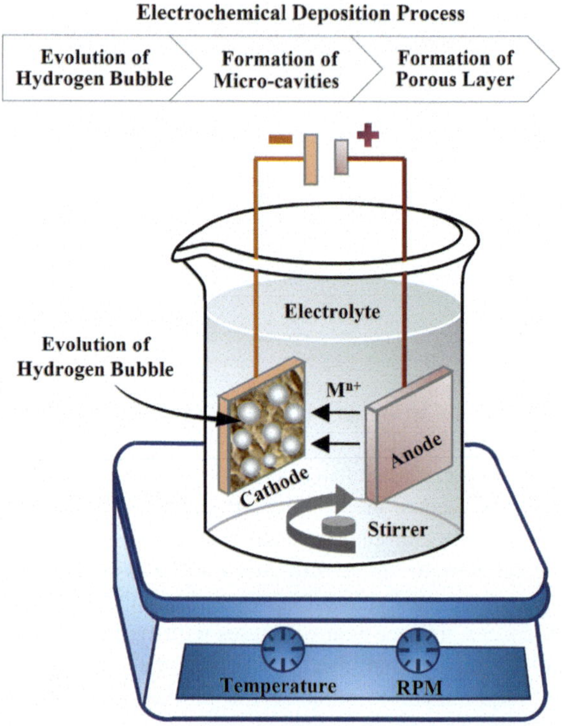

Fig. 5 Schematic diagram of electrochemical deposition process

The electrodeposition arrangement consists of anode, cathode, electrolyte and potentiostat to provide current density and time. This process is based on oxidation and reduction reaction. The metal surface forms a layer of oxide film by applying anodic current in an electrolyte solution. Both anode electrode and cathode electrode are immersed in an electrolyte and an external voltage is applied between the anode and cathode. At the anode, a layer of metal will form, while at the cathode, hydrogen will be generated [38].

Gupta and Mishra [38] investigated the pool boiling heat transfer performance of Cu–TiO_2. The electrochemical deposition process was used to deposit the nanoparticle on the copper substrate, ensuring excellent adhesion. Initially, high current densities were applied for short durations, followed by lowered current density for longer periods in the subsequent stage. The maximum HTC of 15.1 W/cm^2 K and CHF of 198.8 W/cm^2 was achieved with modified surface. This represented a significant increase of 184% in HTC and 85% in CHF compared to the bare copper surface. The improved performance can be attributed to the altered surface morphology and increased porosity. Wu et al. [39] used anodizing method to deposit TiO_2 and SiO_2 on a Cu substrate. The coated surface exhibited a 31% reduction in wall superheat temperature. The CHF of 1800 kW/m^2 and 1330 kW/m^2 was observed with TiO_2 and SiO_2 coated surface respectively. An increase in CHF of 38% and 51% was obtained

Fig. 6 Contact angle measurement on smooth copper, SiO_2 and TiO_2 coated test surfaces. Reprinted with permission from Ref. [39]. Copyright (2010) Elsevier

with FC-72 and water respectively over uncoated surface. Figure 6 shows the surface wettability of different considered surfaces.

3.2.2 Electron Beam Evaporation

Over the past few decades, electron-beam treatment method has seen extensive application in surface modification of metals and alloys. These approaches are renowned for their precision and efficiency, offering several advantages over conventional techniques. These include remarkably short process times, uniform dispersion of electron beam energy enabling precise control over beam parameters and resulting material structure and properties. Furthermore, electron-beam treatment technologies are integral to additive techniques, which are recognized as modern approaches for producing novel materials with distinct functional properties.

Das et al. [40] studied the PBHT performance of TiO_2 coated surface employing the electron beam evaporation method with deposits thickness ranging from 250 to 1000 nm. The nanomaterial coated surface exhibited a maximum reduction in wall superheat temperature of 42.5% for the deposits thickness of 1000 nm as related to the bare surface. Additionally, an increase in HTC of 75% was noticed over bare Cu surface. This enhancement was mainly attributed to the increased nucleation density, surface wettability, and bubble departure frequency. Figure 7 shows the SEM images of TiO_2 coated surface.

Ray and Bhaumik [41] investigated the pool boiling behavior of R-134a on the TiO_2 coated surface. They employed the electron beam evaporation technique to deposit nanoparticles with thicknesses ranging from 100 to 300 nm onto a copper heater surface. The HTC increased by 88% for a coating thickness of 300 nm. Additionally, the incipience wall superheat decreased with greater coating thickness. Yeom et al. [42] investigated bubble behavior of Ti and TiO_2 coated zirconium wire for PBHT enhancement. The CHF increased by 27% and 65% for Ti and TiO_2 coatings, respectively as compared to plain wire. Additionally, at a heat flux of 300 kW/m^2, the nucleation site density on the Ti and TiO_2 coated surfaces was 47% and 45% lower, respectively, as related to bare wire.

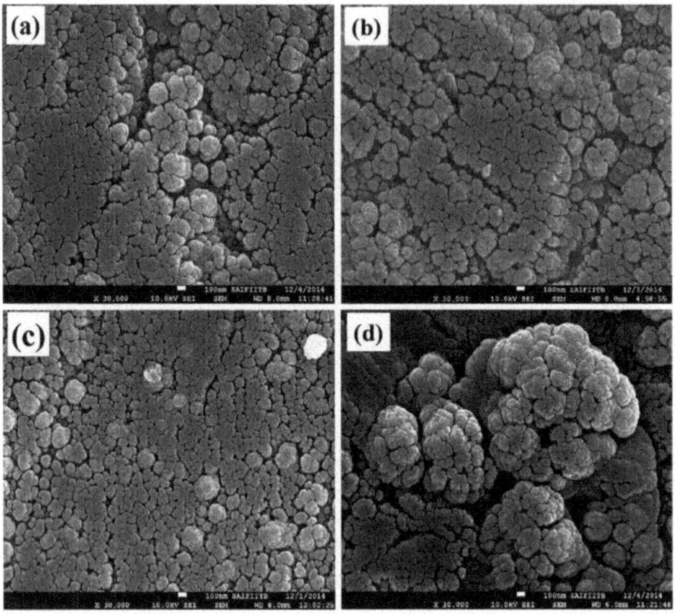

Fig. 7 SEM images of TiO_2 coating of thickness **a** 250 nm, **b** 500 nm, **c** 750 nm and **d** 1000 nm on copper surface. Reprinted with permission from Ref. [40]. Copyright (2017) Elsevier

3.2.3 Vacuum and Glancing Angle Coating Method

Yan et al. [43] created a TiO_2 coated surface on a copper substrate. The nanomaterial coating, with a thickness of 80 nm, was developed using the vacuum deposition method. The coated surface exhibited a 23% reduction in wall superheat compared to the uncoated surface. Additionally, there was a 12% increase in CHF compared to the bare copper surface. Ray et al. [44] modified the boiling surface by depositing the TiO_2 on Cu substrate utilizing the glancing angle coating method to enhance the PBHT characteristics. An improvement in HTC of 35% and 43% were observed with deposits thicknesses of 100 nm and 200 nm, respectively, over bare surface.

3.2.4 Nanofluid Boiling-Induced Coating

The suspension of nanoparticles in base fluid can improve the thermophysical properties of fluids. During pool-boiling conditions, these nanoparticles may experience prolonged boiling durations and adhere to heating surfaces, thereby modifying their inherent features like wettability and surface roughness over time. Changing the tensions between the liquid/vapor interface and solid surface modifies the equilibrium of forces and the extent of contact line, affecting factors such as wettability and the quantity of active nucleation sites. During the nucleate boiling of nanofluid phenomenon, the nanoparticles deposit on the boiling surface, leading to alterations in its characteristics, including surface wettability and roughness. Notably, the roughness of the surface subsequent to nanoparticle coating is affected by various factors such as the nanoparticle fraction and inherent thermal properties, boiling surface roughness, as well as the thickness of deposits. Additionally, the nucleation sites act as a crucial factor in enhancing boiling heat transfer efficiency and can be regulated by variables like surface wettability and roughness and fluid thermal properties.

Watanabe et al. [45] studied the PBHT of TiO_2, SiO_2, and Al_2O_3 nanofluid. They employed a boiling process to deposit the nanoparticles onto a copper substrate. TiO_2 exhibited the highest adhesion strength, while SiO_2 showed the lowest. The CHF and HTC increased for entire considered modified surfaces and an enhancement in CHF was 1.6 times compared to the uncoated surface. Figure 8 shows the heat transfer coefficients for the nanoparticle deposited surfaces.

Jung et al. [22] conducted experiments on the PBHT performance of TiO_2 nanofluid considering an average particle size of 85 nm, and the nanofluid concentration ranged from 0.00001 to 0.1 vol.%. They observed an increase in CHF up to 0.0004 vol.% concentration compared to when the heating surface was unmodified. This improvement was attributed to an increase in nucleation sites and porosity. Naphon and Thongjing [46] studied PBHT on a TiO_2 coated brass surface. The HTC increased as nanoparticle concentration decreased, and boiling pressure increased. Additionally, they noted a decrease in bubbles departure frequency with an increase in nanoparticle concentration, attributed to the reduction in BHT. Ali et al. [47] conducted experimental research on the PBHT performance of TiO_2 nanofluid. For the 15 wt% TiO_2, an enhancement in CHF and HTC were 22% and 38% while for the 12 wt%, an enhancement of 89% and 24%, respectively were observed as related

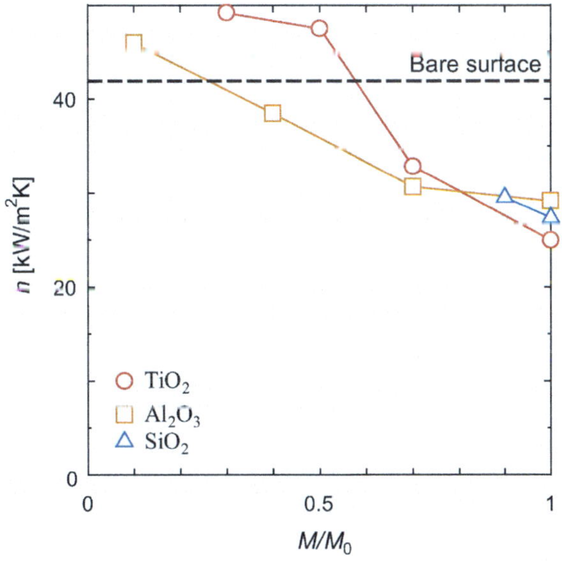

Fig. 8 Heat transfer coefficients for the nanoparticle deposited surfaces. Reprinted with permission from Ref. [45]. Copyright (2019) Elsevier

to base fluid. An enhancement in HTC of 11% was obtained for the 15 wt% TiO_2 nanofluid as compared to the 12 wt% concentration.

3.2.5 Other Coating Methods

The electrolytic deposition, flame spraying, plating, galvanizing, plasma spraying of a polymer, metallic coating of a foam substrate, welding, sintering, or brazing of particles, can all be used to create a homogeneous porous coating surface. In comparison to a plain surface, a uniformly porous layer covering the boiling surface can frequently raise the CHF. A lot of porous materials were used to improve the heat transmission from boiling pools. In recent years, the advancement in porous materials and their potential in many new applications have come to light. The primary goal of the early advancements in nanostructures was to expand the methods for creating nanoporous surfaces through surface preparation. Nanoporous membranes of various materials have demonstrated their promised applicability in several developing domains, mostly due to their straightforward production procedure. It is possible to create porous layers chemically or electrochemically. The application of nanostructured surfaces for pool boiling research necessitates considering several interrelated factors. One way that nanostructures can improve heat transfer performance is by increasing bubble nucleation sites and introducing surface irregularity.

4 Conclusions

- During the electrochemical deposition process, the emergence of hydrogen bubbles from the surface generated a significant quantity of cavities. This phenomenon increased the surface porosity and roughness as related to bare substrate.
- The extent of the dry region is dependent on surface wettability, and it expands as the contact angle increases. Consequently, surfaces with superhydrophilic properties demonstrate the highest critical heat flux because they facilitate the rehydration of the dry area beneath the vapor bubble, thereby postponing the onset of dry-out.
- On the superhydrophilic surface, there are fewer active nucleation sites at the beginning of the boiling process, and the diameter of departing bubbles is larger, causing a delay in the initiation of nucleate boiling compared to the smooth copper surface.
- The hydrophobic surface demonstrated enhanced heat transfer efficiency because it facilitated the early release of nucleated bubbles, thereby generating a greater number of activated bubbles.

- At high heat flux, the hydrophilic surface exhibited outstanding heat transfer performance, primarily attributable to its abundance of active nucleation sites, surface roughness, and enhanced microlayer evaporation facilitated by increased liquid transport.

References

1. Zakir, O., et al.: A review on TiO2 nanotubes: synthesis strategies, modifications, and applications. J. Solid State Electrochem. **27**(9), 2289–2307 (2023). https://doi.org/10.1007/s10008-023-05538-2
2. Chakroborty, S., Nath, N., Soren, S., Barik, A., Kaur, K.: Plasmonic-based TiO2 and TiO2 nanoparticles for photocatalytic CO2 to methanol conversion in energy applications: current status and future prospects. Top. Catal. **67**(1), 232–245 (2024). https://doi.org/10.1007/s11244-023-01816-5
3. Kumar, S., Chanana, A.: TiO2 based nanomaterial: synthesis, structure, photocatalytic properties, and removal of dyes from wastewater. Korean J. Chem. Eng. **40**(8), 1822–1838 (2023). https://doi.org/10.1007/s11814-023-1455-6
4. El-Rahman, A.M.A., Rabia, M., Mohamed, S.H.: Nitrogen doped TiO2 films for hydrogen generation and optoelectronic applications. J. Mater. Sci. Mater. Electron. **34**(14), 1149 (2023). https://doi.org/10.1007/s10854-023-10551-2
5. Yang, J., et al.: Electrochemical activation of oxygen vacancy-rich TiO2@MXene as high-performance electrochemical sensing platform for detecting imidacloprid in fruits and vegetables. Microchim. Acta **190**(4), 146 (2023). https://doi.org/10.1007/s00604-023-05734-x
6. Damoom, M.M., et al.: The role of TiO2 nanoparticles in enhancing the structural, optical, and electrical properties of PVA/PVP/CMC ternary polymer blend: nanocomposites for capacitive energy storage. J. Solgel Sci. Technol. **108**(3), 742–755 (2023). https://doi.org/10.1007/s10971-023-06223-6
7. Mirjamali, F., Samadi, M., Moradlou, O., Zirak, M.: G-C3N4 quantum dots decorated on urchin-like TiO2 nanostructures for the photoelectrochemical water splitting. J. Mater. Sci. **59**(11), 4483–4497 (2024). https://doi.org/10.1007/s10853-024-09501-y
8. Guan, R., et al.: Wettability control of defective TiO2 with alkyl acid for highly efficient photocatalytic ammonia synthesis. Nano Res. **16**(8), 10770–10778 (2023). https://doi.org/10.1007/s12274-023-5779-7
9. Li, Y., Wei, Y., He, W., Tang, Z., Xiong, J., Zhao, Z.: Ordered macroporous structured TiO2-based photocatalysts for CO2 reduction: a review. Chin. Chem. Lett. **34**(12), 108417 (2023). https://doi.org/10.1016/j.cclet.2023.108417
10. Suhan, M.B.K., et al.: Sustainable pollutant removal and wastewater remediation using TiO2-based nanocomposites: a critical review. Nano Struct. Nano Objects **36**, 101050 (2023). https://doi.org/10.1016/j.nanoso.2023.101050
11. Anusha, B., Anbuchezhiyan, M., Deepa, C., Srinivasan Alias Arunsankar, N.: Novel approaches to the degradation of nitrophenols using TiO2-biopolymer-ligand-metal complex as photocatalysts. J. Mater. Sci. Mater. Electron. **35**(8), 544 (2024). https://doi.org/10.1007/s10854-024-12323-y
12. Jia, Z.-M., Zhao, Y.-R., Shi, J.-N.: Adsorption kinetics of the photocatalytic reaction of nano-TiO2 cement-based materials: a review. Constr. Build. Mater. **370**, 130462 (2023). https://doi.org/10.1016/j.conbuildmat.2023.130462
13. Sawal, M.H., Jalil, A.A., Khusnun, N.F., Hassan, N.S., Bahari, M.B.: A review of recent modification strategies of TiO2-based photoanodes for efficient photoelectrochemical water splitting performance. Electrochim. Acta **467**, 143142 (2023). https://doi.org/10.1016/j.electacta.2023.143142

14. Madkhali, N., et al.: Recent update on photocatalytic degradation of pollutants in waste water using TiO2-based heterostructured materials. Results Eng. **17**, 100920 (2023). https://doi.org/10.1016/j.rineng.2023.100920
15. Singh, S.K., Sharma, D.: Review of pool and flow boiling heat transfer enhancement through surface modification. Int. J. Heat Mass Transf. **181**, 122020 (2021). https://doi.org/10.1016/j.ijheatmasstransfer.2021.122020
16. Singh, S.K., Sharma, D.: Experimental investigation on pool boiling heat transfer performance of superhydrophilic, hydrophilic and hydrophobic surface. Int. J. Thermophys. **45**(4), 53 (2024). https://doi.org/10.1007/s10765-024-03350-2
17. Singh, S.K., Sharma, D., Singh, A.K.: Numerical investigation on pool boiling heat transfer of silica and alumina nanofluids. Heat Transf. Eng. **45**(9), 728–747 (2024). https://doi.org/10.1080/01457632.2023.2220471
18. Singh, S.K., Sharma, D.: Pool boiling heat transfer enhancement using nanoparticle coating on copper substrate. In: Shanker, U., Hussain, C.M., Rani, M. (eds.) Handbook of Green and Sustainable Nanotechnology: Fundamentals, Developments and Applications, pp. 1–23. Springer International Publishing, Cham (2022). https://doi.org/10.1007/978-3-030-69023-6_57-1
19. Singh, A.K., Singh, S.K., Sharma, D.: Numerical simulation of nucleate boiling heat transfer performance using SiO_2/water nanofluid. In: IOP Conference Series: Materials Science and Engineering, p. 012012. IOP Publishing (2022). https://doi.org/10.1007/978-3-030-69023-6_57-1
20. Yim, K., Lee, J., Naccarato, B., Kim, K.J.: Surface wettability effect on nucleate pool boiling heat transfer with titanium oxide (TiO2) coated heating surface. Int. J. Heat Mass Transf. **133**, 352–358 (2019). https://doi.org/10.1016/j.ijheatmasstransfer.2018.12.075
21. Kim, J.M., Kim, T., Yu, D.I., Kim, M.H., Moriyama, K., Park, H.S.: Time effect on wetting transition of smart surface and prediction of the wetting transition for critical heat flux in pool boiling. Int. J. Heat Mass Transf. **114**, 735–742 (2017). https://doi.org/10.1016/j.ijheatmasstransfer.2017.06.114
22. Jung, J.-Y., Kim, H., Kim, M.H.: Effect of ionic additive on pool boiling critical heat flux of titania/water nanofluids. Heat Mass Transf. **49**(1), 1–10 (2013). https://doi.org/10.1007/s00231-012-1055-1
23. Ray, M., Deb, S., Bhaumik, S.: Pool boiling heat transfer of refrigerant R-134a on TiO2 nano wire arrays surface. Appl. Therm. Eng. **107**, 1294–1303 (2016). https://doi.org/10.1016/j.applthermaleng.2016.07.080
24. Bourdon, B., Bertrand, E., Di Marco, P., Marengo, M., Rioboo, R., De Coninck, J.: Wettability influence on the onset temperature of pool boiling: experimental evidence onto ultra-smooth surfaces. Adv. Colloid Interface Sci. **221**, 34–40 (2015). https://doi.org/10.1016/j.cis.2015.04.004
25. Zhang, X., Deng, Z., Yan, Z., Wei, C., Zhu, M.: Effect of TiO2 addition on the viscosity of ladle refining slags. Metall. Mater. Trans. B (2024). https://doi.org/10.1007/s11663-024-03055-9
26. Alsaif, N.A.M., et al.: The impact of TiO2 on physical, optical characteristics and shielding qualities against γ-ray features of titanium bismo-borate glasses. Opt. Quant. Electron. **56**(5), 816 (2024). https://doi.org/10.1007/s11082-024-06702-2
27. Ansari, M.A., Shariq, M., Ansari, S.S., Husain, A.: Efficiency assessment of TiO2-based photocatalytic concrete for clean and sustainable construction: a state-of-the-art review. Iran. J. Sci. Technol. Trans. Civ. Eng. (2024). https://doi.org/10.1007/s40996-024-01415-8
28. Çakıroğlu, B.: Graphene quantum dots on TiO2 nanotubes as a light-assisted peroxidase nanozyme. Microchim. Acta **191**(5), 268 (2024). https://doi.org/10.1007/s00604-024-06341-0
29. Saraev, A.A., Kurenkova, A.Yu., Mishchenko, D.D., Trigub, A.L., Gerasimov, E.Yu., Kozlova, E.A. : Cu/TiO2 photocatalysts for CO2 reduction: structure and evolution of the cocatalyst active form. Trans. Tianjin Univ. **30**(2), 140–151 (2024). https://doi.org/10.1007/s12209-024-00384-3

30. Sanad, M.M.S., Meselhy, N.K., Eraky, M.S., Toghan, A.: One-pot synthesis of TiO2-decorated SnSe/ZnS as a novel anode material with promoted storage capacity and cyclability in lithium-ion batteries. J. Mater. Sci. Mater. Electron. **35**(12), 850 (2024). https://doi.org/10.1007/s10854-024-12596-3
31. Sinha, R., Ghosal, P.S.: A comprehensive appraisal on status and management of remediation of DBPs by TiO2 based-photocatalysts: Insights of technology, performance and energy efficiency. J. Environ. Manage. **328**, 117011 (2023). https://doi.org/10.1016/j.jenvman.2022.117011
32. Hu, B., Zhang, S., Wang, D., Chen, X., Si, X., Zhang, D.: Experimental study of nucleate pool boiling heat transfer of self-rewetting solution by surface functionalization with TiO2 nanostructure. Can. J. Chem. Eng. **97**(S1), 1399–1406 (2019). https://doi.org/10.1002/cjce.23399
33. Han, K., Ma, L., Tian, Y., Luo, J.: Photoinduced superlubricity on TiO2 surfaces. Friction **12**(3), 428–438 (2024). https://doi.org/10.1007/s40544-023-0736-8
34. Sun, W., et al.: Research on regulation mechanism of TiO2 photocathodic protection based on cohesive energy, defect patterns, and nanofilm size. J. Mater. Eng. Perform. (2024). https://doi.org/10.1007/s11665-024-09139-6
35. Tanji, K., et al.: Epigrammatic progress on the photocatalytic properties of ZnO and TiO2 based hydroxyapatite@photocatlyst toward organic molecules photodegradation: a review. J. Water Process Eng. **53**, 103682 (2023). https://doi.org/10.1016/j.jwpe.2023.103682
36. Zhang, Y., Yan, J.: Recent advances in the synthesis of defective TiO2 nanofibers and their applications in energy and catalysis. Chem. Eng. J. **472**, 144831 (2023). https://doi.org/10.1016/j.cej.2023.144831
37. Feizpoor, S., Rahim Pouran, S., Habibi-Yangjeh, A.: Recent progress on photocatalytic evolution of hydrogen gas over TiO2-x-based emerging nanostructures. Mater. Sci. Semicond. Process **162**, 107444 (2023). https://doi.org/10.1016/j.mssp.2023.107444
38. Gupta, S.K., Misra, R.D.: An experimental investigation on flow boiling heat transfer enhancement using Cu-TiO2 nanocomposite coating on copper substrate. Exp. Therm. Fluid Sci. **98**, 406–419 (2018). https://doi.org/10.1016/j.expthermflusci.2018.06.012
39. Wu, W., Bostanci, H., Chow, L.C., Hong, Y., Su, M., Kizito, J.P.: Nucleate boiling heat transfer enhancement for water and FC-72 on titanium oxide and silicon oxide surfaces. Int. J. Heat Mass Transf. **53**(9), 1773–1777 (2010). https://doi.org/10.1016/j.ijheatmasstransfer.2010.01.013
40. Das, S., Saha, B., Bhaumik, S.: Experimental study of nucleate pool boiling heat transfer of water by surface functionalization with crystalline TiO2 nanostructure. Appl. Therm. Eng. **113**, 1345–1357 (2017). https://doi.org/10.1016/j.applthermaleng.2016.11.135
41. Ray, M., Bhaumik, S.: Nucleate pool boiling heat transfer of hydro-fluorocarbon refrigerant R134a on TiO2 nanoparticle coated copper heating surfaces. Heat Transf. Eng. **40**(12), 997–1006 (2019). https://doi.org/10.1080/01457632.2018.1450333
42. Yeom, H., Sridharan, K., Corradini, M.L.: Bubble dynamics in pool boiling on nanoparticle-coated surfaces. Heat Transf. Eng. **36**(12), 1013–1027 (2015). https://doi.org/10.1080/01457632.2015.979116
43. Yan, W., Lin-lin, W., Ming-Yan, L.: Antifouling and enhancing pool boiling by TiO2 coating surface in nanometer scale thickness. AIChE J. **53**(12), 3062–3076 (2007). https://doi.org/10.1002/aic.11345
44. Ray, M., Deb, S., Bhaumik, S.: Experimental investigation of nucleate pool boiling heat transfer of R134a on TiO2 coated TF surface. Mater. Today Proc. **4**(9), 10002–10009 (2017). https://doi.org/10.1016/j.matpr.2017.06.310
45. Watanabe, Y., Enoki, K., Okawa, T.: Nanoparticle layer detachment and its influence on the heat transfer characteristics in saturated pool boiling of nanofluids. Int. J. Heat Mass Transf. **125**, 171–178 (2018). https://doi.org/10.1016/j.ijheatmasstransfer.2018.04.072

46. Naphon, P., Thongjing, C.: Pool boiling heat transfer characteristics of refrigerant-nanoparticle mixtures. Int. Commun. Heat Mass Transf. **52**, 84–89 (2014). https://doi.org/10.1016/j.icheatmasstransfer.2014.01.014
47. Ali, H.M., Generous, M.M., Ahmad, F., Irfan, M.: Experimental investigation of nucleate pool boiling heat transfer enhancement of TiO2-water based nanofluids. Appl. Therm. Eng. **113**, 1146–1151 (2017). https://doi.org/10.1016/j.applthermaleng.2016.11.127

Exploring TiO_2-Based Catalysts for Seawater Splitting: Mechanistic Insights and Efficiency in Hydrogen and Oxygen Evolution Reactions

Pulkit Garg, Tarun Mahajan, and Ankit Tyagi

Abstract Hydrogen (H_2) production using seawater is becoming increasingly popular due to the pressing need for sustainable and clean energy sources. Titanium dioxide (TiO_2), known for its stability, low cost, and non-toxicity, exhibits promising performance in both photocatalytic and electrocatalytic processes. In photocatalytic water splitting, TiO_2 utilizes solar or artificial light to initiate redox reactions, facilitating the production of H_2 and oxygen (O_2) from seawater. Various studies have investigated the factors influencing photocatalytic H_2 production, including the concentration of ions and impurities in seawater, and have proposed methods to enhance efficiency. The current flows through the TiO_2 electrode to split the water into its components, H_2 and O_2 in electrocatalytic water splitting. Recent studies have explored the TiO_2-based electrocatalysts and their performance in diverse environments, revealing significant improvements in activity and stability. This chapter encapsulates recent research endeavors to advance TiO_2-based catalytic systems as photocatalysts and electrocatalysts for seawater splitting, highlighting their potential in driving the transition towards an H_2-based economy.

Graphical Abstract

P. Garg · T. Mahajan · A. Tyagi (✉)
Department of Chemical Engineering, Indian Institute of Technology Jammu, Jammu 181221, India
e-mail: ankit.tyagi@iitjammu.ac.in

J. Prakash et al. (eds.), *Titanium Dioxide-Based Multifunctional Hybrid Nanomaterials*, Engineering Materials, https://doi.org/10.1007/978-3-031-80625-4_11

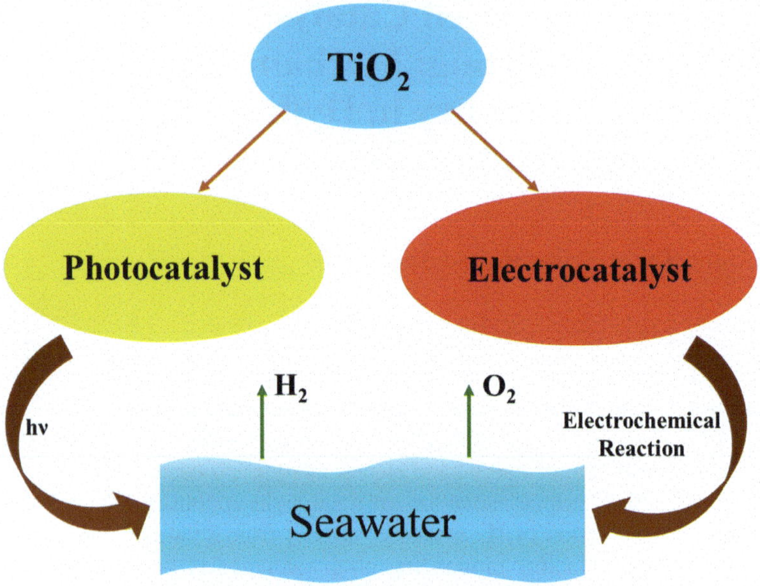

TOC: TiO_2 as photocatalyst and electrocatalyst in seawater splitting.

Keywords Titanium dioxide · Seawater electrolysis · Photocatalyst · Hydrogen evolution reaction · Electrocatalyst · Oxygen evolution reaction

1 Introduction

Green Energy, a fascinating and lucrative term to hear, is the need of the hour. Climate change and energy demand have made it a top priority for every country to develop a clean and green energy source. To date, the energy demand is fulfilled from burning fossil fuels and other conventional energy sources, which has led to an increase in global warming and other hazardous effects on the environment, threatening the habitat of many living organisms. Fossil fuel combustion provides almost 80% of the energy to cater the world's energy requirements [1, 2]. H_2 is considered the best possible solution to address this problem because of its high heating value of 39.4 Wh/kg, which is more than twice that of methane and gasoline, having heating values of 15.4 Wh/kg and 13.19 Wh/kg, respectively [3].

Producing H_2 is still a tedious task. However, producing H_2 from the water splitting can be one way to significantly reduce our dependency on fossil fuels [4, 5]. Currently, commercial water splitting systems utilize high-purity water as the reactant, and to enhance ion conductivity, these systems often require the addition of alkaline, acidic, or buffer solutions [6]. Population growth, Industrial development, and water pollution have led to global water scarcity. If freshwater is used extensively for electrolysis, this problem could worsen significantly. Therefore, water splitting systems utilizing seawater compared to freshwater have numerous advantages and have emerged as an up-and-coming alternative [7]. Seawater, constituting 96.5% of all water on Earth, is abundant and can be readily used as a clean, virtually limitless energy resource. Additionally, the relatively uniform distribution of seawater worldwide allows countries to efficiently leverage this resource for trade and energy development [8].

Recently, photocatalysis and electrocatalysis have been widely used as eco-friendly methods in various fields, such as dye degradation [9], sensors [10, 11], batteries [12, 13] etc. Furthermore, numerous studies have reported photocatalytic and electrocatalytic techniques for H_2 generation through water splitting. However, limited durability and high production cost limits the commericialization of photocatalytic water splitting, thereby requiring further development [14].

Seawater splitting by photocatalysis or electrolysis offers numerous benefits; however, its complex chemical composition poses several challenges. Unlike pure water, water splitting using seawater must address issues such as the deposition of microorganisms and ionic impurities, chlorine evolution reaction (ClER), and corrosion of active site because of chloride (Cl^-) ions [15, 16].

Various semiconductor materials such as Cu_2O [17], TiO_2 [18], Fe_2O_3 [19], ZnO [20], CdS [21], and $BiVO_4$ [22] that accelerate the chemical reactions in the presence of light energy have been widely explored. Among them, TiO_2 stands out as an excellent candidate in photo and electrochemical applications for seawater splitting owing to its superior corrosion resistance in marine ecosystems [23, 24]. The utilization of TiO_2 as a catalyst has garnered substantial attention in the field of catalysis, particularly in the realms of photocatalysis and electrocatalysis [25]. TiO_2 is an extensively studied catalyst because of its non-toxicity, low price, and excellent stability [26]. The photocatalytic activity of TiO_2 arises from its ability to produce electron–hole pairs under solar irradiation and to facilitate redox reactions, such as reducing organic contaminants and H_2 production from water [27]. Furthermore, as a semiconductor material, TiO_2 exhibits remarkable properties that make it highly effective in promoting various chemical reactions simultaneously with light irradiation and electrical potential [28]. Its high stability, abundance, and tunable surface chemistry make TiO_2 an attractive candidate for electrocatalytic applications. Fazil et al. showed that Cu-TiO_2 nanoparticles (NPs) exhibit a band gap reduction from 3.19 to 3.08 eV and increase of specific surface area from 169.9 to 188.2 m^2/g. Within the range of compositions studied, the 2.5% Cu-doped TiO_2 displays notable H_2 production, boasting a quantum yield of 17.67% [29]. One of the main benefits of TiO_2 is its ability to facilitate water oxidation (O_2/H_2O) and the proton reduction (H^+/H_2) simultaneously. Nevertheless, TiO_2 still requires development owing to its low photocatalytic activity arising from its high charge recombination rate and wide band gap under visible light [30].

2 Fundamentals of Seawater Splitting

Using freshwater for H_2 production will lead to further scarcity of already scarce water resources around the globe needed to sustain life. Instead, we can use seawater as a substitute for H_2 production because of its abundant availability and economic feasibility. However, seawater constitutes approx. 96.5% water and 3.5% of ionic impurities such as chloride ions, sodium ions, sulphate ions etc. (Fig. 1), that occlude porous structure and active sites of the membrane and catalysts, respectively, and poison it. We can call seawater a natural electrolyte because it has a conductivity of 5 Sm^{-1} which is considerably larger than the conductivity of pure water (5.5×10^{-6} Sm^{-1}) with 0.5 M NaCl electrolyte [31]. The fundamental mechanism of seawater splitting can be classified in terms of photocatalytic and electrocatalytic seawater splitting.

2.1 Photocatalytic Water Splitting

When semiconductor materials have their band gap energy less than or equal to the energy of the illuminated light, electrons jumps from valence band (VB) to conduction band (CB) leaving behind holes in VB, respectively, as shown in Fig. 2a [16]. When a semiconductor's VB lies below the water oxidation potential (0.81 V, pH = 7) and its CB exceeds the water reduction potential (-0.42, pH = 7), water splitting occurs. Since the CB and VB positions of TiO_2 match this criteria required for water splitting, it is proved to be an excellent semiconductor material for photocatalytic water splitting and can lead to subsequent reactions (Eqs. 1–4) [33, 34]. Appropriate cocatalysts are often loaded onto semiconductor surfaces to mitigate the charge carriers recombination and enhance the surface reaction efficiency. The photocatalytic process occurring in seawater closely resemble to those in pure water,

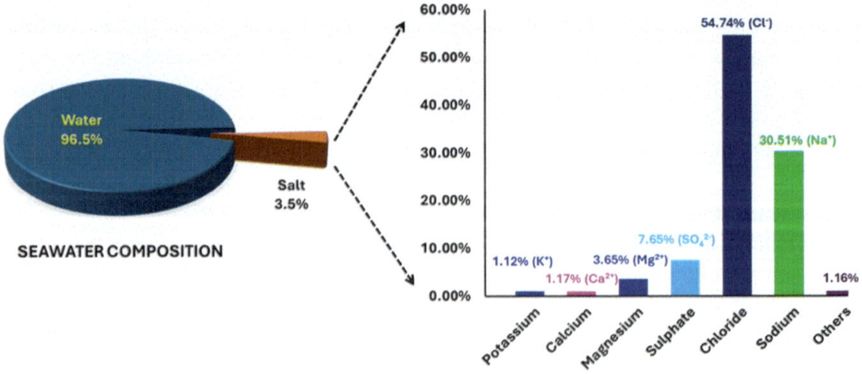

Fig. 1 Chart depicting the average composition of different salts present in seawater [32]

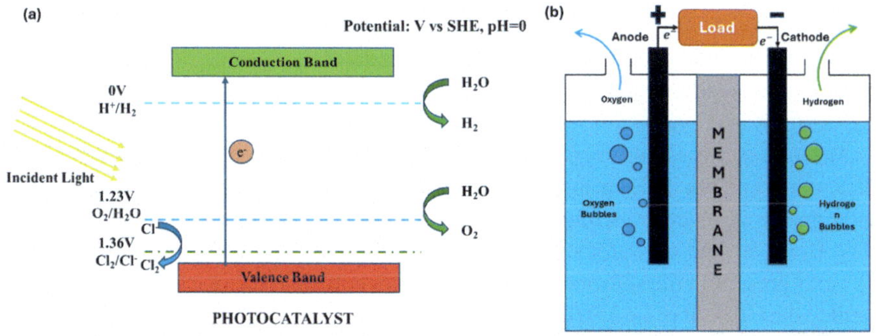

Fig. 2 **a** Energy band diagram for photocatalytic water splitting **b** electrocatalytic seawater splitting

but its saline nature and impurities can influence surface reactions and product formation [35]. For example, Cl^- ions, having a lesser oxidation potential than water, lead to the creation of chlorine-containing oxidation products, highlighting the complex interplay between semiconductor photocatalysis and seawater chemistry [36].

$$TiO_2 + h\vartheta \rightarrow TiO_2\left(e^- + h^+\right) \tag{1}$$

$$2H_2O\ (l) + 4h^+ \rightarrow 4H^+\ (aq) + O_2\ (g), E^o_{ox} = 0.81\ \text{eV} \tag{2}$$

$$2H^+\ (aq) + 2e^- \rightarrow H_2\ (g), E^o_{Red} = -0.42\ \text{eV} \tag{3}$$

Overall mechanism:

$$H_2O\ (l) + TiO_2\ (s) + h\vartheta \rightarrow H_2\ (g) + 1/2O_2\ (g) \tag{4}$$

2.2 Electrocatalytic Water Splitting

Electrocatalytic water splitting entails a sequence of intricate electrochemical reactions happening on the catalyst loaded electrode surface, typically composed of a metal oxide or other conductive material [37]. At the cathode during the HER, protons (H^+) are reduced to form H_2 gas, while concurrently at the anode during the OER, water molecules are oxidized to generate O_2 gas (Fig. 2b) [38]. Electrocatalysts are pivotal in facilitating these reactions by offering water molecule adsorption over active sites, facilitating electron transfer, and stabilizing reaction intermediates. Commonly employed electrocatalysts include metal oxides like ruthenium dioxide (RuO_2), iridium oxide (IrO_2), and manganese oxide (MnO_x), which were chosen for their notable catalytic activity and stability under harsh conditions [39]. The mechanism of electrocatalytic splitting of water necessitates a sequence of steps, including water molecule adsorption, formation of reaction intermediates, transfer of electrons and protons, and eventual release of H_2 and O_2 gases [40].

The primary mechanism of water splitting is shown in (Eqs. 5–11), where the HER undergoes two-electron pathways, and the oxygen evolution reaction (OER) undergoes four electron pathways, making it kinetically sluggish compared to HER [41]. However, the seawater splitting mechanism is slightly different because of the Cl^- ions present in the seawater. During seawater splitting, the reactions at the anode include the ClER at $pH < 3.0$ (Eqs. 9, 10) and hypochlorite (ClO^-) generation at $pH > 7.5$ (Eq. 6), both of which compete with the OER. Notably, Cl_2 and ClO^- formation are two-electron processes, providing a kinetic advantage over the OER, which is a four-electron process. Nonetheless, OER becomes thermodynamically more viable than ClO^- formation when the pH rises above 7.5 [42]. The below reactions occur at anode and cathode in alkaline and acidic mediums during seawater electrolysis.

In Alkaline Media:

Anode:

$$4OH^- \rightarrow 2H_2O + O_2 + 4e^- \quad E_a^\circ = 0.4\ V \tag{5}$$

$$Cl^- + 2OH^- \rightarrow ClO^- + H_2O + 2e^- \quad E_a^\circ = 0.89\ V \tag{6}$$

Cathode:

$$2H_2O + 2e^- \rightarrow 2OH^- + H_2 \quad E_c^\circ = -0.83\ V \tag{7}$$

In Acidic Media:

Anode:

$$2H_2O \rightarrow 4H^+ + O_2 + 4e^- \quad E_a^\circ = 1.23\ V \tag{8}$$

$$2Cl^- \rightarrow Cl_2 + 2e^- \quad E_a^\circ = 1.36\ V \tag{9}$$

or

$$Cl^- + H_2O \rightarrow HClO + H^+ + 2e^- \tag{10}$$

Cathode:

$$2H^+ + 2e^- \rightarrow H_2 \quad E_c^\circ = 0\ V \tag{11}$$

Understanding the thermodynamics of this electrochemical reaction is essential for optimizing energy efficiency and overall performance. The general equation of the Gibbs free energy is presented below:

$$\Delta G = \Delta H - T\Delta S \tag{12}$$

where ΔS is the entropy change, ΔH is the enthalpy change, and ΔT is the temperature. For water electrolysis, the Nernst equation relates the Gibbs-free energy change to the cell potential (E_{cell}) is shown below [43]:

$$\Delta G = -nFE_{cell} \tag{13}$$

Here, 'F' is the Faraday constant having a value of 96,485 C mol^{-1} and the number of electrons taking part in the reaction is represented by '**n**'.

For a standard hydrogen electrode (SHE), the standard reduction potential required for HER $\left(E^{\circ}_{red,cathode}\right)$ are 0 V, and the standard reduction potential required for the OER $\left(E^{\circ}_{red,anode}\right)$ is 1.23 V. Therefore, the standard cell potential E°_{cell} for water electrolysis is [44]:

$$E^{\circ}_{cell} = 0 \quad 1.23 = -1.23 \text{ V}$$

This implies that, under standard conditions, the cell potential for water electrolysis is − 1.23 V.

3 Application of TiO_2 as a Photocatalyst or Photoanode for Seawater Splitting

This section explores the impact of TiO_2 photocatalyst for efficient evolution of O_2 and H_2. As a semiconductor material, TiO_2 has garnered substantial interest for photocatalytic water splitting. However, TiO_2 is more efficient at generating H_2 than O_2 because more energy is required to create electron–hole pairs in TiO_2 than to evolve O_2 [45]. Therefore, several authors reported the utilization of TiO_2 specifically for HER, while some utilized the TiO_2-based photocatalysts or photoanode for overall water splitting, i.e., OER and HER.

3.1 TiO_2-based Photocatalysts for HER

This section delves into the aspects influencing photocatalytic H_2 production and the methods to enhance efficiency (Table 1). One prominent aspect is the impact of seawater ions and impurities on the efficiency and stability of the reaction. Specifically, CO_3^{2-} is identified as having a significant negative impact on the H_2 production rate, necessitating adjustments to the pH of seawater to restore production rates. Under suitable light irradiation, TiO_2 photocatalysts generate electron–hole pairs, which drive the depletion of protons to form H_2 gas at the cathode [35].

Simamora et al. investigated the use of nanostructured TiO_2 promoted with NiO and CuO as photocatalysts to increase H_2 yields from the photocatalytic seawater splitting (PC-SS). The study comprised oxalic acid as a sacrificial reagent on CuO/nano-TiO_2 compared to NiO/nanoTiO_2 photocatalysts. The results indicate that the H_2 formed on CuO/nanoTiO_2 is 8.53 μmol g cat^{-1}. In contrast, on NiO/nanoTiO_2 photocatalyst, the yielded H_2 is comparatively less (i.e., 1.46 μmol g cat^{-1} indicating CuO/nanoTiO_2 outperforms NiO/nanoTiO_2 in generating H_2 from seawater) [46].

Cheng et al. investigated the utilization of a mesoprous brookite/anatase (meso-B/A) TiO_2 photocatalyst for H_2 production from seawater. The research focuses on tackling the issues of side reactions, and the authors engineered the super hydrophilic meso-B/A TiO_2 photocatalyst, confirmed through XRD and RAMAN spectrum, indicating the transitions from the brookite/rutile (B/R) type-I heterojunction to brookite/anatase type-II heterojunction. The photocatalytic H_2 generation rate using brookite/anatase TiO_2 attained 6.59 mmol g^{-1} h^{-1} during seawater splitting, surpassing that anatase/rutile TiO_2 (P25) and of brookite/rutile TiO_2 by 2.9 times and 2.5 times, respectively as shown in Fig. 3 [47].

In another study, Nam et al. aimed to develop Pt/TiO_2 as cathode material by varying Pt precursor ranging 0–0.4 wt%. From the XRD pattern, it was observed that all the samples (Pt precursor ranging from 0 to 0.4 wt% Pt) were initially amorphous. However, annealing the samples in a dry 10% H_2/Ar environment facilitated crystalline phase formation. The XRD peaks corresponding to 2θ values of approximately 25.4° (101), 37.8° (004), 48° (200), and 53° (105), along with 55.1° (211) and 75° (215), were identified as indicative of the TiO_2 crystalline phase (anatase). It also evaluated different seawater electrolytes acquired from combined and single nano-filtration and reverse osmosis processes. The pure TiO_2 cathode and Ti metal showed relatively lesser H_2 generation rates compated to Pt mesh and hydrogenase enzyme immobilized TiO_2 as cathode as shown in Fig. 4a. Specifically, the 0.2 wt% Pt/TiO_2 cathode exhibited superior conduct among Pt/TiO_2 cathodes tested with natural seawater as electrolyte, as shown in Fig. 4b [48].

Li et al. synthesized nanorod arrays by electrodepositing cobalt-phosphate (Co-Pi) particles on TiO_2@g-C_3N_4 nanorod arrays (CCNRs) on FTO substrates with varying g-C_3N_4 mass ratios. A heterojunction was created because the CB and VB of g-C_3N_4 were at a higher level than those of TiO_2. Furthermore, Co-Pi particles are believed to act as rapid redox mediators during water oxidation, facilitating efficient separation of charge and reducing the energy barrier needed for photoelectrochemical water splitting. Across the entire spectrum, incident photon-to-current conversion (IPCE) of TiO_2@g-C_3N_4 NRs surpasses that of pure TiO_2 NRs, indicating effective separation of photogenerated hole pairs in the TiO_2@g-C_3N_4 NRs heterostructure. Following Co-Pi modification, the IPCE value increases, peaking at 41% at 380 nm. Under 100 mW cm^{-2} (AM1.5G) light illumination, a photocurrent density of 1.6 mA cm^2

Table 1 Performance of TiO_2 based photocatalyst for HER

S. No	Catalyst	Electrolyte	Sacrificial agent	H_2 generation (mmol g^{-1} h^{-1})	Spectral range/ source	Ref.
1	TiO_2/CuO	Synthetic seawater	Oxalic acid	1.704	UV–Vis	[46]
2	N-doped TiO_2	Seawater	No sacrificial agent	40	AM 1.5 G, 100 mW cm^{-2}	[52]
3	H- TiO_2/Cu_2O	Simulated seawater, natural seawater	–	45.7, 17.9	365nm	[53]
4	Brookite TiO_2/Pt	Natural seawater	Methanol	1.476	Full solar spectra	[54]
5	2.5% CuO/nanoTiO_2	Synthetic seawater	–	0.0157	UV–Vis	[55]
6	SiO_2/Ag@TiO_2	Synthetic seawater	Glycerol	0.857	UV	[56]
7	$Na_2Ti_3O_7$/Ag/ CdS	Natural seawater	$Na_2S{\cdot}9H_2O$ and Na_2SO_3	1793	Visible	[57]
8	Granular TiO_2/Pt	Natural seawater	Methanol	6.31	$\lambda > 320$ nm	[58]
9	CdS/TiO_2	Artificial seawater	Na_2S, Na_2SO_3	2.640	UV	[59]
10	Fe_2O_3/C-TiO_2	Seawater	–	170.8	Visible	[60]
11	GaP-TiO_2–SiO_2:Rh nanocomposite	Seawater	Methanol	0.0103	–	[61]
12	Cu_2O/TiO_2	Seawater	–	5.1	Xe arc light source	[62]
13	TiO_2-rGO-Cu	Simulated seawater	Methanol	16.08	Xe light source	[63]
14	TiO_2/CdS/CNT nanocomposite	Seawater	–	1.373	Visible	[64]
15	MoS_2/TiO_2	Seawater		200	Visible	[65]
16	Nano TiO_2	Seawater	–	100	Solar spectra	[66]

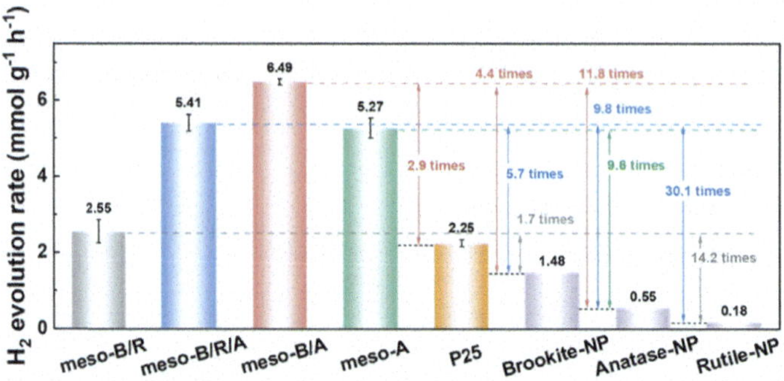

Fig. 3 Rates of evolution of H_2 from TiO_2 photocatalysts. Reprinted with permission from [47], Copyright (2024), Elsevier

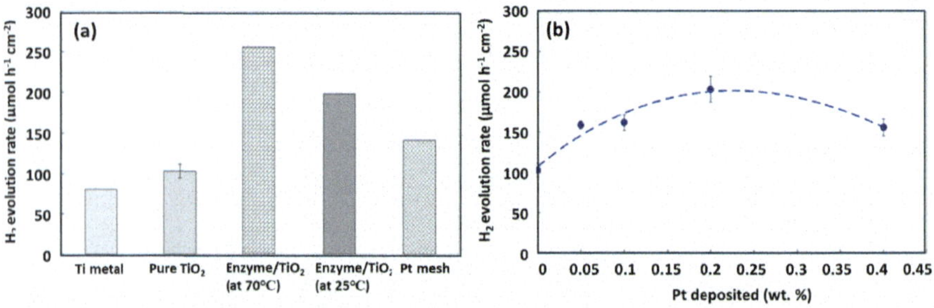

Fig. 4 **a** HER on cathodes with seawater electrolyte **b** various Pt/TiO_2 cathodes effect on HER. Reprinted with permission from [48], Copyright (2011), Elsevier

is achieved at 1.23 V (RHE). Notably, the performance of the CCNRs photoanode remains highly stable during seawater splitting, maintaining its efficiency even after 10 h of continuous light exposure [49].

Dong et al. synthesized nanotube arrays of Bi-BiOI shell-core microsphere-modified TiO_2 (TNA) by solvothermal approach. Notably, the protective Bi shell around the BiOI microspheres prevented seawater corrosion and narrowed the photoanode bandgap, enhancing their catalytic stability. The transient photocurrent densities exhibited by Bi-BiOI/TNA and BiOI/TNA electrodes were 1.42 and 0.19 mA cm^{-2}, respectively, which are 12.9 and 1.7 times superior than bare TNA (0.11 mA cm^{-2}), as the bare TNA responds exclusively to UV light. In 4 h, Bi-BiOI/TNA achieved H_2 yield which is 3.8 and 7.6 times greater than that from BiOI/TNA (36.2 mmol cm^{-2}) and bare TNA (18.2 mmol cm^{-2}), respectively [50].

A critical study conducted by Yaqian et al. demonstrated that ions present in seawater negatively affect the rate of H_2 production rate, among which CO_3^{2-} has the most severe impact. At the concentration of 1.5 mmol L^{-1}, the H_2 generation rate is reduced by nearly 40%. Experiments found that the zeta potential on the surface of TiO_2 particles that had a value of 18.7 mV in pure DI water reduced to 5.5 mV after treating with Simulated seawater and after CO_3^{2-} and SO_4^{2-} treatment, it is reduced to 6.3 and 6.9 mV, which implies that these ions adsorb on TiO_2 surface. Fourier transform Infrared and thermogravimetric analysis (TGA)-mass spectroscopy further confirmed that CO_3^{2-} was the principal ion adsorbed on the surface of TiO_2, as shown in Fig. 5a–d [51].

Li et al. demonstrated a stoichiometric production of H_2 and O_2 through thermal-aided PC-SS, utilizing polarization of charge facilitated by the electrolyte over a N-doped TiO_2 photocatalyst. Vast studies revealed that the photo-polarized facets of opposite charge enabled the selective adsorption of ionic compounds in seawater onto them. This selective adsorption results in a five-times increment in the charge-carrier lifetime, resulting in an overall energy conversion efficiency of about 15.9 at 270 °C and the H_2 production rate of 40 mmol g^{-1} h^{-1} was achieved by utilizing a concentrated light furnace. Numerous ionic species have been investigated, revealing that, in addition to Cl^- anions, cations such as Na^+, typically considered inert in PC-SS, may also apply a significant polarization effect [52].

Xiao et al. synthesized a H-TiO_2/Cu_2O photocatalyst, composed of hydroxyl-rich TiO_2 porous microstructures integrated with Cu_2O NPs. The synthesis of H-TiO_2/

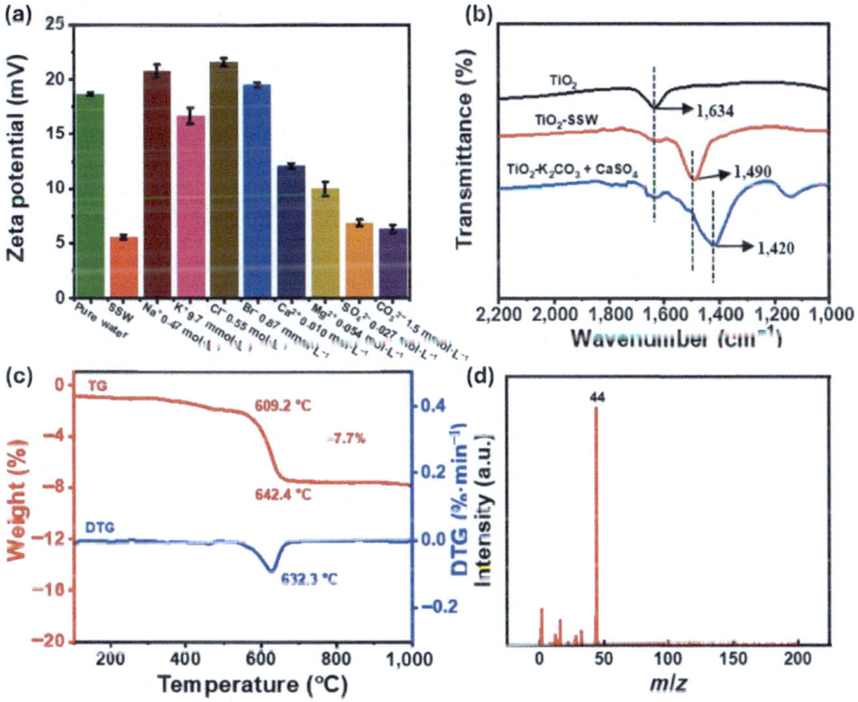

Fig. 5 **a** Zeta Potential of TiO_2 in pure water, simulated seawater, and simulated seawater with CO_3^{2-} and SO_4^{2-}. **b** FTIR spectra for TiO_2 in pure water, simulated seawater, aqueous solution of K_2CO_3 and $CaSO_4$. **c** Derivative thermogravimetric of TiO_2. **d** Mass spectra of efflux from TGA of TiO_2 from simulated seawater. Reprinted with permission from [51], Copyright (2022), Springer Nature

Cu_2O relies on the premise that the hydrophilic surfaces and TiO_2 porous microstructures would support Cu_2O NPs in seawater and facilitate selective H^+ adsorption. In simulated seawater and natural seawater, H-TiO_2/Cu_2O achieved an H_2 generation rate of 45.7 mmol g^{-1} h^{-1} and17.9 mmol g-^{1} h^{-1}, respectively. Additionally, H-TiO_2/Cu_2O maintained over 90% of its H_2 evolution activity for 30 h in natural seawater [53].

3.2 *TiO_2-Based Photocatalysts for OER*

This section discusses the factors that influence photocatalytic O_2 production and how photocatalytic efficiency can be increased. But firstly, let us explore why there is a need to focus on a photocatalyst for OER. OER photocatalysts face specific challenges like low efficiency & stability, dependency on sacrificial agents and cocatalysts, inefficient surface catalytic reactions, and spatial separation of reduction and oxidation reactions [67]. Several investigations attempted to find solution to these challenges. Fu et al. developed a 2D Ti/TiO_2@NiB_x bifunctional electrode for efficient H_2 production. TiO_2 nanorods supported on a anti-corrosive Ti plate (Ti/TiO_2) were initially created via alkaline etching. Subsequently, NiB_x particles were precisely plated onto the Ti/TiO_2 substrate using electroless plating. The Ti plate and Ti/TiO_2 substrate exhibited minimal HER catalytic activity. However, after modification with NiB_x, the interactions between boron and nickel elements facilitated the optimal selection of active sites. Nickel cations contributed to enhanced electrochemical conductivity, while boron anions effectively modified the electronic structure. For HER, the Ti/TiO_2@NiB_x electrode exhibited a 91 mV overpotential at 10 mA cm^{-2} current density and 113 mV dec^{-1} Tafel slope, suggesting that the electrode followed the Volmer-Heyrovsky reaction step. For OER, it demonstrated a 146.5 mV dec^{-1} Tafel slope and 338 mV overpotential at 10 mA cm^{-2}, served as a dual-functional electrode for seawater splitting. Furthermore, the observed volume ratio of H_2 to O_2 during electrolysis matched the theoretical value 2:1, revealing that the Ti/TiO_2@NiB_x electrode achieved a faradaic efficiency close to 100%. The H_2 and O_2 evolution rates are 8.93 μmol s^{-1} and 4.46 μmol s^{-1}, respectively [68].

In another study, Zhao et al. introduce a double-layer assembly composed of stainless steel (SS) and TiO_2 designed to function as a protective layer and cocatalyst for the Si photoanode. The TiO_2 layer safeguards the Si photoanode's surface from plasma damage during the sputtering of SS catalysts. Meanwhile, chromium-incorporated NiFeOOH, derived from the SS film, is an effective cocatalyst for O_2 generation and delays the adsorption of corrosion causing ions from seawater. By employing this strategy, the TiO_2 layer efficiently guards the photoelectrode from the severe plasma sputtering environment during SS coating, thus inhibiting surface damage. The double-layer assembly of Si photoanode prevented the Cl^- ion adsorption and maintained stability for 167 h in Cl^--containing alkaline electrolytes. Figure 6a shows the schematic diagram of the heterojunction of Si/TiO_2/SS, and Fig. 6b shows the J-V curves obtained for different electrolytes using simulated and

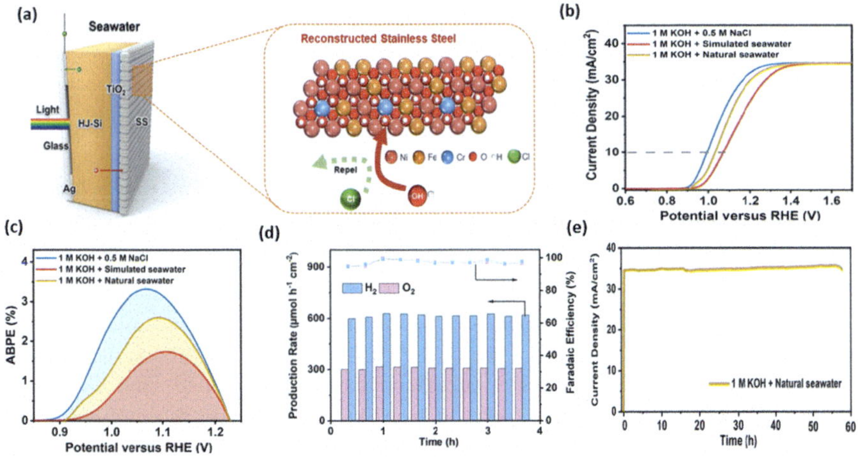

Fig. 6 **a** A schematic illustration of heterojunction of $Si/TiO_2/SS$ utilized in seawater splitting **b** J–V curves and **c** the equivalent ABPE curves **d** of H_2 and O_2 production rate and faradaic efficiency at 1.5 V versus RHE and **e** the stability tests in alkaline natural seawater. Reprinted with permission from 2024 [69], Copyright (2024), Springer Nature

natural seawater. Moreover, Fig. 6c–e shows that this photoanode achieved an applied bias photon-to-current efficiency of 2.62% for over 50 h in alkaline natural seawater and H_2 and O_2 generation rate of up to 600 and 300 μmol $h^{-1}cm^{-2}$ respectively, suggesting the total utilization of charge carriers [69].

4 Application of TiO_2 as an Electrocatalyst for Seawater Electrolysis

TiO_2-based materials are advantageous as an electrocatalyst in water splitting because of its diverse properties for example, high surface area, high chemical and thermal stability etc. [70, 71]. However, splitting seawater presents challenges because catalysts must remain stable at high salinities and at all pH ranges while retaining high catalytic efficiency [72]. Additionally, the effectiveness of HER and OER electrocatalysts in seawater electrolysis can be hindered by the generation of precipitates on electrode surfaces [73].

4.1 TiO_2-Based Electrocatalysts for HER

TiO_2, due to its abundant, inexpensive, and environmentally friendly nature, is a sustainable option for large-scale applications like fuel cells, water splitting applications, catalysis, etc. TiO_2 exhibits excellent stability and corrosion resistance, ensuring long-term performance in electrochemical environments [74]. However, TiO_2 has inherently exceptionally low electrical conductivity (10^{-12}–10^{-7} S cm^{-1}) and limited catalytic activity compared to noble metal catalysts like Au, Ag, Pt, etc. [75, 76], resulting in fewer studies in this area. Overall, TiO_2 holds significant potential as an electrocatalyst for HER and remains an active research area in renewable energy conversion. Due to their similar redox potentials, the primary problem in seawater electrolysis is simultaneous evolution of O_2 and Cl_2 at the anode. Chlorine's multistep electro-oxidation is further becomes complex by its significant dependence on temperature, pH, and applied voltage. Consequently, some researchers have proposed using alkaline electrolyte-based water for the HER instead of seawater. For instance, the investigation by Danilov et al. delved into examining the electrocatalytic properties of Fe-TiO_2 combined coatings for the HER under alkaline conditions. Incorporating TiO_2 particles into the Fe matrix enhances the coating's microhardness such that the catalyst Fe/TiO_2 (5 wt%) exhibited microhardness of 665 HV at 10 A dm^{-2} current density, 1.3 pH, and 298 K. The cyclic voltammetry curves (CV) indicated that the current at potential $E < -1.2$ V and $E > 0.7$ V relates to HER and OER, respectively. Furthermore, from the CV curves, it can be inferred that as the TiO_2 concentration in the coatings increases, the overpotential and Tafel slope values for both the HER and OER decrease. The Tafel slope values determined for the samples Fe, Fe/TiO_2 (2 wt%), and Fe/TiO_2 (5 wt%) analyzed for HER are 115, 108, 106 mV dec^{-1}, respectively, implying that the HER mechanism followed the sequence Volmer-Heyrovsky reactions. The electrochemical impedance spectroscopy curves exhibited incurvated semicircles, indicating that the charge transfer step controls the HER [77].

Further, Chen et al. deposited Pt NPs on Mo-doped TiO_2 ($Ti_{0.9}Mo_{0.1}O_2$), which was used as support for the HER. The catalyst comprises 5 wt% Pt supported on $Ti_{0.9}Mo_{0.1}O_2$ displayed a 4.4-fold mass activity at an overpotential of 50 mV compared to 20% Pt/C due to the H_2 spillover effect (i.e., migration of H_2 atoms from the Pt catalyst onto the $Ti_{0.9}Mo_{0.1}O_2$ support). It exhibited superior stability over time when correlated to the 20% Pt@C catalyst. The synthesized $Ti_{0.9}Mo_{0.1}O_2$ support was found to exhibit an amorphous structure, with molybdenum doping being successful. The Pt/$Ti_{0.9}Mo_{0.1}O_2$ catalyst demonstrated excellent durability, with minimal current density loss after 10,000 cycles [78].

Simulated seawater containing alkaline and chloride solutions has also been used for TiO_2-based electrocatalysts. For example, Hamed et al. synthesized Ni nanoparticle-grafted TiO_2 nanotube electrodes for the HER in simulated seawater. This electrode is created by electrochemically grafting Ni nanoparticles onto anodic TiO_2 nanotubes under various conditions. The s-Ni/TiO_2/Ti electrode achieves an overpotential of 126 mV, outperforming the conventional Pt/C catalyst in 1M KOH.

Additionally, in an electrolyte mixture of 1.0 M KOH and 0.5 M NaCl, the electrode requires an overpotential of only 127 mV for HER. Notably, the s-Ni/TiO_2/Ti electrode maintains high performance over 100 h, highlighting its long-term stability. This makes it a highly effective and cost-efficient alternative to precious metal electrocatalysts, particularly in corrosive environments such as seawater [79].

4.2 TiO_2-Based Electrocatalysts for OER

The high surface area of TiO_2 nanoparticles offers numerous active sites for OER. TiO_2 exhibits compatibility with various electrolytes like aqueous (e.g., $MgCl_2.6H_2O$), solid, gel, etc., which is crucial for diverse electrochemical environments [80–82]. Further, when combined with other materials in composite structures, such as metal nanoparticles or carbon-based materials, TiO_2 demonstrates synergistic effects, augmenting its catalytic efficiency. Thus, combining various metals in mutual bonding can enhance the adsorption-free energy of intermediates involved in the OER. This optimization lowers the energy barriers during seawater splitting. Despite its modest OER activity compared to some alternatives, ongoing research studies aim to enhance TiO_2's performance, unlocking its full potential in sustainable energy applications.

For instance, Haq et al. discussed the design and performance of an electrocatalyst, Au-$GdCo_2B/TiO_2$, for seawater electrolysis, highlighting its high activity, specific O_2 selectivity, and improved practical viability. The Au-Gd-Co_2B/TiO_2 electrode exhibited the lowest Tafel slope at 43 mV dec^{-1} among other synthesized materials such as Gd-Co_2B/TiO_2, Gd-Co_3O_4/TiO_2, and IrO_2/TiO_2. In addition, the Au-Gd-Co_2B/TiO_2 catalyst attained current densities of 100, 500, and 1000 mA cm^{-2} in 1 M KOH by utilizing overpotentials of 260, 300, and 320 mV, respectively. Furthermore, In alkaline and alkaline along with seawater electrolytes, the continuous operation for 200 h and identical polarization curves of the Au-Gd-Co_2B/TiO_2 electrode after 1000 cycles demonstrate the catalyst's superior chemical and mechanical stability, making it highly suitable for large-scale electrolysis of seawater [83]. Other factors such as the branched TiO_2 nanosheets formed in-situ over Ti foil, enhanced electronic conduction in Gd, the relativistic influence of Au on Co oxidation, and increased interfacial link among the Au NCs and Gd-Co_2B hydrophilic surfaces facilitated by microwave treatments also contributed to the catalyst effectiveness [84].

To mitigate the generation of byproducts and the release of toxic and corrosive Cl_2 gas during seawater splitting, incorporating buffer media (borate buffer) is pivotal. This approach effectively addresses these challenges by stabilizing the pH at approximately 8.5. For instance, Shaban et al. synthesized ternary mixed oxides of Co, Ni and Fe on carbon-doped TiO_2 (C-TiO_2), which were fabricated using the spray pyrolysis process on 3-D porous structures of Cu foam and Ni–Fe foam (NFF). Subsequently, the performance of the electrodes were estimated using buffered natural seawater, 1.0 M KOH, and natural seawater as an electrolyte. Adding a borate buffer reduced the potential required to achieve 80 mA cm^{-2} current density to just 968 mV in buffered

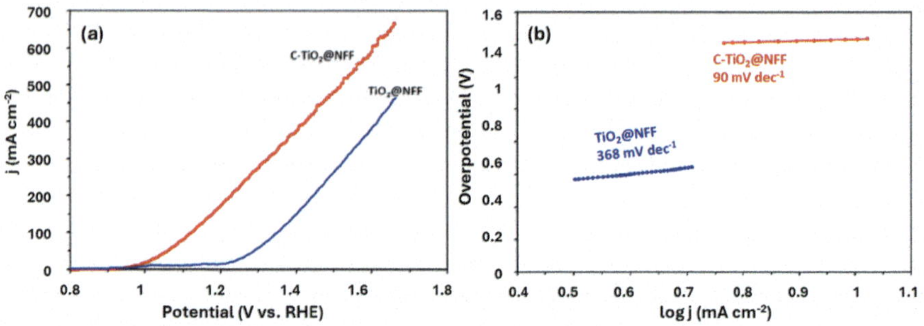

Fig. 7 **a** j-V curves for C-TiO_2@NFF and TiO_2@NFF as working electrode in three-electrode configurations. **b** The equivalent Tafel slopes extracted from the j-V curves. Reprinted with permission from [85], Copyright (2024), Elsevier

seawater as compared to 1.16 V versus RHE in natural seawater. The current density notably increased to 200.8 mA cm^{-2} in buffered seawater from 93.3 mA cm^{-2} in seawater at 1.3 V versus RHE.

Further, an increase in current density in the case of C-TiO_2@NFF (668 mA cm^{-2}) was observed as compared to TiO_2@NFF (463 mA cm^{-2}) at 1.62 V versus RHE and low Tafel slope of 90 mV dec^{-1} was observed for C-TiO_2@NFF electrodes as compared to TiO_2@NFF electrode (368 mV dec^{-1}) as shown in Fig. 7a, b respectively. Particularly, the two-electron processes involved in ClO^- production and ClER have a kinetic advantage over OER. Nonetheless, OER becomes thermodynamically more viable than ClO^- production as the pH goes above 7.5 [85].

5 Conclusion

The urgency to adopt green energy solutions has become increasingly apparent, considering climate change and the depletion of traditional energy sources. With nearly 80% of energy demand met by burning fossil fuels, the detrimental environmental effects are undeniable. H_2 emerges as a promising alternative due to its high energy density, surpassing methane and gasoline.

TiO_2-based catalysts hold immense potential in driving the transition towards green energy solutions, with specific advancements showcasing impressive H_2 and O_2 production and stability. While TiO_2-based catalysts have shown potential as a clean energy source with improved H_2 yields, challenges persist. These include limitations in harnessing sunlight, production costs, stability, and the need for further development to integrate TiO_2-based catalysts into real-world seawater splitting systems.

This chapter discussed various strategies to overcome these challenges and enhance the catalytic activity of TiO_2 under visible light, including semiconductor coupling, composites, and doping. Additionally, it examines the impact of different factors on TiO_2-based photocatalysts and electrocatalysts. Ultimately, the vast information presented suggests a promising future with the potential to improve photocatalyst performance and develop more efficient systems.

References

1. Tao, X., Zhao, Y., Wang, S., et al.: Recent advances and perspectives for solar-driven water splitting using particulate photocatalysts. Chem. Soc. Rev. **51**, 3561–3608 (2022). https://doi.org/10.1039/d1cs01182k
2. Kalair, A., Abas, N., Saleem, M.S., et al.: Role of energy storage systems in energy transition from fossil fuels to renewables. Energy Storage **3** (2021). https://doi.org/10.1002/est2.135
3. Abe, J.O., Popoola, A.P.I., Ajenifuja, E., Popoola, O.M.: Hydrogen energy, economy and storage: review and recommendation. Int. J. Hydrogen Energy **44**, 15072–15086 (2019). https://doi.org/10.1016/j.ijhydene.2019.04.068
4. Lei, Y., Wang, Y., Liu, Y., et al.: Designing atomic active centers for hydrogen evolution electrocatalysts. Angew Chemie Int. Ed **59**, 20794–20812 (2020). https://doi.org/10.1002/anie.201914647
5. Tong, W., Forster, M., Dionigi, F., et al.: Electrolysis of low-grade and saline surface water. Nat. Energy **5**, 367–377 (2020). https://doi.org/10.1038/s41560-020-0550-8
6. Hu, H., Wang, X., Attfield, J.P., Yang, M.: Metal nitrides for seawater electrolysis. Chem. Soc. Rev. **53**, 163–203 (2024). https://doi.org/10.1039/D3CS00717K
7. Fan, R., Liu, C., Li, Z., et al.: Ultrastable electrocatalytic seawater splitting at ampere-level current density. Nat. Sustain. **7**, 158–167 (2024). https://doi.org/10.1038/s41893-023-01263-w
8. Gupta, P., Rathore, V.: Study of TiO_2 material: a photocatalyst for contrary pollutants. Mater. Today Proc. **42**, 1345–1352 (2020). https://doi.org/10.1016/j.matpr.2020.12.1198
9. Garg, P., Prajapati, A., Khare, P., et al.: Electrochemical degradation of metronidazole-containing wastewater using Ni-foam as an electrode. Mater. Today Proc. (2023). https://doi.org/10.1016/j.matpr.2023.01.163
10. Sun, X., Wang, C., Su, D., et al.: Application of photocatalytic materials in sensors. Adv. Mater. Technol. 5 (2020). https://doi.org/10.1002/admt.201900993
11. Kempahanumakkagari, S., Vellingiri, K., Deep, A., et al.: Metal–organic framework composites as electrocatalysts for electrochemical sensing applications. Coord. Chem. Rev. **357**, 105–129 (2018). https://doi.org/10.1016/j.ccr.2017.11.028
12. Han, J., Meng, X., Lu, L., et al.: Single-atom Fe-N x -C as an efficient electrocatalyst for zinc–air batteries. Adv. Funct. Mater. 29 (2019). https://doi.org/10.1002/adfm.201808872
13. Jia, C., Zhang, F., She, L., et al.: Ultra-large sized siloxene nanosheets as bifunctional photocatalyst for a Li-O_2 battery with superior round-trip efficiency and extra-long durability. Angew Chemie **133**, 11357–11361 (2021). https://doi.org/10.1002/ange.202101991
14. Young, J.L., Steiner, M.A., Döscher, H., et al.: Direct solar-to-hydrogen conversion via inverted metamorphic multi junction semiconductor architectures. Nat. Energy **2**, 17028 (2017). https://doi.org/10.1038/nenergy.2017.28
15. Ruan, X., Li, S., Huang, C., et al.: Catalyzing artificial photosynthesis with TiO2 heterostructures and hybrids: emerging trends in a classical yet contemporary photocatalyst. Adv. Mater. 36 (2024). https://doi.org/10.1002/adma.202305285
16. Zhang, J., Hu, W., Cao, S., Piao, L.: Recent progress for hydrogen production by photocatalytic natural or simulated seawater splitting. Nano Res. **13**, 2313–2322 (2020). https://doi.org/10.1007/s12274-020-2880-z
17. Sun, J., Song, P., Zhou, H., et al.: A surface configuration strategy to hierarchical Fe-Co-S/Cu2O/Cu electrodes for oxygen evolution in water/seawater splitting. Appl. Surf. Sci. **567**, 150757 (2021). https://doi.org/10.1016/j.apsusc.2021.150757

18. Zhang, Y.X., Wu, S.M., Tian, G., et al.: Titanium vacancies in TiO2 nanofibers enable highly efficient photodriven seawater splitting. Chem. A Eur. J. **27**, 14202–14208 (2021). https://doi.org/10.1002/chem.202101817
19. Liu, C., Zhang, N., Li, Y., et al.: Long-term durability of metastable β-Fe2O3 photoanodes in highly corrosive seawater. Nat. Commun. **14**, 4266 (2023). https://doi.org/10.1038/s41467-023-40010-9
20. Sahoo, P., Sharma, A., Padhan, S., Thangavel, R.: Visible light driven photosplitting of water using one dimensional Mg doped ZnO nanorod arrays. Int. J. Hydrogen Energy **45**, 22576–22588 (2020). https://doi.org/10.1016/j.ijhydene.2020.06.173
21. Zhu, C., Liu, C., Fu, Y., et al.: Construction of CDs/CdS photocatalysts for stable and efficient hydrogen production in water and seawater. Appl. Catal. B Environ. **242**, 178–185 (2019). https://doi.org/10.1016/j.apcatb.2018.09.096
22. Dabodiya, T.S., George, T., Mathew, F.A., Vadivel Murugan, A.: Ultrastable, high photoelectrocatalytic performance of altervalent cation-doped BiVO4 photoanode and effect of interfacial contact with nanoporous carbon for seawater splitting using a 3D-printed flow device. ACS Appl. Energy Mater. **7**, 2309–2328 (2024). https://doi.org/10.1021/acsaem.3c03068
23. Crossland, E.J.W., Noel, N., Sivaram, V., et al.: Mesoporous TiO2 single crystals delivering enhanced mobility and optoelectronic device performance. Nature **495**, 215–219 (2013). https://doi.org/10.1038/nature11936
24. Lu, Q., Yu, Y., Ma, Q., et al.: 2D transition-metal-dichalcogenide-nanosheet-based composites for photocatalytic and electrocatalytic hydrogen evolution reactions. Adv. Mater. **28**, 1917–1933 (2016). https://doi.org/10.1002/adma.201503270
25. Soundarya, T.L., Harini, R., Manjunath, K., et al.: Pt-doped TiO2 nanotubes as photocatalysts and electrocatalysts for enhanced photocatalytic H2 generation, electrochemical sensing, and supercapacitor applications. Int. J. Hydrogen Energy **48**, 31855–31874 (2023). https://doi.org/10.1016/j.ijhydene.2023.04.289
26. Rafique, M., Hajra, S., Irshad, M., et al.: Hydrogen production using TiO2-based photocatalysts: a comprehensive review. ACS Omega **8**, 25640–25648 (2023). https://doi.org/10.1021/acsomega.3c00963
27. Fazil, M., Alshehri, S.M., Mao, Y., Ahmad, T.: Enhanced photo/electrocatalytic hydrogen evolution by hydrothermally derived Cu-doped TiO2 solid solution nanostructures. Langmuir **40**, 4063–4076 (2024). https://doi.org/10.1021/acs.langmuir.3c02860
28. Zhu, J., Zäch, M.: Nanostructured materials for photocatalytic hydrogen production. Curr. Opin. Colloid Interface Sci. **14**, 260–269 (2009). https://doi.org/10.1016/j.cocis.2009.05.003
29. Chang, J., Yang, Y.: Advancements in seawater electrolysis: progressing from fundamental research to applied electrolyzer application. Renewables **1**, 415–454 (2023). https://doi.org/10.31635/renewables.023.202300034
30. Tong, M.-H., Wang, T.-M., Lin, S.-W., et al.: Ultra-thin carbon doped TiO2 nanotube arrays for enhanced visible-light photoelectrochemical water splitting. Appl. Surf. Sci. **623**, 156980 (2023). https://doi.org/10.1016/j.apsusc.2023.156980
31. Iqbal, T., Riaz, A., Zafar, M.: An overview of photocatalytic water splitting on semiconductor oxides for hydrogen production. Int. J. Environ. Anal. Chem. **103**, 8080–8090 (2023). https://doi.org/10.1080/03067319.2021.1981304
32. Seawater composition. https://www.britannica.com/science/seawater/Dissolved-inorganic-substances
33. Horikoshi, S., Serpone, N.: Can the photocatalyst TiO2 be incorporated into a wastewater treatment method? Background and prospects. Catal. Today **340**, 334–346 (2020). https://doi.org/10.1016/j.cattod.2018.10.020
34. Tran, P.D., Wong, L.H., Barber, J., Loo, J.S.C.: Recent advances in hybrid photocatalysts for solar fuel production. Energy Environ. Sci. **5**, 5902 (2012). https://doi.org/10.1039/c2ee02849b
35. Ray, S.S., Gusain, R., Kumar, N.: Water purification using various technologies and their advantages and disadvantages. In: Carbon Nanomaterial-Based Adsorbents for Water Purification. Elsevier, pp. 37–66 (2020)

36. Sen, R., Das, S., Nath, A., et al.: Electrocatalytic water oxidation: an overview with an example of translation from lab to market. Front. Chem. **10** (2022). https://doi.org/10.3389/fchem.2022.861604
37. Gao, G., Sun, Z., Chen, X., et al.: Recent advances in Ru/Ir-based electrocatalysts for acidic oxygen evolution reaction. Appl. Catal. B Environ. **343**, 123584 (2024). https://doi.org/10.1016/j.apcatb.2023.123584
38. Roger, I., Shipman, M.A., Symes, M.D.: Earth-abundant catalysts for electrochemical and photoelectrochemical water splitting. Nat. Rev. Chem. **1**, 0003 (2017). https://doi.org/10.1038/s41570-016-0003
39. Zhang, F., Yu, L., Wu, L., et al.: Rational design of oxygen evolution reaction catalysts for seawater electrolysis. Trends Chem. **3**, 485–498 (2021). https://doi.org/10.1016/j.trechm.2021.03.003
40. Khatun, S., Hirani, H., Roy, P.: Seawater electrocatalysis: activity and selectivity. J. Mater. Chem. A **9**, 74–86 (2021). https://doi.org/10.1039/d0ta08709b
41. Ursúa, A., Gandía, L.M., Sanchis, P.: Hydrogen production from water electrolysis: current status and future trends. Proc. IEEE **100**, 410–426 (2012). https://doi.org/10.1109/JPROC.2011.2156750
42. Ebbesen, S.D., Jensen, S.H., Hauch, A., Mogensen, M.B.: High temperature electrolysis in alkaline cells, solid proton conducting cells, and solid oxide cells. Chem. Rev. **114**, 10697–10734 (2014). https://doi.org/10.1021/cr5000865
43. Sohail, M., Rauf, S., Irfan, M., et al.: Recent developments, advances and strategies in heterogeneous photocatalysts for water splitting. Nanoscale Adv. **6**, 1286–1330 (2024). https://doi.org/10.1039/d3na00442b
44. Chen, X., Shen, S., Guo, L., Mao, S.S.: Semiconductor-based photocatalytic hydrogen generation. Chem. Rev. **110**, 6503–6570 (2010). https://doi.org/10.1021/cr1001645
45. Pan, Q., Zhao, P., Gao, L., et al.: In-depth study on the effects of impurity ions in saline wastewater electrolysis. Molecules **28**, 4576 (2023). https://doi.org/10.3390/molecules28124576
46. Simamora, A.J., Chang, F.C., Wang, H.P., et al.: H2 fuels from photocatalytic splitting of seawater affected by nano-TiO2 promoted with CuO and NiO. Int. J. Photoenergy **2013**, 1–5 (2013). https://doi.org/10.1155/2013/419182
47. Cheng, Z., Zhang, X., Bo, C., et al.: Precise design of TiO2 photocatalyst for efficient photocatalytic H2 production from seawater splitting. Int. J. Hydrogen Energy **55**, 542–549 (2024). https://doi.org/10.1016/j.ijhydene.2023.11.211
48. Nam, W., Oh, S., Joo, H., Yoon, J.: Preparation of Pt deposited nanotubular TiO2 as cathodes for enhanced photoelectrochemical hydrogen production using seawater electrolytes. J. Solid State Chem. **184**, 2920–2924 (2011). https://doi.org/10.1016/j.jssc.2011.08.041
49. Li, Y., Wang, R., Li, H., et al.: Efficient and stable photoelectrochemical seawater splitting with TiO2@g-C3N4 nanorod arrays decorated by Co-Pi. J. Phys. Chem. C **119**, 20283–20292 (2015). https://doi.org/10.1021/acs.jpcc.5b05427
50. Li, F., Dong, B., Feng, S.: Bi shell-BiOI core microspheres modified TiO2 nanotube arrays photoanode: improved effect of Bi shell on photoelectrochemical hydrogen evolution in seawater. Int. J. Hydrogen Energy **44**, 29986–29999 (2019). https://doi.org/10.1016/j.ijhydene.2019.09.210
51. He, Y., Li, P., Liu, W., et al.: The critical factors of photocatalytic H2 production from seawater by using TiO2 as photocatalyst. Nano Res. **16**, 4620–4624 (2023). https://doi.org/10.1007/s12274-022-5078-8
52. Li, Y., Zhou, H., Cai, S., et al.: Electrolyte-assisted polarization leading to enhanced charge separation and solar-to-hydrogen conversion efficiency of seawater splitting. Nat. Catal. (2024). https://doi.org/10.1038/s41929-023-01069-1
53. Xiao, S.T., Wu, S.M., Wu, L., et al.: Confined heterojunction in hollow-structured TiO2 and its directed effect in photodriven seawater splitting. ACS Nano **17**, 18217–18226 (2023). https://doi.org/10.1021/acsnano.3c05174

54. Zhang, J., Lei, Y., Cao, S., et al.: Photocatalytic hydrogen production from seawater under full solar spectrum without sacrificial reagents using TiO2 nanoparticles. Nano Res. **15**, 2013–2022 (2022). https://doi.org/10.1007/s12274-021-3982-y
55. Simamora, A.J., Hsiung, T.L., Chang, F.C., et al.: Photocatalytic splitting of seawater and degradation of methylene blue on CuO/nano TiO2. Int. J. Hydrogen Energy **37**, 13855–13858 (2012). https://doi.org/10.1016/j.ijhydene.2012.04.091
56. Gao, M., Connor, P.K.N., Ho, G.W.: Plasmonic photothermic directed broadband sunlight harnessing for seawater catalysis and desalination. Energy Environ. Sci. **9**, 3151–3160 (2016). https://doi.org/10.1039/c6ee00971a
57. Xiong, G., Shen, S., Xiao, S., et al.: Well-defined Z-scheme Na2Ti3O7/Ag/CdS multidimensional heterojunctions with enhanced H2 production from seawater under visible light. Int. J. Hydrogen Energy **47**, 30503–30516 (2022). https://doi.org/10.1016/j.ijhydene.2022.07.031
58. Sakurai, H., Kiuchi, M., Jin, T.: Pt/TiO2 granular photocatalysts for hydrogen production from aqueous glycerol solution: durability against seawater constituents and dissolved oxygen. Catal. Commun. **114**, 124–128 (2018). https://doi.org/10.1016/j.catcom.2018.06.013
59. Ji, S.M., Jun, H., Jang, J.S., et al.: Photocatalytic hydrogen production from natural seawater. J. Photochem. Photobiol. A Chem. **189**, 141–144 (2007). https://doi.org/10.1016/j.jphotochem.2007.01.011
60. Yang, Y., Liu, L., Qi, Q., et al.: A low-cost and stable Fe2O3/C-TiO2 system design for highly efficient photocatalytic H2 production from seawater. Catal. Commun. **143**, 106047 (2020). https://doi.org/10.1016/j.catcom.2020.106047
61. Van, D.H., Wang, Y.H., Wu, J.C.S.: Z-scheme photocatalyst Pt/GaP-TiO2-SiO2: Rh for the separated H2 evolution from photocatalytic seawater splitting. Appl. Catal. B Environ. **296**, 120339 (2021). https://doi.org/10.1016/j.apcatb.2021.120339
62. Lv, S., Wang, Y., Zhou, Y., et al.: Oxygen vacancy stimulated direct Z-scheme of mesoporous Cu2O/TiO2 for enhanced photocatalytic hydrogen production from water and seawater. J. Alloys Compd. **868**, 159144 (2021). https://doi.org/10.1016/j.jallcom.2021.159144
63. Xia, L., Lu, Y., Li, Y.Z., et al.: TiO2-rGO-Cu complex: a photocatalyst possessing an interfacial electron transport mechanism to enhance hydrogen production from seawater. Chem. Phys. Lett. **822**, 140498 (2023). https://doi.org/10.1016/j.cplett.2023.140498
64. Sampath, S., Sellappa, K.: Visible-light-driven photocatalysts for hydrogen production by water splitting. Energy Sourc. Part A Recover Util. Environ. Eff. **42**, 719–729 (2020). https://doi.org/10.1080/15567036.2019.1602194
65. Guo, L., Yang, Z., Marcus, K., et al.: MoS2/TiO2 heterostructures as nonmetal plasmonic photocatalysts for highly efficient hydrogen evolution. Energy Environ. Sci. **11**, 106–114 (2018). https://doi.org/10.1039/C7EE02464A
66. Deepanprakash, D., Premnath, V., Raghu, C., et al.: Harnessing power from sea water using nano material as photocatalyst and solar energy as light source: the role of hydrocarbon as dual agent. Int. J. Energy Res. **38**, 249–253 (2014). https://doi.org/10.1002/er.3017
67. Lin, S., Huang, H., Ma, T., Zhang, Y.: Photocatalytic oxygen evolution from water splitting. Adv. Sci. 8 (2021). https://doi.org/10.1002/advs.202002458
68. Fu, C., Weng, S., Fan, J., et al.: Boron-based materials modified on the surface of TiO2 nanorods via electroless plating toward high-efficient solar-driven water splitting. Chem. Eng. J. **430**, 132881 (2022). https://doi.org/10.1016/j.cej.2021.132881
69. Zhao, S., Liu, B., Li, K., et al.: A silicon photoanode protected with TiO2/stainless steel bilayer stack for solar seawater splitting. Nat. Commun. 15 (2024). https://doi.org/10.1038/s41467-024-47389-z
70. Hu, W., Chen, S., Xia, Q.: IrO2/Nb–TiO2 electrocatalyst for oxygen evolution reaction in acidic medium. Int. J. Hydrogen Energy **39**, 6967–6976 (2014). https://doi.org/10.1016/j.ijhydene.2014.02.114
71. Nie, M., Sun, H., Liao, J., et al.: Study on the catalytic performance of Pd/TiO2 electrocatalyst for hydrogen evolution reaction. Int. J. Hydrogen Energy **46**, 6441–6447 (2021). https://doi.org/10.1016/j.ijhydene.2020.11.135

72. Fang, F., Wang, Y., Shen, L., et al.: Interfacial carbon makes nano-particulate RuO2 an efficient, stable, ph-universal catalyst for splitting of seawater. Small **18** (2022). https://doi.org/10.1002/smll.202203778
73. Pan, S., Li, R., Wang, J., et al.: Floating seawater splitting device based on NiFeCrMo metal hydroxide electrocatalyst and perovskite/silicon tandem solar cells. ACS Nano **17**, 4539–4550 (2023). https://doi.org/10.1021/acsnano.2c10477
74. Das, R., Ambardekar, V.B.P.: Titanium dioxide—advances and applications. In: Muhammad Ali, H. (ed.) Titanium Dioxide—Advances and Applications. IntechOpen (2021)
75. Nandedkar, D.: Analysis of conductivity of noble metals near room temperature. Recent Adv. Prop. 'Physics/Electrical Eng. Mater. **1**, 2–24 (2022). https://doi.org/10.9734/bpi/mono/978-93-5547-658-6/ch1
76. Earle, M.D.: The electrical conductivity of titanium dioxide. Phys. Rev. **61**, 56–62 (1942). https://doi.org/10.1103/PhysRev.61.56
77. Danilov, F.I., Tsurkan, A.V., Vasil'Eva, E.A., Protsenko, V.S.: Electrocatalytic activity of composite Fe/TiO2 electrodeposits for hydrogen evolution reaction in alkaline solutions. Int. J. Hydrogen Energy **41**, 7363–7372 (2016). https://doi.org/10.1016/j.ijhydene.2016.02.112
78. Chen, K., Deng, S., Lu, Y., et al.: Molybdenum-doped titanium dioxide supported low-Pt electrocatalyst for highly efficient and stable hydrogen evolution reaction. Chinese Chem. Lett. **32**, 765–769 (2021). https://doi.org/10.1016/j.cclet.2020.05.030
79. Cheshideh, H., Chen, H.Y., Liao, K.W., et al.: Reactive surface intermediates over Ni-grafted TiO2 nanotube arrays towards hydrogen evolution reaction in alkaline and chloride media. Int. J. Hydrogen Energy **48**, 31479–31490 (2023). https://doi.org/10.1016/j.ijhydene.2023.04.305
80. Wang, X., Zhao, Y., Mølhave, K., Sun, H.: Engineering the surface/interface structures of titanium dioxide micro and nano architectures towards environmental and electrochemical applications. Nanomaterials **7**, 382 (2017). https://doi.org/10.3390/nano7110382
81. Paul, S., Rahman, M.A., Bin, S.S., et al.: TiO2 as an anode of high-performance lithium-ion batteries: a comprehensive review towards practical application. Nanomaterials **12**, 2034 (2022). https://doi.org/10.3390/nano12122034
82. Sun, C., Huang, G., Xu, C., et al.: Improved compatibility of aqueous electrolyte with TiO2(B) toward high-voltage aqueous rechargeable Mg-ion batteries. Energy Storage Mater. **66**, 103197 (2024). https://doi.org/10.1016/j.ensm.2024.103197
83. Haq, T.U., Haik, Y., Hussain, I., et al.: Gd-doped Ni-oxychloride nanoclusters: new nanoscale electrocatalysts for high-performance water oxidation through surface and structural modification. ACS Appl. Mater. Interfaces **13**, 468–479 (2021). https://doi.org/10.1021/acsami.0c17216
84. Haq, T., Pasha, M., Tong, Y., et al.: Au nanocluster coupling with Gd-Co2B nanoflakes embedded in reduced TiO2 nanosheets: seawater electrolysis at low cell voltage with high selectivity and corrosion resistance. Appl. Catal. B Environ. **301**, 120836 (2022). https://doi.org/10.1016/j.apcatb.2021.120836
85. Shaban, Y.A., Khan, S.U.M., Salam, M.A., et al.: Synergistic 3D architecture of Fe Co Ni@CF/C-TiO2@NFF for high-performance electrochemical seawater splitting for hydrogen generation. J. Taiwan Inst. Chem. Eng. **162**, 105578 (2024). https://doi.org/10.1016/j.jtice.2024.105578

Biomedical Applications of TiO_2-based Nanomaterials

Suresh Babu Naidu Krishna

Abstract Titanium dioxide (TiO_2) has been utilized for decades in a variety of applications and is thought to be a safe and harmless substance. TiO_2-based nanomaterials have emerged as promising candidates for various biomedical applications due to their unique properties, including biocompatibility, photocatalytic activity, and large surface area. These nanomaterials have been extensively studied for their potential use in drug delivery, tissue engineering, cancer therapy, and antibacterial applications. In this chapter, we will discuss the recent advances in the development of TiO_2-based nanomaterials for biomedical applications and highlight their potential impact on healthcare. Furthermore, the study will also encompass the examination of composite nanostructures containing titanium dioxide, with a focus on their surface and bulk characteristics, to analyze the existing difficulties and potential opportunities.

Keywords TiO_2-based nanomaterials · Biomedical applications · Drug delivery · Tissue engineering · Cancer therapy · Antibacterial applications

1 Introduction

Nanobiotechnology is one of the most recent and rising branches of exploration. It is the interface between biology and nanotechnology, and it involves both the study and the development of unique nano biosystems that are functionalized [1]. There is a significant amount of opportunity for this multidisciplinary study field to contribute to the development of enhanced healthcare technology.

William Gregor is recognized with the initial discovery of the element titanium in 1791. Titanium dioxide (TiO_2) is a white substance that has limited solubility. It is

S. B. N. Krishna (✉)
Faculty of Health Sciences, Durban University of Technology, Durban 4000, South Africa
e-mail: sureshk@dut.ac.za

School of Life Sciences, University of KwaZulu-Natal, Westville Campus, Durban 4000, South Africa

J. Prakash et al. (eds.), *Titanium Dioxide-Based Multifunctional Hybrid Nanomaterials*, Engineering Materials, https://doi.org/10.1007/978-3-031-80625-4_12

widely used in various biomedical applications, particularly in the sectors of personal care products, healthcare, and pharmaceuticals [2, 3]. Titania, also known as Titanium dioxide (TiO_2), is an inorganic compound that is commercially significant and has a diverse variety of uses. According to Ziental et al. [4], it has enhanced performance in the field of photocatalytic applications for eliminating different organic pollutants from air and water. Additionally, it is extensively employed as a dielectric and pigment in paints. Furthermore, it is considered a feasible choice to produce photoelectrochemical energy. Titania is utilized in biomedical sciences, specifically in the fields of bone tissue engineering and medicines, due to its biocompatible and nontoxic properties [5, 6]. Biomedical applications for these intriguing TiO_2 nanoparticles can be divided into four categories: biosensing [7], drug transport, antibacterial activity [8], and implant uses. The potential use of titanium dioxide (TiO_2) in electrochemical and photoelectrochemical (PEC) biosensors for the detection of a wide range of analytes has been the subject of extensive research in recent years [9]. PEC biosensors are believed to be a very sensitive approach that has low background signals and good analytical performance. This is because they possess the two ways for excitation, which are light and current, respectively when it comes to detection. On the other hand, this approach has a few drawbacks or shortcomings. Despite having a broad bandgap (3.2 eV for anatase and 3.0 eV for rutile phase), titanium dioxide (TiO_2) only absorbs a modest percentage of the ultraviolet area of the solar spectrum [10].

Nanostructured TiO_2 seems to be a harmless and inert substance when it encounters humans, as stated in the literature. An in-depth comprehension of nanoscale phenomena will facilitate the development of innovative biomedical devices and the construction of antibacterial surfaces, implant devices, and more. The goal of this chapter is to explicitly investigate the recent advancements that have been made in the biomedical application of TiO_2 nanoparticles. We will focus on the nanotoxicity of these particles as well as the critical criteria that determine their biocompatibility. In addition to this, it covers the issues that will arise in these areas in the future.

2 Overview of Structure, Synthesis, Stabilization of Titanium Dioxide Nanoparticles for Biomedical Applications

TiO_2 is classified as a member of the transition metal oxides group. In nature, there exist four well recognized polymorphs of titanium dioxide (TiO_2), namely anatase (tetragonal), brookite (orthorhombic), rutile (tetragonal), and TiO_2 (B) (monoclinic). Anatase and rutile are the two most prevalent tetragonal crystallographic polymorphs (Fig. 1) of titanium dioxide, and they get their names from two of the most numerous minerals in the world. Brookite is the specimen that possesses the third and more uncommon orthorhombic crystal structure [11, 12].

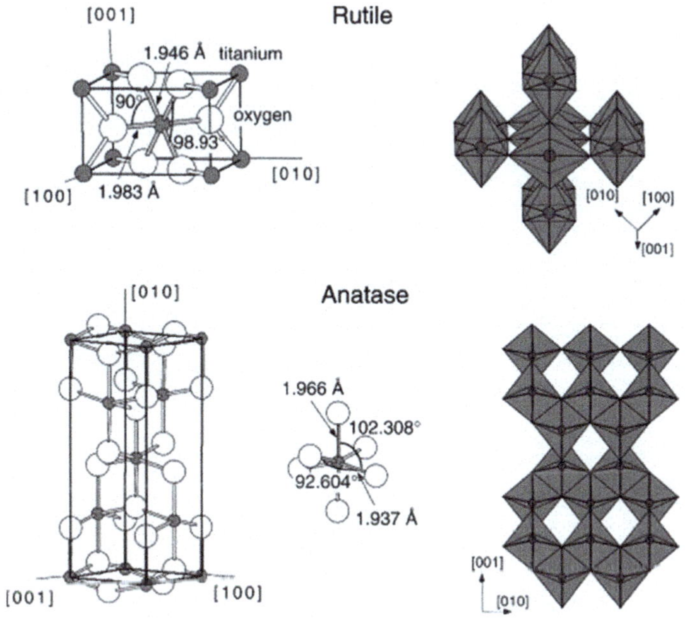

Fig. 1 Chemical structure of rutile and anatase [12]. Copyright (2003), Elsevier

Specifically, the top-down technique and the bottom-up approach are two fundamental approaches that can be utilized in the synthesis of TiO_2 nanostructures. Each of these approaches has its own set of characteristics and uses that are distinct from the other.

I. The top-down approach is a method that involves the reduction in size of larger structures to build smaller ones. Using this technique, smaller structures are constructed out of larger ones. The fabrication of integrated circuits is a good illustration of the top-down technique. In this process, the circuitry is produced layer by layer, with each successive step resulting in a smaller size. When this procedure is finished, there is no need for any additional action to be taken [13].
II. Bottom-up technique: On the other hand, the bottom-up technique involves the construction of intricate nanostructures through the methodical assembly of molecular precursors. The use of this technique makes it possible to create nanostructures that are both detailed and clearly defined. The physical, chemical, hybrid, and biological approaches are the four primary sub-divisions that fall under the umbrella of the bottom-up methodology. According to Wan et al. [14], these techniques are intended to offer a methodical and regulated approach to the formation of nanostructures.
III. Physical methods: The bottom-up technique is a physical method that relies on physical processes to build nanostructures. These processes include deposition and self-assembly.

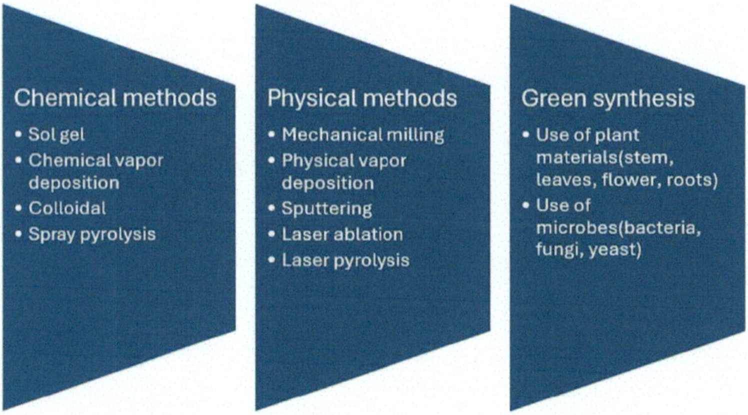

Fig. 2 Illustration of different synthesis methods of TiO_2 nanoparticles

IV. Chemical methods: Chemical methods (Fig. 2) involve the utilization of chemical reactions to construct nanostructures. Sol–gel synthesis and chemical vapour deposition are two examples of the processes that fall within this category.

V. Hybrid methods: Hybrid methods utilize both physical and chemical processes to produce nanostructures. This allows for improved control and qualities to be achieved in many cases.

VI. Green methods: Biological approaches also known as "green approach" make use of biological things, such as proteins or organisms, to aid the controlled production of nanostructures [15].

Most of the titanium(IV) oxide is created by the purification of rutile mineral or through the application of ilmenite ($FeTiO_3$) to either the chloride or sulfate processes, both of which ultimately result in the production of pure titania when completed. The amorphous TiO_2 may undergo a transformation into anatase or brookite by a process known as calcination, which takes place at approximately 400 degrees Celsius [16–18]. This transformation might take place when the thermal treatment is given. However, when heated to temperatures that are higher than 600 degrees Celsius, these polymorphs undergo a transformation into rutile.

Several investigations have demonstrated that titanium dioxide (TiO_2) has significantly higher efficacy as a photocatalyst when present in the form of nanoparticles compared to its bulk powder form [19]. It is necessary to make use of synthetically produced TiO_2 nanoparticles to make the shift from two-dimensional to three-dimensional nanostructures. A plethora of one-of-a-kind characteristics and potential uses present themselves when the shape and size of TiO_2 nanoparticles are precisely controlled [8]. This is especially true in the field of biomedical sciences. TiO_2 nanoparticles have the unique capacity to transition effortlessly from dimensions ranging from nanometers down to the molecular scale, and from spherical forms to various morphologies such as nanorods, nanotubes, and nanowires. This incredible

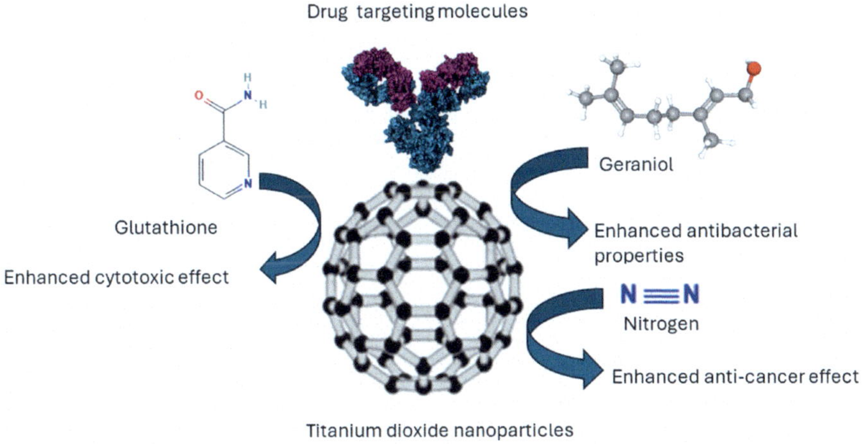

Fig. 3 Surface modification of titanium dioxide nanoparticles for biomedical applications

ability is made possible by the precise manipulation of size and shape [20]. This level of control over the morphological properties of TiO_2 nanoparticles is particularly fascinating since it shows promise for a wide variety of applications in the fields of biomedical and technology application [7].

Numerous different approaches are utilized in the production of titanium dioxide (TiO_2) in general, as well as titania nanoparticles (NPs). The sol–gel synthesis and hydrothermal methods are the most used techniques [21, 22]. Green chemistry and microwave methods are becoming increasingly popular [23, 24]. With careful planning and modification of the process parameters (i.e., substrates used, ratio of solvents, temperature, and process time), it is possible to obtain the desired materials with varying specific physicochemical properties. These properties include surface area, morphology and form, NP size and uniformity in size distribution, crystallinity, and crystal phase, photoactivity, and a great deal of other properties. Further modifications to these features can be made during the synthesis of NPs through the incorporation of a variety of surfactants or dopants, or through alterations that occur after the synthesis has been completed, such as doping, surface functionalization, or interaction with organic molecules (Fig. 3).

3 Applications

With its fascinating potential, nanotechnology empowers scientists to manipulate and harness materials at the nanometer scale [25]. Biotechnology has played a crucial role in bridging the gap between biological systems and industry. In addition, the unique physical and chemical characteristics of nanomaterials have led to the application of nanotechnology in various healthcare fields [7, 26]. Exploring the cutting-edge realm of nanobiotechnology, this field delves into the fascinating intersection of biology and nanotechnology. It involves the study and development of innovative nano biosystems with enhanced functionality. The multidisciplinary research has immense potential in advancing medical engineering.

3.1 Antimicrobial Activity of TiO_2 Nanoparticles

Using nanotechnology, several nanostructures that possess distinctive characteristics have been manufactured for use in biotechnological applications. Among these nanostructures, nano-sized titanium dioxide (TiO_2) has been utilized extensively [16]. Unique aspects of nanoparticles, such as their morphology, size, chemistry, and nanostructure, have a significant impact on the antimicrobial activity of these extremely small particles. To be more specific, the antibacterial activity of titanium dioxide nanoparticles is based on the photocatalytic performance of TiO_2, which is dependent on the morphological, structural, and textural features of the material [10, 17].

Multiple studies demonstrate the inhibitory effect of TiO_2 by its photocatalytic activity [18], resulting in the generation of reactive oxygen species (ROS) such as O_2, OH, and H_2O_2. Reactive oxygen species (ROS) are generated through redox reactions that take place between an adsorbent, such as water or oxygen, and electrons of titanium dioxide (TiO_2) when exposed to ultraviolet (UV) radiation with a wavelength below 385 nm [27]. The schematic diagram of bacterial killing by TiO_2 is depicted in Fig. 4. One significant limitation associated with TiO_2 is its reliance on UV light irradiation for the successful activation of the photocatalyst. Nevertheless, current research has effectively addressed this issue by the utilization of visible light-activated photocatalysts, such as $Ag/AgBr/TiO_2$, as well as doped-TiO_2 nanostructures [28].

The study conducted by Lalabadi et al. examined the antibacterial properties of TiO_2 nanoparticles, zinc oxide (ZnO), and a combination of TiO_2/ZnO nanoparticles supported by 4A zeolite (4A z). The diameter of 4A z particles ranged from 400 to 600 nm, while the average size of the nanoparticles' crystals was roughly 50 nm. The XRD, SEM, MAP, and TEM investigation revealed that the crystallographic plate, size, shape, and morphology of the nanocomposite closely resemble those of 4A z nanoparticles. Furthermore, the antimicrobial test by disc diffusion method (Fig. 5) have indicated a significant antibacterial activity against both gram-positive as well gram-negative bacteria of the nanoparticles supported in 4A z. The antibacterial effectiveness of nanoparticles incorporated into 4A z was contingent upon the specific type of nanoparticles and the classification of the microorganism as either gram positive or gram negative. *Pseudomonas fluorescens* and *Escherichia coli* O157:H7 had the highest level of sensitivity among the microorganisms tested. The incorporation of nanoparticles into 4A z enables the regulation of their release and boosts their antibacterial efficacy. Therefore, utilizing these composites can be a

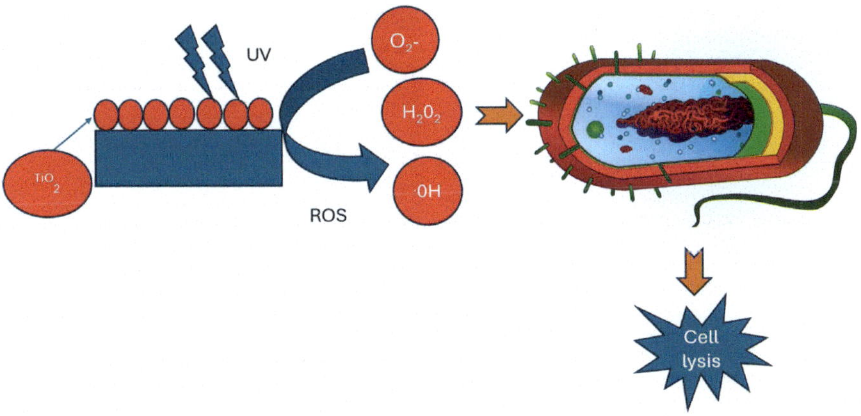

Fig. 4 Illustration of bacterial lysis by ROS response

favourable strategy for developing innovative active packaging in the food business [29].

The rise of antibiotic resistance in pathogenic bacteria has become a global risk, prompting the development of efficient and effective alternatives to traditional antibiotics and advocating for their usage as substitutes. Utilizing natural chemicals can significantly improve the inherent physical and chemical characteristics of NPs [30, 31]. Recently, polyamide 12 powder has incorporated TiO_2 as an antibacterial component. Furthermore, nitrogen-doped titanium dioxide nanoparticles have shown antimicrobial capabilities when exposed to visible light [32, 33]. The TiO_2 nanoparticles exhibit enhanced compatibility with polymer materials due to their tiny size, high surface energy, and strong hydrophilicity resulting from surface hydroxylation [34]. The integration of inorganic nanoparticles into organic polymeric nanocomposites is a promising approach to improve their antibacterial characteristics [35]. For instance, the combination of monoterpenes with TiO_2 resulted in a more pronounced antibacterial impact [36]. Monoterpenes are organic molecules found in nature that contain biologically active substances including phenols, flavonoids, and terpenoids. Geraniol (GER) is a specific kind of terpenoid compound that has minimal toxicity and is very soluble [8]. When it is coupled with TiO_2 nanoparticles, it may show improved antibacterial characteristics. The findings indicate that TiO_2 nanoparticles with germicidal efficacy rate (GER) present a favourable substitute for antibiotics, specifically in the management of antibiotic-resistant strains. Geraniol-induced surface alteration of TiO_2 nanoparticles led to increased antibacterial efficacy against various bacterial strains, including methicillin-resistant Staphylococcus aureus (MRSA) [37]. Further investigation is required to acquire a comprehensive understanding of the possible applications of modified TiO_2 nanoparticles in the fields of biomedicine and the environment.

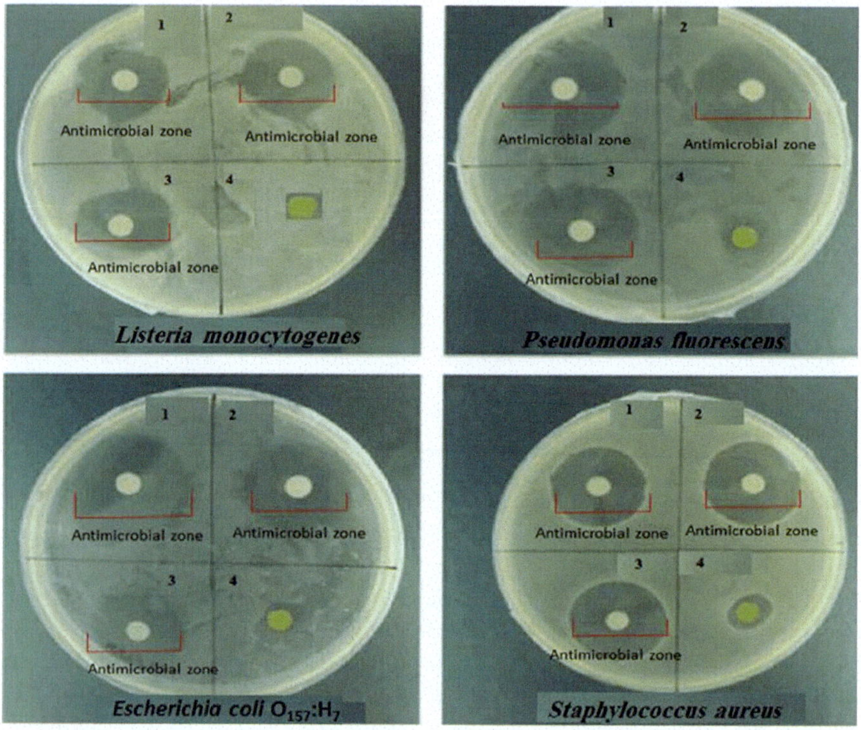

Fig. 5 Antibacterial activity of nanoparticles doped in 4A z. Reprinted with permission from [29]. Copyright (2019), Springer

3.2 TiO_2 Based Chemotherapeutics for Cancer Therapy

In cancer treatment, light is now being utilized to activate photosensitizers (PSs) in photodynamic therapy (PDT) and to trigger the release of drugs from delivery systems [38]. This approach is now being more commonly used in clinical practice alongside radiotherapy and chemotherapy. Light is a highly desirable external stimulus for drug delivery and activation of PS due to its ability to offer precise, temporal, and spatial control over the activation and/or release of therapeutic agents. With precision, this method can be used in a localized manner, avoiding any diffusion effects that would occur with thermal or chemical stimuli. It presents simple transitions between 'on' and 'off' states and allows for fine-tuning of the desired response based on the wavelength and intensity of light [39].

As potential photoactive drug delivery carriers, TiO_2 nanostructures are garnering a lot of attention presently. TiO_2-based nanostructures, in addition to their photoactivity, possess a wide range of beneficial qualities, such as a high surface area, stability, availability, and the possibility of surface modification. These properties make them an ideal carrier for the attachment of a variety of pharmaceuticals. Recently, in 2014,

Wang et al. [40] reported the discovery of a new multifunctional porous TiO_2 nanoparticles loaded with paclitaxel. These nanoparticles have the capacity to activate drug release using ultraviolet (UV) light. As a cancer-targeting agent, they introduced polyethyleneimine (PEI) into TiO_2 nanoparticles and changed them. The scientists believed this new breakthrough could potentially provide more information regarding the treatment of skin cancer and light-controlled medication release.

In a recent study by Maja Nesic et al. [38] metalloid- drug delivery system (NCS), the use of colloidal TiO_2 nanoparticles as a delivery system to deliver the ruthenium complex to the melanoma cell line in a regulated manner. The experimental setup demonstrated a reduced rate of complex release when subjected to visible light, but an increased rate of release when exposed to UV light (Fig. 6). Consequently, the system's ability to cause cancer in amelanotic melanoma cells was shown to be dependent on light. The presence of the nanocomposite system enhances cell death when exposed to UV light and reduces it when exposed to red light. Both elements of the system have the potential to function as photosensitizers through the production of reactive oxygen species, hence facilitating cellular apoptosis (Fig. 7). Hence, the mechanism exhibits dual functionality, serving as both a photodynamic therapeutic agent and a light-tunable system for metalloid-drug administration. Conse quently, it holds potential for the advancement of novel and enhanced drug delivery methodologies through the utilization of light as a stimulant from the outside.

It has been established that TiO_2, is a good choice for the fabrication of effective biosensors. A considerable degree of chemical inertness is exhibited by it, and it demonstrates resistance to heat. Additionally, it is a chemical that is biocompatible and provides a high level of electrical conductivity. Titania has been utilized in the production of biosensors in a variety of forms, including nanosheets (2D), nanotubes (1D), nanoparticles (OD), and porous gel matrices. Biosensors that are based on titanium dioxide as their constituent can identify a variety of biomarkers and measure the concentration of those biomarkers [41, 42]. For TiO_2-based biosensors to be effective, the shape of the sensor that has been developed is of utmost importance. Having a geometry that is appropriate results in a better incorporation of the biomolecules within the network of the sensor, which in turn makes it easier for TiO_2 molecules to interact with biomarkers in vitro (Table 1) [43].

Titanium has several applications, including sensing applications, and it can also be utilized in cell capture methods [43]. Zhang et al. [44] achieved a successful diagnosis of cancer through the capture of malignant cells through the utilization of titania nanofiber that was manufactured through electrospinning. Due to the fact that the cells were captured, cancer drugs were able to start therapy and diagnostic procedures early [45]

TiO_2 platforms, which include label-free microfluidic immunosensors, photoelectrochemical biosensors, field-effect transistors, and amperometrics, have been utilized to make significant advancements in the detection of small biomolecules in clinical blood samples, as well as the detection of circulating cancer cells and pathogens [46, 47]. When compared to other nanoplatforms, the band gap modification of TiO_2 nanostructures using a straightforward doping or deposition procedure is very possible and efficient [48, 49]. This is done with the intention of improving

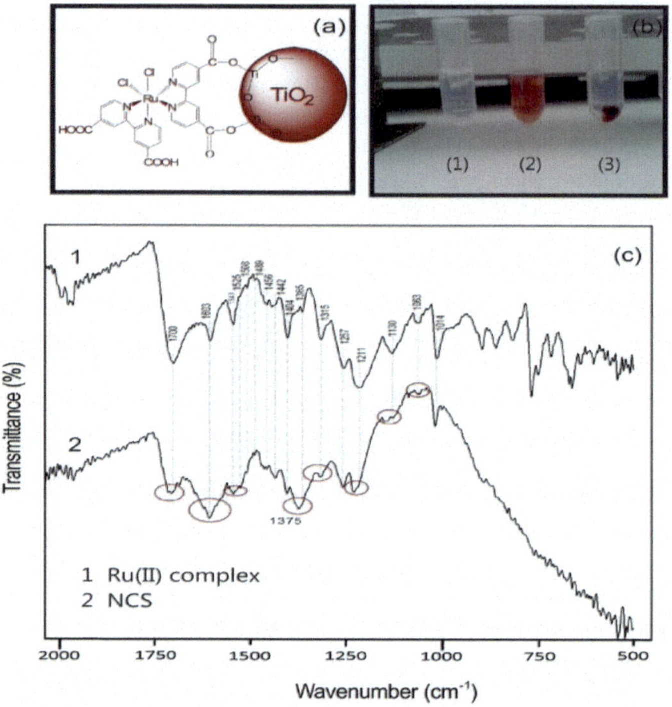

Fig. 6 The diagram illustrates the link that is established between the Ru-complex and the surface of TiO_2 nanoparticles. A photograph depicting the test tubes containing NPs (1), complex (2), and NCS (3) after the process of centrifugation. **c** The Fourier transform infrared (FTIR) of the complex and NCS [38]. Copyright (2017). Elsevier

the detection performance. Because of this, it is extremely expected that reusable and high-precision biosensors will be introduced to the market with the assistance of improved and cost-effective TiO_2 nanoparticles. These nanomaterials have a broad band gap and a high surface activity [50]. As a result, we predict the emergence of reusable, high-precision biosensors in the market, leveraging cost-effective TiO_2 nanoparticles. These nanomaterials, characterized by a broad band gap and high surface activity, hold great promise for advancing diagnostic technologies.

3.3 *Titanium Dioxide-Based Biomaterials in Orthopaedic Applications*

TiO_2 nanostructures have been effectively established into medical applications as implants due to the nanotopographical features that differentiate them from steel implants in medical applications [60]. This has allowed them to cover some of the

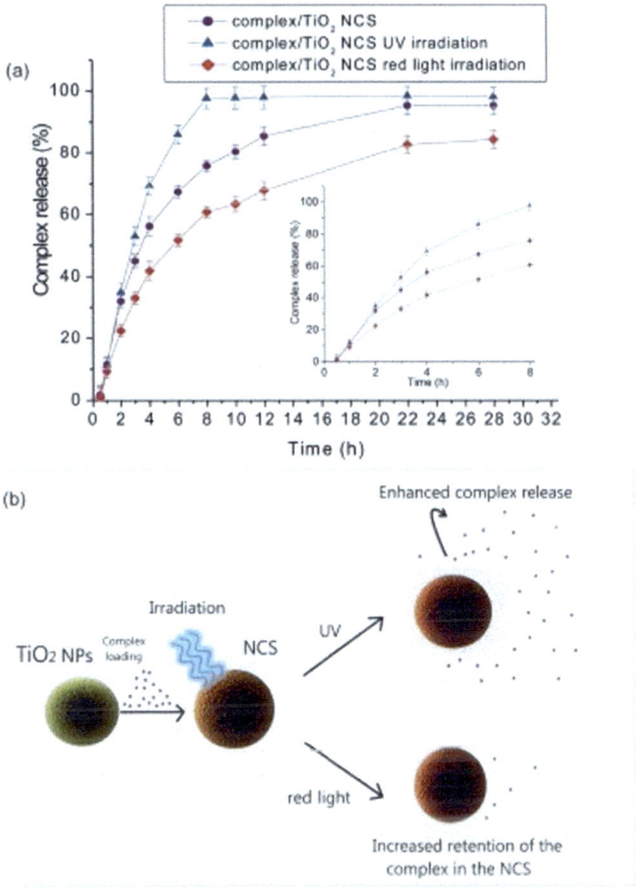

Fig. 7 The complex release profiles of NCS in the dark versus NCS treated with continuous UV and red light [29], respectively (**a**). The inset shows the magnified first half of the graph (up to 8 h of complex release). Process of complex loading onto TiO_2 nanoparticles (NPs) and subsequent irradiation (**b**). The following sketch provides insight into the UV and red light-tunable release rate of the complex from the particle surface [38]. Copyright (2017). Elsevier

shortcomings that other implants have. Considering the breadth of the chapter, this section will focus on the utilization of TiO_2 nanostructures that are valuable in the fields of bone and dentistry.

Bone is a naturally occurring and highly organized organic–inorganic composite material composed of collagen fibrils that have been strengthened with hydroxyapatite (HA) nanocrystals that are evenly distributed. Bone is a unique human tissue that possesses an inorganic phase that serves the purpose of mechanical reinforcing [61].

The skeletal system plays a crucial role in providing support, facilitating mobility, and safeguarding the interior organs. During physical activities such as ambulation

Table 1 TiO_2 nanomaterial-based cancer therapy

Nanomaterial	Cell line	Cancer type	Therapeutic application	Reference
CPE/TiO_2 and Au nanocomposite + chitosan	Carcinoembryonic antigen	Multiple	Early cancer diagnosis	[51]
TiO_2-coated gold nano bipyramid (NBP@TiO_2)	Human glioblastoma (U-87 MG) cells	Lung cancer	Photothermal and photocatalytic cancer therapy	[52]
Au NR–TiO_2 NCs	HeLa cells	Cervical cancer	Photothermal and photocatalytic cancer therapy	[53]
TiO_2 NP	MCF-7 and MDA-MB-468 breast cancer epithelial cells	Breast cancer	Increased apoptotic effect	[54]
TiO_2-APTES-Ce6 NPs	U87 cells		Anticancer photodynamic therapy	[55]
TiO_2	T-24 human bladder cancer cells	Bladder cancer	Anticancer photodynamic therapy	[56]
Mn-doped TiO_2 rhombic nanocomposites	Breast cancer cells	Breast cancer	PDT	[57]
TiO_2 nanoplatforms (DOX@TiO_2x@PDA-Cy5.5	Breast cancer cells	Breast cancer	PTT, PDT	[58]
TiO_2 NP	HeLa cells	Antitumor	Antineoplastic effect	[59]

and mastication, substantial forces are transmitted to the osseous tissue, necessitating the load-bearing capacity of artificial implants. The mechanical characteristics of bone tissue are upheld by a perpetual process of bone creation and resorption, commonly referred to as bone turnover, which is governed by "Wolff's Law" [62]. Furthermore, synthetic materials used in bone tissue must possess non-toxic, non-immunogenic, non-thrombogenic, and non-carcinogenic properties, in addition to their mechanical performance [63]. Titanium and its alloys are favoured materials because of their exceptional specific strength, low elastic modulus that aligns with the elastic modulus of bone tissue, and most importantly, their ability to create a thin yet highly stable oxide layer (known as passivation) on the surface, which contributes to its inertness (Table 2) [62]. Furthermore, they have capacity to form a thin, stable oxide (TiO_2) layer on the surface that is resistant to corrosion. However, there are drawbacks such as limited interaction with the surrounding tissue and their impact on bioactivity. Coating the surface of this metal family with biocompatible compounds is a potential solution to address these challenges [64].

Table 2 A comparison of different implants used in bone tissue engineering

Material	Tensile strength (Mpa)	Yield strength (MPa)	Elastic modulus (GPa)	Reference
Bone (cortical)	70–150	30–70	15–30	[65]
Titanium based alloys	690–1100	585–1060	55–110	[64]
Stainless steel	490–1350	190–690	200–210	[66]
Co based alloys	655–1793	310–1586	210–253	[67]
Dense HA ceramics	40–100	–	70–120	[67]
Bioglass 45S5	42	–	35	[68]

Improving the biological functionality of bone implants relies heavily on their surfaces, since they are the points of contact with the surrounding living tissue [62, 69]. Implants with enhanced biological performance that can alter and control the response of living tissue are the goal of surface engineering. Surface engineering often involves changing the chemical (coating) and topographical (roughness) properties of a medical device. The goal of topographical alterations to titanium and its alloys was to make implant surfaces rougher, so that they might have a larger surface area than bigger, smoother surfaces. The biomechanical interaction between the implant and bone tissue is improved, and cell attachment is enhanced, thanks to the larger surface area. Many methods, including as acid etching and grit-blasting, have been devised to achieve this goal. To achieve grit-blasting, implant surfaces are bombarded with particles of silica, hydroxyapatite, alumina, or TiO_2. Common acids used in acid etching procedures include hydrofluoric, nitric, and sulfuric acid. Dohan Ehrenfest et al. [70] conducted an in-depth study on the topic of topographical implant alteration and its practical implications. The advent of nanotechnology in recent years has increased the range of topographical changes from the micro- to the nanoscale, with consequential effects on ions, cells, and biomolecules at this microscopic level. Materials with at least one dimension sized between 1 and 100 nm are the primary focus of nanotechnology, which encompasses the production, characterization, and application of these materials. Coatings having a thickness below 100 nm or nanostructured coatings composed of nanoparticles or nanocrystals with dimensions lower than 100 nm are examples of how surface engineers are increasingly interested in this nanoscale dimension [71] Implant surfaces can have their mechanical qualities maintained through coating deposition, and their usefulness enhanced by adding (bio)chemical substances that stimulate better bone regeneration. The coating should allow for optimal integration into the surrounding tissues, while the substrate is responsible for the load-bearing function of the implant once these compounds have been successfully immobilized onto the implant surfaces.

Titanium alloys have several advantages as implant materials; however, they are prone to biofilm formation [72], corrosion [73], and even hypersensitivity [74]

leading to revision surgeries which are complicated to the original replacement surgeries done on patients. To prevent infections caused by bacteria, one of the most prevalent procedures that has been used is to coat titanium implants with antimicrobial agents (Fig. 8). It has been shown that the Gram-positive *Staphylococcus aureus* is the most extensively researched of all the pathogenic bacteria that are responsible for implant associated infections (IAIs) [75]. It is possible for *Staphylococcus aureus* to produce multicellular biofilms, which serve to protect the bacterial cells from the immune system of the host environment [76]. In addition, *Staphylococcus aureus* has developed a wide range of defence mechanisms to circumvent the immune system of the host, and its biofilms have demonstrated an increased resistance to antibiotics. Further, Gram-negative bacteria, such as *Pseudomonas aeruginosa*, *Porphyromonas gingivalis*, and *Escherichia coli*, as well as *Staphylococcus epidermidis* and *Streptococcus sanguinis*, can cause infections that are invasive and potentially life-threatening [77]. Additionally, the coatings should be biocompatible, non-cytotoxic, and encourage tissue integration [78]. This is in addition to the fact that they should avoid microbial attacks. The process of applying coatings often involves the addition of a new layer to the surface of the material, which does not interfere with the characteristics and characteristics of the bulk material. This can be accomplished by using variety of techniques, such as electrochemical deposition [79], ionized jet deposition (IJD) [80], sol–gel approach [81], micro-arc oxidation [82], and many more. The coatings are loaded with antimicrobial agents, such as inorganic elements, antibiotics, antimicrobial peptides, polymers, and hybrid inorganic–organic moieties [60], which successfully prevent the formation of biofilms and increase tissue integration. This is done to prevent IAIs.

4 Conclusion

Within the scope of this chapter, an attempt was made to investigate the fundamental characteristics of titanium dioxide nanostructure applications in the field of biomedical sciences. Beginning in the 1960s, titanium alloys have been utilized as materials for implant applications. They have endured the test of time and continue to be utilized as widely as possible in a variety of orthopaedic and dental applications even in the present day. Since that time, ongoing research has been conducted with the goal of modifying the composition of the alloy as well as the surface characteristics to create a material that possesses the optimal mix of chemical and mechanical capabilities. To achieve this objective, the surfaces of bone implants will be progressively enhanced with biomolecules to speed up bone healing. A diverse range of biomolecules, including growth factors, bioactive proteins, enzymes, and non-viral genes (DNAs, RNAs), are presently undergoing pre-clinical evaluation for this objective.

To achieve maximum therapeutic effectiveness, it is imperative to regulate the destiny of these biomolecules. This entails the development of efficient immobilization techniques, whether through covalent or non-covalent means. This chapter will

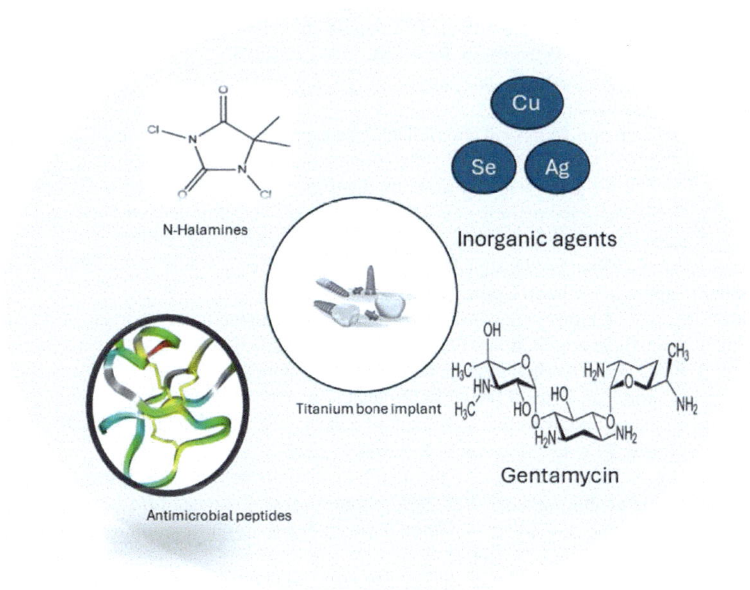

Fig. 8 Different antibacterial coatings on titanium implants

provide valuable insights for future researchers and academicians, allowing them to explore various synthesis methods and applications of TiO_2 nanomaterials. This knowledge will inspire them to develop new and innovative protocols with a wide range of novel applications.

References

1. Krishna, S.B.N., et al., ZnO based 0–3D diverse nano-architectures, films and coatings for biomedical applications. J. Mater. Chem. B (2024)
2. Hashemzadeh, H., et al.: PDMS nano-modified scaffolds for improvement of stem cells proliferation and differentiation in microfluidic platform. Nanomaterials **10**(4), 668 (2020)
3. Bendre, A., et al.: Recent developments in microfluidic technology for synthesis and toxicity-efficiency studies of biomedical nanomaterials. Mater. Today Adv. **13**, 100205 (2022)
4. Ziental, D., et al.: Titanium dioxide nanoparticles: prospects and applications in medicine. Nanomaterials **10**(2), 387 (2020)
5. Jafari, S., et al.: Biomedical applications of TiO(2) nanostructures: recent advances. Int. J. Nanomed. **15**, 3447–3470 (2020)
6. Sungur, Ş., Titanium dioxide nanoparticles. In: Handbook of Nanomaterials and Nanocomposites for Energy and Environmental Applications, pp. 713–730 (2021)
7. Javed, R., et al.: Diverse biotechnological applications of multifunctional titanium dioxide nanoparticles: an up-to-date review. IET Nanobiotechnol. **16**(5), 171–189 (2022)
8. Thakur, N., et al.: A critical review on the recent trends of photocatalytic, antibacterial, antioxidant and nanohybrid applications of anatase and rutile TiO2 nanoparticles. Sci. Total. Environ. **914**, 169815 (2024)

9. Baranowska-Wójcik, E., et al.: Effects of titanium dioxide nanoparticles exposure on human health—a review. Biol. Trace Elem. Res. **193**, 118–129 (2020)
10. Jafari, S., et al., Biomedical applications of TiO2 nanostructures: recent advances. Int. J. Nanomed. 3447–3470 (2020)
11. Gupta, S.M., Tripathi, M.: A review of TiO 2 nanoparticles. Chin. Sci. Bull. **56**, 1639–1657 (2011)
12. Diebold, U.: The surface science of titanium dioxide. Surf. Sci. Rep. **48**(5), 53–229 (2003)
13. Gazit, E.: Self-assembled peptide nanostructures: the design of molecular building blocks and their technological utilization. Chem. Soc. Rev. **36**(8), 1263–1269 (2007)
14. Wan, Y., Shi, Y., Zhao, D.: Designed synthesis of mesoporous solids via nonionic-surfactant-templating approach. Chem. Commun. **9**, 897–926 (2007)
15. Palithya, S., et al.: Green synthesis of silver nanoparticles using flower extracts of Aerva lanata and their biomedical applications. Part. Sci. Technol. **40**(1), 84–96 (2022)
16. Khashan, K.S., et al.: Antibacterial activity of TiO2 nanoparticles prepared by one-step laser ablation in liquid. Appl. Sci. **11**(10), 4623 (2021)
17. Liu, J., et al.: Nano-modified titanium implant materials: a way toward improved antibacterial properties. Front. Bioeng. Biotechnol. **8**, 576969 (2020)
18. Sakalauskaite, S., et al.: Influence of the metabolic activity of microorganisms on disinfection efficiency of the visible light and P25 TiO2 photocatalyst. Catalysts **11**(12), 1432 (2021)
19. Carp, O., Huisman, C.L., Reller, A.: Photoinduced reactivity of titanium dioxide. Prog. Solid State Chem. **32**(1–2), 33–177 (2004)
20. Thakur, N., Thakur, N.: Removal of organic dyes and free radical assay by encapsulating polyvinylpyrrolidone and Tinospora Cordifolia in dual (Co–Cu) doped TiO2 nanoparticles. Environ. Pollut. **335**, 122229 (2023)
21. Noman, M.T., Ashraf, M.A., Ali, A.: Synthesis and applications of nano-TiO2: a review. Environ. Sci. Pollut. Res. **26**(4), 3262–3291 (2019)
22. Chen, X., Mao, S.S.: Titanium dioxide nanomaterials: synthesis, properties, modifications, and applications. Chem. Rev. **107**(7), 2891–2959 (2007)
23. Muniandy, S.S., et al.: Green synthesis of mesoporous anatase TiO2 nanoparticles and their photocatalytic activities. RSC Adv. **7**(76), 48083–48094 (2017)
24. Falk, G.S., et al.: Microwave-assisted synthesis of TiO2 nanoparticles: photocatalytic activity of powders and thin films. J. Nanopart. Res. **20**(2), 23 (2018)
25. Prakash, J., et al.: Recent advances on metal oxide based nano-photocatalysts as potential antibacterial and antiviral agents. Catalysts **12**(9), 1047 (2022)
26. Razavi, H., Janfaza, S.: Ethosome: a nanocarrier for transdermal drug delivery. Arch. Adv. Biosci. **6**(2), 38–43 (2015)
27. Koizumi, R., et al.: Visible-light-induced antibacterial activity of carbon and nitrogen co-doped rutile TiO2 films. Thin Solid Films **780**, 139944 (2023)
28. Li, S., et al.: Durable antibacterial and UV-protective Ag/TiO2@ fabrics for sustainable biomedical application. Int. J. Nanomed. **12**(null), 2593–2606 (2017)
29. Azizi-Lalabadi, M., et al.: Antimicrobial activity of Titanium dioxide and Zinc oxide nanoparticles supported in 4A zeolite and evaluation the morphological characteristic. Sci. Rep. **9**(1), 17439 (2019)
30. Badeggi, U.M., et al.: Green synthesis of gold nanoparticles capped with procyanidins from Leucosidea sericea as potential antidiabetic and antioxidant agents. Biomolecules **10**(3), 452 (2020)
31. du Maire, P., et al.: Characterisation of the mechanical properties of polyamide 12 powder when using titanium dioxide as antimicrobial additive. Materialwiss. Werkstofftech. **54**(4), 385–390 (2023)
32. Ou, B., et al.: Mechanical and antibacterial properties of polypropylene/polyamide 6 blends-TiO2 nanocomposites. Polym. Plast. Technol. Eng. **51**(8), 849–853 (2012)
33. Altan, M., Yildirim, H.: Mechanical and antibacterial properties of injection molded polypropylene/TiO2 nano-composites: effects of surface modification. J. Mater. Sci. Technol. **28**(8), 686–692 (2012)

34. Younis, A.B., et al.: Synthesis and characterization of TiO2 nanoparticles combined with geraniol and their synergistic antibacterial activity. BMC Microbiol. **23**(1), 207 (2023)
35. Zhao, Y., et al.: Fabrication of biomaterial/TiO2 composite photocatalysts for the selective removal of trace environmental pollutants. Chin. J. Chem. Eng. **27**(6), 1416–1428 (2019)
36. Chen, W., Viljoen, A.M.: Geraniol—a review of a commercially important fragrance material. S. Afr. J. Bot. **76**(4), 643–651 (2010)
37. Bahmani, M.: Synthesis and evaluation of the antibacterial effect of titanium dioxide nanoparticles in comparison with ampicillin, colistin, and ertapenem on Staphylococcus aureus. J. Pharm. Negative Results **10**(1), 16–20 (2019)
38. Nešić, M., et al.: Light controlled metallo drug delivery system based on the TiO2-nanoparticles and Ru-complex. J. Photochem. Photobiol. A **347**, 55–66 (2017)
39. Allison, R.R., et al.: Photosensitizers in clinical PDT. Photodiagn. Photodyn. Ther. **1**(1), 27–42 (2004)
40. Wang, C.Y., et al.: The study of using TiO2 nanotube arrays as a drug delivery for alendronate. Adv. Mater. Res. **335**, 1469–1472 (2011)
41. Ghosh, S., Das, A.: Modified titanium oxide (TiO2) nanocomposites and its array of applications: a review. Toxicol. Environ. Chem. **97**(5), 491–514 (2015)
42. Eskandarinezhad, S., et al.: Metal and metal oxide nanoparticles/nanocomposites as electrochemical biosensors for cancer detection. J. Electrochem. Soc. **169**(4), 047504 (2022)
43. Yin, Z.F., et al.: Recent progress in biomedical applications of titanium dioxide. Phys. Chem. Chem. Phys. **15**(14), 4844–4858 (2013)
44. Zhang, N., et al.: Electrospun TiO2 nanofiber-based cell capture assay for detecting circulating tumor cells from colorectal and gastric cancer patients. Adv. Mater. **24**(20), 2756–2760 (2012)
45. Cleeton, C., et al.: Electrospun nanofibers for drug delivery and biosensing. ACS Biomater. Sci. Eng. **5**(9), 4183–4205 (2019)
46. Zanghelini, F., et al.: Biosensing breast cancer cells based on a three-dimensional TIO2 nanomembrane transducer. Biosens. Bioelectron. **92**, 313–320 (2017)
47. Mun, K.-S., et al.: A stable, label-free optical interferometric biosensor based on TiO2 nanotube arrays. ACS Nano **4**(4), 2070–2076 (2010)
48. Nah, Y.C., Paramasivam, I., Schmuki, P.: Doped TiO2 and TiO2 nanotubes: synthesis and applications. ChemPhysChem **11**(13), 2698–2713 (2010)
49. Zaleska, A.: Doped-TiO2: a review. Recent Pat. Eng. **2**(3), 157–164 (2008)
50. Wang, X., et al.: One-dimensional titanium dioxide nanomaterials: nanowires, nanorods, and nanobelts. Chem. Rev. **114**(19), 9346–9384 (2014)
51. Aslan, S., Anik, Ü.: Development of TiO2 and Au nanocomposite electrode as CEA immunosensor transducer. Electroanalysis **26**(6), 1373–1381 (2014)
52. Jang, D., et al.: Direct deposition of anatase TiO2 on thermally unstable gold nanobipyramid: morphology-conserved plasmonic nanohybrid for combinational photothermal and photocatalytic cancer therapy. Appl. Mater. Today **27**, 101472 (2022)
53. Lee, J., et al.: Gold nanorods-conjugated TiO 2 nanoclusters for the synergistic combination of phototherapeutic treatments of cancer cells. J. Nanobiotechnol. **16**, 1–12 (2018)
54. Lagopati, N., et al.: Photo-induced treatment of breast epithelial cancer cells using nanostructured titanium dioxide solution. J. Photochem. Photobiol. A **214**(2), 215–223 (2010)
55. Youssef, Z., et al.: Titania and silica nanoparticles coupled to Chlorin e6 for anti-cancer photodynamic therapy. Photodiagn. Photodyn. Ther. **22**, 115–126 (2018)
56. Kubota, Y., et al.: Photokilling of T-24 human bladder cancer cells with titanium dioxide. Br. J. Cancer **70**(6), 1107–1111 (1994)
57. Mushtaq, A., et al.: Facile synthesis of Mn doped TiO2 rhombic nanocomposites for enhanced T1-Magnetic resonance imaging and photodynamic therapy. Mater. Res. Bull. **144**, 111481 (2021)
58. Hasanzadeh Kafshgari, M., Goldmann, W.H.: Insights into theranostic properties of titanium dioxide for nanomedicine. Nanomicro. Lett. **12**(1), 22 (2020)
59. Fujishima, A., Rao, T.N., Tryk, D.A.: Titanium dioxide photocatalysis. J. Photochem. Photobiol. C **1**(1), 1–21 (2000)

60. Akshaya, S., et al.: Antibacterial coatings for titanium implants: recent trends and future perspectives. Antibiotics **11**(12), 1719 (2022)
61. Wu, S., et al.: Hydrogels for bone organoid construction: from a materiobiological perspective. J. Mater. Sci. Technol. **136**, 21–31 (2023)
62. Bosco, R., et al.: Surface engineering for bone implants: a trend from passive to active surfaces. Coatings **2**(3), 95–119 (2012)
63. Williams, D.F.: On the mechanisms of biocompatibility. Biomaterials **29**(20), 2941–2953 (2008)
64. Kaur, M., Singh, K.: Review on titanium and titanium based alloys as biomaterials for orthopaedic applications. Mater. Sci. Eng. C **102**, 844–862 (2019)
65. Ratner, B.D., et al.: Biomaterials science: an introduction to materials in medicine. Elsevier (2004)
66. Wagner, W.R., et al.: Biomaterials science: an introduction to materials in medicine. Academic Press (2020)
67. Ratner, B.D.: Biomaterials: been there, done that, and evolving into the future. Annu. Rev. Biomed. Eng. **21**, 171–191 (2019)
68. Lubas, M., et al.: Spectroscopic and microstructural investigation of novel Ti–10Zr–45S5 bioglass composite for dental applications. J. Mol. Struct. **1221**, 128545 (2020)
69. Moura, C.C., et al.: The effect of a nanothickness coating on rough titanium substrate in the osteogenic properties of human bone cells. J. Biomed. Mater. Res. Part A Official J. Soc. Biomater. Jpn. Soc. Biomater. Aust. Soc. Biomater. Korean Soc. Biomater. **94**(1), 103–111 (2010)
70. Ehrenfest, D.M.D., et al.: Classification of osseointegrated implant surfaces: materials, chemistry and topography. Trends Biotechnol. **28**(4), 198–206 (2010)
71. Mendonça, G., et al.: Advancing dental implant surface technology–from micron-to nanotopography. Biomaterials **29**(28), 3822–3835 (2008)
72. Busscher, H., et al.: Biofilm formation on dental restorative and implant materials. J. Dent. Res. **89**(7), 657–665 (2010)
73. Prestat, M., Thierry, D.: Corrosion of titanium under simulated inflammation conditions: clinical context and in vitro investigations. Acta Biomater. **136**, 72–87 (2021)
74. Egusa, H., et al.: Suspected association of an allergic reaction with titanium dental implants: a clinical report. J. Prosthet. Dent. **100**(5), 344–347 (2008)
75. Silva-Santana, G., et al.: Staphylococcus aureus biofilms: an opportunistic pathogen with multidrug resistance. Rev. Res. Med. Microbiol. **32**(1), 12–21 (2021)
76. Minkiewicz-Zochniak, A., et al.: Biofilm formation on dental implant biomaterials by Staphylococcus aureus strains isolated from patients with cystic fibrosis. Materials **14**(8), 2030 (2021)
77. Cerioli, M., et al.: Pseudomonas aeruginosa implant-associated bone and joint infections: experience in a regional reference center in France. Front. Med. **7**, 513242 (2020)
78. Trindade, R., et al.: Osseointegration and foreign body reaction: titanium implants activate the immune system and suppress bone resorption during the first 4 weeks after implantation. Clin. Implant Dent. Relat. Res. **20**(1), 82–91 (2018)
79. Luo, J., et al.: Multi-metal ions doped hydroxyapatite coatings via electrochemical methods for antibacterial and osteogenesis. Colloid Interface Sci. Commun. **43**, 100435 (2021)
80. Graziani, G., et al.: Ionized jet deposition of antimicrobial and stem cell friendly silver-substituted tricalcium phosphate nanocoatings on titanium alloy. Bioact. Mater. **6**(8), 2629–2642 (2021)
81. Nichol, T., et al.: The antimicrobial activity and biocompatibility of a controlled gentamicin-releasing single-layer sol-gel coating on hydroxyapatite-coated titanium. Bone Joint J. **103**(3), 522–529 (2021)
82. Hu, Y., et al.: Preparation of coating on the titanium surface by micro-arc oxidation to improve corrosion resistance. Coatings **11**(2), 230 (2021)

Titanium Dioxide-Based Multifunctional Hybrid Nanomaterials: A Comprehensive Study of Their Biological Applications

Rajpal Tyagi and Anuj Maurya

Abstract Titanium dioxide (TiO_2) nanoparticles (NPs) are among the most extensively utilized metallic oxide nanoparticles. Whether in their pure state or modified with polymers or with other metals, TiO_2 NPs plays an essential role. Nevertheless, the modification of size and shape through doping leads to enhancements in the physical, chemical, and biological characteristics of TiO_2 nanoparticles. TiO_2 NPs have attracted significant interest in biological research because of their distinct characteristics and wide-ranging applications. Researchers have established unique impacts of diverse TiO_2 nanostructures, such as nanoparticles, nanopores, nanorods, and nanotubes, in many areas of biotechnology. There has been a remarkable increase in studies on TiO_2 NPs due to their chemical stability, low toxicity, and diverse characteristics. These nanoparticles have an essential function in advancing the field of nanomedicine by enhancing nanotechnology through immobilizing biomolecules. They have applications in drug delivery systems, cancer therapy, bone/dental implants, and antibacterial agents. They have also shown potential in biosensors, detecting specific biomolecules with exceptional sensitivity. Therefore, this chapter presents continuing studies about the multifaceted roles of TiO_2 NPs in biological systems, emphasizing their impact on human health and potential for advancing nanomedicine have been comprehensively discussed.

Keywords TiO_2 NPs · Nanomedicine · Nanotechnology · Drug delivery · Antibacterial · Biosensors · Phototherapy

R. Tyagi (✉) · A. Maurya
Department of Chemistry, Multanimal Modi College Modinagar, Ghaziabad, Uttar Pradesh, India
e-mail: tyagirajpal25@gmail.com

A. Maurya
e-mail: 2017pcy5308@mnit.ac.in

J. Prakash et al. (eds.), *Titanium Dioxide-Based Multifunctional Hybrid Nanomaterials*, Engineering Materials, https://doi.org/10.1007/978-3-031-80625-4_13

1 Introduction

Titanium dioxide (TiO_2) nanoparticles (NPs) are very versatile materials with very substantial potential for utilization in biotechnological applications. TiO_2 nanoparticles provide the impression of being a harmless and inert substance when they are exposed to the human body. This chapter examines the diverse functions of TiO_2 NPs emphasizing their substantial influence on human health.

Nanotechnology is the discipline of science that deals with materials on the nanometer scale. Nanotechnology, one of the most fascinating fields of science, enables scientists to build, manipulate, and use materials on the nanometer scale [1–4]. Nanobiotechnology is a fast-expanding area that combines nanotechnology with biology. This is a multidisciplinary discipline with applications in nearly every area of biotechnology [1, 2]. Biotechnology has facilitated the transmission of biological system knowledge to business. Furthermore, the diverse physical, chemical, and biological characteristics of nanoparticles have resulted in the application of nanotechnology across many healthcare environments [5–10].

This multidisciplinary study has significant potential for improving medical engineering [11, 12]. Nanotechnology has created several nanostructures with unique features for biotechnological applications, [7, 9, 13–16] including nano-sized titanium dioxide. TiO_2 is a White, poorly soluble TiO_2 is used in cosmetics, medications, and pharmaceuticals [6, 17, 18]. Additionally, it has a pair of crystalline forms called Anatase and Rutile which are used in many important industrial processes. Mixed polymorphs of TiO_2 (such as Anatase 80% and Rutile 20%) are more biomedically effective than one crystal. Anatase form of TiO_2 is more photocatalytic and cytotoxic than Rutile [18].

Most metal-oxide nanoparticles are harmful and expensive to scale up [8, 15] whereas, TiO_2 NPs are cheap, non-toxic, and easy to synthesize in labs and factories. Sol–gel, hydrothermal, and electrochemical anodization can produce substantial amount of TiO_2 NPs [19]. Simple and adaptable anodic oxidation (or anodization) synthesizes TiO_2 NPs with regulated structures and shapes that align perpendicularly and readily to the Ti substrate [20, 21]. Sol–gel, hydrothermal, and chemical vapour deposition are the most prevalent techniques for synthesizing TiO_2 NPs with desired properties [22, 23]. With their low toxicity, nanosized features, intrinsic properties, good biocompatibility, and versatile fabrication techniques, nanostructured TiO_2 holds immense potential for a wide variety of applications [13, 14]. For example, nanostructured arrays made up of TiO_2 NT wall, along with Ag_2O and Cu NPs, are excellent choices for biomedical applications like antibacterial surface coating. In addition, they possess excellent cytocompatibility and are highly beneficial for promoting osteoblast cell spreading, proliferation, and differentiation [24, 25]. In addition, TiO_2 NPs possess a diverse array of uses, such as cancer therapy, drug delivery systems, bone/dental implants, and antibacterial agents (Fig. 1).

Recently, there has been significant progress in the development of different NP systems for delivering chemotherapeutic agents. The aim is to successfully eradicate tumour cells. However, many of these delivery systems still struggle to specifically

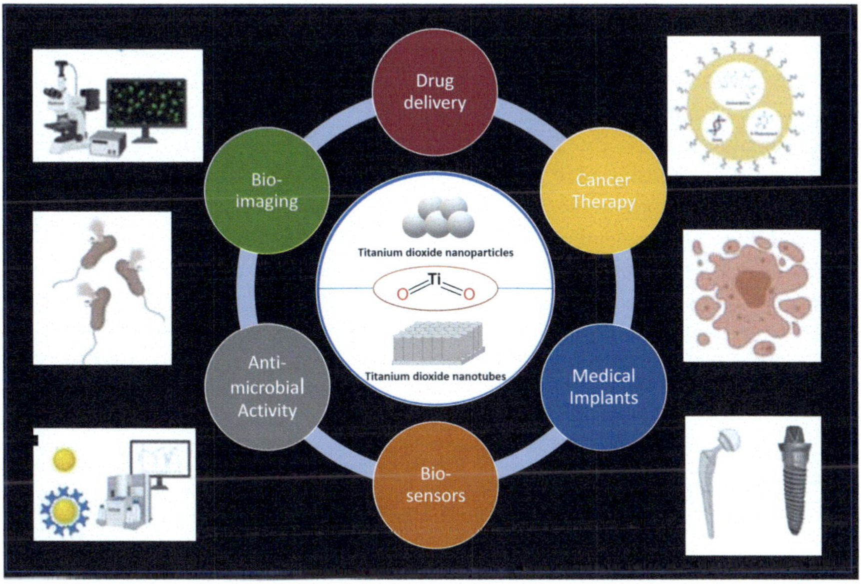

Fig. 1 Various applications of TiO_2-based hybrid nanomaterials in the field of biomedicine

target tumor cells and have limited control over the drug release process. As a result, there are often serious side effects [26, 27]. Thus, achieving targeted accumulation of drug at the tumor site remains a significant issue. As a result, we need to focus on making the best drug delivery system possible, one that can release drugs in response to triggers and target cancer cells. With their exceptional properties, porous TiO_2 NPs have the potential to address this issue. They possess a large functional surface and exhibit impressive photocatalytic activity [28]. For example, TiO_2 NTs are preferable structures for surface coating of implants compared to implants made of polymers. TiO_2 NTs exhibit excellent stability, resisting disintegration and swelling. TiO_2 NTs are tubular and self-organized nanostructures that have gained significant interest in the field of implant manufacturing due to their impressive mechanical stability, cost-effectiveness in preparation, and enhanced biocompatibility when compared to TiO_2 film [29–31]. Furthermore, TiO_2 NPs have the potential to function as an intelligent delivery system for regulating drug release in implants. These features of TiO_2 nanostructures deal with the key issues of the previously designed implant [32]. Research has also been conducted to enhance the performance of TiO_2 NTs implants by immobilizing biomolecules on the surface of these nanotubular matrices. Administering growth hormones, such as bone morphogenetic protein 2 (BMP2), in TiO_2 NTs implants can enhance their ability to induce bone formation [33, 34]. TiO_2 NTs can serve as matrices for immobilizing proteins and enzymes in biosensors. For example, the enzyme glucose oxidase (GOx) can be immobilized on the surface of TiO_2 NTs to create an enzymatic biosensor that can detect glucose [35].

This study aims to update readers on the latest developments in the comprehensive utilization of TiO_2 NPs in the field of medical biotechnology, namely in bioimaging, drug transport, phototherapy, and antibacterial activity. Based on the literature, TiO_2 NPs are considered to be an inert and non-toxic substance when they come into contact with the human body. A thorough understanding of nanoscale phenomena will enable advancing cutting-edge biomedical devices and producing surfaces capable of neutralizing microorganisms. It will also enable the exploration of techniques for immobilizing biomolecules in TiO_2 NPs, from synthesis to enhancing the effectiveness of these functionalized TiO_2 NPs in biomedicine, whether as implants or biosensors. This study specifically examines the recent progress in the biomedical use of TiO_2 NPs, highlighting its nanotoxicity and key parameters that influence its biocompatibility. It also addresses the challenges and future prospects of TiO_2 NPs in biomedical fields.

2 TiO_2 NPs—An Advancement in Nanomedicine

TiO_2 NPs have a special photoactivity that makes them promising for use in nanomedicine. When exposed to ultraviolet light, titanium dioxide produces reactive oxygen species (ROS), which have the ability to preferentially target cancer cells. The use of TiO_2 NPs in photodynamic therapy (PDT) has shown promise as a possible treatment for a wide range of diseases, including cancer and psoriasis, and it can also be utilised in genetic engineering as a nucleic acid endonuclease. TiO_2 NPs are currently being extensively studied due to their intriguing characteristics, including little toxicity, biological and chemical responsiveness, transparency in thin films, and strong chemical stability. In addition to their other applications, TiO_2 NPs have prominent medical applications that have the potential to considerably improve the field of medicine. There is a great potential for TiO_2 NPs in the field of nanomedicine. TiO_2 NPs improve the effectiveness of treatment while limiting the harm to healthy tissues [36, 37].

In recent times, many areas of biomedicine have recently shown an increasing interest in using metallic oxide NPs [38]. Furthermore, nanostructures and nanocomposites, based on TiO_2 nanoparticles can serve as potential instruments for a wide range of biological applications [39]. Until now, a substantial amount of scientific work that focuses on the biological applications of TiO_2 NPs has been outlined. The following section discusses their possible use in medication delivery, phototherapy, and as antibacterial agents.

2.1 Application in Drug Delivery

TiO_2 NPs are being considered as possible alternatives for delivering drugs because of their distinctive physicochemical features. These NPs offer a versatile platform for selective drug delivery, enhancing the potency of treatment while reducing potential adverse effects. In this section, we explore the application of TiO_2 NPs in administration of medicines, highlighting examples of their successful utilization.

Traditionally, drugs are administered by oral, nasal, and parenteral methods to distribute them throughout the whole body. This hampers the efficacy of the medicine and also gives rise to potential adverse effects. Therefore, it is necessary to develop medication delivery methods that are more effective in order to overcome any limitations. The integration of nanotechnology into biomedicine has been facilitated by advancements in biotechnology, material science, and engineering techniques, specifically for the purpose of targeted medication delivery. TiO_2 NPs have been used in drug delivery procedures due to their distinctive electrochemical characteristics, tunable structures, and biocompatibility [30]. TiO_2 NPs with surface modifications are effective drug transporters. Therapeutic compounds can be released regulated by their surface modification. Controlled release enhances medication stability and precise distribution. TiO_2 NPs focus therapeutic substances, minimizing adverse effects and improving treatment success. Polymers, peptides, and antibodies can modify TiO_2 NP surfaces to increase drug loading. TiO_2 nanoparticles coated with polyethylene glycol (PEG) improve drug solubility and circulation.

The TiO_2 NP-mediated drug delivery in living cells via targeted and non-targeted approach is shown in Fig. 2a. The targeted delivery technique involves TiO_2 NPs tailored medication with photocatalytic breakdown of organic molecules by UV light stimulation. When UV energy exceeds TiO_2 NPs band gap, valence electrons are excited to the conduction band, forming an electron (e^-) and hole (h^+) pair and active free radicals ($OH^·$ and $O2^·$) that decompose organic compounds on the surface, laying the groundwork for light-triggered drug release at a targeted site [40].

TiO_2 NPs are responsive to environmental elements such as light, pH, and temperature. Drug release is triggered by photothermal effects when it is subjected to near-infrared (NIR) light. For example, when subjected to laser exposure, TiO_2 NPs that contain doxorubicin have the ability to release the medicine. It is possible to engineer TiO_2 NPs in a way that allows them to specifically target disease areas by attaching active targeting ligands, such as antibodies, to the nanoparticles. For instance, cancer cells that have an excessive amount of folate receptors are targeted by TiO_2 NPs that have been modified with folate [41].

It is important to mention that TiO_2 NPs have been reported to have potential anticancer properties as they can generate active free radicals when exposed to UV light [28, 31]. To improve the biocompatibility of TiO_2 NPs and enhance their ability to release drugs in a controlled manner, they are coated with PEG. Using TiO_2 NPs as nanocarriers, we successfully achieved targeted drug delivery of paclitaxel, an anticancer drug. Enhancing the biocompatibility of TiO_2 NPs involves attaching PEG to their surface and grafting folic acid (FA) as a ligand for targeted drug delivery. The characterization studies demonstrated the successful PEGylation and grafting of the ligand onto these NPs. In addition, the NPs loaded with the drug exhibited a significantly higher adsorption capacity. In addition, the release studies conducted on paclitaxel from FA–PEG–TiO_2 NPs demonstrated an initial rapid release, followed by a sustained release phase [42]. Overall, the use of TiO_2 NPs presents exciting

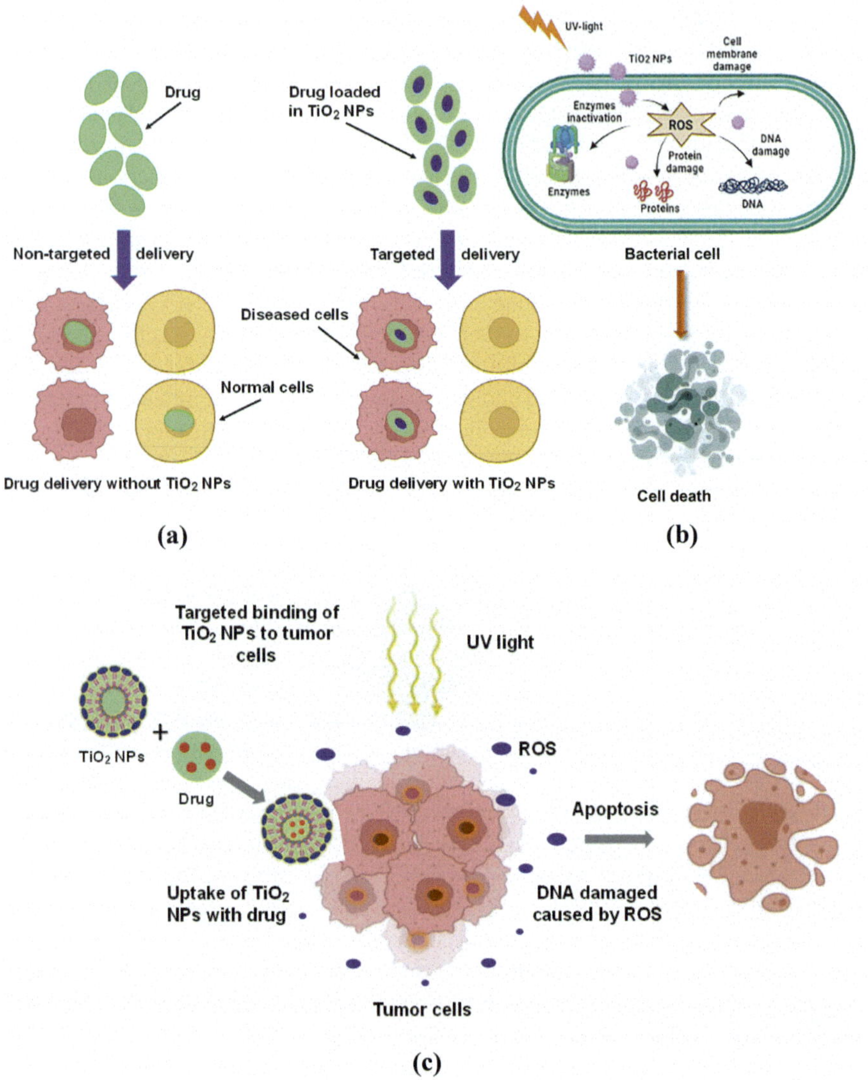

Fig. 2 **a** An illustration of the general method of administering drugs to living cells, both with and without the presence of TiO_2 NPs. **b** A simplified illustration of the mechanism of action of TiO_2 NPs against bacteria. **c** The irradiated TiO_2 NPs as diagnostic and therapeutic agents for detecting and causing DNA damage and inducing apoptotic pathway [36] (open access)

prospects for precise drug delivery, potentially revolutionizing the field of medicine. With ongoing research, we are eagerly awaiting significant advancements that will revolutionize patient care and enhance treatment results.

2.2 Antimicrobial Activity

In antimicrobial applications, the overuse of antibiotics has led to the formation of multidrug resistant (MDR) bacterial strains, a development that raises concerns for human health and food safety [43]. While searching for new antibacterial chemicals, researchers have focused on metal oxide NPs. Biofilms contribute to antibiotic resistance [44]. As a result, the focus of the study has turned to investigating the antibacterial properties of metal and metal oxide nanoparticles [45]. TiO_2 NPs are a promising antibacterial agent due to their photocatalytic properties, chemical stability, non-toxicity, affordability, and generally recognised as safe [46]. Additionally, due to their antibacterial action, TiO_2 NPs are among the metal oxide NPs that are increasingly used in biomedicine [47].

TiO_2 NPs have shown remarkable bactericidal properties, effectively combating bacteria that are resistant to antibiotics. They have natural antimicrobial properties. These substances have the potential to be utilized for targeted drug administration in areas affected by infection. Antibacterial infections can be effectively treated with TiO_2 NPs loaded with antibiotics. With their impressive photocatalytic effects, they have the ability to break down organic pollutants, making them incredibly useful for purifying water and disinfecting air. With the help of TiO_2 NPs, wound healing is enhanced by preventing infections and promoting tissue regeneration. TiO_2 NPs demonstrate photocatalytic antimicrobial activity when exposed to UV light with a wavelength of less than 385 nm (Fig. 2b). The antimicrobial activity of TiO_2 NPs' effectiveness relies on the thickness of the microbial cell surface, with the order of effectiveness being virus > bacterial wall > bacterial spore. Photocatalysis by TiO_2 NPs induces the generation of hydroxyl radicals, which in turn causes oxidative stress on the cell membrane of bacteria, resulting in antibacterial activity. TiO_2 NPs increase the oxidation of unsaturated phospholipids present in the plasma membrane, leading to harm to the bacterial membrane. It interferes with vital biological activities such as breathing, energy production, and the ability of substances to pass across cell membranes [48].

The methicillin-resistant staphylococcus aureus (MRSA), a major nosocomial infection, is known for causing broad-spectrum infections and contributing to the rise of bacterial resistance. As per the literature, the researchers investigated the antibacterial effects of TiO_2 NPs on MRSA. A study was conducted to assess the impact of TiO_2 NPs on the formation of biofilm by MRSA using the tissue culture plate method. Out of the total 30 isolates, 22 exhibited a strong ability to form biofilms, while 2 showed a weaker capacity for biofilm formation. The analysis suggested that the TiO_2 NPs (500 μg/ml) effectively suppressed the growth of both highly resistant and less resistant MRSA strains. These findings indicate that the TiO_2 NPs can serve as antibacterial agents [49]. In a separate study, scientists employed TiO_2 NPs in combination with antibiotics like ceftazidime and cefotaxime to counteract the presence of multi-drug-resistant Pseudomonas aeruginosa. The specimens were collected from sputum, pus, bronchoalveolar lavage, and the endotracheal tract. Upon 1-h exposure to UV radiation, a noticeable bactericidal impact was detected at doses exceeding

350 μg/ml. The TiO_2 NPs demonstrated minimum inhibitory concentration values that were six times greater than those of the antibiotics. Hence, the antibacterial efficacy was augmented through the amalgamation of antibiotics and TiO_2 NPs [50].

Adding antibacterial agents to synthetic polymer-based nanocomposites creates versatile materials for packaging, biomedical, and general-purpose applications. However, thermodynamic and kinetic barriers hinder inorganic and hydrophilic nanoparticles from dispersion in hydrophobic polymer matrices, making TiO_2 NP-based products challenging to make. Titania-based antibacterial nanocomposites have garnered interest recently [51]. For instance, Petica and colleagues reported the synthesis of Ag-TiO_2 nanocomposites using the electrochemical method to improve their photocatalytic properties and antifungal and antibacterial activity [52]. Using TiO_2 alone or in conjunction with polymers lowers microbiological growth, reduces food deterioration and extends shelf life.

2.3 Application in Cancer Treatment

Cancer remains a significant global cause of death, presenting considerable challenges in terms of treatment. The primary treatment typically involves a combination of surgery, chemotherapy, or radiotherapy. Chemotherapy encounters obstacles like drug resistance or limited access to tumor cells, which can undermine its effectiveness [53]. Titanium dioxide nanoparticles (TiO_2 NPs) have shown great potential in the fight against cancer. With their exceptional properties, such as biocompatibility, photoactivity, and surface reactivity, these substances have become highly sought-after for cutting-edge cancer treatments.

Phototherapy, including PDT and photothermal therapy (PTT), is gaining recognition as a promising treatment option for various types of cancer. This technique offers a range of benefits, such as reduced severity, high effectiveness, and minimal side effects [34]. In recent times, NPs have been employed to transport photosensitisers for PDT. PDT with TiO_2 NPs is a highly targeted and efficient method for eliminating cancer cells (Fig. 2c). Their capacity to produce ROS when exposed to UV radiation makes them valuable partners in the battle against cancer. Numerous studies are currently being conducted to explore the potential of molybdenum oxide, TiO_2, ZnO, and tungsten oxide nanoparticles as photosensitizers in PDT [54]. Several studies have highlighted the encouraging findings of utilizing TiO_2 NPs in PDT and PTT due to their distinct catalytic properties. TiO_2 NPs can serve as photosensitizers or be combined with photosensitizers for enhanced delivery.

Scientists have effectively demonstrated the targeted destruction of cancer cells using a combination of antibody-immobilized TiO_2 NPs and UV irradiation. PDT has gained significant attention in the past 30 years as a therapeutic approach. Within this article, we delve into the diverse uses of TiO_2 NPs in the realm of cancer treatment, supported by compelling research examples that showcase their immense promise.

Both titanium dioxide and zinc oxide have proven to be highly effective photosensitizers for PDT applications. Interestingly, studies have shown that their efficiency in PDT is comparable [55].

In a study conducted by Yurt et al. [56], the synthesis of the Zinc phthalocyanine (ZnPc) molecule and its integration into titanium dioxide nanoparticles was explored. The researchers then examined the potential of these molecules for photodynamic therapy (PDT) in breast and cervical tumors. The researchers showed that the use of ZnPc-TiO_2 resulted in a much higher level of photodynamic activity compared to using ZnPc alone. Additionally, the incorporation of ^{131}I into ZnPc-TiO_2, it exhibited a significant increase in the absorption of this compound by cervical and breast cancers. This suggests that ZnPc-TiO_2 could be a promising agent for photodynamic therapy [56]. In a separate study conducted by Yurt et al., it was found that ZnPc-TiO_2 exhibited a greater phototoxic effect compared to ZnPc in colon tumors [57].

Radiotherapy is another method for treating cancer. These tiny particles of titanium dioxide have the ability to enhance the impact of ionizing radiation on cancer cells, making them more susceptible to treatment. Several studies have examined the ability of TiO_2 NPs to improve the efficiency of cancer detection and treatment by boosting radiosensitivity [58]. Radiotherapy is a widely utilized technique in cancer treatment, applying ionizing radiation to specifically target tumor cells. Advancements in radiotherapy techniques lead to the discovery of innovative methods. The primary objective behind exploring these methods is to minimize or prevent harm to the healthy tissues surrounding the tumors. Thus, understanding tumour radiosensitivity is crucial in determining the optimal radiation dose to effectively eliminate cancer cells while minimizing harm to healthy cells. Radiosensitive medicines, for example, fluorouracil, have the capability of targeting cancer cells in a selective manner [59]. A study by Youkhana et al. examined the use of TiO_2 NPs as radiosensitizers on two cultured cell lines: human keratinocyte (HaCaT) and prostate cancer (DU145) cells. Precise calculations of radiation dosage were conducted for the experiment involving anatase TiO_2 NPs. Cell survival rates were assessed using cell viability assay and clonogenic assay. Based on the cytotoxicity test results, it was observed that TiO_2 NPs exhibited no toxicity at concentrations up to 4 mM. Furthermore, when the concentrations exceeded 4 mM, the cell viability remained constant. Ultimately, the use of TiO_2 NPs shows promise in advancing the development of more precise and efficient cancer therapics. With ongoing research, the potential of TiO_2-based nanomaterials in cancer therapy is being further explored, promising to shape the future of this field [59]

3 Developing Nanotechnology for Biological Systems

Titanium dioxide nanoparticles (TiO_2 NPs) are increasingly being utilized in nanotechnology, specifically for their effectiveness in biological systems. Due to their distinct characteristics, including as size-dependent behaviour, surface reactivity, and biocompatibility, nanoparticles are highly desirable for a range of biological applications, including cellular tracking in biological imaging and biosensing of particular biomolecules. TiO_2 NPs have been utilized in the field of pharmacy, particularly in pharmaceutical chemistry and technology. They have found applications in several sectors such as dentistry, surgery, and medical implants [60]. This article explores the progress of TiO_2 nanoparticles and their significant contribution to the advancement of biological research and healthcare. TiO_2 NPs are leading the way in nanotechnology, bringing about a significant transformation in biological systems. The subsequent paragraph delineates the prospective function of TiO_2 NPs in the domains of bioimaging, biosensors, and medical implants.

3.1 Bioimaging

Scientific research and biomedical purposes utilize a diverse range of imaging techniques. These include spectroscopy techniques like NIR spectroscopy and Raman spectroscopy, nuclear magnetic resonance (MRI), radio imaging using specific nuclides, computed tomography (CT) scanning, and more advanced scanning techniques such as laser ablation, ICP-MS, MALDI-MS, and others [61]. Advancements in diagnostic procedures result in prompt treatment and recovery of patients. NPs, especially TiO_2 NPs, are extensively researched and utilized in diagnostic procedures such as magnetic resonance imaging (MRI) and computed tomography (CT) as contrast agents. TiO_2 NPs are stimulated by exposure to radiation, enabling them to function as both diagnostic and therapeutic agents simultaneously. To enable cellular imaging and tracking, TiO_2 NPs can be modified with fluorescent dyes or quantum dots, enabling the use of fluorescent probes. Their small size enables efficient infiltration into deep tissues.

A study evaluated the image contrast properties of TiO_2 NPs using MRI and CT scanner. There was a noticeable difference in imaging between the control and TiO_2 NP samples on T2-weighted images. This suggests that TiO_2 NPs have the potential to be used as a cutting-edge theragnostic drug with the ability to enhance radio sensitivity and provide radiological diagnostic capabilities. These benefits are likely due to the chemical group modification on the surface of the NPs [62, 63]. In addition, TiO_2 NPs possess several advantageous photo-physical properties, notably a photocatalytic attribute. Extensive research has been conducted on the photocatalytic activity of TiO_2 NPs for their potential application in diagnostic assays. According to reports, a thorough study was conducted on the in-situ labelling approaches for fluorescence microscopy to stain the TiO_2 NPs that were absorbed by the cells. In the first approach, fluorescent biotin and fluorescent streptavidin were utilized to label the NPs prior to and following cellular uptake. In the second approach, the copper-catalyzed azide-alkyne cycloaddition method was employed to label and detect azide-conjugated TiO_2 NPs. Using synchrotron X-ray fluorescence microscopy (XFM), researchers were able to detect TiO_2 NPs. The XFM results showed a remarkable match with the location of optical fluorescence observed through confocal microscopy [42]. It is quite simple to prepare and modify TiO_2 NPs, including the addition of europium(III). Additionally, TiO_2 hollow nano shells can serve as effective two-photon nanoprobes. They adhere to HeLa cervical cancer cells when coated with poly(ethyleneimine) and are subsequently detected [64]. Based on the literature, a nanoarchitecture called $Ag@SiO_2@mTiO_2$ was created by coating mesoporous titania onto silver-silica core–shell NPs and the fluorescence was enhanced by the metal core. The particles were equipped with the fluorescent cancer treatment doxorubicin along with fluorescent flavin mononucleotide, and they were utilized for the simultaneous delivery of drugs and imaging using fluorescence and surface-enhanced Raman spectroscopy (SERS) [29].

3.2 TiO_2 Based Implants

The unique nanoscale surface features of TiO_2 NPs have facilitated their effective integration into medical applications as implants, effectively mitigating some limitations associated with titanium implants. We shall explore the utilization of these important nanostructures in the fields of bone, dentistry, and drug-release. The introduction of TiO_2 NPs into medical applications as implants has been facilitated by their nano topographical properties, effectively overcoming some limitations of titanium implants. Here, we will explore the utilization of these important nanostructures in bone, dental, and drug-release implants.

3.2.1 Bone Implants

Bone is a frequently transplanted tissue [65]. There are three primary methods to address bone defects. One option is to use an autologous bone graft, where the bone is taken from the patient's own body. Despite being considered the most reliable method, this approach has its drawbacks, such as temporary instability, minor flaws, and restricted availability. Using bone allografts, which involves transplanting bone from a donor, is a second method that helps to address some of the limitations of the first method. Nevertheless, the clinical failure rate of this method is quite high, ranging from 30 to 60% over a span of 10 years. The third approach involves the use of synthetic implants and the development of new bone through engineering. This method is rapidly gaining popularity as it surpasses the limitations of previous methods.

Bone is a fascinating natural material that combines an organic matrix of collagen fibrils with an inorganic reinforcement of Hap nanocrystals [66]. The natural bone acquires its distinctive mechanical characteristics through the arrangement of hierarchical structures, ranging from collagen fibres at the nanoscale to the dense cortical bone and the macroporous bone network at the macroscopic level [65]. The ordered assembly of bone contributes to both its tensile strength and toughness. Therefore, a bone scaffold that is effective must replicate the intricate structure of bone by combining different materials with diverse mechanical properties at various length scales. Furthermore, the significance of the surface properties is very vital in the interaction between the extracellular matrix and the implant, as well as in the process of osseointegration [65]. Osseointegration is the establishment of a direct and strong link between living bone and the surface of an implant that carries a load, both structurally and functionally. Osseointegration refers to the process of freshly generated bone growing around an artificial implant. Consequently, the success or failure of an implant is, then, highly dependent on the way tissue and the implant interact at the interface.

As we know, titanium and its various alloys have proven to be excellent choices for bone implants. This is due to their impressive tensile strength, steady support, low elastic modules that seamlessly align with tissue, and their capacity to generate a durable TiO_2 layer on the surface, which prevents corrosion. Nevertheless, there are drawbacks such as limited interaction with neighbouring tissue and their impact on biological processes [66]. In a recently conducted study, a highly effective approach was outlined to improve the binding of TiO_2 NTs. This technique utilized high-pressure torsional treatment to enhance the refinement of grains in the underlying titanium layer (Fig. 3). This technique improved the bonding of TiO_2 NTs with bone tissue and their ability to interact with living organisms by enhancing the stiffness of the surface. The research findings indicate that TiO_2 NTs measuring 0.4 μm in length have greater adhesion strength than lengthier NTs measuring 2 μm. This is attributed to the reduced interfacial stress in the shorter NTs [68].

3.2.2 Dental Implants

When disorders, diseases, or injuries cause organ loss, there is a need to restore the affected organs, such as joints or teeth and typically done by replacing them with artificial materials. For optimal performance, it is crucial that these artificial materials establish a strong connection with the surrounding tissues through fibrous connective tissues [69, 70]. Unlike other organ replacements, the continued effectiveness of dental implants relies heavily on the speedy recovery process and secure implantation into the jaw bone and enamel. The surface treatment, topography, macro design, and geometry of a dental implant system play crucial roles in ensuring the durability of the implants, both in the short and long term [71, 72].

With a deep understanding of the intricate processes involved in osseointegration, researchers have dedicated considerable time and effort to studying and enhancing implant surfaces. These surfaces play a crucial role in ensuring the stability of

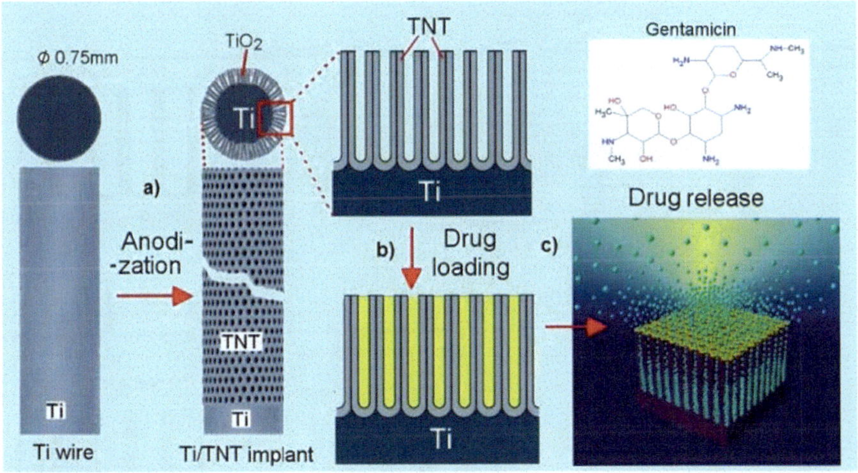

Fig. 3 A schematic showing bone implant fixation prepared by fabrication of TiO_2 NTs on Ti wires followed by **a** electrochemical anodization, **b** loading of local therapeutics and the release of drug molecules [67] (open access)

implants and facilitating a quicker and more successful osseointegration process in newly formed bone. The surface characteristics of the dental implant are determined at different scales, including the macro, micro, and nano levels. Nano roughness is the most frequently observed type of nano features found on dental implant surfaces [73]. Thus, the surface topography of dental implants plays a vital role in achieving proper osseointegration. The porous surface plays a vital role in this field. With its porous surface, a dental implant covered by TiO_2 NTs arrays offers a larger area for drug loading and improved biocompatibility, making it an ideal choice (Fig. 4).

A study confirmed the fact mentioned above by growing TiO_2 NTs on dental implant surfaces using the anodic oxidation method. Using TiO_2 NTs measuring 60 nm in diameter and 10 μm in length, the incorporation of bone morphogenetic protein 2 (BMP2) resulted in enhanced osseointegration [75]. The interaction between implants and blood is crucial for peri-implant healing. Under certain circumstances, the immune system can react to the presence of titanium implants, resulting in inflammation or fibrosis, which can ultimately lead to the failure of the implant. Research has shown that TiO_2 NTs exhibit excellent blood compatibility, suggesting their great promise for enhancing the surfaces of implants that come into contact with blood. Immobilizing certain biological molecules on TiO_2 NTs has the potential to enhance osseointegration in dental implants [76].

In a study, Huang et al. explored the impact of various structures of TiO_2 NTs on the behaviour and adhesion of blood platelets. The research findings indicate that various factors such as surface chemistry, structure, wettability, and crystalline phase had a significant impact on the responses of platelet-rich plasma. Specifically, it was noted that amorphous TiO_2 NTs with a larger diameter had the ability to decrease platelet adhesion and activation. In addition, there have been reports

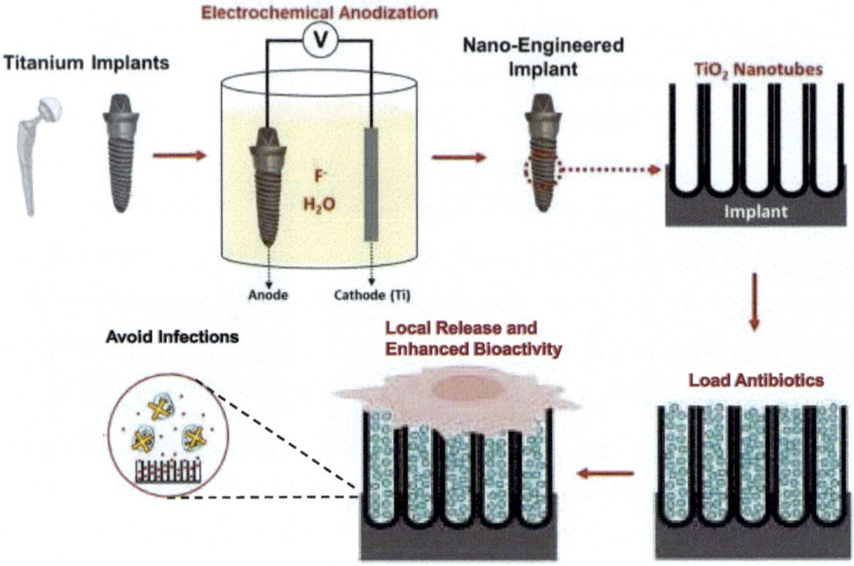

Fig. 4 A diagram showing anodized TiO_2 NTs dental implants for enhanced bioactivity and drug release [74] (open access)

suggesting that the combination of various characteristics of TiO_2 NTs, including wettability, surface topography, and chemistry, can enhance cell adhesion, migration, proliferation, and differentiation. For this particular scenario, creating distinct wettability patterns on the surface of TiO_2 nanostructures has shown great potential in enhancing the practicality of biomedical applications, like implant scaffolds [77, 78].

3.3 *Biosensors Based on TiO_2 NPs*

The research and development of biosensors has been quite significant in recent years, thanks to the remarkable progress in nanotechnology [7, 19]. There has been a growing fascination with the sensing application of TiO_2 for use in biosensors in recent years [17, 79]. With their exceptional properties, nanomaterial-based biosensors offer highly sensitive and fast biological detection. With their well-defined nanostructured platforms, TiO_2 NPs offer excellent transport pathways, strong adhesion to the substrate, and a high surface area. This means they have a large number of atoms on their surface, which can interact with numerous biomolecules. As a result, they are ideal for use as electrodes in the production of biosensors [80].

This section of the chapter focuses solely on nano biosensors that utilize TiO_2 NPs, despite the various nanomaterials that have been employed in medical biosensors [81–83]. These biosensors have the ability to detect specific biomolecules such as DNA and proteins with exceptional sensitivity. These biosensors have a wide range of applications in disease diagnosis and drug screening. Recent research in the field of biomolecules and TiO_2 nanostructures has yielded significant achievements in the development of bio-nano hybrid devices, including biomolecule-sensitized solar cells (BSSCs) and photoelectrochemical cells (PECs) [84]. In addition to the devices mentioned above, the combination of biomolecules and TiO_2 nanostructured thin film has paved the way for advancements in this field, particularly in the realm of biosensors. With the ability to detect biological analytes, there are exciting possibilities for exploring how patient response to medical or surgical therapy can be monitored. In order for a biosensor to be successful in the market, it needs to be cost-effective, easy to use, highly sensitive and accurate, fast, and able to be manufactured efficiently with high selectivity [85]. In addition, TiO_2 nanostructures have been used to create a range of detection devices, including sensors for humidity, oxygen, and hydrogen. These nano-sized semiconductors have shown great potential as electrode materials in biosensors because of their unique properties, including a porous structure, a large specific surface area, and exceptional biocompatibility. Using TiO_2 as an immobilizing matrix helps preserve the biocatalytic activity of enzymes by reacting with their amine and carboxyl groups [86].

The PEC biosensors have drawn significant attention compared to the other electrochemistry-based biosensors as they allow for both optical and electrochemical detection. The PEC biosensors utilize the PEC process, where the conversion of photons to electricity takes place as shown in Fig. 5. In the PEC process, the semiconductor absorbs photons, creating electron–hole pairs. These pairs are then separated, with electrons going to the working electrode and holes going to the electrolyte. When the system detects the target analyte, it causes a shift in the photocurrent. The magnitude of this change is exactly proportional to the amount of the desired substance present in the sample. In a recent study, Fan et al. reported on application of a PEC biosensor based on CdS-sensitized Fe-doped TiO_2 to detect the antigen for squamous cell cancer. Doping TiO_2 with iron increases the composite's absorption in the visible light spectrum. Moreover, carboxyl functionalized CdS NPs bond adequately to the rough surfaces of Fe-TiO_2 NPs [87]. To develop biosensors with specific features and actions, PEC biosensors may be combined with various sensing techniques. Bernacka-Wojcik et al. created a disposable biosensor using dye-sensitized solar cells (DSSCs) and TiO_2 nanoparticulate sheets as photoanodes and colorimetric DNA detection to identify specific Mycobacterium tuberculosis DNA sequences [88].

4 TiO_2 NPs Immobilized with Biomolecules

TiO_2 NPs have attracted considerable attention in the field of nanotechnology because of their distinct characteristics and wide-ranging uses. By immobilizing these nanoparticles purposefully with proteins, a potent synergy arises, allowing for groundbreaking advancements in biological systems and presenting promising

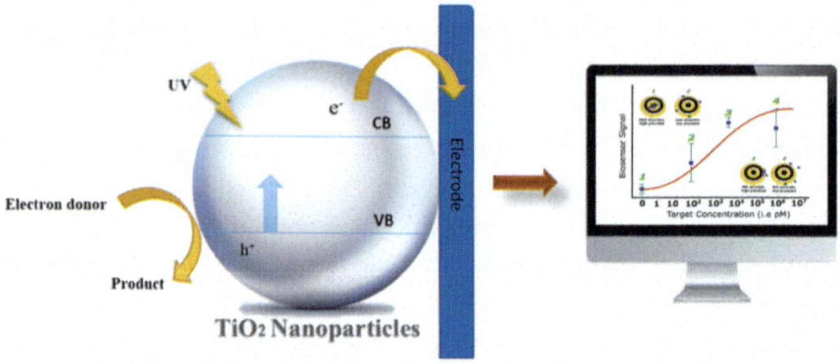

Fig. 5 A schematic diagram illustrating the working principle of PEC biosensors based on TiO_2 [41] (open access)

opportunities in merging nanotechnology and biology. Titanium (Ti) and its alloys are widely utilized in implant manufacturing due to their biocompatibility. To create a nanostructured surface, TiO_2 NTs can be used as they have dimensions similar to the tiny components of human bone and can enhance the effectiveness of implants. Efforts are currently being made to enhance the performance of these TiO_2 NTs by applying a coating of biomolecules to their nanotubular matrices [89, 90].

Enhancing the biocompatibility of the functionalized TiO_2 NTs involves facilitating quick osseointegration, inhibiting bacterial adhesion on the surfaces, and ensuring a controlled and prolonged release of loaded drugs. These nanotubular matrices have also been utilized in the production of high-performance biosensors. These biosensors have the ability to immobilize enzymes on their surfaces, which could potentially be used for disease diagnosis. The biomolecules can be trapped within the nanotubular matrices by applying a layer on the surface of the TiO_2 NTs and/or by filling them. Various methods can be employed for the adsorption of biomolecules in TiO_2 NTs, including physical interactions like hydrogen bonding, hydrophobic interactions, and electrostatic interactions, as well as chemical interactions like covalent binding with the formation of ether, amide, and thioether linkages. There are two types of sol–gel coating methods: spin coating and dip coating [91].

These methods are now employed to improve the layering of TiO_2 NTs with biological molecules, specifically by using chitosan biopolymer on these nanotubular matrices [92, 93].

Biomolecules can be covalently immobilized on TiO_2 NTs through a covalent bond, BMP2 has the capability to functionalize TiO_2 NTs by adhering to the catechol and quinine groups that are formed from dopamine polymerization [94]. Using TiO_2 NTs, bioactive compounds can be loaded to facilitate precise release into a specific anatomical region. The process of lyophilization has frequently been used to load titanium dioxide nanotubes with connective tissue growth factor (CNN2) and the antimicrobial peptide protein B (CecB) [95, 96]. For the purpose of enhancing the compatibility of TiO_2 NTs used in implants, researchers have modified these matrices with biomolecules that are compatible with living cells. These biomolecules have the potential to encourage cell migration to the implant site, improve cell adhesion, facilitate the development of bone-forming cells from mesenchymal stem cells, or even promote the growth of osteoblastic cells if the biomolecule possesses mitogenic properties. These nanotubular matrices have proven to be highly effective in offering a large surface area that can be used to immobilize biomolecules, particularly enzymes, in the production of biosensors. Similar to a biochemist, lectins could be utilized for immobilization in TiO_2 NTs, enhancing the compatibility of implants with living organisms and potentially being employed in the production of biosensors. These biosensors could be used to identify glycosylated biomarkers, which are crucial for diagnosing diseases [97].

5 Summary and Perspectives

Titanium dioxide (TiO_2) based nanoparticles (NPs) represents a promising frontier in biological applications due to their unique properties and wide range of potential uses. These nanoparticles have been thoroughly researched due to their low toxicity, chemical stability, and multifaceted properties. They have applications in drug delivery systems, bone/drug implants, antibacterial agents and so on. Porous TiO_2 NPs can address targeted drug release at tumor sites, while their high functional surface and photocatalytic activity make them ideal for surface coating implants. TiO_2 NPs can also serve as matrices for immobilizing proteins and enzymes in biosensors. Their unique photoactivity, which produces reactive oxygen species (ROS) when exposed to ultraviolet light, makes them promising alternatives for drug delivery. They can be engineered to target disease areas by attaching active targeting ligands, such as antibodies. TiO_2 NPs have shown remarkable bactericidal properties, effectively combating bacteria resistant to antibiotics.

TiO_2 NPs have shown great potential in the fight against cancer due to their exceptional properties such as biocompatibility, photoactivity, and surface reactivity. TiO_2 NPs have been found to be highly effective photosensitizers for photodynamic therapy (PDT) applications, with antibody-immobilized TiO_2 NPs and UV irradiation being highly effective. Also in radiotherapy, they can enhance the impact of ionizing radiation on cancer cells, making them more susceptible to treatment. TiO_2 NTs have found applications in various sectors such as dentistry, surgery, and medical implants due to their size-dependent behaviours, surface reactivity, and biocompatibility. They have been successfully integrated into medical applications as implants due to their unique nanoscale surface features, making them ideal for bone, dentistry, and drug-release implants. TiO_2 NTs have also shown potential in biosensors, detecting specific biomolecules with exceptional sensitivity. They have been used in bio-nano hybrid devices such as PECs.

Future perspectives encompass personalized medicine, combination therapy, utilizing drug carriers based on TiO_2, personalized medicine shows great potential for both diagnostics and therapies. To ensure the safety of TiO_2 NPs in biological systems requires conducting toxicity assessment and biocompatibility studies. Regulatory approval and clinical translation are of utmost importance for fully harnessing the potential of TiO_2 NPs in cancer therapy. Efforts are being made to develop biodegradable TiO_2 NPs for sustainable applications, showcasing a commitment to environmental responsibility.

References

1. Jamil, B., et al.: Nanomaterials: toxicity, risk management and public perception. In: Nanomaterials: Ecotoxicity, Safety, and Public Perception, Chapter 14. Springer (2019)
2. Javed, R., et al.: Role of capping agents in the application of NPs in biomedicine and environmental remediation: recent trends and future prospects. J. Nanobiotechnol. **18**, 1–15 (2020)
3. Ali, I., et al.: Recent advances in syntheses, properties and applications of TiO2 nanostructures. RSC Adv. **8**, 30125–30147 (2018)
4. Reghunath, S., Pinheiro, D., Devi, S.: A review of hierarchial nanostructures of TiO2: advances and applications. Appl. Surf. Sci. Adv. **3**, 100063 (2021)
5. Esfandyari, J., et al.: Capture and detection of rare cancer cells in blood by intrinsic fluorescence of a novel functionalized diatom. Photodiagn. Photodyn. Ther. **30**, 101753 (2020)
6. Garimella, R., Eltorai, A.E.M.: Nanotechnology in orthopedics. J. Orthop. **14**, 30–33 (2017)
7. Janfaza, S. et al.: A selective chemiresistive sensor for the cancer-related volatile organic compound hexanal by using molecularly imprinted polymers and multiwalled carbon nanotubes. Mikrochim. Acta **186** (2019)
8. Razavi, H., Darvishi, M.H., Janfaza, S.: Silver sulfadiazine encapsulated in lipid-based nanocarriers for burn treatment. J. Burn Care Res. **1** (2017) https://doi.org/10.1097/bcr.000000000000 0602
9. Hashemzadeh, H., et al.: Gold nanowires/fibrin nanostructure as microfluidics platforms for enhancing stem cell differentiation: bio-AFM study. Micromachines **11**, 50 (2019)
10. Hashemzadeh, H., Javadi, H., Darvishi, M.H.: Study of structural stability and formation mechanisms in DSPC and DPSM liposomes: a coarse-grained molecular dynamics simulation. Sci. Rep. **10** (2020)
11. Darvishi, M.H., et al.: Targeted DNA delivery to cancer cells using a biotinylated chitosan carrier. Biotechnol. Appl. Biochem. Online/Biotechnol. Appl. Biochem. **64**, 423–432 (2017)
12. Jha, R.K., Jha, P.K., Chaudhury, K., Rana, S.V.S., Guha, S.K.: An emerging interface between life science and nanotechnology: present status and prospects of reproductive healthcare aided by nano-biotechnology. Nano Rev. **5**, 22762 (2014)
13. Naseri, N., Janfaza, S., Irani, R.: Visible light switchable bR/TiO 2 nanostructured photoanodes for bio-inspired solar energy conversion. RSC Adv. **5**, 18642–18646 (2015)
14. Molaeirad, A., Janfaza, S., Karimi-Fard, A., Mahyad, B.: Photocurrent generation by adsorption of two main pigments of Halobacterium salinarum on TiO2 nanostructured electrode. Biotechnol. Appl. Biochem. Online/Biotechnol. Appl. Biochem. **62**, 121–125 (2014)
15. Hashemzadeh, H., Allahverdi, A., Peter, E.: Comparison between three-dimensional spheroid and two-dimensional monolayer in a549 lung cancer and pc9 normal cell lines under treatment of silver nanoparticles. Modares J. Biotechnol. **10**, 573–580 (2019)
16. Esfandyari, J., Shojaedin-Givi, B., Mozafari-Nia, M., Hashemzadeh, H., Naderi-Manesh, H.: Diatom biosilica shell manipulation with gold, spion nanoparticles and trastuzumab with aims of diagnostics of her2 cells. Modares J. Biotechnol. **10**, 581–588 (2019)
17. Baan, R., et al.: Carcinogenicity of carbon black, titanium dioxide, and talc. Lancet Oncol. **7**, 295–296 (2006)
18. Bavykin, D.V., Friedrich, J.M., Walsh, F.C.: Protonated titanates and TiO_2 nanostructured materials: synthesis, properties, and applications. Adv. Mater. **18**, 2807–2824 (2006)
19. Filho, J.T., Rocco, A.M.: Titanium oxide nanotubes: synthesis of anatase phase, characterization and photocatalytic application. Revista Virtual De Química **5** (2013)

20. Xiao, P., et al.: Nanostructured electrode with titania nanotube arrays: fabrication, electrochemical properties, and applications for biosensing. J. Nanosci. Nanotechnol. **9**, 2426–2436 (2009)
21. Minagar, S., Berndt, C.C., Wang, J., Ivanova, E., Wen, C.: A review of the application of anodization for the fabrication of nanotubes on metal implant surfaces. Acta Biomater. **8**, 2875–2888 (2012)
22. Nyamukamba, P., et al.: Synthetic methods for titanium dioxide nanoparticles: a review. In: Yang, D. (ed.) Titanium Dioxide—Material for a Sustainable Environment, pp. 151–175 (2018)
23. Samriti, N., Gupta, R., Ruzimuradov, O., Prakash, J.: Synthesis of multifunctional hybrid semiconductor nanomaterials. In: Advances in Material Research and Technology, pp. 31–51 (2023). https://doi.org/10.1007/978-3-031-39481-2_2
24. Gao, A., et al.: The effects of titania nanotubes with embedded silver oxide nanoparticles on bacteria and osteoblasts. Biomaterials **35**, 4223–4235 (2014)
25. Hang, R., et al.: Antibacterial activity and cytocompatibility of Cu–Ti–O nanotubes. J. Biomed. Mater. Res. Part A **102**, 1850–1858 (2013)
26. Lammers, T., et al.: Polymeric nanomedicines for image-guided drug delivery and tumor-targeted combination therapy. Nano Today **5**, 197–212 (2010)
27. Zahednezhad, F., Zakeri-Milani, P., Mojarrad, J.S., Valizadeh, H.: The latest advances of cisplatin liposomal formulations: essentials for preparation and analysis. Expert Opin. Drug Deliv. **17**, 523–541 (2020)
28. Wang, T., et al.: Potential application of functional porous TiO2 nanoparticles in light-controlled drug release and targeted drug delivery. Acta Biomater. **13**, 354–363 (2015)
29. Zhang, W., et al.: Mesoporous titania based yolk–shell nanoparticles as multifunctional theranostic platforms for SERS imaging and chemophotothermal treatment. Nanoscale **6**, 14514–14522 (2014)
30. Wang, Q., et al.: TiO2 nanotube platforms for smart drug delivery: a review. Int. J. Nanomed. **11**, 4819 (2016)
31. Xu, J., et al.: Photokilling cancer cells using highly cell-specific antibody–TiO2 bioconjugates and electroporation. Bioelectrochemistry **71**, 217–222 (2007)
32. Xu, Z., Jiang, X.: Rapid fabrication of TiO2 coatings with nanoporous composite structure and evaluation of application in artificial implants. Surf. Coat. Technol. **381**, 125094 (2020)
33. Liu, P., et al.: Concurrent photothermal therapy and photodynamic therapy for cutaneous squamous cell carcinoma by gold nanoclusters under a single NIR laser irradiation. J. Mater. Chem. B **7**, 6924–6933 (2019)
34. Lucky, S.S., Soo, K.C., Zhang, Y.: Nanoparticles in photodynamic therapy. Chem. Rev. **115**, 1990–2042 (2015)
35. Youssef, Z., et al.: The application of titanium dioxide, zinc oxide, fullerene, and graphene nanoparticles in photodynamic therapy. Cancer Nanotechnol. **8**, 1–62 (2017)
36. Javed, R., et al.: Diverse biotechnological applications of multifunctional titanium dioxide nanoparticles: an up-to-date review. IET Nanobiotechnol. **16**, 171–189 (2022)
37. Çeşmeli, S., Avci, C.B.: Application of titanium dioxide (TiO2) nanoparticles in cancer therapies. J. Drug Target. **27**, 762–766 (2018)
38. McNamara, K., Tofail, S.A.M.: NPs in biomedical applications. Adv. Phys. X. **2**, 54–88 (2017)
39. Yin, Z.F., et al.: Recent progress in biomedical applications of titanium dioxide. Phys. Chem. Chem. Phys. **15**, 4844–4858 (2013)
40. Tyagi, R., Maurya, A., Samriti, N., Prakash, J., Kohli, S.: Role of TiO2-based photocatalysts in waste water treatment. In: Advances in Material Research and Technology, pp. 201–220 (2023). https://doi.org/10.1007/978-3-031-39481-2_9
41. Jafari, S., et al.: Biomedical applications of TiO_2 nanostructures: recent advances. Int. J. Nanomed. **15**, 3447–3470 (2020)
42. Venkatasubbu, G.D., et al.: Folate targeted PEGylated titanium dioxide NPs as a nanocarrier for targeted paclitaxel drug delivery. Adv. Powder Technol. **24**, 947–954 (2013)
43. Khan, M., et al.: Controlled grafted poly (quaternized-4-vinylpyridineco- acrylic acid) brushes attract bacteria for effective antimicrobial surfaces. J. Mater. Chem. B. **6**, 3782–3791 (2018)

44. Reygaert, W.C.: An overview of the antimicrobial resistance mechanisms of bacteria. AIMS Microbiol. **4**, 482–501 (2018)
45. Ahmad, M.A., et al.: Appraisal of comparative therapeutic potential of undoped and nitrogen-doped titanium dioxide NPs. Molecules **24**, 3916 (2019)
46. de Dicastillo, C.L.: Antimicrobial effect of titanium dioxide nanoparticles. In: Antimicrobial Resistance—A One Health Perspective. Intech Open (2020)
47. Sobha, K., et al.: Emerging trends in nanobiotechnology. Biotechnol. Mol. Biol. Rev. **4**, 1–12 (2010)
48. Hajipour, M.J., et al.: Antibacterial properties of NPs. Trends Biotechnol. **30**, 499–511 (2012)
49. Jesline, A., et al.: Antimicrobial activity of zinc and titanium dioxide NPs against biofilm-producing methicillin-resistant Staphylococcus aureus. Appl. Nanosci. **5**, 157–162 (2015)
50. Arora, B., Murar, M., Dhumale, V.: Antimicrobial potential of TiO2 NPs against MDR Pseudomonas aeruginosa. J. Exp. Nanosci. **10**, 819–827 (2015)
51. Kubacka, A., et al.: Understanding the antimicrobial mechanism of TiO2-based nanocomposite films in a pathogenic bacterium. Sci. Rep. **4**, 4134 (2014)
52. Petica, A., et al.: Synthesis and characterization of silver-titania nanocomposites prepared by electrochemical method with enhanced photocatalytic characteristics, antifungal and antimicrobial activity. J. Mater. Res. Technol. **8**, 41–53 (2019)
53. Javed, R., Ahmad, M.A., Ao, Q.: Theranostic applications of nanobiotechnology in cancer. In: Nanotheranostics, pp. 277–295. Springer (2019)
54. Sun, J., et al.: Recent progress in metal-based nanoparticles mediated photodynamic therapy. Molecules **23**, 1704 (2018)
55. Zhang, H., Shan, Y., Dong, L.: A comparison of TiO₂ and ZnO nanoparticles as photosensitizers in photodynamic therapy for cancer. J. Biomed. Nanotechnol. **10**, 1450–1457 (2014)
56. Yurt, F., et al.: Photodynamic therapy and nuclear imaging activities of zinc phthalocyanine-integrated TiO2 nanoparticles in breast and cervical tumors. Chem. Biol. Drug Des. **91**, 789–796 (2017)
57. Yurt, F., et al.: Investigation of in vitro PDT activities of zinc phthalocyanine immobilised TiO_2 nanoparticles. Int. J. Pharm. **524**, 467–474 (2017)
58. Babu, B., et al.: Inorganic nanoparticles as radiosensitizers for cancer treatment. Nanomaterials **13**, 2873 (2023)
59. Youkhana, E.Q., Feltis, B., Blencowe, A., Geso, M.: Titanium dioxide nanoparticles as radiosensitisers: an in vitro and phantom-based study. Int. J. Med. Sci. **14**, 602–614 (2017)
60. Kurzmann, C., et al.: In vitro evaluation of experimental light activated gels for tooth bleaching. Photochem. Photobiol. Sci. **18**, 1009–1019 (2019)
61. Wolfbeis, O.S.: An overview of nanoparticles commonly used in fluorescent bioimaging. Chem. Soc. Rev. **44**, 4743–4768 (2015)
62. Akasaka, H., et al.: Investigation of the potential of using TiO2 NPs as a contrast agent in computed tomography and magnetic resonance imaging. Appl. Nanosci. **10**, 3143–3148 (2020)
63. Jalvo, B., et al.: Antimicrobial and antibiofilm efficacy of self-cleaning surfaces functionalized by TiO2 photocatalytic NPs against Staphylococcus aureus and Pseudomonas putida. J. Hazard. Mater. **340**, 160–170 (2017)
64. Sandoval, S., et al.: Europium-doped TiO2 hollow nanoshells: twophoton imaging of cell binding. Chem. Mater. **24**, 4222–4230 (2012)
65. Saiz, E., Zimmermann, E.A., Lee, J.S., Wegst, U.G.K., Tomsia, A.P.: Perspectives on the role of nanotechnology in bone tissue engineering. Dent. Mater. **29**, 103–115 (2013). https://doi.org/10.1016/j.dental.2012.08.001
66. Bosco, R., Van Den Beucken, J., Leeuwenburgh, S., Jansen, J.: Surface engineering for bone implants: a trend from passive to active surfaces. Coatings **2**, 95–119 (2012). https://doi.org/10.3390/coatings2030095
67. Gulati, K., Aw, M.S., Losic, D.: Drug-eluting Ti wires with titania nanotube arrays for bone fixation and reduced bone infection. Nanoscale Res. Lett. **6** (2011)

68. Hu, N., et al.: Enhanced interfacial adhesion and osseointegration of anodic TiO2 nanotube arrays on ultra-fine-grained titanium and underlying mechanisms. Acta Biomater. **106**, 360–375 (2020). https://doi.org/10.1016/j.actbio.2020.02.009
69. Oshima, M. et al.: Functional tooth restoration by next-generation bio-hybrid implant as a bio-hybrid artificial organ replacement therapy. Sci. Rep. **4** (2014). https://doi.org/10.1038/srep06044
70. Larsson, L., et al.: Regenerative medicine for periodontal and peri-implant diseases. J. Dent. Res. **95**, 255–266 (2015). https://doi.org/10.1177/0022034515618887
71. Shibli, A., Ehrenfest, D.D.: In dental implant surfaces, nanowar has begun . . . but nanoquest is still at stake! Poseido **1**, 131–140 (2013)
72. Richard, C.: Innovative surface treatments of titanium alloys for biomedical applications. In: Paper Presented at: Materials Science Forum (2017)
73. Mohammadi, M., Hesaraki, S., Hafezi-Ardakani, M.: Investigation of biocompatible nanosized materials for development of strong calcium phosphate bone cement: comparison of nano-titania, nano-silicon carbide and amorphous nano-silica. Ceram. Int. **40**, 8377–8387 (2014). https://doi.org/10.1016/j.ceramint.2014.01.044
74. Hakim, L.K., et al.: The current applications of nano and biomaterials in drug delivery of dental implant. BMC Oral Health **24** (2024)
75. Shim, S.C., Choe, B.H., Jang, I.S., et al.: Effect of tio2 nanotube arrays on osseointegration for dental implant. In: Paper Presented at: Advanced Materials Research (2014)
76. Kim, G.-H., et al.: Evaluation of osteoblast-like cell viability and differentiation on the Gly-Arg-Gly-Asp-Ser peptide immobilized titanium dioxide nanotube via chemical grafting. J. Nanosci. Nanotechnol. **16**, 1396–1399 (2016). https://doi.org/10.1166/jnn.2016.11916
77. Li, H., et al.: Multifunctional wettability patterns prepared by laser processing on superhydrophobic TiO2 nanostructured surfaces. J. Mater. Chem. B **3**, 342–347 (2015). https://doi.org/10.1039/C4TB01814A
78. Lai, Y., et al.: Bioinspired patterning with extreme wettability contrast on TiO2 nanotube array surface: a versatile platform for biomedical applications. Small **9**, 2945–2953 (2013). https://doi.org/10.1002/smll.201300187
79. Abdullah, M., Kamarudin, S.K.: Titanium dioxide nanotubes (TNT) in energy and environmental applications: an overview. Renew. Sustain. Energy Rev. **76**, 212–225 (2017). https://doi.org/10.1016/j.rser.2017.01.057
80. Lee, H.-C., Zhang, L.-F., Lin, J.-L., Chin, Y.-L., Sun, T.-P.: Development of anodic titania nanotubes for application in high sensitivity amperometric glucose and uric acid biosensors. Sensors **13**, 14161–14174 (2013)
81. Ma, F., Zhang, Q., Zhang, C.-Y.: Nanomaterial-based biosensors for DNA methyltransferase assay. J. Mater. Chem. B **8**, 3488–3501 (2020). https://doi.org/10.1039/C9TB02458A
82. Sibel, A.O., Shah, A., (eds.): Nanomaterials-based enzyme biosensors for electrochemical applications: recent trends and future prospects. In: New Developments in Nanosensors for Pharmaceutical Analysis, pp. 381–408. Elsevier (2019)
83. Su, S., et al.: Two-dimensional nanomaterials for biosensing applications. TrAC. Trends Anal. Chem. **119**, 115610 (2019). https://doi.org/10.1016/j.trac.2019.07.021
84. Mahyad, B., Janfaza, S., Hosseini, E.S.: Bio-nano hybrid materials based on bacteriorhodopsin: potential applications and future strategies. Adv. Coll. Interface Sci. **225**, 194–202 (2015). https://doi.org/10.1016/j.cis.2015.09.006
85. Estevez, M.-C., Otte, M.A., Sepulveda, B., Lechuga, L.M.: Trends and challenges of refractometric nanoplasmonic biosensors: a review. Anal. Chim. Acta **806**, 55–73 (2014). https://doi.org/10.1016/j.aca.2013.10.048
86. Kennedy, J.F., Kay, I.M.: Hydrous titanium oxides–new supports for the simple immobilisation of enzymes. J. Chem. Soc. Perkin Trans. I 329–335 (1976) https://doi.org/10.1039/p19760000329
87. Fan, D., et al.: An ultrasensitive label-free immunosensor based on CdS sensitized Fe–TiO2 with high visible-light photoelectrochemical activity. Biosens. Bioelectron. (Online) **74**, 843–848 (2015)

88. Bernacka-Wojcik, I., et al.: Inkjet printed and "doctor blade" TiO2 photodetectors for DNA biosensors. Biosens. Bioelectron. (Online) **25**, 1229–1234 (2010)
89. Saji, V.S., Choe, H.C., Yeung, K.W.K.: Nanotechnology in biomedical applications: a review. Int. J. Nano Biomater. **3**, 119 (2010)
90. Oliveira, W.F., et al.: Functionalization of titanium dioxide nanotubes with biomolecules for biomedical applications. Mater. Sci. Eng. C Biomimetic Mater. Sens. Syst. **81**, 597–606 (2017)
91. Bakri, A.S., Sahdan, M.Z., Adriyanto, F., Said, N.D.M., Raship, A.: Influences of deposition layer on the properties of titanium dioxide thin films fabricated by dip coating technique. ARPN J. Eng. Appl. Sci. **11**, 8834–8839 (2016)
92. Chen, X., et al.: Fabrication of selenium-deposited and chitosan-coated titania nanotubes with anticancer and antibacterial properties. Colloids Surf. B Biointerfaces **103**, 149–157 (2013)
93. Kumeria, T., et al.: Advanced biopolymer-coated drug-releasing titania nanotubes (TNTs) implants with simultaneously enhanced osteoblast adhesion and antibacterial properties. Colloids Surf. B Biointerfaces **130**, 255–263 (2015)
94. Ma, Y., et al.: Nanotubes functionalized with BMP2 knuckle peptide improve the osseointegration of titanium implants in rabbits. J. Biomed. Nanotechnol. **11**, 236–244 (2015)
95. Shen, X., et al.: Cecropin B loaded TiO2 nanotubes coated with hyaluronidase sensitive multilayers for reducing bacterial adhesion. Mater. Des. **92**, 1007–1017 (2016)
96. Wei, H., et al.: Increased fibroblast functionality on CNN2-loaded titania nanotubes. Int. J. Nanomed. **1091** (2012) https://doi.org/10.2147/ijn.s28694
97. Mi, L., et al.: A novel photoelectrochemical immunosensor by integration of nanobody and TiO 2 nanotubes for sensitive detection of serum cystatin C. Anal. Chim. Acta **902**, 107–114 (2016)

Antibacterial Applications of TiO_2 Based Hybrid Semiconductor Nanomaterials: Recent Advances and Future Prospects

Junghyun Cho

Abstract Low-temperature solution growth of TiO_2 films provides unique surface functionalities for base materials. It involves multiple stages of crystal growth at various length scales and is of particular help for the substrates that may not survive high processing temperatures. Having a fundamental understanding of the mechanisms involved in those stages is essential to obtain desirable TiO_2 phases and functional performances. In this review, we report four major phases of TiO_2 for their nanostructures and photocatalytic activities that are essential in antimicrobial properties. To further improve the photocatalytic activities through more efficient electron–hole separation, the TiO_2-based hybrid structures are adopted for creating heterojunction interfaces between heterophases of TiO_2 (i.e., rutile, anatase, brookite, TiO_2 (B)), between TiO_2 and other semiconductor oxides (e.g., ZnO, SnO_2, Ce_2O_3, NiO), or between TiO_2 and nanometals (e.g., Cu, Ag, Ni). For such hybrid structures, 1D TiO_2 rutile or TiO_2 (B) nanowires, nanorods, or nanotubes can be fabricated as a 'core' structure using a hydrothermal method or other solution grown processes. To create a hybrid configuration of TiO_2 with other semiconductor oxides, the energy level diagrams of oxides are proposed. Metal nanoparticles have been extensively used as an antimicrobial agent, so the hybrid configuration between TiO_2 and metal nanoparticles can provide a synergistic effect on the antibacterial performance with additional toxicity modes different from the photocatalytic reactions via generation of reactive oxygen species (ROS). The size and shape of the surface (or 'shell') feature in addition to its chemistry seem to play an important role in inactivating the bacterial pathogens, for which nanostructured surface can be optimally designed for the TiO_2 core structure. Especially, the brookite and TiO_2 (B) phases are much less studied than their rutile and anatase counterparts, thereby providing unique opportunities for further development of antimicrobial surfaces that can be nonselective to types of microorganisms.

J. Cho (✉)
Materials Science and Engineering Program, Department of Mechanical Engineering, Thomas J. Watson College of Engineering and Applied Science, Binghamton University (State University of New York), Binghamton, NY 13902, USA
e-mail: jcho@binghamton.edu

J. Prakash et al. (eds.), *Titanium Dioxide-Based Multifunctional Hybrid Nanomaterials*, Engineering Materials, https://doi.org/10.1007/978-3-031-80625-4_14

Keywords Titanium oxides · Rutile · Anatase · Brookite · TiO_2 (B) · Hybrid · Nanostructures · Photocatalytic · Antibacterial · Nanoparticles

1 Introduction

A bacterial infection spreads faster than the speed at which new antibiotics are being developed. Further, its antimicrobial resistance (AMR) has emerged as one of the global health issues of the twenty-first century as the drugs used to treat bacteria no longer become effective [1]. It was reported that AMR could kill 10 million people per year by 2050 [2, 3]. Albeit controversial for its magnitude, [4] it is clear that the spread of AMR could be significant and lethal in the near future and that there is a strong demand to subdue its potential impact. The bacterial spread often begins with bacterial colonization on high touch surfaces found in biomedical tools, hospitals, public buildings, and smart phones and devices, which is grown into biofilms. One of the effective ways to mitigate the spread of bacterial pathogens and AMR is to have such surfaces covered with antibacterial coating [5–10]. On bacterial contact, the surfaces covered with an antibacterial material that generate metal ions, [11, 12] penetrate the bacteria cell wall, produce reactive oxygen species (ROS), and cause gene modification and metabolic change may provide mechanisms to inactivate the bacterial spread [13]. This functional surface can be made from nanoparticles and nanostructures of metals or metal oxides.

Among them, titanium oxide (TiO_2) has been extensively studied for its various properties originated from its unique band gap structure and widely applied to solar energy conversions, sensors, high-k dielectrics, and environmental protection applications [14–19]. As electron–hole pairs are generated upon absorption of incoming photons, ROS are generated in aqueous or air environment and react with dye molecules or cell surfaces for their decomposition. Therefore, TiO_2 provides a very promising solution for antibacterial requirements in many applications in various environments including air, water, and space. Especially in space application, where water supply is limited, re-circulation of clean water is essential [20, 21].

TiO_2 has been reported commonly with four polymorphs (anatase, rutile, brookite and TiO_2 (B)), [22–24] but the rutile is the most stable phase while the other three are metastable phases. The nanostructured anatase is generally considered the most photoactive phase [9, 10]. Although some literature suggested the selectivity of exposing surface planes on the photocatalytic property, their relation with surface free energy was not clearly established [25–27]. For the anatase structure, surface energies were measured to be 0.44 J/m^2, 0.53 J/m^2, and 0.90 J/m^2 for (101), (100), and (001), respectively [28–30]. Despite TiO_2 and other oxide nanostructures acting as a strong biocide, their efficacy has been limited and inconsistent due to the difficulty in its scaling up for large area coatings, sensitivity to the active surface area, and diffusion length of charge carriers from photocatalytic reactions. The last aspect results in the high recombination rate of electrons and holes that will reduce the quantum efficiency and decrease the photocatalytic activity [31]. One way to overcome this difficulty is to use the hybrid oxide structures (or nanocomposites) to connect one phase of TiO_2 with its another phase [32] or other oxide phases such as ZnO and SnO_2 [33, 34]. It will effectively stimulate electron transfer from one phase to another while maximizing the multiple controlling mechanisms for photocatalytic and antibacterial performance.

Metal oxide coatings such as TiO_2 and its hybrid phases can be processed through vapor phase deposition or solution grown processes [35–42]. In the past two decades, solution grown methods for thin oxide films have received special attention owing to their potential to be processed via low temperature synthetic routes that can also provide a high crystalline structure with various functional properties [43–45]. Compared with other competitive thin film deposition techniques, solution processes not only have a capability to conformally form thin film on the complex base structures, but also to obtain innovative nanostructures with minimal investment for capital equipment [46, 47]. Among them, hydrothermal method has been recognized as a very effective technique to produce highly crystalline structures unlike other solution-based processes. As shown in literature, crystal phase formation and growth kinetics in hydrothermal system can be well controlled by solution parameters including precursor concentration, temperature, pH and ionic species [48–52].

Specifically, titanium oxide precursor species involving SO_4^{2-} ions could interfere with the ligand construction during the solution process, whereas some other species such as Cl^- ions could modify the surface energy for certain crystal faces [53–55]. While Cl^- ions have been shown to contribute to the rutile TiO_2 growth in the previous studies, [56, 57] some of the earlier reports indicated that SO_4^{2-} ions could effectively induce an anatase TiO_2 phase in solution process [58, 59]. Several mechanisms have been established so far to explain the formation of various anatase microstructures from the solution processes [60]. For example, anatase TiO_2 surface morphologies composed of randomly oriented, nanotextured, or single-crystalline nanoparticles were prepared on the fluorine-doped tin oxide (FTO)-coated glass substrates [9]. Among these morphologies, the highly textured anatase film with a strong [001] out-of-plane texture on (001) nanofacets exhibited strong photocatalytic dye degradation and anti-bacterial performance under both UV light and dark environment [9].

Such work provides a possible pathway for the developments of TiO_2 nanostructures and its hybrid configurations for generating the antibacterial surface. It can start with nanoparticles whose spontaneous self-organization process with neighboring particles by sharing a common crystallographic orientation, which enables these particles to bond each other and to continuously grow. The driving force for this phenomenon is the overall energy reduction during the self-organization process of these nanoparticles. This oriented attachment (OA) mechanism has been in fact observed on the highest surface energy planes, but it has also been seen on other surfaces such as (100) and (110) planes, [61] which offers the basis of the structural organization observed. With this non-classical crystallization pathway (instead of the classical crystallization that is based on ion-by-ion growth), it offers a process scheme that can make the hybrid oxide structures at more sustainable low-temperature, low-energy consumed processes.

2 Different Polymorphs of TiO_2

The crystal structure of TiO_2 is one of the major factors determining its functional properties such as photocatalysis and antimicrobial performance. TiO_2 has four major polymorphs in nature including rutile, anatase, brookite, and TiO_2 (B). Table 1 explains their crystal structure and relevant properties [31]. All four types of TiO_2 are constructed based upon a TiO_6 octahedral unit while differing in its organization through sharing edges and corners in different fashions (Fig. 1). The characteristics of the Ti–O bond determine the variations in structural and functional properties of the above TiO_2 polymorphs [62].

The rutile phase is thermodynamically the most stable one while the other three phases are metastable. Therefore, rutile can be normally obtained by annealing the other three phases at elevated temperatures for sufficient time. It was also reported that such transformation may depend on the crystallite size, anatase having the smallest particle size that transforms into brookite (> 11 nm) or rutile (> 35 nm) when they grow [63]. The anatase phase shows the most photocatalytic performance than the other TiO_2 phases. In addition to the structural differences and density, the bandgap energies and band structures of TiO_2 phases exhibit some difference, where anatase is a semiconductor with an indirect bandgap while rutile and brookite with a direct bandgap [17, 22]. It results in much higher photocatalytic activity for the indirect bandgap anatase phases than that of the other phases not only because of a longer lifetime of photoexcited electrons and holes but also of their lighter average effective mass [22] and increased surface area [64].

The photocatalytic activity of brookite is reported to be highly dependent upon its point defects [65] and dopants [66]. Recently TiO_2 (B) phase received more attention due to its unique crystallographic structure with fibrous morphology [67] and its photocatalytic performance [68]. TiO_2 (B) nanowires can be converted from layered hydrogen titanates upon heating, [69] while TiO_2 (B) nanoparticles can be obtained from dissolved Ti metal through hydrothermal process [70]. In particular,

Table 1 Crystal structure, lattice parameters, and properties of four major TiO_2 polymorphs [31]

Phase	Crystal systems, space group	Lattice parameters (Å)	Volume (Å^3)	Density (g/cm^3)	Bandgap energy (eV)
Rutile	Tetragonal, $P4_2/mnm$	a = 4.594 c = 2.959	62.45	4.25	3.02–3.04
Anatase	Tetragonal, $I4_1/amd$	a = 3.784 c = 9.515	136.24	3.89	3.20–3.23
Brookite	Orthorhombic, Pbca	a = 9.148 b = 5.447 c = 5.145	257.38	4.12	3.14–3.31
TiO_2 (B)	Monoclinic, C2/m	a = 12.179 b = 3.741 c = 6.525 β = 107.054°	284.22	3.73	3.09–3.22

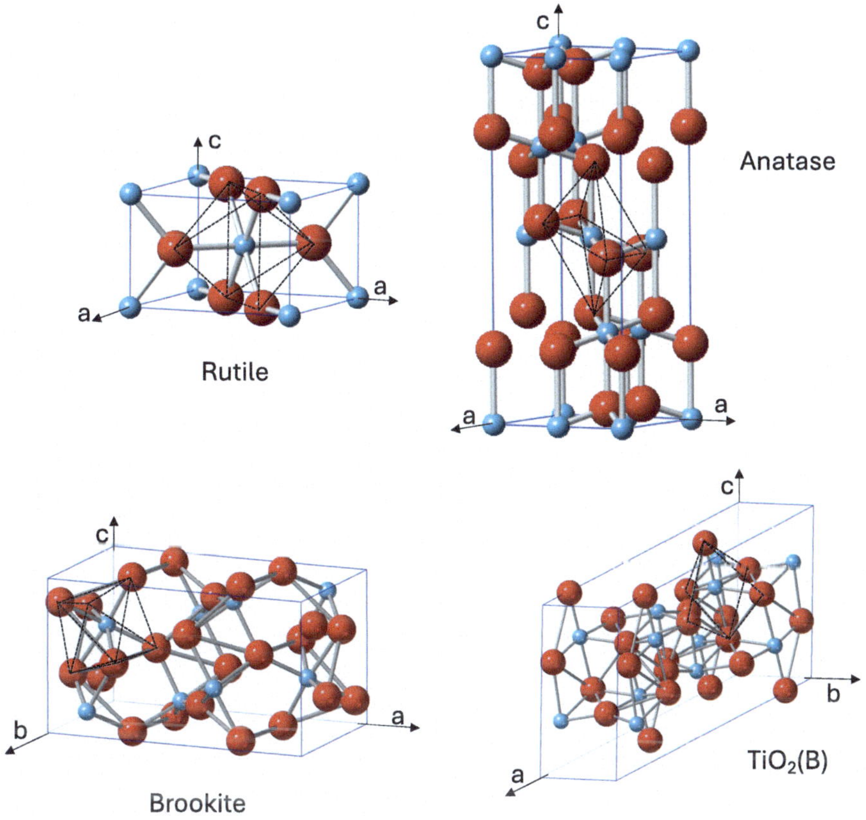

Fig. 1 Crystal models of four major polymorphs of TiO_2. The unit cell is shown by a solid rectangular box, in which a TiO_6 octahedral unit is indicated as dotted lines

TiO_2 (B) has an advantage over the other TiO_2 polymorphs because of its open structure that can act as an excellent electrode for energy storage electrode [70].

3 Antibacterial Behavior of TiO_2 Photocatalysts

The wide bandgap semiconductor oxides are well known as a photocatalytic agent for decomposing organic contaminants and inactivating various types of microrganisms. It starts with excited electrons of the oxide when irradiated by the photon energy (ultraviolet or visible light) higher than its bandgap energy, creating the electron (e^-)—hole (h^+) pairs. Chemical reactions then ensue to produce active radicals and ionic species through the oxidation and reduction reactions.

$$\textbf{\textit{Oxidation}}: h^+ + OH^- \rightarrow {}^{\cdot}\mathbf{OH}\ (\textit{hydroxyl radical}) \quad (1)$$

$$\textbf{\textit{Reduction}}: e^- + O_2 \rightarrow {}^{\cdot}\mathbf{O_2}^- \text{ (superoxide radical anion)} \quad (2)$$

The superoxide radical anion then undergoes successive reactions as shown below:

$$^{\cdot}O_2^- + H^+ \rightarrow HOO^{\cdot} \quad (3)$$

$$HOO^{\cdot} + HOO^{\cdot} \rightarrow \mathbf{H_2O_2} + O_2 \quad (4)$$

$$H_2O_2 + e^- \rightarrow {}^{\cdot}\mathbf{OH} + OH^- \quad (5)$$

In aqueous [9, 71–73] or air [74–76] environment, electrons and holes generated from the photon irradiation can either recombine to release the dissipated energy, or can further react with the species in water or air to generate the reactive radicals and species such as $^{\cdot}OH$, $^{\cdot}O_2^-$, and H_2O_2, as shown in the above oxidation and reduction reactions. These species are a strong oxidizer for organic and biomolecular contaminants. Furthermore, holes have high oxidative potential that can directly oxidize organic species. In order to be more reactive under visible light, sub-bands created within the bandgap due to impurities [77, 78] and defects [79] can also play an important role in the photocatalytic reactions.

Some of these nanostructured oxides have shown promising photocatalytic responses. Recent examples of nanostructured coatings of TiO_2 (rutile, anatase) and ZnO are shown in Fig. 2 for their photocatalytic deactivation of methylene blue (MB) dyes under UV light. In particular, these coatings are based upon highly 001-textured anatase nanoclusters and ZnO nanorods. Complete bleaching of the MB dye was observed after 2 h of UV irradiation for ZnO nanorods (or 1 h for anatase), where the dye degradation rate depended on specific microstructures and crystalline planes generated from those secondary nanostructures. On the other hand, TiO_2 rutile nanorods exhibited a very slow photocatalytic degradation of the MB dye.

Regarding the dye degradation capabilities among various facets of anatase films, it was shown that anatase (004) contributes most to the photocatalytic performance. This behavior is attributed to a nanotextured (004) facet that has more atomic density of Ti cations compared to that of a (101) facet. An atomic density of Ti on a (004) plane is $0.0698/Å^2$ while that on a (101) plane is $0.0484/Å^2$. More Ti atoms with broken bonds provide more active surface state. These (004) facets in the anatase film can be easily activated via absorbing the UV light and release radical agents to decompose organic dye molecules [26].

Photocatalytic activity gives a good indication for the antimicrobial property required for prevention of the biofilm formation and spread. The aforementioned TiO_2 (anatase) and ZnO nanorod surfaces indeed displayed promising antibacterial behavior under UV light [9, 80–87]. For the nanostructured TiO_2 (anatase) coating, Teflon, known to proliferate the bacteria, was used as a substrate. The antibacterial test was then performed using the carrier test under two conditions, with bacterial strains being deposited directly onto the surface: with UV light and with no light. As

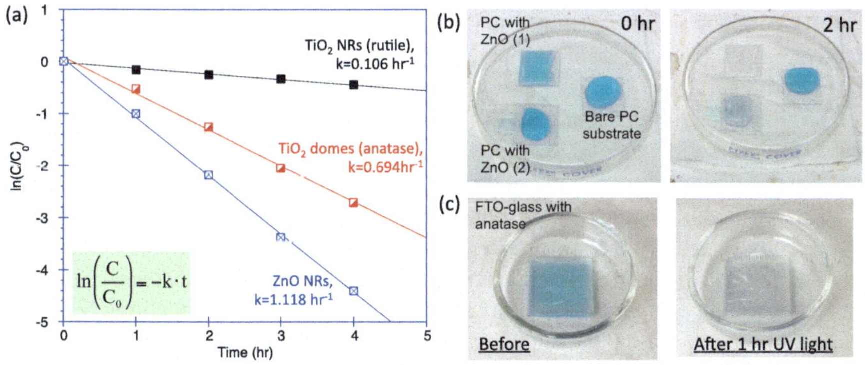

Fig. 2 Photocatalytic activities of nanostructured oxides under UV light; **a** MB dye degradation rates measured from UV/Vis spectroscopy; **b** MB dye degradation on ZnO surfaces; **c** MB degradation on anatase surface. PC: polycarbonate, FTO: fluorine-doped tin oxide. Reprinted with permission from [10]. Copyright 2023 Springer Nature

shown in Fig. 3a, the number of both bacteria was reduced on the anatase surface, compared to that on the control surface (Teflon) [9]. In particular, the anatase-coated surface was more effective with the Gram-positive bacteria (*Staphylococcus aureus*) exhibiting about one log reduction (i.e., 90% reduction) under UV light for 10 min.

The antibacterial performance of the TiO_2 (anatase) surface was even stronger in a dark environment. Again, the antibacterial performance was measured with the *S. aureus* bacteria, for which it exhibited about a 4-log reduction (i.e., 99.99%) relative to Teflon surfaces during four hour exposure in a dark environment (Fig. 3b) [9]. The anatase surface, however, had a reduced effect on the viability of the Gram-negative bacteria, *Pseudomonas aeruginosa,* only showing a 1-log reduction in the viability relative to Teflon surfaces (Fig. 3b). This behavior is often observed from other antimicrobial treatment studies that report more difficulty in inactivating the Gram-negative bacteria than the Gram-positive bacteria [88, 89]. The Gram-negative

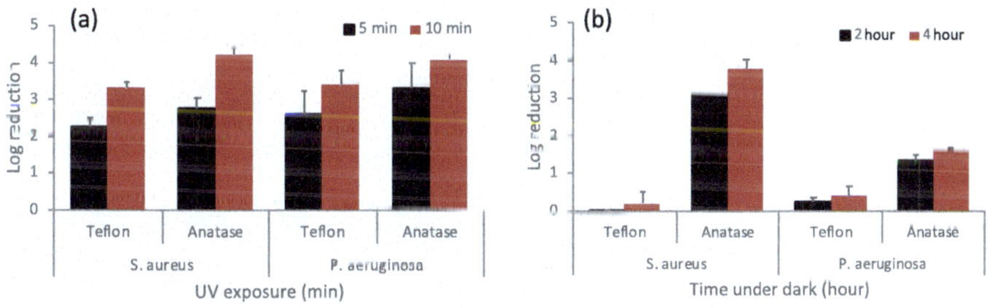

Fig. 3 **a** Antibacterial activity of TiO_2 (anatase) surface under UV light for up to 10 min. **b** Antibacterial activity of anatase surfaces under dark environment for up to 4 h. Note that all experiments were done in triplicate. Reprinted with permission from [9]. Copyright 2019 Elsevier

bacteria have an additional outer membrane (which is not present for the Gram-positive bacteria), so they normally tend to be more resistant to the anti-biotics and antibacterial surfaces.

The contact killing mechanisms of bacteria from metal oxides have been studied and several potential inactivation mechanisms for their 'cytotoxicity' were proposed [82, 90]. First, the presence of metal ions or other metal species can break the balance of bacterial metabolism and be responsible for a significant reduction of the cell's overall activity and proliferation. Secondly, the physical contact between nanoscale material and bacteria can damage the outer wall or membrane of bacteria leading to a leakage, which will be enhanced if the size of contact surface features becomes smaller, eventually killing the damaged bacteria. This phenomenon was detected via SEM observation by Bai et al. [83]. Therefore, the antibacterial performance of the anatase surface under no light can be attributed to the nanocube facets created from the highly textured (004) planes. These nanostructures of anatase must be "toxic" to bacteria even without the effect of photocatalytic reactions under UV light.

The antibacterial activity of nanostructured surfaces also needs to be tested in a liquid environment as numerous bacterial cell proliferations and biofilm formations occur in liquid. The antibacterial test was explored following the bacterial attachment to the respective target surfaces [91]. In order to visualize surface attached bacteria, the *P. aeruginosa* PAO1 strain was used, and the red fluorescent protein mCherry was expressed to enable detection by fluorescence microscopy. The number of attached bacterial cells were then counted on the nanofilm coated and control surfaces using ImageJ. Figure 4 shows microscope images from the attachment assays in the dark for 6 h using *P. aeruginosa.* As seen here, both ZnO and TiO_2 coatings show no trace of bacteria strains unlike their control surfaces.

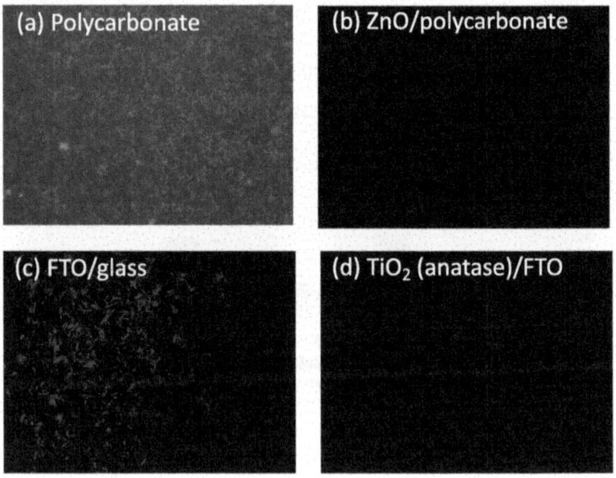

Fig. 4 Effective deactivation of *P. aeruginosa* in liquid: **a** polycarbonate surface; **b** ZnO surface; **c** FTO/glass surface; **d** TiO_2 (anatase) surface. Reprinted with permission from [10]. Copyright 2023 Springer Nature

4 Antibacterial Behavior of TiO_2 Hybrid Structure Between Its Polymorphs

The photocatalytic process requires an efficient electron–hole charge separation. Therefore, to further improve the antimicrobial efficacy of TiO_2 through its photocatalytic performance, the hybrid (nanocomposite) structure has some advantages as it offers a pathway for an efficient electron–hole separation. For example, a vertically grown array of rutile nanorods can transform its surface into another phase (e.g., anatase) to garner the strengths of each component. One potential scheme is to create a core–shell structure by using the oxide nanostructures as a very efficient outer functional surface (shell) that complements more robust TiO_2 nanorod framework (core) that can be grown on the surface of any base materials. Such a core–shell hybrid surface, as shown in Fig. 5, garners the benefits from each component for greater photocatalytic activities than either of the material alone. Besides more efficient electron–hole separation by moving electrons from anatase nanostructures to rutile, and holes from rutile and anatase, it can take advantage of the oxide nanostructures as a very efficient outer functional surface, which may not only better respond under UV light but also extending its enhanced photoactivity into visible wavelengths or room light [92] so that the nanocomposite coating becomes *smarter*. This is attributed to a possibility of creating some sub-band levels within the band structure of anatase due to tailored defects (e.g., Ti^{3+}, Ti^{4+}, O^{2-}, oxygen vacancies, OH^-, etc.), disordered interfaces, and lattice distortion [93]. Such defect levels trap electrons near the conduction band while trapping holes near the valence band.

Furthermore, this configuration will reduce the recombination of electrons within the oxidized species in the environment while holes accumulated in the outer layer generate the *hydroxyl radicals* in the presence of moisture, which are a very strong oxidative agent. The high oxidative potential of the hole in the outer shell permits the direct oxidation of organic species. The bandgap energies of both rutile (3.0 eV) and anatase (3.2 eV) will also enable the hybrid structure to cover a wider wavelength

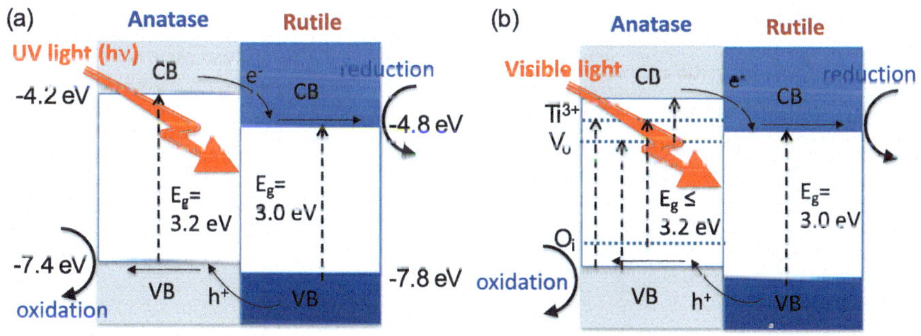

Fig. 5 Schematic illustration of electron–hole generations from the hybrid system upon light illumination: **a** anatase–rutile. Reprinted with permission from [10]. Copyright 2023 Springer Nature; **b** anatase–rutile with defect levels inside the bandgap

range of the light by the creation of the sub-bands within the band structure due to impurities and defects (e.g., oxygen vacancies, interfaces), thus making it more efficient.

Other benefits include increase of the outer surface area that is in contact with the organic species and microbes through nanostructure development. Various shapes and sizes of nanowires and nanoclusters can be possible with the hydrothermally grown oxide nanostructures. This will make much greater surface area for the catalytic reactions, compared to that of a monolithic nanorod array that is made with TiO_2 alone. All these make the rutile/anatase structure very efficient to achieve the antibacterial behavior that does not solely depend on UV light or hydroxyl radicals.

The multiphase heterostructures of TiO_2 with brookite or TiO_2 (B) have been much less reported in the literature, compared to that between rutile and anatase due to the difficulty in fabrication and its phase stability. The anatase-brookite heterojunction TiO_2 has, however, shown good photocatalytic activities, where both the conduction band edge (CBE) and valence band edge (VBE) of brookite are located above the corresponding ones of anatase [94]. Therefore, the brookite phase can work with either anatase or rutile [95] as a shell layer. In addition, the anatase/TiO_2 (B) heterostructure shows improved photoactivities compared to each single phase of anatase or TiO_2 (B) due to the synergistic effects of high specific surface area and more efficient separation of the photogenerated electron–hole pairs [96, 97]. In particular, the 1D TiO_2 (B) core—anatase shell heterostructure was fabricated through an alkaline hydrothermal process, followed by ion exchange and calcination [96]. It showed enhanced photocatalytic degradation of organic species compared to its individual phases and Degussa P25. Another example by Liu et al. shows that the hybrid TiO_2 (B)—anatase nanowires (NWs) transformed from the hydrogen titanate NWs ($H_2Ti_2O_5{\cdot}H_2O$) through the calcination at elevated temperatures exhibited the best photocatalytic performance, compared to NWs made of TiO_2 (B) or anatase phase, as shown in Fig. 6 [98].

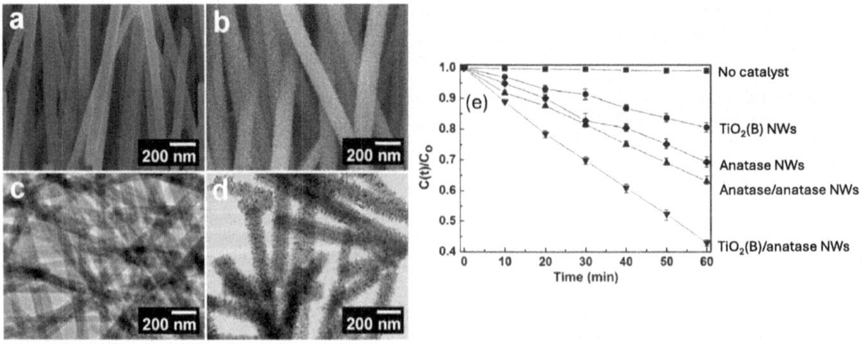

Fig. 6 SEM image of **a** TiO_2 (B) nanowires and **b** TiO_2 (B)/anatase nanowires; TEM images of **c** TiO_2 (B) nanowires and **d** TiO_2 (B)/anatase nanowires; **e** photocatalytic degradation of dyes for the corresponding nanowires, compared to the control sample. Reprinted with permission from [98]. Copyright 2011 American Chemical Society

Studies on antibacterial properties of the TiO_2 hybrid heterojunction that involves brookite or TiO_2 (B) are less commonly observed [99, 100]. However, it is expected that such hybrid structures as explained above will provide the ROS species and nanostructured surfaces that should provide a pathway to inactivate the bacteria. One example is shown for a carbon-doped anatase—brookite hybrid that forms the heterojunction photocatalysts [101]. Figure 7 shows the bacterial growth is significantly suppressed with the presence of the anatase-brookite heterojunction nanoparticles under visible light, compared to that of the commercially available P-25 Degussa nanoparticles (70% anatase + 30% rutile) [101]. The carbon doping makes the hybrid system respond under visible light. In this example, the photogenerated electrons are transferred from the conduction band of brookite to that of anatase, thereby resulting in inactivating the bacteria such as *S. aureus* while avoiding the electron–hole recombination that is a common issue with non-metal doped TiO_2 due to the additional energy levels created from oxygen vacancies in the bandgap [102, 103].

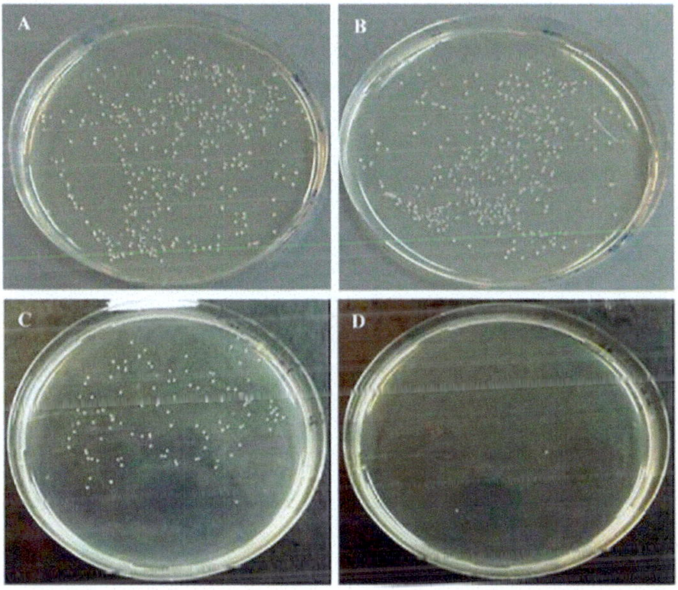

Fig. 7 Bacterial colony growth in the presence of TiO_2 nanoparticles; **a** Degussa P-25, under dark, **b** C-doped anatase brookite nanoparticles, under dark; **c** Degussa P-25 with visible light illumination; **d** C-doped anatase-brookite nanoparticles with visible light illumination. Reprinted with permission from [101]. Copyright 2013 American Chemical Society

5 Antibacterial Behavior of TiO_2 Hybrid Structure with Other Oxides

To further improve antimicrobial efficacy of TiO_2, the hybrid (nanocomposite) structure was suggested with vertically grown 1D TiO_2 nanorods or nanowires (e.g., rutile, TiO_2 (B)), each of which is covered with another oxide (e.g., ZnO, SnO_2, etc.) to achieve the advantages of each component. The main goal is to maximize the photogenerated electron–hole charge separation with the other oxide nanostructures that can also serve as a very efficient outer functional surface. There have been studies on TiO_2, ZnO, Cu_2O, SnO_2, WO_3, and Fe_2O_3 as a photocatalyst for decomposing organic contaminants. Figure 8 show the energy positions of the CBE and VBE along with the bandgap indicated between these levels, [10, 104, 105] which provide a means of designing a functional shell layer for the 1D TiO_2 nanorods as a core. As the photocatalytic reaction favors the degradation of organic pollutants including microorganism, more efficient photocatalysis should provide a mechanism for antibacterial properties. For this efficient photocatalysis, the CBE of an outer layer (shell) of the semiconductor oxides must be positioned above the redox potential of H_2/H^+ while its VBE must be below the redox potential of OH^-/O_2. This criterion gives an insight for selecting an outer shell oxide material including ZnO, Cu_2O, Ce_2O_3, and NiO for the rutile nanorods (core).

Such a hybrid surface can not only be better responded under UV light, but it is also possible to make the nanocomposite coating respond under visible light or room light. This is due to the potential of creating the aforementioned sub-bands within

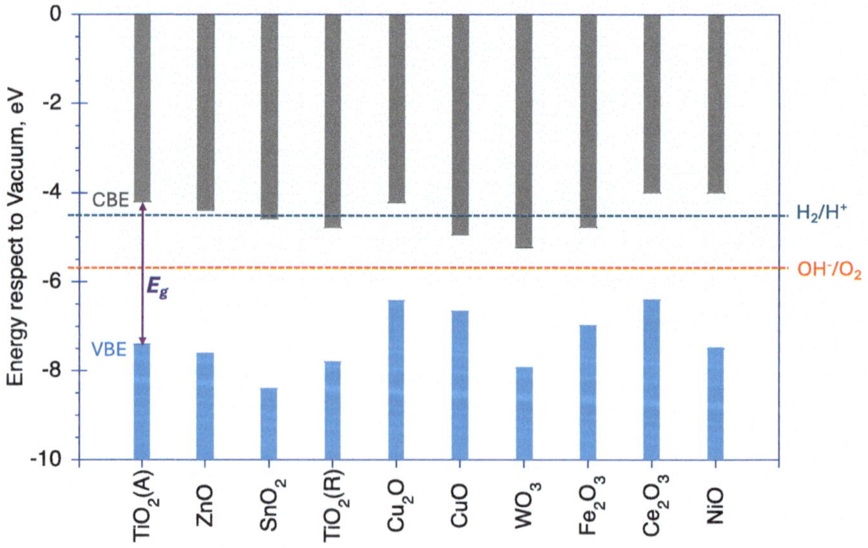

Fig. 8 Energy positions of the conduction band edge (CBE) and valence band edge (VBE) for several semiconducting oxides, along with the water oxidation/reduction potential levels

the band structure of TiO_2 with the presence of impurities and/or defects, thereby providing a means of reducing its bandgap energy for visible light response.

Figure 9 shows two examples of such hybrid structure containing TiO_2 by introducing the rutile core or the anatase shell. It will offer very efficient charge separation mechanisms under light illumination since it moves excited electrons from ZnO shell to rutile core (or from anatase shell to SnO_2 core), and holes from rutile core to ZnO shell (or from SnO_2 core to anatase shell). This process is favored because the bottom of the conduction band edge (lowest unoccupied molecular orbital (LUMO)) of ZnO and anatase is higher than that of rutile and SnO_2, respectively and in addition, the top of their valence band level (highest occupied molecular orbital (HOMO)) is higher than that of rutile and SnO_2, respectively [106–110]. This configuration will also generate the *hydroxyl radicals* accumulated in the outer layer in the presence of moisture, which are very strong oxidative agent. Furthermore, the high oxidative potential of the holes in the outer shell allows the direct oxidation of organic species. One example of the ZnO nanorod shell created on the TiO_2 rutile nanorod core is shown in Fig. 10. The advantage of the core–shell structure is that only the surface can be adjusted for the specific functional properties by a surface dopant or a shell layer while the nanorod core structure remains.

The photoelectrochemical inactivation of microbial cells when exposed to semiconductor materials can involve multiple mechanisms [111, 112]. The majority of these theories are related to the photocatalytic reactions of TiO_2 [105]. It includes oxidative damage to cell membrane that can lead to discharge of intracellular components and eventual demise of the cell. The damage is mainly made from the attach of ROS produced during the photocatalytic reactions. Further, the hydroxyl radical is a strong oxidative species that is nonselective, so can be very effective to any type of biomolecules.

With such operating mechanisms, the TiO_2 hybrid structures have also been investigated for antibacterial properties. Among them, TiO_2/ZnO nanocomposites were

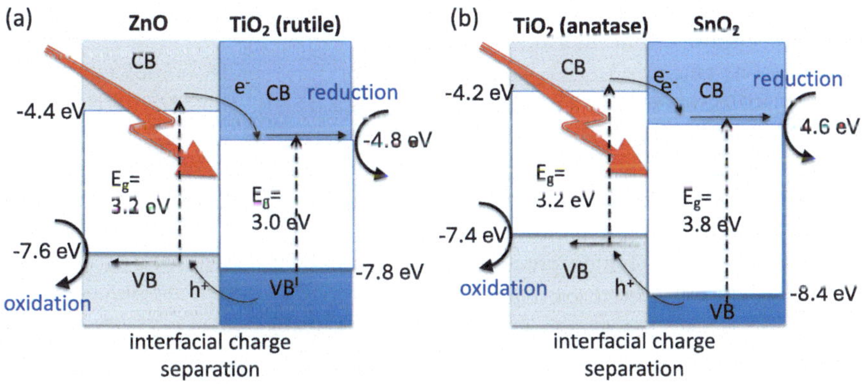

Fig. 9 Schematic illustration of electron–hole separation from the hybrid system: **a** ZnO shell–TiO_2 rutile core. Reprinted with permission from [10]. Copyright 2023 Springer Nature; **b** TiO_2 anatase shell–SnO_2 core

Fig. 10 Cross-sectional SEM images of vertically grown TiO_2 nanorods: **a** pristine rutile TiO_2 nanorods; **b** ZnO-covered rutile nanorods. Reprinted with permission from [10]. Copyright 2023 Springer Nature

shown as bactericidal agents against Gram-positive bacterial strains due to the generation of ROS and the role of Zn^{2+} ions [113]. Further, TiO_2 (anatase)/SnO_2 nanocomposite also showed the potential for antibacterial properties on *S. aureus* under visible light irradiation, compared to TiO_2 or SnO_2 alone [114]. The Cu_2O–TiO_2 nanocomposites with a p-n heterojunction was synthesized via a supercritical solvothermal route and showed a photocatalytic inactivation of various bacteria including *Acinetobacter baumannii, P. aeruginosa, E. coli,* and *S. aureus* under visible light [115]. Figure 11 shows the SEM images of the *A. baumannii* morphologies, where the well preserved plump rod shape is clearly visible before the light illumination. However, with the light illumination on the Cu_2O–TiO_2 nanocomposite surface, the change of the bacteria size and the bacteria damage are observed with time. The bacterial killing behavior was attributed to due to the leakage of K^+ ions and critical damage of the cell structure.

The Ce_2O_3/TiO_2 hybrid nanofibers through electrospinning of sol–gel solution also showed a strong bactericidal activity against food pathogens such as *S. aureus* (Gram positive) and *Salmonella typhimurium* (Gran negative) [116]. Figure 12a, b show the SEM images of the Ce_2O_3/TiO_2 hybrid nanofibers. The growth rates of the bacteria, measured by optical density (OD) at 600 nm (0.1 OD_{600nm} corresponding to 10^8 cells per milliliter) through time scales are shown in Fig. 12c, d [116]. The highest concentration of the Ce_2O_3/TiO_2 solution (40 μg/ml) was found to completely inhibit the growth of the pathogens. The inactivation of bacteria is attributed to the release of ROS from the hybrid nanofiber surfaces that caused cell membrane damage, followed by the leakage of internal molecules. In addition, the antibacterial activity might have been contributed from the cerium ion release that penetrates through the bacterial cell wall and the interruption of the enzymatic activity.

The same electrospinning process was done to fabricate the NiO/TiO_2 hybrid nanofibers that also showed a strong bactericidal activity against food pathogens such as *S. aureus, E. coli, S. typhimurium,* and *Klebsiella pneumoniae* [117]. This hybrid structure inhibits the growth of tested strains through disruption of cell membranes and depressed activity of some enzymes. The presence of Ni that resulted in Ni doping in TiO_2 also seems to help bind the bacteria to the surface of hybrid nanofibers.

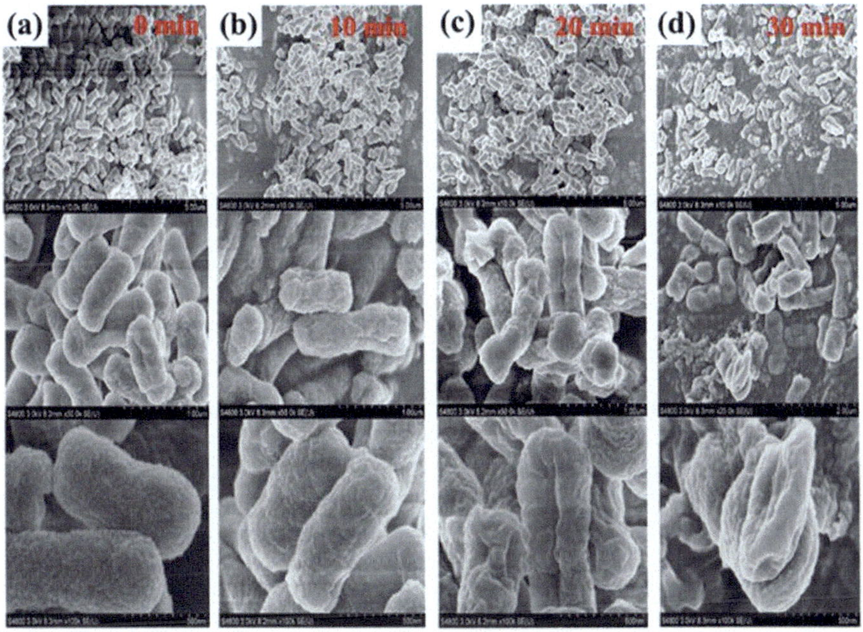

Fig. 11 FE-SEM images of A. baumannii on 5%Cu_2O–TiO_2 catalyst surface: **a** under no visible light; **b–d** under visible light for 10 min, 20 min, and 30 min, respectively, at different magnifications (from top to bottom). Reprinted with permission from [115]. Copyright 2019 American Chemical Society

6 Antibacterial Behavior of TiO_2 Hybrid Structure with Nanometals

Copper [118–120], silver [121–123], zinc [124, 125], and other metals [126] have been extensively used and studied as antimicrobial agents with long history. With the development of formulations that lead to nanometals, these metallic materials are now being tried to form more efficient heterojunction interface with semiconductor metal oxides. The promising surface coatings can then be designed with the hybrid structure from these metal nanoparticles and oxides, which may trigger more efficient photocatalytic activation as well as the contact interactions with various bacteria. For example, 1D TiO_2 nanostructures covered with noble metal nanoparticles (e.g., Au, Ag, Pt) can provide an effective heterojunction interface as it forms the Schottky barrier between metal and TiO_2 that enhances the segregation of charges and reduces the charge recombination [127]. Further, such a heterojunction can benefit from surface plasmon resonance (SPR) effect from metal components. It has been demonstrated that nanometals placed on 1D TiO_2 structure extends the optical band edge of the TiO_2 from UV to visible light because of the SPR effect given by metal components [128, 129].

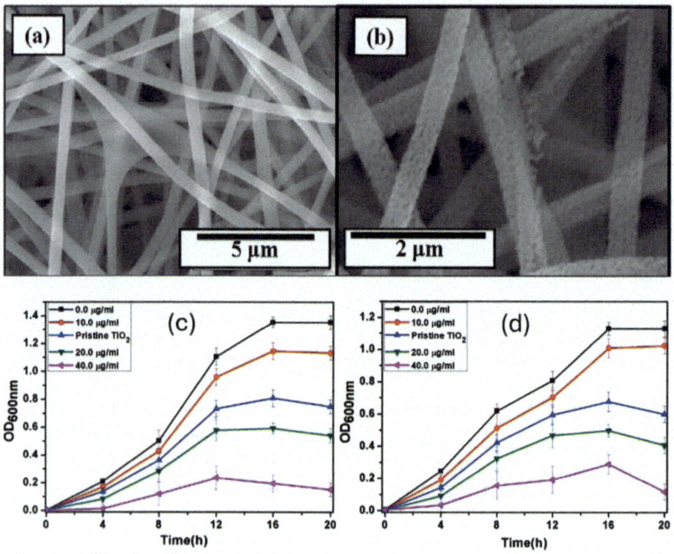

Fig. 12 **a**, **b** FE-SEM images of the Ce_2O_3/TiO_2 hybrid nanofibers at different magnifications. **c**, **d** Growth rates of *S. aureus* and *S. typhimurium*, respectively, exposed to different concentration of Ce_2O_3/TiO_2, along with a control sample and a pristine TiO_2. Reprinted with permission from [116]. Copyright 2012, Springer Nature

Surface coatings that combine TiO_2 with other metals demonstrated a higher efficiency in deactivating *E. coli* and *bacteriophage MS2*, compared to a TiO_2 coating alone. In this example, loading TiO_2 nanowire membrane with both Cu and Ag (Cu–Ag–TiO_2) improved the disinfection of drinking water compared to that from TiO_2, Cu–TiO_2, or Ag–TiO_2 [130]. The improved performance was attributed to the enhanced photocatalytic property by loading silver and copper on the membrane along with the positive effect from the free Ag and Cu ions in water. In addition, the antimicrobial activities of Ag-doped TiO_2 were tested against *E. coli* and other viruses [131]. It exhibited very high antimicrobial activities greater than 99.99% against *E. coli* under UVA illumination, thereby outperforming a pristine TiO_2.

The 'nanosize' of Cu has also been shown to be more effective than their ionic or microsized counterparts, in which Cu and CuO nanoparticles are more toxic to *Escherichia coli* and *Lactobacillus brevis* than their microsized Cu particles and are with toxicities approaching to those of the ionic Cu species [132]. Interestingly, these Cu and CuO nanoparticles establish a different toxicity mode, where nanoparticles are internalized within *E. coli*. This behavior provides a unique genotoxicity potential inside the cells as they generate significant ROS and cause damage to DNA. Figure 13 demonstrates Cu nanoparticles (n-Cu, n-CuO, n-$Cu(OH)_2$) are capable of crossing the cell membrane intact.

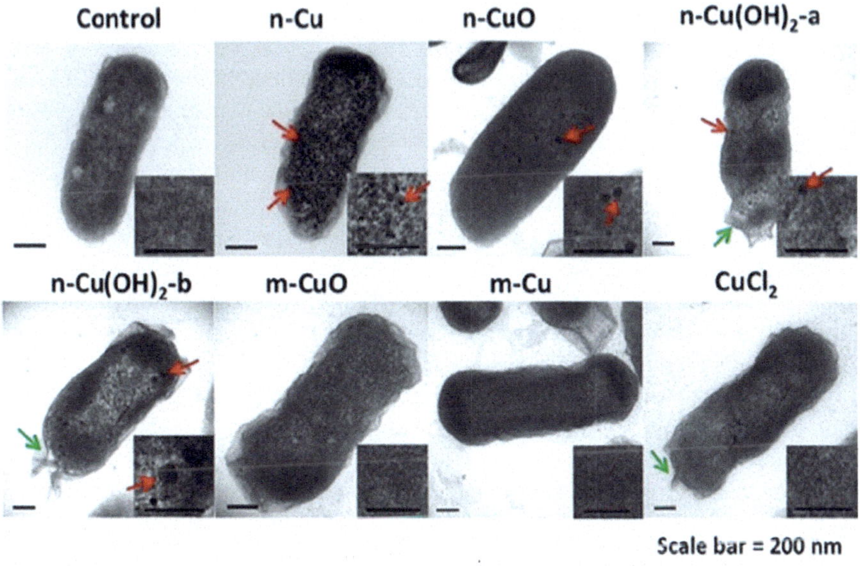

Fig. 13 TEM images of *E. coli* treated with various Cu species. Red arrows indicate Cu particles and green arrows indicate membrane damage. Reprinted with permission from [132]. Copyright 2015, American Chemical Society

7 Conclusion

Titania (TiO_2) is a wide bandgap semiconductor that possesses numerous functionalities including very active photocatalytic properties. A high degree of nanocrystallinity and nanoshape can be easily obtained via solution grown processes including hydrothermal methods. Therefore, it provides an ideal surface coating material for the base structure that needs protection from bacterial infection. Among various TiO_2 crystalline phases, anatase is the most photoactive phase, so becomes one of the most promising antibacterial agents. Its applicability is, however, limited and inconsistent for different bacterial pathogens since a single phase of anatase coatings or nanoparticles not only provide limited surface area, but is also prone to rapid charge recombination.

Therefore, TiO_2-based hybrid structure provides excellent opportunities to mitigate the spread of the harmful pathogens of bacteria and their antimicrobial resistance (AMR) due to its unique bandgap engineering design through the heterojunction interface. One example is the core–shell framework, where one of the TiO_2 phases can serve as a core, a shell, or both. Such a novel design of nanocomposites enables multiple modes of inactivating the harmful pathogens besides the photocatalytic reactions of ROS (e.g., hydroxyl radicals, superoxide, hydrogen peroxide, etc.) and holes, which can be non-selective regardless of their bacterial types. Further, through the hybrid scheme, defects and lattice distortion created across the heterojunction interface result in 'self-doping' that expands the optical absorption edge to visible light.

In particular, one-dimensional framework can be obtained from rutile or TiO_2 (B) by forming nanorods, nanotubes, or nanowires from a low-temperature, low-power consumption process of hydrothermal methods. Then various types of the hybrid structure can be engineered by priming its core surface with the most photoactive phases or toxic materials to bacteria: (i) nanostructured anatase or brookite; (ii) other semiconductor oxides such as ZnO, SnO_2, Cu_2O, Ce_2O_3, and NiO; (iii) metal nanoparticles such as Cu, Ag, and Zn. While the main mechanism of the bacteria killing is from the creation of ROS under light illumination, other modes of bacterial killing will help further improve the antibacterial activity. Besides, the TiO_2 hybrids will be reactive even under room light or without the light at all.

Even though the hybrid structure indeed provides good opportunity to further develop antibacterial properties, the main challenge is to apply the coating that can be more universal and non-selective so that it can provide more generic protection. Further, the bacterial killing needs to be done on an 'immediate' contact rather than waiting for a few hours or even minutes. It can be possible if more bacterial killing mechanisms work simultaneously. In fact, the film approach may be more beneficial as there have been some concerns on utilization of nanoparticles in everyday applications. In this case, the film needs to incorporate the nanostructures to effectively destroy the cell membranes, as well as to increase the surface area. It is also important to increase the crystallinity of the nanostructures as the functionalities will depend on the degree of crystallinity while reducing the random defects resulting from low-temperature processes compared to vapor phase deposition techniques. But tailored defects can be beneficial as they can narrow the bandgap energies so that it can be better responded under visible or room light.

Many vapor phase deposition processes commonly used in the semiconductor industry suffer from a significant carbon footprint. For example, chemical vapor deposition processes that provide more conformal coverage on complex shapes of the substrate consume high energy and toxic precursors while generating large waste and greenhouse gases. Therefore, more sustainable solution-based processes will play a critical role in developing more efficient process through recycling and reusing resources and byproducts.

Acknowledgements This review article was written partly based upon our previous studies and supported by NASA SBIR Program (Contract #: NNX16CJ42P) through N2 Biomedical LLC, Bedford, MA (Arash Aslani, Ward Halverson). Prof. Karin Sauer's group provided the antibacterial tests in her lab at Binghamton Biofilm Research Center. Some of the characterization work presented here was performed at the Analytical and Diagnostics Lab (ADL) of the S^3IP Center. I would also like to acknowledge contributions of the research work from my former graduate students: Jong Hyun Shim, Tao Tao, Biplab Roy, and Guangneng Zhang.

References

1. Murray, C.J.L., Ikuta, K.S., Sharara, F., Swetschinski, L., Robles Aguilar, G., Gray, A., Han, C., Bisignano, C., Rao, P., Wool, E., Johnson, S.C., Browne, A.J., Chipeta, M.G., Fell, F., Hackett, S., et al.: Global burden of bacterial antimicrobial resistance In: 2019: A Systematic Analysis. Lancet **399**(10325), 629–655 (2022)
2. O'neill, J.: Antimicrobial resistance: tackling a crisis for the health and wealth of nations. Rev. Antimicrob. Resist. (2014)
3. O'neill, J.: Tackling drug-resistant infections globally: final report and recommendations (2016)
4. De Kraker, M.E., Stewardson, A.J., Harbarth, S.: Will 10 million people die a year due to antimicrobial resistance by 2050? Plos Med **13**(11), E1002184 (2016)
5. Cloutier, M., Mantovani, D., Rosei, F.: Antibacterial coatings: challenges, perspectives, and opportunities. Trends Biotechnol. **33**(11), 637–652 (2015)
6. Chen, X., Zhou, J., Qian, Y., Zhao, L.: Antibacterial coatings on orthopedic implants. Mater. Today Bio. **19**, 100586 (2023)
7. Wang, N., Yu, J., Yan, J., Hua, F.: Recent advances in antibacterial coatings for orthodontic appliances. Front. Bioeng. Biotechnol. **11** (2023)
8. Prakash, J., Krishna, S.B.N., Kumar, P., Kumar, V., Ghosh, K.S., Swart, H.C., Bellucci, S., Cho, J.: Recent advances on metal oxide based nano-photocatalysts as potential antibacterial and antiviral agents. Catalysts **12**(9), 1047 (2022)
9. Tao, T., Bae, I.-T., Woodruff, K.B., Sauer, K., Cho, J.: Hydrothermally-grown nanostructured anatase TiO_2 coatings tailored for photocatalytic and antibacterial properties. Ceram. Int. **45**(17, Part B), 23216–23224 (2019)
10. Cho, J.: Application of nanostructured metal oxides and its hybrids for inactivation of bacteria and viruses. In: Prakash, J., Cho, J., Campos Janegitz, B., Sun, S. (eds.) Multifunctional Hybrid Semiconductor Photocatalyst Nanomaterials: Application on Health, Energy And Environment, pp. 53–80. Springer International Publishing, Cham (2023)
11. Yasuyuki, M., Kunihiro, K., Kurissery, S., Kanavillil, N., Sato, Y., Kikuchi, Y.: Antibacterial properties of nine pure metals: a laboratory study using Staphylococcus Aureus and Escherichia Coli. Biofouling **26**(7), 851–858 (2010)
12. Vaidya, M.Y., Mcbain, A.J., Butler, J.A., Banks, C.E., Whitehead, K.A.: Antimicrobial efficacy and synergy of metal ions against Enterococcus Faecium, Klebsiella Pneumoniae and Acinetobacter Baumannii in Planktonic And Biofilm phenotypes. Sci. Rep. **7**(1), 5911 (2017)
13. Slavin, Y.N., Asnis, J., Hńfeli, U.O., Bach, H.: Metal nanoparticles: understanding the mechanisms behind antibacterial activity. J. Nanobiotechnol. **15**, 1–20 (2017)
14. Carp, O.: Photoinduced reactivity of titanium dioxide. Prog. Solid State Chem. **32**(1–2), 33–177 (2004)
15. Pelaez, M., Nolan, N.T., Pillai, S.C., Seery, M.K., Falaras, P., Kontos, A.G., Dunlop, P.S.M., Hamilton, J.W.J., Byrne, J.A., O'shea, K., Entezari, M.H., Dionysiou, D.D.: A review on the visible light active titanium dioxide photocatalysts for environmental applications. Appl. Catal. B Environ. **125**, 331–349 (2012)
16. Gaya, U.I., Abdullah, A.H.: Heterogeneous photocatalytic degradation of organic contaminants over titanium dioxide: a review of fundamentals, progress and problems. J. Photochem. Photobiol. C Photochem. Rev. **9**(1), 1–12 (2008)
17. Roy, B.K., Zhang, G., Cho, J.: Titanium oxide nanoparticles precipitated from low-temperature aqueous solutions: III. Thin film properties. J. Am. Ceram. Soc. **95**(2), 676–683 (2012)
18. Roy, B.K., Zhang, G., Magnuson, R., Poliks, M., Cho, J.: Electrodeposition of titania thin films on metallic surface for high-K dielectric applications. J. Am. Ceram. Soc. **93**(3), 774–781 (2010)
19. Roy, B.K., Zhang, G., Yoo, M., Bae, I.-T., Cho, J.: Developments of low-temperature solution processing for nanostructured titania dielectric films. Sci. Adv. Mater. **2**(1), 90–101 (2010)

20. Pickett, M.T., Roberson, L.B., Calabria, J.L., Bullard, T.J., Turner, G., Yeh, D.H.: Regenerative water purification for space applications: needs, challenges, and technologies towards 'closing the loop'. Life Sci. Space Res. **24**, 64–82 (2020)
21. Grigoriev, A.I., Sinyak, Y.E., Samsonov, N.M., Bobe, L.S., Protasov, N.N., Andreychuk, P.O.: Regeneration of water at space stations. Acta Astronaut. **68**(9), 1567–1573 (2011)
22. Zhang, J., Zhou, P., Liu, J., Yu, J.: New understanding of the difference of photocatalytic activity among anatase, rutile and brookite TiO_2. Phys. Chem. Chem. Phys. **16**(38), 20382–20386 (2014)
23. Diebold, U.: The surface science of titanium dioxide. Surf. Sci. Rep. **48**(5), 53–229 (2003)
24. Vittadini, A., Casarin, M., Selloni, A.: Structure and stability of TiO_2-B surfaces: a density functional study. J. Phys. Chem. C **113**(44), 18973–18977 (2009)
25. Daghrir, R., Drogui, P., Robert, D.: Modified TiO_2 for environmental photocatalytic applications: a review. Ind. Eng. Chem. Res. **52**(10), 3581–3599 (2013)
26. Zhang, D., Li, G., Yang, X., Yu, J.C.: A micrometer-size TiO_2 single-crystal photocatalyst with remarkable 80% level of reactive facets. Chem. Commun. (Camb) (29), 4381–4383 (2009)
27. Yang, H.G., Sun, C.H., Qiao, S.Z., Zou, J., Liu, G., Smith, S.C., Cheng, H.M., Lu, G.Q.: Anatase TiO_2 single crystals with a large percentage of reactive facets. Nature **453**(7195), 638–641 (2008)
28. Imani, S.M., Ladouceur, L., Marshall, T., Maclachlan, R., Soleymani, L., Didar, T.F.: Antimicrobial nanomaterials and coatings: current mechanisms and future perspectives to control the spread of viruses including sars-cov-2. ACS Nano **14**(10), 12341–12369 (2020)
29. Zhang, B., Wei, F., Wu, Q., Piao, L., Liu, M., Jin, Z.: Formation and evolution of the high-surface-energy facets of anatase Tio_2. J. Phys. Chem. C **119**(11), 6094–6100 (2015)
30. Lazzeri, M., Vittadini, A., Selloni, A.: Structure and energetics of Stoichiometrictio2anatase surfaces. Phys. Rev. B **63**(15) (2001)
31. Eddy, D.R., Permana, M.D., Sakti, L.K., Sheha, G.A.N., Solihudin, Hidayat, S., Takei, T., Kumada, N., Rahayu, I.: Heterophase polymorph of TiO_2 (Anatase, Rutile, Brookite, TiO_2 (B)) for efficient photocatalyst: fabrication and activity. Nanomaterials **13**(4), 704 (2023)
32. Cao, Y., He, T., Chen, Y., Cao, Y.: Fabrication of rutile TiO_2-Sn/Anatase TiO_2-N heterostructure and its application in visible-light photocatalysis. J. Phys. Chem. C **114**(8), 3627–3633 (2010)
33. Rangel-Mendez, J.R., Matos, J., Cházaro-Ruiz, L.F., González-Castillo, A.C., Barrios-Yáñez, G.: Microwave-assisted synthesis of C-doped TiO_2 and ZnO hybrid nanostructured materials as quantum-dots sensitized solar cells. Appl. Surf. Sci. **434**, 744–755 (2018)
34. Huo, J., Hu, Y., Jiang, H., Huang, W., Li, C.: SnO_2 Nanorod@TiO_2 hybrid material for dye-sensitized solar cells. J. Mater. Chem. A **2**(22), 8266–8272 (2014)
35. Ghanashyam Krishna, M., Vinjanampati, M., Dhar Purkayastha, D.: Metal oxide thin films and nanostructures for self-cleaning applications: current status and future prospects. Eur. Phys. J. Appl. Phys. **62**(3), 30001 (2013)
36. Lu, Z., Jiang, X., Zhou, B., Wu, X., Lu, L.: Study of effect annealing temperature on the structure, morphology and photocatalytic activity of Si doped Tio2 thin films deposited by electron beam evaporation. Appl. Surf. Sci. **257**(24), 10715–10720 (2011)
37. Nejand, B.A., Sanjabi, S., Ahmadi, V.: The effect of sputtering gas pressure on structure and photocatalytic properties of nanostructured titanium oxide self-cleaning thin film. Vacuum **85**(3), 400–405 (2010)
38. Zhao, L., Han, M., Lian, J.: Photocatalytic activity of TiO_2 films with mixed anatase and rutile structures prepared by pulsed laser deposition. Thin Solid Films **516**(10), 3394–3398 (2008)
39. Maruyama, T., Arai, S.: Titanium dioxide thin films prepared by chemical vapor deposition. Solar Energy Mater. Solar Cells **26**(4), 323–329 (1992)
40. Alotaibi, A.M., Sathasivam, S., Williamson, B.A.D., Kafizas, A., Sotelo-Vazquez, C., Taylor, A., Scanlon, D.O., Parkin, I.P.: Chemical vapor deposition of photocatalytically active pure Brookite TiO_2 thin films. Chem. Mater. **30**(4), 1353–1361 (2018)

41. Zhang, G., Roy, B.K., Allard, L.F., Cho, J.: Titanium oxide nanoparticles precipitated from low-temperature aqueous solutions: II. Thin-film formation and microstructure developments. J. Am. Ceram. Soc. **93**(7), 1909–1915 (2010)
42. Cai, N., Cho, J.: Low temperature processed SnO_2 films using aqueous precursor solutions. Ceram. Int. **39**(1), 143–151 (2013)
43. Liu, B., Aydil, E.S.: Growth of oriented single-crystalline rutile TiO_2 nanorods on transparent conducting substrates for dye-sensitized solar cells. J. Am. Chem. Soc. **131**(11), 3985–3990 (2009)
44. Sutiono, H., Tripathi, A.M., Chen, H.-M., Chen, C.-H., Su, W.-N., Chen, L.-Y., Dai, H., Hwang, B.-J.: Facile synthesis of [101]-oriented rutile TiO_2 nanorod array on fto substrate with a tunable anatase–rutile heterojunction for efficient solar water splitting. Acs Sustain. Chem. Eng. **4**(11), 5963–5971 (2016)
45. Noh, H.W., Jeong, S.M., Cho, J., Hong, J.-I.: Ultrahigh photosensitivity of the polar surfaces of single crystalline ZnO nanoplates. Nanoscale **10**(15), 6801–6805 (2018)
46. Dong, L., Cheng, K., Weng, W., Song, C., Du, P., Shen, G., Han, G.: Hydrothermal growth of rutile TiO_2 nanorod films on titanium substrates. Thin Solid Films **519**(15), 4634–4640 (2011)
47. Zhang, G., Howe, J.Y., Coffey, D.W., Blom, D.A., Allard, L.F., Cho, J.: A biomimetic approach to the deposition of ZrO_2 films on self-assembled nanoscale templates. Mater. Sci. Eng. C **26**(8), 1344–1350 (2006)
48. Huang, Q., Zhou, G., Fang, L., Hu, L., Wang, Z.-S.: TiO_2 nanorod arrays grown from a mixed acid medium for efficient dye-sensitized solar cells. Energy Environ. Sci. **4**(6), 2145 (2011)
49. Roy, B.K., Cho, J.: Dielectric properties of solution-deposited crystalline barium titanate thin films. J. Am. Ceram. Soc. **95**(4), 1189–1192 (2012)
50. Amin, G., Asif, M.H., Zainelabdin, A., Zaman, S., Nur, O., Willander, M.: Influence of Ph, precursor concentration, growth time, and temperature on the morphology of Zno nanostructures grown by the hydrothermal method. J. Nanomaterials **2011**(1), 269692 (2011)
51. Zhang, Q., Gao, L.: Preparation of oxide nanocrystals with tunable morphologies by the moderate hydrothermal method: insights from rutile TiO_2. Langmuir **19**(3), 967–971 (2003)
52. Whittingham, M.S., Guo, J.-D., Chen, R., Chirayil, T., Janauer, G., Zavalij, P.: The hydrothermal synthesis of new oxide materials. Solid State Ionics **75**, 257–268 (1995)
53. Perales-Martínez, I.A., Rodríguez-González, V.: Towards the hydrothermal growth of hierarchical cauliflower-like TiO_2 anatase structures. J. Sol-Gel Sci. Technol. **81**(3), 741–749 (2016)
54. Ding, K., Miao, Z., Hu, B., An, G., Sun, Z., Han, B., Liu, Z.: Shape and size controlled synthesis of anatase nanocrystals with the assistance of ionic liquid. Langmuir **26**(7), 5129–5134 (2010)
55. Feng, X., Shankar, K., Varghese, O.K., Paulose, M., Latempa, T.J., Grimes, C.A.: Vertically aligned single crystal TiO_2 nanowire arrays grown directly on transparent conducting oxide coated glass: synthesis details and applications. Nano Lett. **8**(11), 3781–3786 (2008)
56. Shim, J.H., Cho, J.: Hierarchical organization of TiO_2 nanostructures in low-temperature solution processes. J. Am. Ceram. Soc. **99**(2), 431–439 (2016)
57. Shim, J.H., Bae, I.T., Cho, J.: Microstructure development of hydrothermally grown TiO_2 thin films with vertically aligned nanorods. J. Am. Ceram. Soc. **101**(1), 50–60 (2018)
58. Zhao, Z., Sun, Z.C., Zhao, H.F., Zheng, M., Du, P., Zhao, J.L., Fan, H.Y.: Phase control of hierarchically structured mesoporous anatase TiO_2 microspheres covered with {001} facets. J. Mater. Chem. **22**(41), 21965–21971 (2012)
59. Zhao, Z., Tan, H.Q., Zhao, H.F., Li, D., Zheng, M., Du, P., Zhang, G.Q., Qu, D., Sun, Z.C., Fan, H.Y.: Orientated anatase TiO_2 nanocrystal array thin films for self-cleaning coating. Chem. Commun. **49**(79), 8958–8960 (2013)
60. Ban, T., Nakatani, T., Uehara, Y., Ohya, Y.: Microstructure of six-pointed starlike anatase aggregates. Cryst. Growth Des. **8**(3), 935–940 (2008)
61. Cho, K.-S., Talapin, D.V., Gaschler, W., Murray, C.B.: Designing Pbse nanowires and nanorings through oriented attachment of nanoparticles. J. Am. Chem. Soc. **127**(19), 7140–7147 (2005)

62. Nosheen, S., Galasso, F.S., Suib, S.L.: Role of Ti–O bonds in phase transitions of TiO_2. Langmuir **25**(13), 7623–7630 (2009)
63. Zhang, H., Banfield, J.F.: Understanding polymorphic phase transformation behavior during growth of nanocrystalline aggregates: insights from TiO_2. J. Phys. Chem. B **104**(15), 3481–3487 (2000)
64. Žerjav, G., Žižek, K., Zavašnik, J., Pintar, A.: Brookite Vs. rutile Vs. anatase: what's behind their various photocatalytic activities? J. Environ. Chem. Eng. **10**(3), 107722 (2022)
65. Khan, S., Je, M., Kim, D., Lee, S., Cho, S.-H., Song, T., Choi, H.: Mapping point defects of brookite TiO_2 for photocatalytic activity beyond anatase and P25. J. Phys. Chem. C **124**(19), 10376–10384 (2020)
66. Zhuang, B., Shi, H., Zhang, H., Zhang, Z.: Sodium doping in brookite TiO_2 enhances its photocatalytic activity. Beilstein J. Nanotechnol. **13**, 599–609 (2022)
67. Marchand, R., Brohan, L., Tournoux, M.: TiO_2 (B) a new form of titanium dioxide and the potassium octatitanate K2ti8o17. Mater. Res. Bull. **15**(8), 1129–1133 (1980)
68. Chakraborty, A.K., Qi, Z., Chai, S.Y., Lee, C., Park, S.-Y., Jang, D.-J., Lee, W.I.: Formation of highly crystallized TiO_2 (B) and its photocatalytic behavior. Appl. Catal. B Environ. **93**(3), 368–375 (2010)
69. Armstrong, A.R., Armstrong, G., Canales, J., Bruce, P.G.: Angew. Chem. Int. Ed. **43**, 2286 (2004)
70. Ren, Y., Liu, Z., Pourpoint, F., Armstrong, A.R., Grey, C.P., Bruce, P.G.: Nanoparticulate TiO_2 (B): an anode for lithium-ion batteries. Angew. Chem. Int. Ed. **51**(9), 2164–2167 (2012)
71. Konstantinou, I.K., Albanis, T.A.: TiO_2-assisted photocatalytic degradation of azo dyes in aqueous solution: kinetic and mechanistic investigations: a review. Appl. Catal. B Environ. **49**(1), 1–14 (2004)
72. Torane, A.P., Ubale, A.B., Kanade, K.G., Pagare, P.K.: Photocatalytic dye degradation study of TiO_2 material. Mater. Today Proc. **43**, 2738–2741 (2021)
73. Chakraborty, A., Samriti, Ruzimuradov, O., Gupta, R.K., Cho, J., Prakash, J.: TiO_2 nanoflower photocatalysts: synthesis, modifications and applications in wastewater treatment for removal of emerging organic pollutants. Environ. Res. **212**, 113550 (2022)
74. Verbruggen, S.W.: TiO_2 photocatalysis for the degradation of pollutants in gas phase: from morphological design to plasmonic enhancement. J. Photochem. Photobiol. C Photochem. Rev. **24**, 64–82 (2015)
75. Binas, V., Venieri, D., Kotzias, D., Kiriakidis, G.: Modified Tio2 based photocatalysts for improved air and health quality. J. Materiomics **3**(1), 3–16 (2017)
76. Obee, T.N., Brown, R.T.: TiO_2 photocatalysis for indoor air applications: effects of humidity and trace contaminant levels on the oxidation rates of formaldehyde, toluene, and 1,3-butadiene. Environ. Sci. Technol. **29**(5), 1223–1231 (1995)
77. Samsudin, E.M., Abd Hamid, S.B.: Effect of band gap engineering in anionic-doped TiO_2 photocatalyst. Appl. Surf. Sci. **391**, 326–336 (2017)
78. Zaleska, A.: Doped-TiO_2: a review. Recent Pat. Eng. **2**(3), 157–164 (2008)
79. Naldoni, A., Allieta, M., Santangelo, S., Marelli, M., Fabbri, F., Cappelli, S., Bianchi, C.L., Psaro, R., Dal Santo, V.: Effect of nature and location of defects on bandgap narrowing in black TiO_2 nanoparticles. J. Am. Chem. Soc. **134**(18), 7600–7603 (2012)
80. Tao, T., Shim, J.H., Joshi, J., Sauer, K., Cho, J.: Tailoring of solution grown TiO_2 and ZnO nanostructures for photoactive performances. In: 2016 Mrs Fall Meeting (Poster Presentation; Boston, MA, Nov 30, 2016). Mrs, Boston, MA (2016)
81. Tao, T., Bae, I.T., Woodruff, K.B., Sauer, K., Cho, J.: Solution grown nanostructured ceramic coatings for photocatalytic and antibacterial surface. In: Techconnect Briefs 2019, Washington, DC, pp. 74–77 (2019). ISBN: 978-0-9988782-8-7
82. Yadav, H.M., Kim, J.-S., Pawar, S.H.: Developments in photocatalytic antibacterial activity of nano TiO_2: a review. Korean J. Chem. Eng. **33**(7), 1989–1998 (2016)
83. Bai, H., Liu, Z., Liu, L., Sun, D.D.: Large-scale production of hierarchical TiO_2 nanorod spheres for photocatalytic elimination of contaminants and killing bacteria. Chem. Eur. J. **19**(9), 3061–3070 (2013)

84. Qi, K., Cheng, B., Yu, J., Ho, W.: Review on the improvement of the photocatalytic and antibacterial activities of Zno. J. Alloy. Compd. **727**, 792–820 (2017)
85. Akhavan, O., Mehrabian, M., Mirabbaszadeh, K., Azimirad, R.: Hydrothermal synthesis of ZnO nanorod arrays for photocatalytic inactivation of bacteria. J. Phys. D Appl. Phys. **42**(22), 225305 (2009)
86. Djurišić, A.B., Leung, Y.H., Ng, A.M.C., Xu, X.Y., Lee, P.K.H., Degger, N., Wu, R.S.S.: Toxicity of metal oxide nanoparticles: mechanisms, characterization, and avoiding experimental artefacts. Small **11**(1), 26–44 (2015)
87. Kadiyala, U., Kotov, N.A., Vanepps, J.S.: Antibacterial metal oxide nanoparticles: challenges in interpreting the literature. Curr. Pharm. Des. **24**(8), 896–903 (2018)
88. Kühn, K.P., Chaberny, I.F., Massholder, K., Stickler, M., Benz, V.W., Sonntag, H.-G., Erdinger, L.: Disinfection of surfaces by photocatalytic oxidation with titanium dioxide and uva light. Chemosphere **53**(1), 71–77 (2003)
89. Nakano, R., Hara, M., Ishiguro, H., Yao, Y., Ochiai, T., Nakata, K., Murakami, T., Kajioka, J., Sunada, K., Hashimoto, K., Fujishima, A., Kubota, Y.: Broad spectrum microbicidal activity of photocatalysis by TiO_2. Catalysts **3**(1), 310–323 (2013)
90. Chen, Y., Ng, T.W., Lu, A., Li, Y., Yip, H.Y., An, T., Li, G., Zhao, H., Gao, M., Wong, P.K.: Comparative study of visible-light-driven photocatalytic inactivation of two different wastewater bacteria by natural sphalerite. Chem. Eng. J. **234**, 43–48 (2013)
91. Petrova, O.E., Sauer, K.: Sags contributes to the motile-sessile switch and acts in concert with Bfisr to enable pseudomonas aeruginosa biofilm formation. J. Bacteriol. **193**(23), 6614–6628 (2011)
92. Liu, L., Chen, X.: Titanium dioxide nanomaterials: self-structural modifications. Chem. Rev. **114**(19), 9890–9918 (2014)
93. Xu, Y., Wu, S., Wan, P., Sun, J., Hood, Z.D.: Introducing Ti^{3+} defects based on lattice distortion for enhanced visible light photoreactivity in TiO_2 microspheres. Rsc Adv. **7**(52), 32461–32467 (2017)
94. Song, W., Liu, Y., Dong, Y., Han, X., Mu, M., Chen, Y., Wang, W., Wang, P., Li, W.: Fabrication of Brookite@Anatase heterojunction TiO_2 via phase transformation from metal organic frameworks for enhanced photocatalytic hydrogen evolution and Tch degradation. Catal. Sci. Technol. **13**(11), 3292–3303 (2023)
95. Luo, B., Li, Z., Xu, Y.: The positive effect of anatase and rutile on the brookite-photocatalyzed degradation of phenol. Rsc Adv. **5**(128), 105999–106004 (2015)
96. Wang, C., Zhang, X., Wei, Y., Kong, L., Chang, F., Zheng, H., Wu, L., Zhi, J., Liu, Y.: Correlation between band alignment and enhanced photocatalysis: a case study with anatase/ TiO_2 (B) nanotube heterojunction. Dalton Trans. **44**(29), 13331–13339 (2015)
97. Zhang, J., Xiao, G., Xiao, F.-X., Liu, B.: Revisiting one-dimensional TiO_2 based hybrid heterostructures for heterogeneous photocatalysis: a critical review. Mater. Chem. Front. **1**(2), 231–250 (2017)
98. Liu, B., Khare, A., Aydil, E.S.: TiO_2–B/anatase core–shell heterojunction nanowires for photocatalysis. Acs Appl. Mater. Interfaces **3**(11), 4444–4450 (2011)
99. Tinoco Navarro, L.K., Jaroslav, C.: Enhancing photocatalytic properties of TiO_2 photocatalyst and heterojunctions: a comprehensive review of the impact of biphasic systems in aerogels and xerogels synthesis, methods, and mechanisms for environmental applications. Gels **9**(12), 976 (2023)
100. Kumaravel, V., Nair, K.M., Mathew, S., Bartlett, J., Kennedy, J.E., Manning, H.G., Whelan, B.J., Leyland, N.S., Pillai, S.C.: Antimicrobial Tio2 nanocomposite coatings for surfaces, dental and orthopaedic implants. Chem. Eng. J. **416**, 129071 (2021)
101. Etacheri, V., Michlits, G., Seery, M.K., Hinder, S.J., Pillai, S.C.: A highly efficient $TiO_{2-x}C_x$ nano-heterojunction photocatalyst for visible light induced antibacterial applications. Acs Appl. Mater. Interfaces **5**(5), 1663–1672 (2013)
102. Etacheri, V., Seery, M.K., Hinder, S.J., Pillai, S.C.: Highly visible light active $TiO_{2-x}N_x$ heterojunction photocatalysts. Chem. Mater. **22**(13), 3843–3853 (2010)

103. Etacheri, V., Seery, M.K., Hinder, S.J., Pillai, S.C.: Nanostructured $Ti_{1-x}S_xO_{2-y}N_y$ heterojunctions for efficient visible-light-induced photocatalysis. Inorg. Chem. **51**(13), 7164–7173 (2012)
104. Xu, Y., Schoonen, M.A.A.: The absolute energy positions of conduction and valence bands of selected semiconducting minerals. Am. Mineral. **85**(3–4), 543–556 (2000)
105. Yemmireddy, V.K., Hung, Y.-C.: Using photocatalyst metal oxides as antimicrobial surface coatings to ensure food safety—opportunities and challenges. Compr. Rev. Food Sci. Food Saf. **16**(4), 617–631 (2017)
106. Yu, X., Zhao, Z., Zhang, J., Guo, W., Qiu, J., Li, D., Li, Z., Mou, X., Li, L., Li, A., Liu, H.: Rutile nanorod/anatase nanowire junction array as both sensor and power supplier for high-performance, self-powered, wireless UV photodetector. Small **12**(20), 2759–2767 (2016)
107. Sutiono, H., Tripathi, A.M., Chen, H.-M., Chen, C.-H., Su, W.-N., Chen, L.-Y., Dai, H., Hwang, B.-J.: Facile synthesis of [101]-oriented rutile TiO_2 nanorod array on fto substrate with a tunable anatase-rutile heterojunction for efficient solar water splitting. ACS Sustain. Chem. Eng. **4**(11), 5963–5971 (2016)
108. Zhao, Y., Li, C., Liu, X., Gu, F., Du, H.L., Shi, L.: Zn-doped Tio2 nanoparticles with high photocatalytic activity synthesized by hydrogen–oxygen diffusion flame. Appl. Catal. B Environ. **79**(3), 208–215 (2008)
109. Xiong, Z., Wu, H., Zhang, L., Gu, Y., Zhao, X.S.: Synthesis of TiO_2 with controllable ratio of anatase to rutile. J. Mater. Chem. A **2**(24), 9291–9297 (2014)
110. Pei, C.C., Leung, W.W.-F.: Enhanced photocatalytic activity of electrospun TiO_2/ZnO nanofibers with optimal anatase/rutile ratio. Catal. Commun. **37**, 100–104 (2013)
111. Matsunaga, T., Tomoda, R., Nakajima, T., Wake, H.: Photoelectrochemical sterilization of microbial cells by semiconductor powders. Fems Microbiol. Lett. **29**(1–2), 211–214 (1985)
112. Dizaj, S.M., Lotfipour, F., Barzegar-Jalali, M., Zarrintan, M.H., Adibkia, K.: Antimicrobial activity of the metals and metal oxide nanoparticles. Mater. Sci. Eng. C **44**, 278–284 (2014)
113. Harun, N.H., Mydin, R.B.S.M.N., Sreekantan, S., Saharudin, K.A., Basiron, N., Aris, F., Wan Mohd Zain, W.N., Seeni, A.: Bactericidal capacity of a heterogeneous TiO_2/ZnO nanocomposite against multidrug-resistant and non-multidrug-resistant bacterial strains associated with nosocomial infections. ACS Omega **5**(21), 12027–12034 (2020)
114. Gnanasekaran, L., Rajendran, S., Senthil Kumar, P., Priya, A.K., Gracia, F., Habila, M.A., Saravanakumar, K.: Visible light stimulated binary nanostructure and defect enriched TiO_2-SnO_2 for photocatalysis and antibacterial activity. Mater. Lett. **316**, 131998 (2022)
115. Li, H., Zhong, J., Zhu, H., Yang, Y., Ding, M., Luo, L., Huo, Y., Li, H.: Hybrid Cu_2O/TiO_2 nanocomposites with enhanced photocatalytic antibacterial activity toward acinetobacter baumannii. ACS Appl. Bio Mater. **2**(11), 4892–4903 (2019)
116. Hassan, M.S., Amna, T., Al-Deyab, S.S., Kim, H.-C., Oh, T.-H., Khil, M.-S.: Toxicity of Ce_2O_3/TiO_2 composite nanofibers against S. Aureus and S. Typhimurium: a novel electrospun material for disinfection of food pathogens. Colloids Surf. A Physicochem. Eng. Asp. **415**, 268–273 (2012)
117. Amna, T., Hassan, M.S., Yousef, A., Mishra, A., Barakat, N.A.M., Khil, M.-S., Kim, H.Y.: Inactivation of foodborne pathogens by NiO/TiO_2 composite nanofibers: a novel biomaterial system. Food Bioproc. Technol. **6**(4), 988–996 (2013)
118. Vincent, M., Duval, R.E., Hartemann, P., Engels-Deutsch, M.: Contact killing and antimicrobial properties of copper. J. Appl. Microbiol. **124**(5), 1032–1046 (2018)
119. Dollwet, H.: Historic uses of copper compounds in medicine. Trace Elem. Med. **2**, 80–87 (1985)
120. Poole, K.: At the nexus of antibiotics and metals: the impact of Cu and Zn on antibiotic activity and resistance. Trends Microbiol. **25**(10), 820–832 (2017)
121. Rai, M., Deshmukh, S.D., Ingle, A.P., Gupta, I.R., Galdiero, M., Galdiero, S.: Metal nanoparticles: the protective nanoshield against virus infection. Crit. Rev. Microbiol. **42**(1), 46–56 (2016)
122. Lara, H.H., Ayala-Nuñez, N.V., Ixtepan-Turrent, L., Rodriguez-Padilla, C.: Mode of antiviral action of silver nanoparticles against Hiv-1. J. Nanobiotechnol. **8**, 1 (2010)

123. Li, W.-R., Xie, X.-B., Shi, Q.-S., Zeng, H.-Y., Ou-Yang, Y.-S., Chen, Y.-B.: Antibacterial activity and mechanism of silver nanoparticles on Escherichia Coli. Appl. Microbiol. Biotechnol. **85**(4), 1115–1122 (2010)
124. Korant, B.D., Kauer, J.C., Butterworth, B.E.: Zinc ions inhibit replication of rhinoviruses. Nature **248**(449), 588–590 (1974)
125. Gupta, P., Rapp, F.: Effect of zinc ions on synthesis of herpes simplex virus Type 2-induced polypeptides. Proc. Soc. Exp. Biol. Med. **152**(3), 455–458 (1976)
126. Frei, A., Verderosa, A.D., Elliott, A.G., Zuegg, J., Blaskovich, M.A.T.: Metals to combat antimicrobial resistance. Nat. Rev. Chem. **7**(3), 202–224 (2023)
127. Nguyen Tri, P., Nguyen, T.A., Nguyen, T.H., Carriere, P.: Chapter 7—antibacterial behavior of hybrid nanoparticles. In: Mohapatra, S., Nguyen, T.A., Nguyen-Tri, P. (eds.) Noble Metal-Metal Oxide Hybrid Nanoparticles, pp. 141–155. Woodhead Publishing (2019)
128. Liu, K., Bi, Y., Qu, S., Tan, F., Chi, D., Lu, S., Li, Y., Kou, Y., Wang, Z.: Efficient hybrid plasmonic polymer solar cells with Ag nanoparticle decorated Tio 2 nanorods embedded in the active layer. Nanoscale **6**(11), 6180–6186 (2014)
129. Leong, K.H., Gan, B.L., Ibrahim, S., Saravanan, P.: Synthesis of surface plasmon resonance (Spr) triggered Ag/TiO_2 photocatalyst for degradation of endocrine disturbing compounds. Appl. Surf. Sci. **319**, 128–135 (2014)
130. Rao, G., Brastad, K.S., Zhang, Q., Robinson, R., He, Z., Li, Y.: Enhanced disinfection of Escherichia Coli and bacteriophage MS2 in water using a copper and silver loaded titanium dioxide nanowire membrane. Front. Environ. Sci. Eng. **10**(4), 11 (2016)
131. Moongraksathum, B., Chien, M.Y., Chen, Y.W.: Antiviral and antibacterial effects of silver-doped TiO_2 prepared by the peroxo sol-gel method. J. Nanosci. Nanotechnol. **19**(11), 7356–7362 (2019)
132. Kaweeteerawat, C., Chang, C.H., Roy, K.R., Liu, R., Li, R., Toso, D., Fischer, H., Ivask, A., Ji, Z., Zink, J.I., Zhou, Z.H., Chanfreau, G.F., Telesca, D., Cohen, Y., Holden, P.A., Nel, A.E., Godwin, H.A.: Cu nanoparticles have different impacts in Escherichia Coli and Lactobacillus Brevis than their microsized and ionic analogues. ACS Nano **9**(7), 7215–7225 (2015)

TiO_2 Hybrid Nanomaterials for Electrochemical Sensing and Biosensing Applications

K. S. Shalini Devi, Xiong, and Seiya Tsujimura

Abstract TiO_2 (TiO_2) nanomaterials such as nanotubes, nanobelts and nanowires, or fibres possess the superhydrophilicity and capillary activity for constructing sensor/ biosensor platforms due to their large surface area, special structural and functional characteristics. Moreover, they exhibit additional features mainly nontoxic, abundance and affordable nature. These properties provided path for numerous applications that includes energy storage, photovoltaics, pharmaceuticals, gas sensors, biosensors, and immunosensors. The hybrid TiO_2 nanomaterials with carbon and metal doping showed progress in electrochemical detection of different analytes with and without enzymatic activity. The detection analytes include glucose, organophosphorus pesticides, pharmaceutical drugs, proteins and various biomarkers. This book chapter elaborates the recent advances and role of TiO_2 nanomaterials, synthesis procedures, detection mechanisms of environmental toxicants mainly gas sensors i.e., volatile organic compounds (VOCs) and health care biomarkers. The last section discusses the prospects and problems for the use of TiO_2 nanomaterials in the future as well as the several initiatives being taken to improve their performance for sensing/ biosensing applications.

Keywords Titanium oxide nanomaterials · Biosensors · Multiple analyte detection · Environmental toxicants · Biomarker detection

1 Introduction

Titania (TiO_2) nanoparticles have been more popular in recent decades due to their high refractive index, dielectric constant, catalytic characteristics, and photoelectric conversion efficiency [1]. TiO_2 polymorphs found in nature include anatase

K. S. Shalini Devi (✉)
Pure and Applied Sciences, University of Tsukuba, Ibaraki 305-5358, Japan
e-mail: shalu2403@gmail.com

K. S. Shalini Devi · Xiong · S. Tsujimura
Division of Material Science, Faculty of Pure and Applied Science, University of Tsukuba, 1-1-1, Tennodai, Tsukuba, Ibaraki 305-5358, Japan

J. Prakash et al. (eds.), *Titanium Dioxide-Based Multifunctional Hybrid Nanomaterials*, Engineering Materials, https://doi.org/10.1007/978-3-031-80625-4_15

(tetragonal), rutile (tetragonal), brookite (orthorhombic) and monoclinic. Anatase is expected to be the most thermodynamically persistent TiO_2 polymorph for nanoparticles smaller than 10–20 nm. For particles bigger than 35 nm, rutile is the most lasting phase. When the size of the TiO_2 material is reduced to the nanoscale scale, enormous physical and chemical characteristics arise [2]. Massive efforts have been expended on TiO_2 material research, which has resulted in the potential applications ranging from photovoltaics [3] and photocatalysis [4] to photo-/electrochromics [5] and sensors/biosensors [6, 7]. Recently, several outstanding studies and papers on the synthesis and characteristics of TiO_2 based nanomaterials have been published [8]. As the size of the TiO_2 nanomaterial decreases, it increases the specific surface area and surface to volume ratio. Therefore, it enhances the interaction between the substrate and its surrounding media, which depends extensively on the material's surface area. The different forms adsorbed in the biosensor with the synthesis of TiO_2 nanomaterials such as porous gel matrices, nanotubes, nanowires, nanorods and nanoparticles (0D, 1D). Compared to TiO_2 nanomaterials-based sensing, it exhibits much superior performance with dopants such as metals/metal oxides [9, 10], quantum dots, polymers [11], and carbon nanomaterials [12, 13] by increasing surface area and conductivity for constructing a stable biosensor. The number of articles on commonly used metals and metal oxides in different sensor fields in the past five years and the most relevant usage of TiO_2 biosensor field shows the moderate usage (Table 1).

The use of TiO_2 nanomaterials in the field of sensors/biosensors are increasing gradually because of its non-toxic nature, low-cost, excellent biocompatibility and provides high surface area for detecting the targeted analytes [14]. There are different detection techniques involved with TiO_2 incorporated detection which involves photoelectrochemical, field effect transistor, fluorescent, surface plasmon resonance, electrochemical etc., This book chapter primarily focuses on the advancements of TiO_2 nanomaterials in electrochemical sensing and biosensing applications. It also discusses the difficulties experienced using this nanomaterial and the potential for the future. We are certain that this study will provide a thorough overview of several doped TiO_2 nanomaterials-based bio/immunosensors and their potential, opening the way for the development of new point-of-care (POC) diagnostic devices.

Table 1 Data from web of science core collection

Materials	Photoelectrochemical sensor	Electrochemical sensor	Gas sensor	Biosensor
Au	331	1934	1128	2240
Pt	45	747	864	447
ZnO	135	818	2724	472
CuO	65	557	522	181
MnO_2	3	229	37	160
TiO_2	**405**	**661**	**880**	**494**

#TiO_2 appears to be particularly notable in the field of photoelectrochemical sensors, with substantial research interest in gas sensors as well. It is also moderately studied in electrochemical sensors and biosensors compared to other materials

2 Synthesis Routes of TiO_2 Nanomaterials

Preparation procedures of TiO_2 nanomaterials have been achieved by numerous techniques such as sol–gel method [15], electrodeposition [16], hydrothermal method [17], oxidation method, microwave synthesis [18], sonochemical method [19], Physical/chemical vapor deposition [20], solvothermal method [21], impregnation [22], and calcination [23] etc., Each method of preparation showed difference in TiO_2 structures in electrochemical sensing application. In brief, the dopants like carbon (C), nitrogen (N) and sulfur (S) with TiO_2 has their own advantages. For instance, pure TiO_2 films possess smaller holes and undergoes shrinking after the calcination process. In the case S-doped TiO_2, sulfur precursors are Whereas, S-TiO_2 films provides a higher surface area comparable to undoped films results in improved photocatalytic activity. This is because the S dopant stimulates the creation of well-ordered mesostructured films, resulting in a greater porosity, larger pores, more hydrophilic surface, and a red shift in the bandgap [24]. Similarly, C-doped TiO_2 enhances the electrical conductivity with high surface area and bandgap. Various carbon source precursors doped TiO_2 includes banana stem, poly ethyclene glycol, melamine borate, glucose, cellulose, oleylamine etc. [25]. Likewise, in the case N-TiO_2 nanomaterials preparation, nitrogen precursors are urea, ammonium chloride, hydrazine, 4-nitrophenol, carbamide, diaminohexane etc., Nitrogen doping can change the chemical properties of the TiO_2 surface, such as the number and nature of surface hydroxyl groups, thereby affecting its interaction with analytes. Not only that, TiO_2 can form heterojunctions with other substances to improve the properties of TiO_2. For example, by mixing Mo-MOF with Ti_3C_2Tx and utilizing Mo-MOF as a precursor, the heterostructure of TiO_2 with abundant defect sites is formed through in situ coordination and calcination processes, thus improving its gas sensing properties [26]. Meanwhile, the crystalline structure of TiO_2 can also be regulated, and the formation of TiO_2 (001) crystalline surface was enhanced by utilizing $NaBF_4$ as a crystalline surface controlling agent under acidic hydrothermal conditions, and TiO_2 heterostructures with excellent properties were prepared by hydrothermal reaction and subsequent drying process [27]. In addition to these nanostructured TiO_2, the synthesis and application of porous titanium dioxide have also attracted much attention. The template method is often used to prepare porous titanium dioxide, such as using self-assembled poly(methyl methacrylate) spheres as templates to prepare high-quality hierarchical-structured three-dimensionally ordered microporous TiO_2 [28], and utilizing the gelation properties of alginate to synthesize hierarchical porous TiO_2 nanomaterials with tuneable pore volumes and nanoparticle sizes [29]. Each synthesis method has its unique advantages and scope of application, and the selection of an appropriate synthesis method depends on the specific application needs and performance requirements of the target titanium dioxide materials.

3 TiO_2 Based Electrochemical Detection Strategies

3.1 TiO_2 Nanomaterials Based Glucose Sensors

A non-enzymatic glucose sensor was recently described employing a nanocomposite film of polypyrrole, chitosan (CS-PPy), and TiO_2. In this art-of-work, the use of thermal plasma-deposited TiO_2 nanoparticles with excellent purity and tiny particle sizes improved the sensitivity and selectivity of glucose biosensors [30]. Similarly, a recent report on non-enzymatic electrochemical glucose sensor was carried out by Bittu et al. By simple anodization of Ti_6Al_4V alloy plate produced conductive TiO_2 nanotubes (TNTs) which was then coupled with Cu_2O using a chemical bath deposition method to obtain a TNT-Cu_2O NPs. In general, it requires a high level of experiment to attain the non-enzymatic biosensors must be reproducible, stable, and repeatable. Highly ordered nanostructure of TiO_2 and its homogeneity with Cu_2O Nps minimises the charge transport barrier and diffusion resistance between electrodes. Thus the electrode showed 85% stability and repeatability was calculated with standard deviation of about 1.9 [31]. Another work on TiO_2 nanoparticles and graphene oxide (GO) showed better performance towards catalytic performance for glucose this might be due to the green synthesis method of aerosol assisted self-assembly (AASA). TiO_2-GO composite had the maximum current flow and obvious oxidation, with lowered peaks at − 0.5 and − 0.6 V [13]. Guo et al. showed the TiO_2 hollow nanofibers (HNF-TiO_2) immobilized with GOx for direct electron transfer investigation of glucose sensing using a sol–gel method of preparation of TiO_2 crystals [32] (Fig. 1a). The strong connection between GOD and HNF-TiO_2 improves sensing stability. Also, it demonstrated the excellent sensing capabilities both in O_2-free and O_2-containing situations, with good sensitivity, reasonable selectivity, long-term stability, and sound dependability. Amino functionalized carbon nanotubes (CNT-TiO_2) nanocomposite was prepared and attempted for glucose sensing by immobilizing glucose oxidase (GOx) enzyme over the prepared nanomaterials as in Fig. 1b. TiO_2-coated CNT were functionalized with amine groups possessed the ability of direct transfer of GOx and the electron transfer rate constant was determined to be 3.5 s^{-1}. The fabricated biosensor exhibited superior anti-interference effect towards the uric acid and ascorbic acid [33].

3.2 TiO_2 Nanomaterials Based Detection of Pharmaceutical Drugs

Antibiotics are widely used in medical, agriculture, and animal husbandry, leading to worries about antibiotic resistance and environmental damage. The identification of antibiotics in environmental samples is extremely important these days. Nida et al. constructed a electrochemical sensor using TiO_2/MWCNT-COOH modified electrode for the detection of linezolid (LNZ) (commonly used to treat bacterial

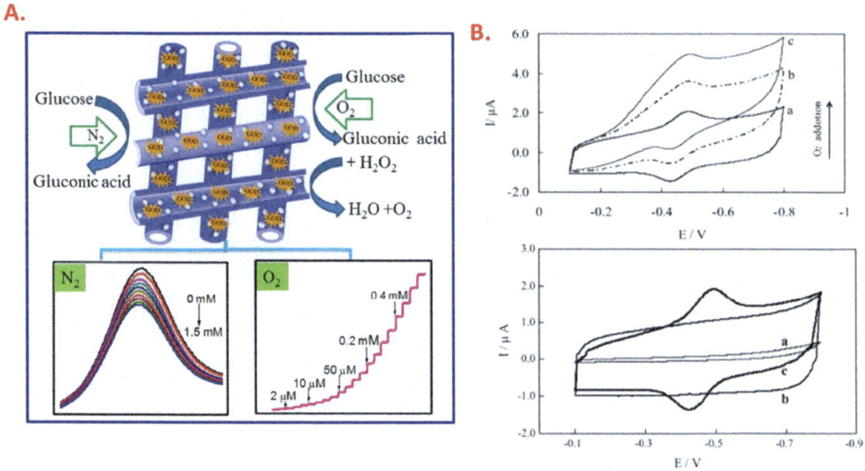

Fig. 1 **a** Detection strategy representation of HNF-TiO_2 based GOx involved enzymatic biosensing of glucose (**a**). Reprinted with permission from [32], Copyright (2017), Elsevier; **b** GOx/NH_2–TiO_2–CNT/GC electrode in 0.1 M PBS (pH 7.4). The CVs from down to up were obtained at N_2 saturated, 2 and 4 min O_2 bubbling in PBS and different control experiments with redox and without response **a** GC bare, **b** NH_2–TiO_2–CNT/GC and **c** GOx/NH_2–TiO_2–CNT/GC electrodes in 0.1 M N_2-saturated PBS. Reprinted with permission from [33], Copyrights (2011), Elsevier

infections) as shown in Fig. 2a. Excessive amount of this LNZ causes resistance that is harmful to the environment. The detection of LNZ at varied concentrations in spiked soil samples was determined using TiO_2-MWCNT-COOH modified electrode [34]. Akthar et al. demonstrated the acetaminophen (AMP) detection using electrochemical strategy with titanium oxide-aluminum oxide nanocomposites modified gold electrode (TiO_2 Al_2O_3 NC/AuE). The nanocomposite was a phytoextract based synthesis using fagonia plant extracts, which is an environment friendly and energy-efficient process that takes place under mild conditions. The fabricated electrode was effectively used to detect AMP in river water and hospital waste effluents, with recoveries above 90% [35]. Another important antibiotic amoxicillin (AMX) was detected using cobalt-doped TiO_2 nanocrystals reported by Souza et al. Stencil made screen printed electrode has been adopted and by simple drop casting of the mixture TiO_2-1%Co nanocrystals over the carbon substrate exhibited efficient AMX detection as in Fig. 2b. The recovery values for AMX determination in pharmaceutical samples were examined in TiO_2-1%Co/SPCE by utilising the standard addition technique and compared with the conventional spectroscopic methods [36]. In other works, ganciclovir (GCV) detection was detected using Fe-Cu/TiO_2/MWCNT modified electrode. The medicine GCV might be administrated and used to treat infections caused by Cytomegalovirus (CMV). The combination of these nanostructure materials has synergistic effects in GCV analysis. The developed sensor was efficiently utilized based on the differential pulse voltammetry (DPV) method to measure this substance in urine and human serum samples [37]. TiO_2 decorated with GO have

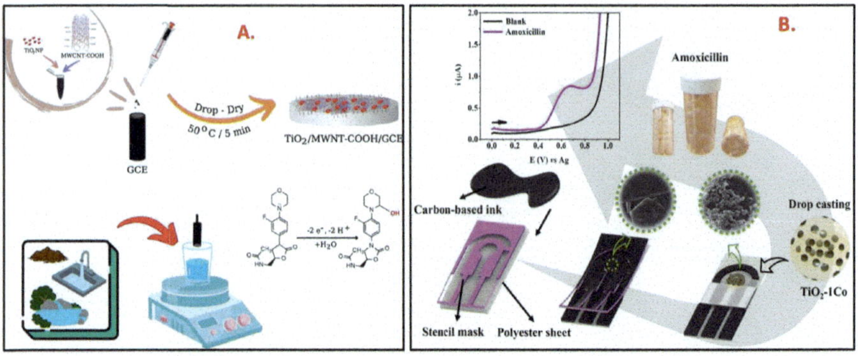

Fig. 2 **a** Representation of TiO_2 Nanoparticle incorporated functionalized carbon nanotube modified GCE for linezolid detection. Reprinted with permission from [34], Copyrights (2024), Elsevier. Screen printed carbon electrode modified with carbon ink made of Fe-Cu/TiO_2/MWCNT nanocomposite for detection of amoxicillin (**b**). Reprinted with Permission from [36], Copyrights (2023), Elsevier

been employed for the determination of chloroquine, a derivative of quinine and also an antimalarial drug, which was investigated by Zoubir et al. this sensor provided a wide linear range and lower detection limit as low as 10^{-8} M [38].

3.3 TiO_2 Nanomaterials for Pesticide Detection

There are variety of pesticides used widely in agriculture to control the insects, pest that destroys the crops in the field and to improve the effective production. Several nanocomposites have been developed in that TiO_2 based detection are as follows, dicapthon (DCP) an organophosphorus pesticide posing a risk to human health when eaten contaminants that are accumulated in soil and water. To detect this pesticide, dai et al. utilized the silver-doped TiO_2 nanoparticles using electrodeposition method. Ag's synergistic action on TiO_2 enhances electron transport, electrocatalytic activity, and selectivity of the developed DCP sensor. The fabricated electrode was tested with fresh cabbage samples and obtained good recovery values [39]. Another detection of fungicide was investigated by prabhu et al. Ru-doped TiO_2 decorated reduced graphene oxide (rGO) have been constructed to detect the 2-phenylphenol (2-PHP) (a phenolic fungicide that acts as a disinfectant, and preservative). TiO_2 utilized in this work was due to their fascinating ability to grab charge carriers, allowing effective doping with various transition metals. The suggested approach was chosen based on its stability and longevity, guaranteeing that the electrode's performance is sustained throughout time [40]. Similar fashion, nickel doped zinc oxide incorporated with TiO_2 have been prepared as a photocatalyst for the electrochemical degradation of an organophosphate namely chlorpyrifos (widely used in animal products, buildings, kill the insects in agricultural farms). Ionic metals are used in TiO_2-ZnO composites to boost their photocatalytic efficiency. TiO_2 may form heterogeneous links between n and p type dopant NiO. As a result, bonds (p-n) can improve electron–hole pair separation via the electric coupling field, promoting surface charge transfer. The produced electron–hole pairs and reactive species, such as hydroxyl radicals ($^{\cdot}OH$) and superoxide radicals (O^{2-}), begin the degradation processes with the adsorbed chlorpyrifos molecules. These radicals' stabbing chlorpyrifos' chemical bonds, breaking them down into smaller, less dangerous molecules [41]. Profenofos, a family of organophosphorus pesticide was investigated electrochemically by Azis et al. This work utilized the carbon paste electrode (CPE) paired with TiO_2 semiconductor for the direct and efficient detection of profenofos. A repeatability test with the functioning electrode until the 30th day was optimal and also found to be maximum when compared to conventional systems [42]. Carbendazim (CRZ), another fungicide, reduces fungal and bacterial infections that affect the fruits and vegetables but in turn causes harmful process with its long-time existence to living beings and the environment. TiO_2 was found to have good electrical activity, enhanced selectivity, and dependability, including the detection of CRZ towards the real time soil samples as in Fig. 3a [43]. Molecularly imprinted polyaniline modified graphene/TiO_2 nanocomposite electrode was fabricated by Nurdin et al. for the direct determination of fipronil (phenylpyrazole insecticide) contamination (Fig. 3b). MIP-PANi coatings on the Gr/TiO_2 surface carried out via electro polymerization. The effective production of MIP-PANi on the Gr/TiO_2 surface was characterised by an increase in the peak current of aniline during both oxidation and reduction. It also demonstrated the high selectivity property, which suggests the potential of selectively oxidising the fipronil in the presence of profonefos interference compound [44].

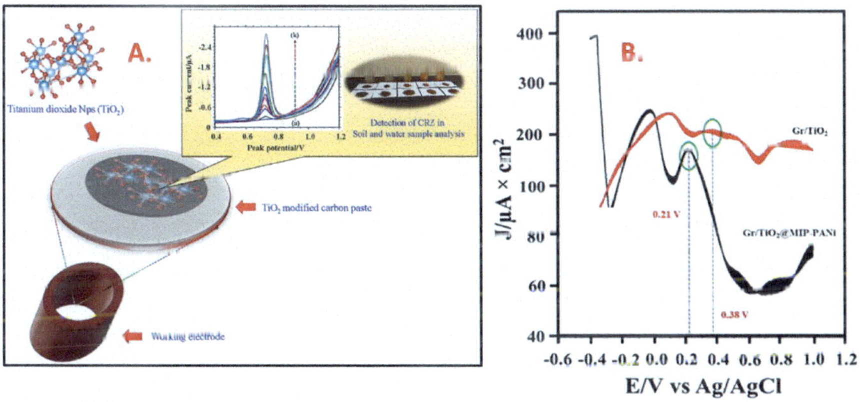

Fig. 3 Illustration of TiO_2 nanoparticles incorporated with carbon paste electrode (CPE) for carbendazim detection (**a**). Reprinted with permission from [43], Copyrights (2022), Elsevier; cyclic voltammogram of fipronil oxidation at Gr/TiO_2@MIP-PANi. **b** SWV voltammogram of fipronil oxidation at Gr/TiO_2@MIP-PANi (black line) and Gr/TiO_2 (red line). Fipronil concentration is 3 mM in 0.1 M of a mixture of supporting electrolytes Na_2SO_4 and HCl (**b**). Reprinted with permission from [44], copyrights (2022), Elsevier

3.4 TiO_2 Nanomaterials for the Detection of Infectious Biomarkers

Recent viral epidemics include SARS, Zika virus, dengue, MERS-CoV, and influenza virus infections worldwide. In findings the solutions, researchers used the various strategies of diagnosis before prevention. A sol–gel technique was utilised to design the nano TiO_2-guanine-based composite materials that can detect the H1N1 influenza virus in clinical samples with high sensitivity and speed. Electrochemical impedance spectroscopy (EIS) analysis revealed the target (ds) DNA-TG-SPGE modified electrode with the largest semicircle exhibited the greatest Rct compared to other modified and unmodified electrodes, indicating absolute difussion. The modified biosensor was tested for practical usage by recovering H1N1 swine flu HA gene from spiked nasal swab samples [45]. Alam et al. designed the biosensor strip for rapid and highly sensitive detection of coronavirus 2 (SARS-CoV-2). The strips were constructed with the monoclonal antibody (mAb) specific to SARS-CoV-2 on the surface of a commercially screen-printed carbon electrode (SPE) enhanced with polyaniline-titania nanotubes (PANi-TNT). The sensor strip's capacity to identify SARS-CoV-2 has been proven, differentiating it from human coronavirus-OC43 (HCoV-OC43), HCoV-NL63, HCoV-229E, and adenovirus. SARS-CoV-2 have been tested precisely in the saliva or nasopharyngeal fluid of infected patients as in Fig. 4 [46]. Another genosensor designed with TiO_2-adenine nanocomposites (NCs) modified screen-printed gold electrode surface (TA-SPGE) was immobilised with haemagglutinin (HA) gene specific probe containing the 5′ amine group and examined for swine flu virus (H1N1). Methylene blue (MB) was used as an indicator for identification of single-stranded viral genome. The suggested genosensor possess a high sensitivity of 10.6 μA/ng cm^2, a lower detection limit of 6.7×10–8 ng/mL, and a wide linear range of 3×10^{-8}–3×10^{-3} ng/mL, for swine flu viral DNA [47].

3.5 TiO_2 Nanomaterials Based Gas Sensors

Electrochemical gas sensors have a wide range of applications in food detection, environmental monitoring, etc. TiO_2 has been widely used for gas sensing due to its environmental friendliness and excellent photocatalytic properties. Zhang et al. introduced a high-performance gas sensor based on (001)$TiO_2/Ti_3C_2T_x$, which can detect trace amounts of ammonia gas under UV-assisted excitation, as in Fig. 5a. The ammonia detection limit is as low as 156 ppt, and its ability to assess food spoilage in practical applications is demonstrated [48]. Seekaew et al. designed a TiO_2/Graphene-carbon nanotube (TiO_2/G-CNT)-based gas sensor, which can detect toluene at room temperature, with a response of 42%–500 ppm, and the toluene sensing mechanism is based on the formation of Schottky metal–semiconductor junctions between a metallic 3D graphene-carbon nanotube structure and n-type semiconductor TiO_2 nanoparticles (Fig. 5b) [49]. Li et al. prepared a WO_3-TiO_2 composite

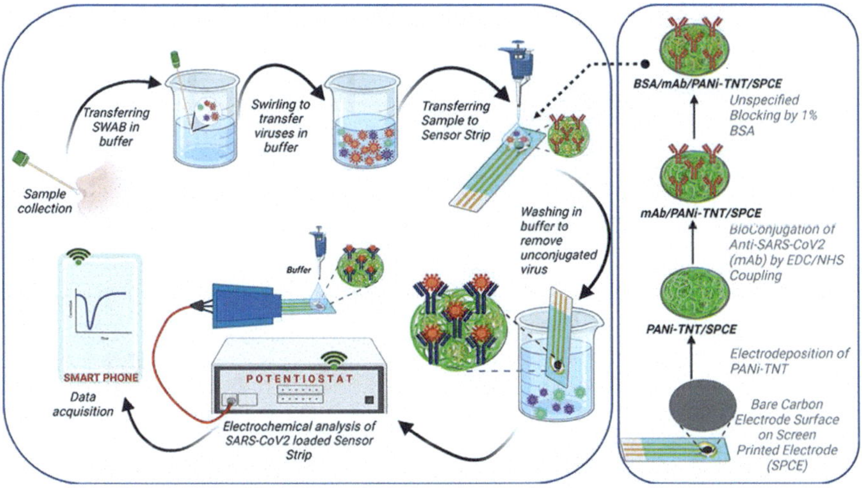

Fig. 4 Schematic illustration of sample analysis from collection to data acquisition using the developed strip and portable data reader. Reprinted with permission from [46], Copyrights (2023), ACS

as a hydrogen sensing element and utilized the heterojunction effect generated by the combination of WO_3 and TiO_2 in the hydrogen sensing at room temperature. The surface area of WO_3-TiO_2 composites was as high as 109.8 m^2/g, which could show a high sensitivity response to 5.26–10,000 ppm H_2 [50].

3.6 TiO_2 Nanomaterials for Protein Detection

Protein detection can be applied in a variety of fields, such as early disease diagnosis, food regulation, etc. Wang et al. introduced a novel and sensitive PEC (photoelectrochemical) aptamer sensor for the detection of cytochrome c. The sensor was integrated on a TiO_2 nanotube array substrate, and the cadmium sulfide and copper indium selenide quantum dots were sequentially integrated into the TiO_2 nanotubes via gold nanoparticles, forming a graded band edge structure favorable for electron/hole transfer, enabling the sensor to perform sensitive detection of cytochrome c in the range of 5 pmol (pM) to 100 nmol (nM) with a detection limit as low as 5 pM (Fig. 6a) [51]. Soomro et al. exploited the high photocatalytic activity and biocompatibility of $BiVO_4$ with in-situ formed TiO_2 on Ti_3C2T_x nanosheets to construct a viable heterojunction (TiO_2/MX*$BiVO_4$) as in Fig. 6b. The excellent photoactivity and robust heterojunction arrangement allowed the sensor to be used for the detection of soluble CD44 protein. This sensor has a wide operating range from 2.2 $\times$ 10^{-4} to 3.2 ng mL^{-1} and a detection limit of 1.4 $\times$ 10^{-2} pg mL^{-1} [52]. Li et al. proposed an electrochemiluminescent (ECL) biosensor for the detection of Carnovan A (ConA). They enhanced the ECL intensity of an organic luminescent probe (PTC,

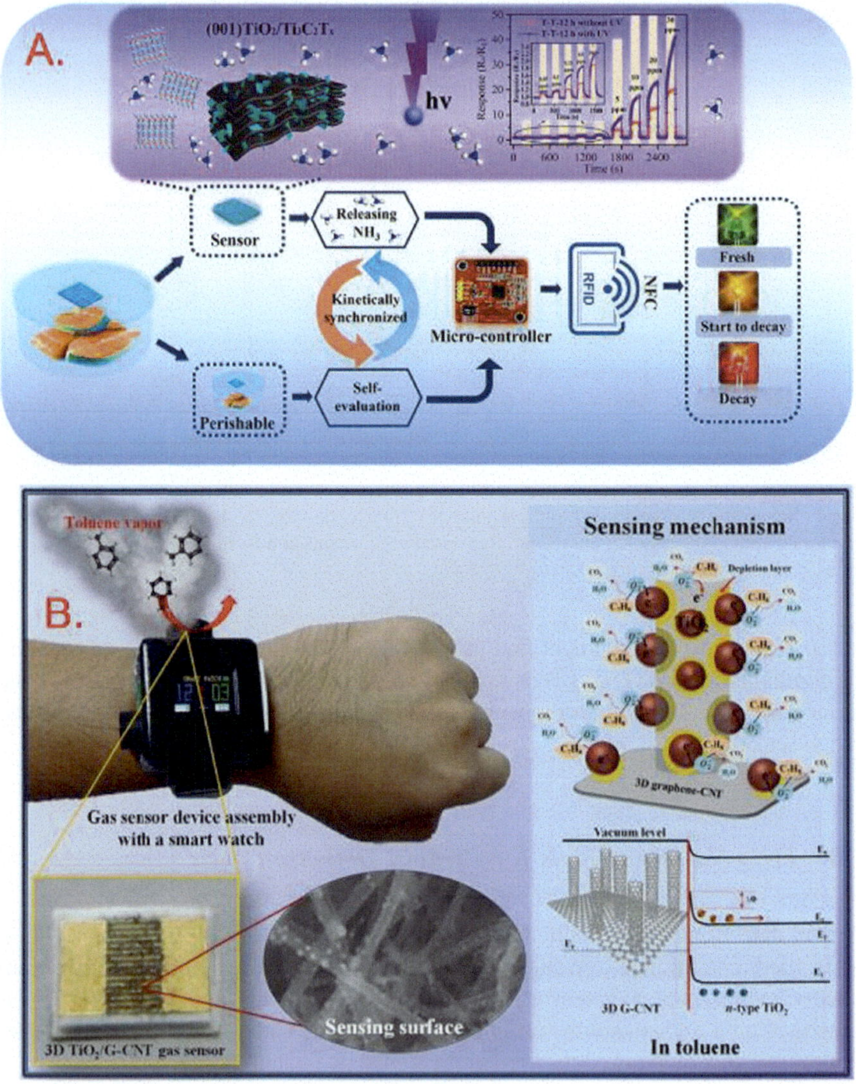

Fig. 5 Ammonia sensor based on (001) TiO_2/Ti_3C2T_x heterostructure and circuit alarm system containing NFC molecules and microcontroller (**a**). Reprinted with permission from [48], Copyrights (2022), Elsevier. Highly sensitive room temperature gas sensor based on titanium dioxide/graphene-carbon nanotube (TiO_2 /G-CNT) (**b**). Reprinted with permission from [49], Copyrights (2019), Elsevier

an amidated product of PTCDA) with platinum-modified inorganic TiO_2 (Pt-TiO_2) material. Glucose oxidase (GOx) was then loaded on the electrode to specifically recognize ConA, and the sensor showed a wide detection range of $1 \times 10^{4\text{-}1} \times 10^3$ ng/mL and a detection limit as low as 1.8×10^5 ng/mL [53]. The comparison of TiO_2 nanomaterial based sensors are clearly listed in Table 2.

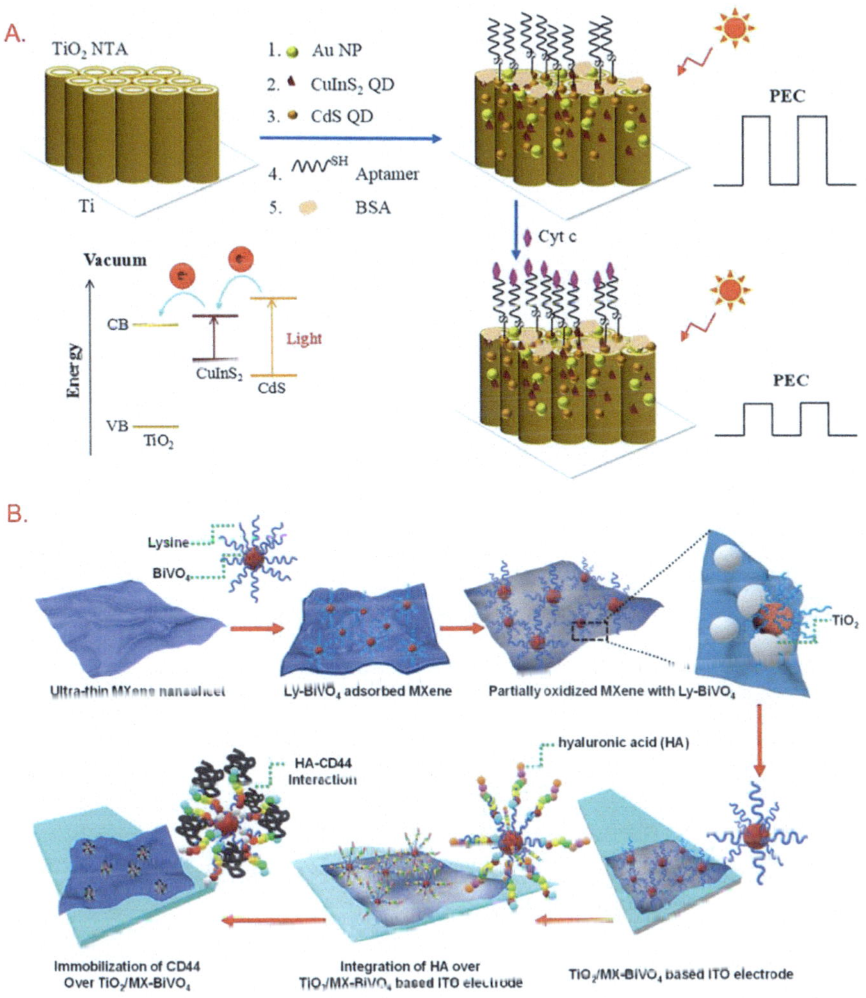

Fig. 6 Preparation and mechanism schematic of quantum dot-based PEC aptasensor (**a**). Reprinted with permission from [51], Copyrights (2019), Elsevier. Fabrication flow chart of CD44 protein PEC biosensor based on TiO_2/MX-$BiVO_4$ (**b**). Reprinted with permission from [52], Copyrights (2020), Elsevier

Table 2 Overview of TiO_2 nanomaterial-based sensors and its analytical performance

Nanomaterial type	Bio component	Technique used	LOD	Linearity range	Refs.
Macro-mesoporous TiO_2	H_2O_2	CV and CA	1.65 μM	5 μM–8 mM	[63]
TiO_2 films with macro/mesoporous texture	Dopamine	CV	–	–	[64]
Porous TiO_2 hollow nanofibers	Glucose	CV	2 μm (O_2-free) and 0.8 μm (O_2-containing)	0.006–1.5 mM (O_2-free) and 0.002–3.17 mM (O_2-containing)	[32]
Au-porous TiO_2 nanotubes	Glutathione	CV and CA	1.3 μM and 0.15 mM	–	[65]
Porous TiO_2–Pt nanowhiskers	Glutathione	CV	0.1 μM	0.5–40 μM	[66]
Au-TiO_2 nanotube arrays	Glucose	CA	0.068 mM and 0.626 mM	0.1–0.9 mM and 1–14 mM	[67]
TiO_2 nanotubes	Glucose	DPV	8.5 μM	0.03–1.0 mM	[68]
BiOBr/TiO_2 nanotube array	Glucose	PEC	10 nM	5×10^2–3×10^7 nM	[69]
TiO_2 nanotubes alginate hydrogel scaffold	Lactate and Glucose	Colorimetric	–	0.1–1 mM (lactate), 0.1–0.8 mM (glucose)	[70]
TiO_2 particles	Phosphopeptides	DPV	0.24 pg/L	(1 pg/L^{-1} mg/L	[71]
TiO_2/polyvinyl alcohol nanocomposite	Ciprofloxacin	DPV	0.04 μM	10–120 μM	[72]
Au/TiO_2 particles	Glucose	ECL	0.22 μM	7–100 μM	[73]

Au—Gold; BiOBr—bismuth oxybromide; Pt—platnium; CV—cyclic voltamogram; CA—chronoamperometry; DPV—differential pulse voltamogram; ECL—electrochemiluminscence; PEC—photoelectrochemical sensor; TiO_2—titanium dioxide; H_2O_2—hydrogen peroxide

4 Challenges and Future Prospects of TiO_2 Nanomaterials

TiO_2's fascinating physicochemical stability, biocompatibility, and optoelectronic properties as well as its reasonable price and abundant reserves, have led it to be used by researchers to design a variety of photoelectrochemical and biosensors [54]. However, there are still some difficulties associated with TiO_2 nanomaterials that need to be overcome in order to exploit the potential of TiO_2 nanomaterials further. How to fully utilize the biocompatibility of TiO_2 nanomaterials is a key technical challenge for their application in disease detection and health monitoring [55]. Exploring the enhancement of the electrocatalytic activity and electron transport rate of TiO_2 nanomaterials should be the primary goal of current research [56].

The large-scale production of special morphology TiO_2 nanomaterials is another difficulty for practical applications. It has been shown that some TiO_2 nanomaterials with high-activity crystalline (001) facets tend to lose high-energy facets due to high surface energy [57]. Therefore, the production of these TiO_2 nanomaterials with high-activity crystal faces often requires doping with F-containing materials to expose the high-activity crystal faces [58]. In addition, the wide band gap of TiO_2 limits its optoelectronic performance in the visible region. For example, anatase materials with high surface energy (001) facets can absorb only 4% of the ultraviolet light from the sun [27]. Although TiO_2 nanomaterials can generate electron–hole pairs to promote redox reactions, their high electron–hole complexation rate limits their catalytic function [59].

POC detection has the advantages of rapid detection, on-site diagnosis, and multi-detection analysis, which are promising tests for potential applications in various fields such as clinical diagnosis and health regulation as well as early disease detection [60]. Electrochemical biosensors with the advantages of fast response, high sensitivity, high selectivity, and easy miniaturization are perfect candidates for POC detection. However, in-situ detection of samples in complex environments in the field is a formidable challenge to the stability and reproducibility of the sensors. Furthermore, electrochemical biosensors have multiple potential advantages in portable, wearable, and implantable devices. It is evident that electrochemical biosensors using TiO_2 have great potential applications, and the design of next-generation electrochemical biosensors doped with TiO_2 nanomaterials with biomedical and environmental diagnostic applications should be explored in the future.

TiO_2 nanomaterials face both challenges and opportunities in electrochemical biosensor applications. TiO_2 nanomaterials, though, need to overcome challenges such as conductivity, sensitivity, and multiple sensing. However, the affordability and customizability of TiO_2 nanomaterials create a unique value. The good biocompatibility of TiO_2 allows it to ensure sensor stability and reliability in the detection of proteins, DNA and other biomolecules [61]. Notably, the electron–hole pairs generated by TiO_2 nanomaterials endow it with electrocatalytic functionality, which is also uniquely useful in the field of electrochemiluminescence. In addition, some of the difficulties faced by TiO_2 can be overcome by compounding materials such as MXene and graphene. For example, the composite of TiO_2 and MXene will form a Schottky barrier at the interface. This self-built electric field facilitates the transfer of photogenerated electrons from TiO_2 to MXene while the photogenerated holes remain in TiO_2, thus inhibiting the compounding of electron–hole pairs and improving the separation efficiency of photogenerated carriers [48]. The composite of TiO_2 and graphene can provide more active sites for molecular adsorption and reaction, thus improving the sensitivity and response speed of the sensor [62].

5 Conclusions

TiO_2 nanomaterials have been utilized to create a wide range of sensing devices for detecting compounds such as pharmaceutical drugs, pesticides, glucose, etc. TiO_2 also behaves as an ideal choice for biosensors because of its high biocompatibility. These biosensors may be integrated into electrochemical/optical systems using various sensing techniques. The majority of innovations have centered on amperometric devices, which may deliver the quickest and most precise readings to date. To construct effective sensors with high accuracy, precision, repeatability, and low detection limits, the characteristics and microstructure of TiO_2 must be optimized prior to the experiments to make it conductive. More simple synthetic approaches should be investigated for the advancement of the healthcare process in order to industrialize the application and fabrication of microminiaturized, integrated, and online monitoring sensors with TiO_2 nanomaterials.

References

1. Li, Z., Wang, S., Wu, J., Zhou, W.: Recent progress in defective TiO2 photocatalysts for energy and environmental applications. Renew. Sustain. Energy Rev. **156**, 111980 (2022)
2. Chen, X., Mao, S.S.: Titanium dioxide nanomaterials: synthesis, properties, modifications, and applications. Chem. Rev. **107**(7), 2891–2959 (2007)
3. Bai, Y., Mora-Seró, I., De Angelis, F., Bisquert, J., Wang, P.: Titanium dioxide nanomaterials for photovoltaic applications. Chem. Rev. **114**(19), 10095–10130 (2014)
4. Schneider, J., Matsuoka, M., Takeuchi, M., Zhang, J., Horiuchi, Y., Anpo, M., Bahnemann, D.W.: Understanding TiO2 photocatalysis: mechanisms and materials. Chem. Rev. **114**(19), 9919–9986 (2014)
5. Joost, U., Šutka, A., Oja, M., Smits, K., Döbelin, N., Loot, A., Järvekülg, M., Hirsimäki, M., Valden, M., Nõmmiste, E.: Reversible photodoping of TiO2 nanoparticles for photochromic applications. Chem. Mater. **30**(24), 8968–8974 (2018)
6. Pourmadadi, M., Rajabzadeh-Khosroshahi, M., Eshaghi, M.M., Rahmani, E., Motasadizadeh, H., Arshad, R., Rahdar, A., Pandey, S.: TiO2-based nanocomposites for cancer diagnosis and therapy: a comprehensive review. J. Drug Deliv. Sci. Technol. **82**, 104370 (2023)
7. Amri, F., Septiani, N.L.W., Rezki, M., Iqbal, M., Yamauchi, Y., Golberg, D., Kaneti, Y.V., Yuliarto, B.: Mesoporous TiO2-based architectures as promising sensing materials towards next-generation biosensing applications. J. Mater. Chem. B **9**(5), 1189–1207 (2021)
8. Han, R., Liu, Y., Mo, Y., Xu, H., Yang, Z., Bao, R., Pan, C.: High anti-jamming flexible capacitive pressure sensors based on core-shell structured AgNWs@TiO2. Adv. Func. Mater. **33**(51), 2305531 (2023)
9. Demir, N., Atacan, K., Ozmen, M., Bas, S.Z.: Design of a new electrochemical sensing system based on MoS2–TiO2/reduced graphene oxide nanocomposite for the detection of paracetamol. New J. Chem. **44**(27), 11759–11767 (2020)
10. Kumar, B., Sinha, S.K.: Nanostructured Cu2O deposited on TiO2 nanotube arrays for ultra-sensitive non-enzymatic cholesterol electrochemical biosensor. Mater. Today Commun. **37**, 107171 (2023)
11. Deng, L., Liu, J., Huang, H., Deng, C., Lu, L., Wang, L., Wang, X.: A molecularly imprinted electrochemical sensor based on TiO2@Ti3C2Tx for highly sensitive and selective detection of chlortetracycline. Molecules **28**(22), 7475 (2023)

12. Shen, Q., You, S.-K., Park, S.-G., Jiang, H., Guo, D., Chen, B., Wang, X.: Electrochemical biosensing for cancer cells based on TiO2/CNT nanocomposites modified electrodes. Electroanalysis **20**(23), 2526–2530 (2008)
13. Jang, H.D., Kim, S.K., Chang, H., Roh, K.-M., Choi, J.-W., Huang, J.: A glucose biosensor based on TiO2–graphene composite. Biosens. Bioelectron. **38**(1), 184–188 (2012)
14. Madhu, S., Ramasamy, S., Manickam, P., Nagamony, P., Chinnuswamy, V.: TiO2 anchored carbon fibers as non-invasive electrochemical sensor platform for the cortisol detection. Mater. Lett. **308**, 131238 (2022)
15. Romero-Arcos, M., Garnica-Romo, M.G., Martínez-Flores, H.E.: Electrochemical study and characterization of an amperometric biosensor based on the immobilization of laccase in a nanostructure of TiO2 synthesized by the sol-gel method. Materials (2016)
16. Jiang, L.-C., Zhang, W.-D.: Electrodeposition of TiO2 nanoparticles on multiwalled carbon nanotube arrays for hydrogen peroxide sensing. Electroanalysis **21**(8), 988–993 (2009)
17. Fan, Y., Lu, H.-T., Liu, J.-H., Yang, C.-P., Jing, Q.-S., Zhang, Y.-X., Yang, X.-K., Huang, K.-J.: Hydrothermal preparation and electrochemical sensing properties of TiO2–graphene nanocomposite. Colloids Surf. B **83**(1), 78–82 (2011)
18. Gijare, M., Chaudhari, S., Ekar, S., Garje, A.: Study of structural, optical and electrochemical properties of GO/TiO2 nanostructures prepared by microwave-assisted technique for its application as a glucose sensor. Adv. Nat. Sci. Nanosci. Nanotechnol. **12**(3), 035015 (2021)
19. Chen, T.-W., Chinnapaiyan, S., Chen, S.-M., Hossam Mahmoud, A., Elshikh, M.S., Ebaid, H., Taha Yassin, M.: Facile sonochemical synthesis of rutile-type titanium dioxide microspheres decorated graphene oxide composite for efficient electrochemical sensor. Ultrason. Sonochem. **62**, 104872 (2020)
20. Khaliq, N., Rasheed, M.A., Cha, G., Khan, M., Karim, S., Schmuki, P., Ali, G.: Development of non-enzymatic cholesterol bio-sensor based on TiO2 nanotubes decorated with Cu2O nanoparticles. Sens. Actuators B Chem. **302**, 127200 (2020)
21. Amri, F., Bonardo, D., Rezki, M., Septiani, N.L.W., Iqbal, M., Syarif, D.G., Nugraha, Yuliarto, B.: Polyvinylpyrrolidone (PVP)-assisted solvothermal synthesis of mesoporous TiO2 nanoparticles as an active material for enzymatic electrochemical glucose sensor. J. Electrochem. Soc. **168**(11), 117503 (2021)
22. Shi, Y., Li, T., Zhao, L., Liu, Y., Ding, K., Li, D., He, P., Jiang, D., Liu, J., Zhou, H.: Ultrathin MXene nanosheet-based TiO2/CdS heterostructure as a photoelectrochemical sensor for detection of CEA in human serum samples. Biosens. Bioelectron. **230**, 115287 (2023)
23. Tan, Q., Yu, Z., Long, R., He, N., Huang, Y., Liu, Y.: Ag3PO4/MXene-TiO2-T: as an all-solid Z-type photocatalytic system with stable and enhanced photocatalytic performance. Opt. Mater. **131**, 112685 (2022)
24. Basavarajappa, P.S., Patil, S.B., Ganganagappa, N., Reddy, K.R., Raghu, A.V., Reddy, C.V.: Recent progress in metal-doped TiO2, non metal doped/codoped TiO2 and TiO2 nanostructured hybrids for enhanced photocatalysis. Int. J. Hydrogen Energy **45**(13), 7764–7778 (2020)
25. Mani, A.D., Reddy, P.M.K., Srinivaas, M., Ghosal, P., Xanthopoulos, N., Subrahmanyam, C.: Facile synthesis of efficient visible active C-doped TiO2 nanomaterials with high surface area for the simultaneous removal of phenol and Cr(VI). Mater. Res. Bull. **61**, 391–399 (2015)
26. Weining Liu, H.L., Huang, D., Tan, X., Zhao, M., Cheng, Q., Yi, M., Ding, Q., Ren, Y., Liu, G.: Engineering α-MoO_3/TiO_2 heterostructures derived from MOFs/MXene hybrids for high-performance triethylamine sensor. Chem. Eng. J. **483**, 149340 (2024)
27. Wei Qiang, X.Q., Chen, C., Zhang, L., Sun, D.: Ti_3C_2 MXene derived (001)TiO_2/Ti_3C_2 heterojunctions for enhanced visible-light photocatalytic degradation of tetracycline. Mater. Today Commun. **33**, 104216 (2022)
28. Zhenxuan Zhao, G.L., Li, B., Guo, L., Fei, C., Wang, Y., Lv, L., Liu, X., Tian, J., Cao, G.: Dye-sensitized solar cells based on hierarchically structured porous TiO_2 filled with nanoparticles. J. Mater. Chem. A **11320** (2015)
29. Chao Yu, X.L., Liu, Z., Yang, X., Huang, Y., Lin, J., Zhang, J., Tang, C.: Synthesis of hierarchically porous TiO_2 nanomaterials using alginate as soft templates. Mater. Res. Bull. **83**, 609–614 (2016)

30. Jeong, H., Yoo, J., Park, S., Lu, J., Park, S., Lee, J.: Non-enzymatic glucose biosensor based on highly pure TiO2 nanoparticles. Biosensors **11**(5), 149 (2021)
31. Kumar, B., Sinha, S.K.: Nanostructured Cu2O deposited on TiO2 nanotube arrays for ultrasensitive non-enzymatic glucose electrochemical biosensor. Ionics **29**(2), 793–805 (2023)
32. Guo, Q., Liu, L., Zhang, M., Hou, H., Song, Y., Wang, H., Zhong, B., Wang, L.: Hierarchically mesostructured porous TiO2 hollow nanofibers for high performance glucose biosensing. Biosens. Bioelectron. **92**, 654–660 (2017)
33. Tasviri, M., Rafiee-Pour, H.-A., Ghourchian, H., Gholami, M.R.: Amine functionalized TiO2 coated on carbon nanotube as a nanomaterial for direct electrochemistry of glucose oxidase and glucose biosensing. J. Mol. Catal. B Enzym. **68**(2), 206–210 (2011)
34. Aydogdu Ozdogan, N., Demır, E., Ozkan, S.A.: Sensitive and selective electrochemical characterization and analytical determination of linezolid in environmental samples using TiO2 nanoparticles and MWCNT-COOH modified glassy carbon electrode. Microchem. J. **199**, 109920 (2024)
35. Akhtar, K., Baig, J.A., Solangi, S.A., Perveen, S., Hussain, S., Kazi, T.G., Afridi, H.I., Abbasi, F.: Phytoextract based synthesis of TiO2.Al2O3 nanocomposites for efficient electrocatalytic detection of acetaminophen from environmental and pharmaceutical samples. Ceram. Int. **50**(7, Part A), 11012–11021 (2024)
36. Souza, L.K.A., Silva-Neto, H.A., Rocha, D.S., de Abreu, F.C., Silva, A.C.A., Dantas, N.O., Coltro, W.K.T., Garcia, P.T.: Disposable sensor based on carbon electrodes modified with cobalt-doped titanium dioxide nanocrystals for electrochemical detection of amoxicillin. J. Electroanal. Chem. **942**, 117587 (2023)
37. Alinejad, T., Chen, C.-S., Shamsipur, M., Bagher Gholivand, M., Paimard, G.: Electrochemical evaluation and determination of antiretroviral drug ganciclovir based on Fe-Cu/TiO2/ multiwalled carbon nanotubes sensor. Measurement **214**, 112846 (2023)
38. Zoubir, J., Daoudi, W., Assabbane, A., Tounsi, A., Bakas, I.: Enhanced in situ electrochemical sensing of trace chloroquine in human urine and serum samples using highly charged TiO2-NPs decorated with reduced graphene oxide. J. Solid State Electrochem. (2024)
39. Dai, Y., Li, H.: Effective electrochemical sensor by using silver doped TiO2 nanoparticles modified glassy carbon electrode for determination of dicapthon pesticide in agricultural food. J. Food Meas. Charact. (2024)
40. Prabhu, K., Malode, S.J., Shetti, N.P., Pandiaraj, S., Alodhayb, A., Muthuramamoorthy, M.: Determination of fungicide at Ru-doped TiO2/reduced graphene oxide decorated electrochemical sensor. Microchem. J. **197**, 109722 (2024)
41. Esfandian, H., Rostamnejad Cherati, M., Khatirian, M.: Electrochemical behavior and photocatalytic performance of chlorpyrifos pesticide decontamination using Ni-doped ZnO-TiO2 nanocomposite. Inorg. Chem. Commun. **159**, 111750 (2024)
42. Thamrin, A., Maulidiyah, M., Muzakkar, M.Z., Ratna, R., Aziza, S.W., Bijang, C.M., Agus Salim, L.O., Prabowo, O.A., Wibowo, D., Nurdin, M.: Examination of carbon paste electrode/ TiO2 nanocomposite as electrochemical sensor for detecting profenofos pesticide. Surf. Eng. Appl. Electrochem. **57**(3), 387–396 (2021)
43. Killedar, L., Ilager, D., Malode, S.J., Shetti, N.P.: Fast and facile electrochemical detection and determination of fungicide carbendazim at titanium dioxide designed carbon-based sensor. Mater. Chem. Phys. **285**, 126131 (2022)
44. Nurdin, M., Arham, Z., Irna, W.O., Maulidiyah, M., Kurniawan, K., Salim, L.O.A., Irwan, I., Umar, A.A.: Enhanced-charge transfer over molecularly imprinted polyaniline modified graphene/TiO2 nanocomposite electrode for highly selective detection of fipronil insecticide. Mater. Sci. Semicond. Process. **151**, 106994 (2022)
45. Yadav, M., Arora, R., Dhanda, M., Singh, G., Mohan, H., Lata, S.: TiO2-guanine as a new amalgamation compound for fabrication of a disposable biosensor with high sensitivity and rapid detection of H1N1 swine flu virus. Microchim. Acta **190**(10), 412 (2023)
46. Alam, A.-M., Uppal, T., Islam, M.S., Misra, M., Verma, S.C.: Polyaniline/titania nanotube-based biosensor strip for sensitive and specific electrochemical detection of SARS-CoV-2. ACS Omega **8**(48), 45700–45707 (2023)

47. Yadav, M., Dhanda, M., Arora, R., Singh, G., Mohan, H., Lata, S.: A TiO2-adenine nanocomposite as modification material for screen-printed gold electrode to detect H1N1 (swine flu) virus: a disposable genosensor. Microchem. J. **197**, 109831 (2024)
48. Dongzhi Zhang, S.Y., Wang, X., Huang, J., Pan, W., Zhang, J., Meteku, B.E., Zeng, J.: UV illumination-enhanced ultrasensitive ammonia gas sensor based on (001)TiO2/MXene heterostructure for food spoilage detection. J. Hazard. Mater. **423**, 127160 (2022)
49. Yotsarayuth Seekaew, A.W., Phokharatkul, D., Wongchoosuk, C.: Room temperature toluene gas sensor based on TiO_2 nanoparticles decorated 3D graphene-carbon nanotube nanostructures. Sens. Actuators B Chem. **279**, 69–78 (2019)
50. Hui Li, C.-H.W., Liu, Y.-C., Yuan, S.-H., Chiang, Z.-X., Zhang, S., Wu, R.-J.: Mesoporous WO3-TiO2 heterojunction for a hydrogen gas sensor. Sens. Actuators B Chem. **341**, 130035 (2021)
51. Lijia Wang, W.G., Sheng, P., Zhang, Z., Zhang, B., Cai, Q.: A label-free cytochrome c photoelectrochemical aptasensor based on $CdS/CuInS_2/Au/TiO_2$ nanotubes. Sens. Actuators B Chem. **281**, 1088–1096 (2019)
52. Razium Ali Soomro, S.J., Kalawar, N.H., Tunesi, M., Karakuş, S., Kilislioğlu, A., Willander, M.: In-situ engineered MXene-TiO_2/ $BiVO_4$ hybrid as an efficient photoelectrochemical platform for sensitive detection of soluble CD44 proteins. Biosens. Bioelectron. **166**, 112439 (2020)
53. Ya Li, Y.L., Chen, Y., Zhang, P., Jia, N.: Fullerene-promoted organic-inorganic hybrid electrochemiluminescence biosensor for sensitive detection of concanavalin A. Sens. Actuators B Chem. **393** (2023)
54. Jingxia Qiu, S.Z., Zhao, H.: Recent applications of TiO_2 nanomaterials in chemical sensing in aqueous media. Sens. Actuators B Chem. **160**, 875–890 (2011)
55. Shu-Juan Bao, C.M.L., Zang, J.-F., Cui, X.-Q., Qiao, Y., Guo, J.: New nanostructured TiO2 for direct electrochemistry and glucose sensor applications. Adv. Funct. Mater. **18**, 591–599 (2008)
56. Shankara, J.S., Kalanur, S., Prashanth, S.N.: Voltammetric sensor for buzepide methiodide determination based on TiO2 nanoparticle-modified carbon paste electrode. Colloids Surf. B **78**, 217–221 (2010)
57. Yang Lu, Y.Z., Zhang, H., Zhang, Y., Wang, G., Zhao, H.: Meaningful comparison of photocatalytic properties of {001} and {101} faceted anatase TiO_2 nanocrystals. Sci. Bull. **61**(13), 1003–1012 (2016)
58. Liuyong Chen, X.Z., Jin, B., Luo, J., Xu, X., Zhang, L., Hong, Y.: Heterojunctions in g-C_3N_4/B-TiO_2 nanosheets with exposed {001} plane and enhanced visible-light photocatalytic activities. Int. J. Hydrogen Energy **41**(18), 7292–7300 (2016)
59. Jiuying Tian, Y.L., Dong, J., Huang, M., Lu, J.: Photoelectrochemical TiO_2 nanotube arrays biosensor for asulam determination based on in-situ generation of quantum dots. Biosens. Bioelectron. **110**, 1–7 (2018)
60. Anoop Singh, A.S., Ahmed, A., Sundramoorthy, A.K., Furukawa, H., Arya, S., Khosla, A.: Recent advances in electrochemical biosensors: applications, challenges, and future scope. Biosensors **11**, 336 (2021)
61. Nan Hao, R.H., Chen, S., Zhang, Y., Zhou, Z., Qian, J., Liu, Q., Wang, K.: Multiple signal-amplification via Ag and TiO_2 decorated 3D nitrogen doped graphene hydrogel for fabricating sensitive label-free photoelectrochemical thrombin aptasensor. Biosens. Bioelectron. **101**, 14–20 (2018)
62. H.-F. Cui, Wu W.W., Li, M.-M., Song, X., Lv, Y., Zhang, T.-T.: A highly stable acetylcholinesterase biosensor based on chitosan-TiO_2-graphene nanocomposites for detection of organophosphate pesticides. Biosens. Bioelectron. **99**, 223–229 (2018)
63. Wu, X., Zhang, H., Huang, K., Zeng, Y., Zhu, Z.: Rose petal and P123 dual-templated macro-mesoporous TiO2 for a hydrogen peroxide biosensor. Bioelectrochemistry **120**, 150–156 2018
64. Yongjun Chen, S.L., Dionysios, D.D.: Photocatalytic activity and electrochemical response of titania film with macro/mesoporous texture. Thin Solid Films **516**(21), 7930–7936 (2008)

65. Mers, S.V.S., Kumar, E.T.D., Ganesh, V.: Gold nanoparticles-immobilized, hierarchically ordered, porous TiO2 nanotubes for biosensing of glutathione. Int. J. Nanomed. **10** (2015)
66. Guihua Chen, J.W., Wu, C., Li, C.-Z., Jiang, H., Wang, X.: Photoelectrocatalytic oxidation of glutathione based on porous TiO2–Pt nanowhiskers. Langmuir 12393–12399 (2012)
67. Xidong Zhang, D.Y., Zhang, L., Lin, S.: Three-dimensional flexible Au nanoparticles-decorated TiO2 nanotube arrays for photoelectrochemical biosensing. J. Mater. Sci. Technol. **56**, 162–169 (2020)
68. Varkani, R.A., Rafiee-Pour, H.-A., Noormohammadi, M.: One step immobilization of glucose oxidase on TiO2 nanotubes towards glucose biosensing. Microchem. J. **170** (2021)
69. Zhigang Wu, J.Z., Yin, Z., Wang, X., Li, Z., Wang, X.: Highly sensitive photoelectrochemical detection of glucose based on BiOBr/TiO2 nanotube array p-n heterojunction nanocomposites. Sens. Actuators B Chem. **312** (2020)
70. Udara Bimendra Gunatilake, S.G.-R., Ojeda, E., Basabe-Desmonts, L., Benito-Lopez, F.: TiO2 nanotubes alginate hydrogel scaffold for rapid sensing of sweat biomarkers: lactate and glucose. ACS Appl. Mater. Interfaces **13**, 37734–37745 (2021)
71. Ying Chen, B.L., Chen, Z., Zuo, X.: Innovative electrochemical sensor using TiO2 nanomaterials to detect phosphopeptides. Anal. Chem. **93**, 10635–10643 (2021)
72. Junwei Zhao, P.H., Jin, W.: Electrochemical sensor based on TiO2/polyvinyl alcohol nanocomposite for detection of ciprofloxacin in rainwater. Int. J. Electrochem. Sci. **16**(10) (2021)
73. Liyan Yu, X.W., Fang, C., Tu, Y.: A disposable biosensor for noninvasive diabetic diagnosis rest on the Au/TiO2 nano-composite intensified electrochemiluminescence. Electrochim. Acta **211**, 27–35 (2016)

The manufacturer's authorised representative in the EU is Springer Nature Customer Service Centre GmbH, Europaplatz 3, 69115 Heidelberg, Germany. If you have any concerns regarding our products, please contact ProductSafety@springernature.com

Printed and bound by CPI Group (UK) Ltd, Croydon, CR0 4YY
01/10/2026
02225962-0012